Ecological Studies, Vol. 135

Analysis and Synthesis

Ecological Studies

Volumes published since 1992 are listed at the end of this book.

Springer

Berlin
Heidelberg
New York
Barcelona
Budapest
Hong Kong
London
Milan
Paris
Singapore
Tokyo

B. Jędrzejewska W. Jędrzejewski

Predation in Vertebrate Communities

The Białowieża Primeval Forest as a Case Study

With 251 Figures and 103 Tables

Springer

Dr. Bogumiła Jędrzejewska
Dr. Włodzimierz Jędrzejewski
Mammal Research Institute
Polish Academy of Sciences
Ul. Gen. Waszkiewicza 1
17-230 Białowieża, Poland

ISBN 978-3-642-08384-6

Library of Congress Cataloging-in-Publication Data
Jędrzejewska, B. (Bogumiła), 1959- . Predation in vertebrate communities: the Białowieża Primeval Forest as a case study / B. Jędrzejewska, W. Jędrzejewski. p. cm. - (Ecological studies; vol. 135) Includes bibliographical references (p.) and index.

1. Predation (Biology) - Bialowieza Forest (Poland and Russia) 2. Population biology - Bialowieza Forest (Poland and Russia) 3. Vertebrates - Ecology - Bialowieza Forest (Poland and Russia) I. Jędrzejewski, W. (Włodzimierz), 1961- . II. Title. III. Series: Ecological studies; v. 135. QL758.J44 1998 596.153-dc21 98-2772 CIP

Cover design: Design & Production GmbH, D-69121 Heidelberg

Acknowledgements

Our work on this book and the 10-year study on Białowieża's predators were, in some ways, the result of team work, and it is a pleasure to acknowledge the help of many friends, colleagues, and teachers. Jacek Goszczyński, Joanna Gliwicz, and Marek Keller, our professors at the Forestry Faculty of Warsaw Agricultural University, fostered our scientific interest in vertebrate ecology and - for years after we had graduated - encouraged and helped us through advice, constructive criticism and discussions. We are grateful to Professor Zdzisław Pucek, the Director of the Mammal Research Institute (Polish Academy of Sciences) in Białowieża, for his unfailing support for our wide-scale research programme, scrupulous reading and correcting of our scientific texts, and encouragement during the work on this book.

Since 1985, when we began the investigations, the research couple has grown into a research team, the Group of Predator Ecology, a group of enthusiasts. Many people who joined us, for shorter or longer periods of cooperation, have contributed enormously to the data synthesised in this book. We thank our colleagues-researchers, Andrzej Zalewski, Krzysztof Schmidt, Henryk Okarma, Rafał Kowalczyk, and Leszek Rychlik, and graduate students, Marcin Brzeziński, Wojciech Nowakowski, Alicja Morylewska, Janusz Kupis, and Marco Musiani. We are very grateful to the technicians who worked on our projects in 1985–1996: Irena Szpakowicz, Nadzieja Wojcieszuk, Swietłana Bogdańska, Alicja Buszko, Edyta Bajko, Jerzy Siemieniuk, Mirosław Szlachciuk, Marek Świć, Eugeniusz Bujko, Stefan Buszko, Maksymilian Szuma, Arkadiusz Arasim, Stanisław Śnieżko, and the late Leonidas Siemieniuk. Their jobs included caring for laboratory rodents and predators, snow tracking, live trapping of animals, radio-tracking, laboratory analysis of scats and pellets, computer work and much more. We are particularly indebted to Lucyna Szymura and Karol Zub for their tremendous work. Lucyna put most of the data into computer files, helped in organising field censuses, and prepared microscope slides of mammalian hair recovered from predator scats and pellets. Karol identified a large part of the prey remnants (especially birds, amphibians, and insects), analysed graphically the intricate data on space use by radio-tracked carnivores, and prepared infinite numbers of figures in the computer programme CorelDraw. We thank them for a creative attitude to all tasks they were fulfilling. We are grateful to all other workers of the Mammal Research Institute, who helped us in many ways.

We appreciate the financial and technical assistance from Earthwatch (USA) and the work of EarthCorps volunteers on our project. Thanks are extended to Linda Wallace-Friedland, Bill Smith, Marylouise Morgan (USA), Don and Betty

Wood (Australia) and other people for helping us in various ways. We also thank many other volunteers from Poland and other countries, for whom no fieldwork was too hard and no time spent in the forest was too long.

The development of our long-term study on predators in Białowieża Forest brought us closer to other investigators and fostered very useful and satisfactory cooperation with Drs. Andrzej L. Ruprecht, Małgorzata Krasińska (Mammal Research Institute), Zbigniew A. Krasiński (Białowieża National Park), Marcin Brzeziński (Warsaw University), Lech Miłkowski (Białowieża State Forestry District), Arkadiusz Szymura (Forestry High School, Białowieża), Eugeniusz Pugacewicz (North-Podlasie Society for Bird Protection), Aleksei N. Bunevich (State National Park Belovezhskaya Pushcha, Belarus Republic), Vadim E. Sidorovich and Mikhail M. Pikulik (Institute of Zoology, Belarussian Academy of Sciences, Minsk). We thank them all for inspiration and sharing their experience and knowledge with us.

Several zoologists have made their yet unpublished findings available for incorporation in this book. They include Drs. Zdzisław Pucek, Marcin Brzeziński, Vladimir A. Dackevich, Aleksei N. Bunevich, Lech Miłkowski, Vadim E. Sidorovich, Mikhail M. Pikulik, and Wiesław Walankiewicz. To all these people we are grateful for their help. We thank Lech Z. Nowacki, the author of line drawings, and the photographers whose pictures of wildlife and Białowieża Forest enliven this book.

The birth and growth of many ideas presented in this book were stimulated by numerous discussions - in personal contacts and letters - with Drs. Zdzisław Pucek, Jacek Goszczyński, Joanna Gliwicz, Ludwik Tomiałojć, Tomasz Wesołowski, Wiesław Walankiewicz (Poland), Burton L. Gordon, George Batzli, (USA), Lennart Hansson, Sam Erlinge (Sweden), Ilkka Hanski, Hannu Ylonen, Erkki Korpimaki (Finland), Eivind Ostbye, Nils C. Stenseth, Geir A. Sonerud (Norway), and Carolyn King (New Zealand). Drs. Mikael Sandell (Sweden), Urs Breitenmoser (Switzerland), and L. David Mech (USA) taught us how to capture and radiotrack carnivores.

Finally, we express our gratitude to managers and employees of the following institutions that permitted our access to the study area and helped in various ways: Białowieża National Park (Dr. Czesław Okołów, Director, and Prof. Janusz B. Faliński, Chairman of the Scientific Council), State National Park Belovezhskaya Pushcha (Mr. Vasilii P. Zhukov, General Director, Dr. Vyacheslav V. Semakov, Director of the Science Department), State Forestry Districts of Białowieża (Mr. Stanisław Ignatiew), Hajnówka (Mr. Józef Fularczyk), and Browsk (Mr. Andrzej Jaworski), and the Forestry High School in Białowieża.

The work on this book was financed by the grant KBN 4 4416 91 02 from the Committee for Scientific Research (Poland) and the budget of the Mammal Research Institute.

Many individuals and institutions contributed to the good points in this book. For the weak points, the responsibility is ours.

Białowieża, April 1998

B. Jędrzejewska
W. Jędrzejewski

Contents

Introduction

With difficulty we guess at things on earth,
and laboriously find out what lies before our feet...
Book of Wisdom, 9, 16

This book is about the relationships between the entire community of predators and populations of their prey in the Białowieża Primeval Forest. Although the ecological term 'predator' is applicable to taxonomically diverse animals, in this work it is used to mean mammals in the order Carnivora and birds of prey in the orders Strigiformes, Accipitriformes, and Falconiformes. We have studied this unique web of relationships for over a decade (1985–1996). Since Białowieża is among the best preserved but also the most thoroughly studied places in the world, we enjoyed the opportunity of combining our data with the results of numerous botanical, zoological, and ecological studies conducted there during the recent half-century. Furthermore, we exploited the wealth of information from the archives of forestry and game management of the forest.

Our objective was to evaluate and quantify the role of predators in the life, population dynamics and densities of their vertebrate prey, mainly ungulates, rodents, birds, shrews, and terrestrial amphibians. We attempted the community, indeed the ecosystem, approach in a belief that exploration of the complicated web of relations among predators, their prey, and environment might reveal the mechanisms and influences not expected from the analyses of one predator - one prey systems.

The woodland of Białowieża (about 1500 km^2), contemporarily located on the Polish-Belarussian borderland, has long earned the fame of the best-preserved temperate deciduous and mixed forest in European lowlands. The diversity and functioning of animal communities in Białowieża Forest may be regarded as model or control in comparison to other regions of temperate Europe. The forest harbours nearly 40 species of predators ranging in size from weasel to lynx and wolf, and from pygmy owl to white-tailed eagle. Still more impressive than the number of species are the ancestral life styles and foraging habits of predators, their habitat selection, and relationships with enormously rich communities of prey that include vertebrates as diverse in size and life styles as pygmy shrews, moles, bank voles, yellow-necked mice, brown hare, wild boar, red deer, European bison, tits, flycatchers, thrushes, hazel hen, brown frogs, and fish, to list just several representatives of a few hundred species.

Chapter 1 briefly presents the forest, its habitats, climate, and seasonal and multiannual variation in these aspects of plant communities, which appeared essential for shaping vertebrate communities. Chapter 2 then introduces the herbivorous and insectivorous animals, which constitute significant prey to the temperate forest predators. In this fairly detailed survey, the patterns of abundance of prey communities, their seasonal and year-to-year variation are described with the aim of elucidating factors other than predation that affect prey numbers.

Chapter 3 is devoted to presentation of the predator community. Population sizes of each species, breeding parameters, habitat preferences, home ranges, and variation in numbers, especially in response to changes in prey availability are described in order to elucidate what determines predator density and its fluctuations. Diet composition, hunting habits, prey selection, and kill rates by predators are dealt with in Chapter 4. Besides an identification of who eats whom, the questions of the predators' adaptations to hunt a given array of prey, and the consequences ('advantages' and 'disadvantages') of predators relying on a given prey are discussed.

Chapters 2, 3 and 4 contain all the necessary information for calculating the predation impact on prey. This is done in Chapter 5, where data on prey abundance, predator numbers, diet composition and kill rates are used to calculate the seasonal and annual numbers of ungulates, rodents, birds, shrews, and frogs taken by all predators from a unit area. More centrally, we analysed whether those losses mattered when compared with prey population numbers and total mortality. The ultimate goal was to tell how important predation and other factors (food, severe winters etc.) were in shaping prey densities and determining the course of seasonal and multiannual population dynamics of prey.

Straightforward as it may seem, this question is among the most difficult ones in the science of ecology. Reliable counts of prey and predator individuals, proper estimating of predator killing rates and diet composition are usually based not on direct observations but on complicated indirect methods, each inflicted with biases and errors. When various 'pieces of the jigsaw' are put together for calculating the total predation, the errors may neutralise one another or may grow with compound interest.

Even if reasonably accurate results are obtained, they may hardly be conclusive. A figure of 80% of available prey being annually taken by predators does not simply translate into a statement that predation is an essential factor in regulating prey population dynamics. Wolves, martens, hawks, and owls may as well kill animals already predestined to death due to food shortage, drought or lack of territory openings. In other words, is predation additive or compensatory to other factors of mortality? Furthermore, one should answer the question as to whether predator actions keep pace with fast changes in prey numbers or lag behind. Does predation alter the pattern of prey fluctuations or does it simply harvest the population with little influence on its ups and downs being driven by food or other factors?

To tackle these problems, we adopted two approaches. The first is rooted in Andrewartha and Birch's (1954) concept that numbers of animals are limited by external conditions. This view (mechanistic paradigm, Krebs 1995) is oriented towards understanding mechanisms, and its basic conceptual framework is to plot population parameters (such as density, birth and death rates) directly against the factors affecting them (weather, food supply, disease etc.). Wherever the data allowed for such analysis, it was the first step made in this book.

Then, we checked whether any of the factors affecting population numbers acted in a density-dependent way, according to the approach rooted in Nicholson's (1933) and Lack's (1966) idea of the crucial role of the negative feedback mechanism sensitive to the size of the population (density-dependent paradigm, Krebs 1995). The density-dependent conceptual model is particularly useful for explaining the strength of predation versus other mortality factors on prey populations. It is based on plotting the percentage predation against densities of prey. When percentage predation grows with increasing prey densities, i.e. is density-dependent, it is thought to stabilise prey numbers. When percentage predation is stable over a wide range of prey densities, it is density-independent, and if it declines with growing prey numbers, it is called inversely density-dependent or depensatory. Only density-dependent factors can regulate prey population. Factors that are density-independent or depensatory are regarded as limiting (but not regulating) prey population. The questions of density dependence and regulatory versus limiting roles of predation have recently become targets for many observational and experimental studies, as well as conceptual reviews and mathematical modelling (e.g. Sinclair 1989; Hanski et al. 1993; Newton 1993; Krebs 1995; Messier 1995a, b; Norrdahl 1995; Jędrzejewski and Jędrzejewska 1996; Sinclair and Pech 1996). Hopefully, this book will be yet another bit towards solving these problems.

1 Białowieża Primeval Forest – Abiotic Conditions and Habitats

Unbounded forests flowing with the honey of wild bees.
Czesław Miłosz, *Notes*, 1978

No book about animals can dispense with a description of their environment. Białowieża Primeval Forest, located in the heart of the immense European plains, represents the biome of temperate deciduous and mixed forests. Białowieża's woodland stretches over nearly 1500 km^2. About one third of that area are old growths of natural origin (Fig. 1.1). Besides being large and sparsely populated (2–3 people/km^2) Białowieża Forest is spatially integrated with other extensive forest tracts in Poland and Belarus that eventually lead to still other woodlands to the east, north, and south (Fig. 1.2).

The reason why Białowieża Forest has persisted till today in such a good state is both typical for the temperate forests in Europe and unique. The typical story, which Białowieża shares with some other ancient woodlands of Europe, was its status as the royal forest. Since early medieval times, when clearing of forests and stubbing the land were done primarily at the sacrifice of lowland deciduous forests on fertile soils, only the woodlands that were set aside as crown lands got a chance to resist the ever-growing demand for arable land and pastures. A unique feature of Białowieża's history is the perseverance and efficiency of that way of protection. Most of the former royal forests in Europe have disappeared or have been so heavily transformed that

Fig. 1.1. The pristine forests of Białowieża National Park in spring. In the foreground lies a trunk of an oak *Quercus robur*, a few hundred years old. A multilayered and multispecies stand (with oak, spruce *Picea abies*, and hornbeam *Carpinus betulus*) is in the background. Photo: S. Wąsik

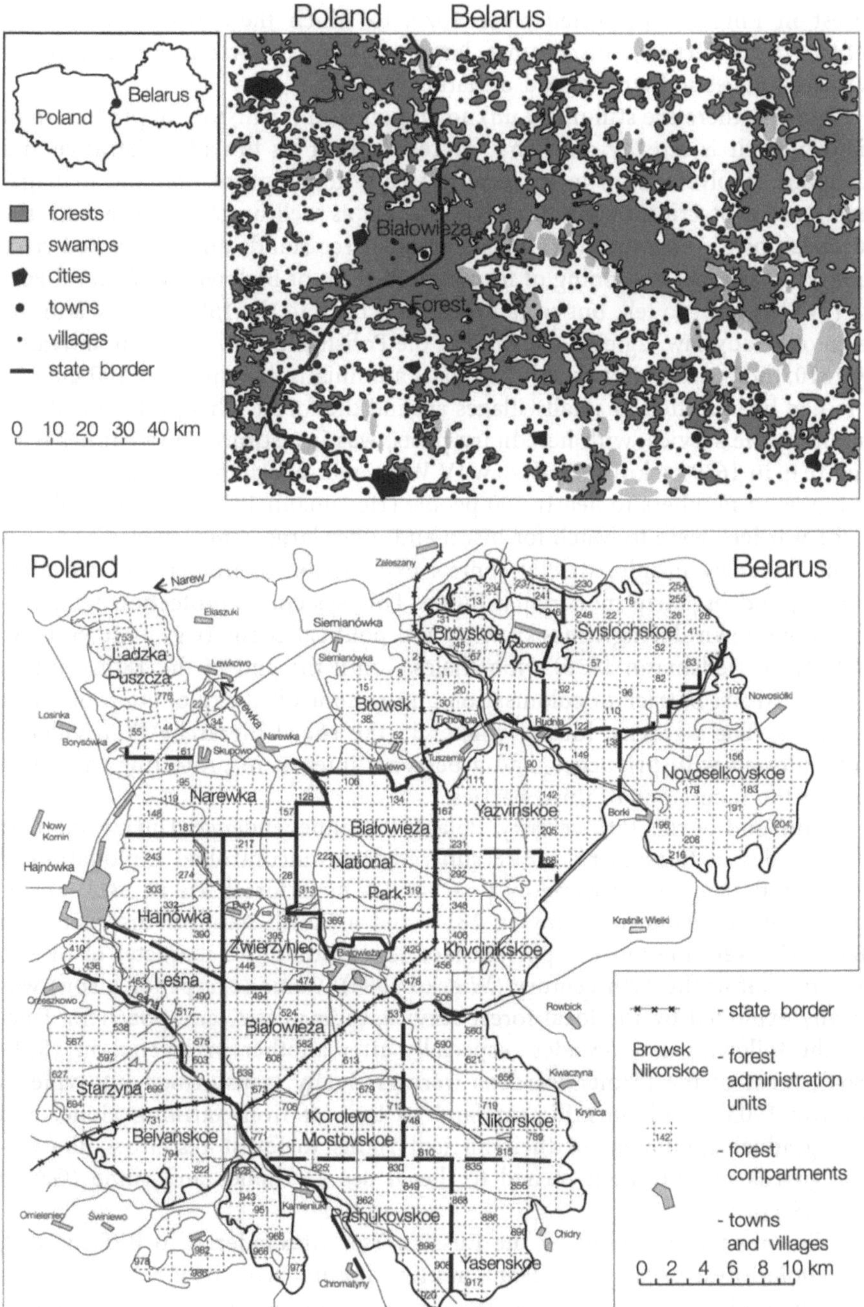

Fig. 1.2. *Upper panel* Białowieża Primeval Forest among the woodlands of eastern Poland and western Belarus. *Lower panel* schematic map of BPF with division into forest administration units (state in 1996)

they have become 'living records of centuries of land use' (such as the New Forest in England, protected since 1079, to which these words written by Tubbs (1986) pertain). In contrast, large parts of Białowieża Forest have remained a majestic living record of nature.

Although there are still many unknown points in the history of protection of Białowieża, it is already clear that success hinged on the administrative and economic involvement of many local people in forest protection and management, and protection of the forest as a whole as well as game (especially the European bison) together with their habitats (tree stands and unforested river valleys). The exact beginning of formal protection of Białowieża Primeval Forest (BPF) is hard to tell, but certainly it dates back to the 15th century and the reign of Władysław Jagiełło, the King of the Polish-Lithuanian Commonwealth. By 1510, there was a well-established local administration and over 100 guards, who had been settled in small villages around the forest. This professional forest and game service, with hereditary posts, was investitured a few times with land (e.g. in 1639 by king Władysław IV Waza and in 1744 by August III Saxon) and grew in numbers to nearly 300 people (Hedemann 1939). The main duties of the wardens were to watch for bison and other large game, supervise access to the forest by people, strictly control exploitation of timber and other resources, and organize monarchial hunts. This system of protection temporarily broke down during times of wars and political chaos (e.g. in 1655–1660, 1700–1710; Hedemann 1936). Nonetheless, until the end of the 18th century, it was a strong barrier to colonisation and uncontrolled exploitation of forest resources. The wardens, located higher in the social hierarchy than villeins, and endowed with privileges, had a personal interest in protecting the forest against the incursions of third parties.

By the end of the 18th century, Poland lost independence. During the first two decades of Russian rule, BPF suffered from uncontrolled exploitation of timber and slaughter of game by poachers. A huge forest fire in 1811, an incursion of the Napoleon army in 1812, and a Polish national insurgence in 1830–1831 were the most important factors moulding the character of BPF in the first half of the 19th century (Więcko 1984). Game (except for bison) was heavily exploited by the local forest service, their guests, and poachers. During the following few decades (especially after another insurgence in 1863–1864), most of the former wardens were gradually transformed into villeins (Karcov 1903). With unsufficient land to live off from farming alone, and with gradual annulment of their ancient rights to forest resources, the former guards of BPF often turned to illegal use of timber and game (Genko 1902–1903).

Following the first hunt by the Russian Tsar Alexander II in BPF in 1860, the whole forest became the Tsar's hunting ground with very restricted timber exploitation, but with an intense 'modern' hunting management based on German models (Karcov 1903). In the second half of the 19th century, pasturing of cattle in BPF reached an unprecedented level and, temporarily, had a damaging effect on the natural regeneration of trees. Cattle grazing in the forest was not fully abandoned until the 1970s.

In the 20th century, the forest experienced the heaviest deterioration in its history. In 1915, the Tsar's court evacuated east before the invasion of the German army. The following 4 years witnessed the slaughter of game by German and Russian soldiers and the local people. By 1919, the population of European bison was eradicated (Pucek 1991a, b). Moreover, industrial scale exploitation of timber (by large clear-cuts) was begun in 1915 by German invaders. After World War I, the Polish state forestry and the English company, The Century European Timber Corporation, continued exploitation. In 1921, the most valuable and nearly untouched old growth located in the heart of BPF was proclaimed the Białowieża National Park. The remaining part of BPF underwent economic exploitation of timber until 1941.

From 1929 until 1939, the Białowieża Forest was a place of representative hunts organised by the Polish president for foreign diplomats. European bison, by a prestigious national and international project, became reintroduced to the forest in 1929, but the bison lived and were bred in an enclosure until the early 1950s, when first specimens were released to the wild (Krasiński 1967; Pucek 1991a, b).

During World War II BPF was first under Soviet rule (1939–1941), and then was a 'Reichjagdgebiet' of the Nazi invaders (1941–1944). In 1945, the new state border between Poland and the Soviet Union (Belarussian SSR) divided the BPF into two parts with different management rules. In 1981, the Soviets built a 2.5-m-high wire fence along the border, and it became a real barrier to ungulate movements. The Polish part of BPF (580 km^2) includes Białowieża National Park (BNP) and the exploited forests administered by the State Forestry. The core of BNP (47 km^2) has been a strict nature reserve (since 1921) where no hunting, timber exploitation or motor transportation is allowed. It embraces the most precious old growths, most of which have never been cut. In the 1970s, the park became one of UNESCO's Man and Biosphere Reserves and World Heritage Sites. In 1996, BNP was enlarged to 100 km^2. The exploited forests on the Polish side of the border are managed by means of small (0.1–6 ha) clear-cuts and selective cutting of large trees, and replantation. Several partial reserves are located in that part of the forest. The Belarussian part of BPF (870 km^2) has been partially protected since 1945. Cutting is small and selective (dead and dying trees only, no clear-cuts). In 1991, the whole Belarussian part of BPF became a State National Park (with some strict reserves but mostly with a partial protection regime) and a UNESCO Man and Biosphere Reserve.

Białowieża Primeval Forest stretches from 52° 30′ to 53° N and from 23°30′ to 24°15′ E. At this latitude, the day length (sunrise to sunset) varies from 16 h 45 min on 21 June to 7 h 42 min on 22 December. The mean annual temperature averages 7 °C (range recorded in 1780–1995: 5.1–9.2 °C, Fig. 1.3). In the last two centuries, i.e. the longest time span covered by the data on animal abundance presented in this book, the 10-year moving average of annual temperature showed that years 1780–1820 and 1920–1970 were markedly cooler than the long-term mean conditions. Warmer than average were the 50-year period between 1820 and 1870 and the last 25 years (1970–1995).

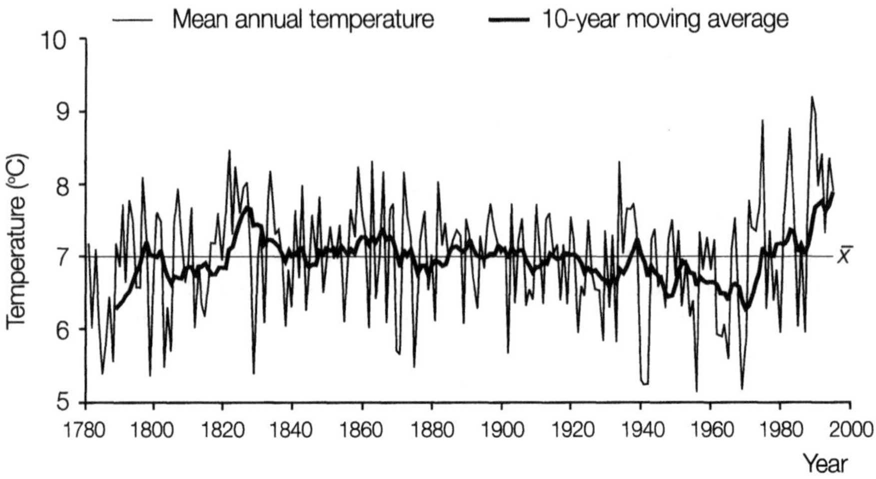

Fig. 1.3. Mean annual temperature of Białowieża in 1780–1995. Records from 1948-1995 come from Białowieża meteorological station. Data from 1780–1947 are reconstructed, based on temperature records in Vilnius. The 10-year moving average gives, for each year, a value averaged for a given year and nine preceding years. From: Jędrzejewska et al. (1997)

During 1870–1920, the annual temperature oscillated around the long-term average. In the whole 215-year series, no consistent trend of either increase or decrease of the mean annual temperature was detected. The mean temperature of July varied from 15.2 to 22.5 °C (on average 18.4 °C) and, from 1780 to 1995, it showed a slight (not significant) trend of decline (by 0.3 °C per 100 years). More variable was the mean temperature of January, the coldest month. During the 215 years, it ranged from –16.8 to 1.8 °C (mean –4.8 °C), and showed a trend of increase at a mean rate of 0.6 °C per 100 years. In consequence, the most pronounced feature of the climate during the 19th and 20th centuries in Białowieża has been a significant decrease in the amplitude between July and January mean temperatures, a slow change from a continental-type to an Atlantic-type climate.

In the temperate zone of Europe, the organizing theme of the environment is seasonality. In the cold seasons (1 November–31 March) of 1948/1949–1995/1996, the mean daily snow cover varied from 1 to 42.5 cm (mean 10 cm) (Fig. 1.4). The deepest snow is usually recorded in February. In some winters (e.g. 1969/1970, 1995/1996), snow with a maximum depth of 95 cm, covered the ground from November until April (Figs. 1.5 and 1.6). In other years (e.g. in 1989/1990, 1994/1995), wet and mild winters occurred, with barely any snow and a temperature that allowed the persistence of some green vegetation on the forest floor (Fig. 1.7). In yet another winters (e.g. 1955/1956), snow cover was poor but frosts as hard as –25 °C lasted for several days and caused deep freezing of the soil.

Mean annual precipitation in 1948–1995 was 639 mm, and it varied from 425 to 940 mm. In March–April, snow-melt makes the rivers rise and, in some years, flood in their wide plain valleys. Rainfall is most abundant in

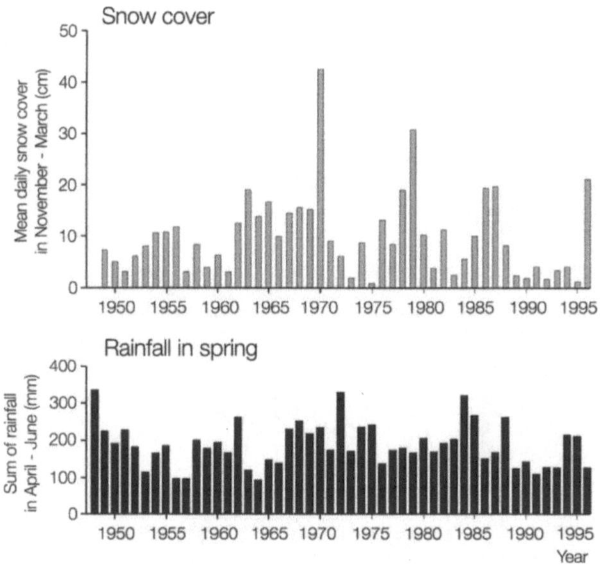

Fig. 1.4. *Upper panel* mean daily snow cover in the cold seasons (1 November–31 March) of 1948/1949–1995/1996 in Białowieża. *Lower panel* sum of rainfall (mm) in April–June in 1948–1996

May to September, but it is the precipitation in April–June that is crucial to plant growth and animal life. This parameter varied among years from the extraordinarily wet springs of 1948, 1972, and 1984 to real droughts in 1956, 1957, 1964, and 1991 (Fig. 1.4). June, July and August are the warmest months, with mean daily temperatures usually around 17 °C, and the highest insolation, on average 7 h/day compared to 48 min/day in December.

The annual cycle of temperature and precipitation shapes the life of plants. The vegetative season (i.e. number of days with mean temperature >5 °C) lasts, on average, 208 days (from 188 days in 1956 to 237 in 1990). Trees and shrubs wake up from winter dormancy in March and, with few exceptions, they immediately enter the flowering phase either before or concurrently with leaf development. Leaves of trees are fully developed in May/June, they begin to turn yellow, reddish and brown in September, and are shed completely in October/November. Seed ripening in deciduous trees usually lasts until autumn and their disseminating in many species only begins in late summer/early autumn and may continue throughout winter. Leafless lime, hornbeam and alder trees still retain seeds and shed them in masses in late autumn and, in decreasing

---

Fig. 1.5. (*see page 10*) Mean temperature and snow cover (for each day) and mean daily precipitation (averaged for each month) in BPF. The winter of 1969/1970 was the most severe and that of 1989/1990 the mildest winter on record (in 1948–1996). Years 1985–1996 are the period of the studies on predation synthesized in this book

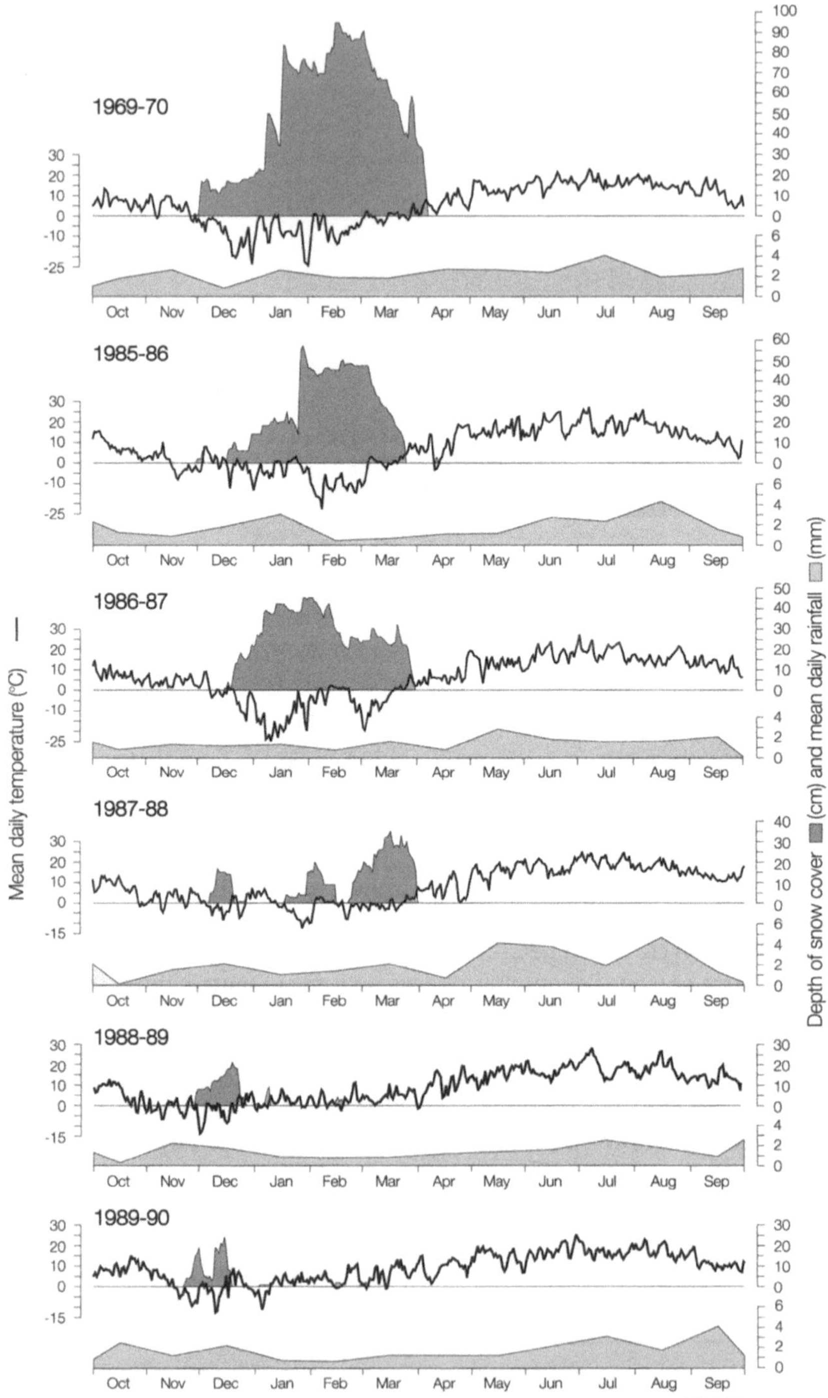

Fig. 1.5. (Legend see page 9)

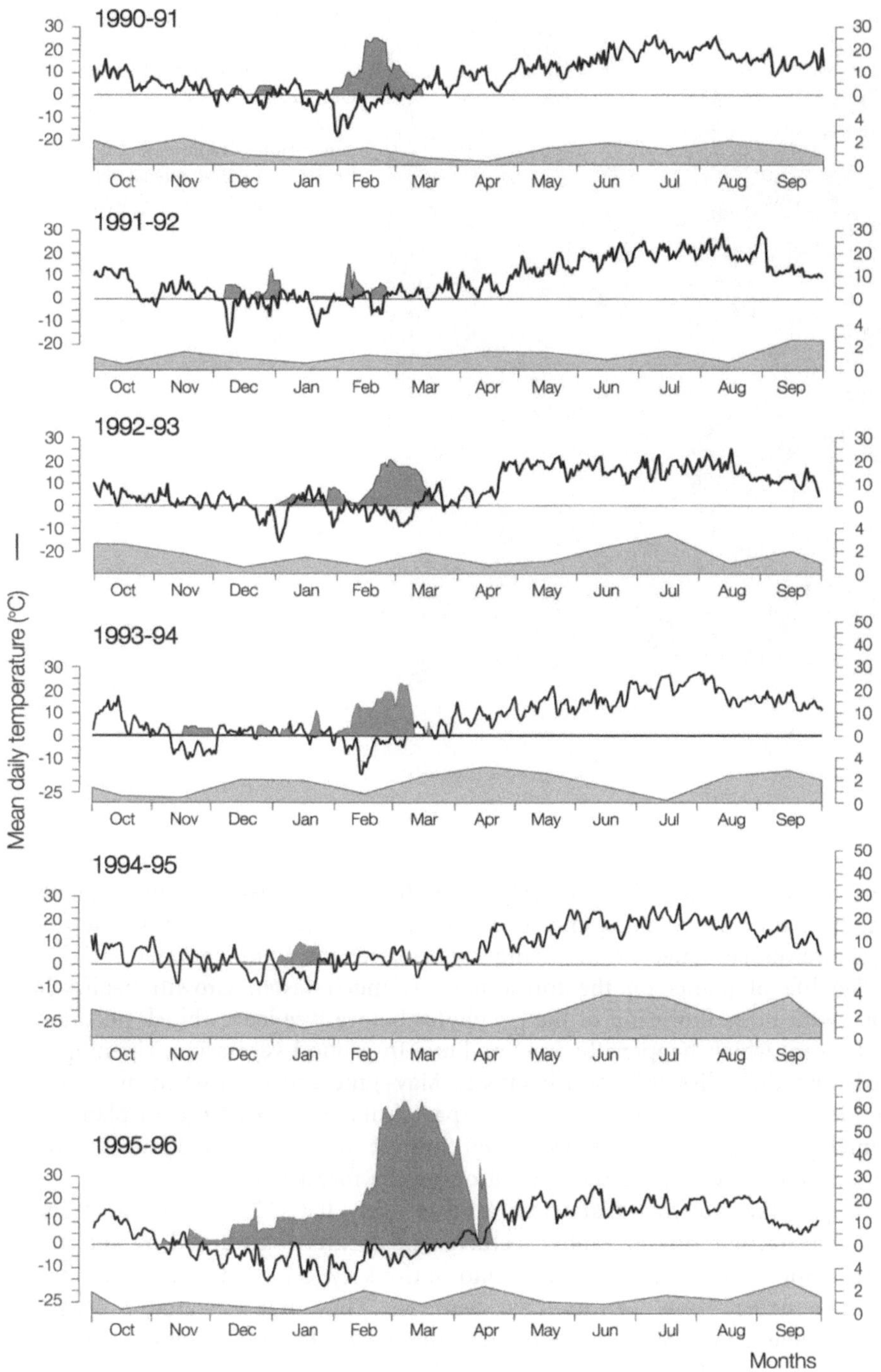

Fig. 1.5. (Legend see page 9)

Fig. 1.6. The severe winter of 1995/1996 (February 1996). Thick snow covers all layers of the forest, from the ground to tree tops. Old growth forest (mature limes *Tilia cordata* and spruces, young generation dominated by hornbeam) in Białowieża National Park. Photo: W. Jędrzejewski

amounts, until spring (Falińska 1971). Coniferous trees need an exceptionally long time for cone ripening and shedding; spruce cones fall in spring and pine seeds are disseminated in the spring 2 years after their pollination.

The life of plants on the forest floor is much faster. Growth begins in April with mass flowering of the geophytes before tree leaves block out most of the sun, which happens by May or June. In ground vegetation, ripening of seeds and their dissemination begins in May–June and ends when plants enter their winter rest. In late October, apart from some wintergreen plants in coniferous forests and semievergreen minute herbs in deciduous stands, there is very little herbaceous vegetation on the forest floor.

Despite the fact that Białowieża land is fairly flat (135–202 m a.s.l.) and lacks spectacular physiographic features, the forests – governed by soil and water conditions – are organized into various types, so distinct that their common names have been used for centuries. They first appeared in a written document in 1550, and were later adopted by the 19th century rules of forest inventory (Genko 1902–1903). Interestingly, these types easily translate into alliances in Braun-Blanquet's scale. Information about plant communities of BPF, given below, is based on the book by Faliński (1986).

Fig. 1.7. The mild winter of 1994/1995 (February 1995). Note snowless ground and ice-free water. Riverside ash *Fraxinus excelsior*-alder *Alnus glutinosa* forest (with admixtures of oaks and spruce) near the Hwoźna river, Białowieża National Park. Photo: W. Jędrzejewski

Oak-lime-hornbeam forest (*Tilio-Carpinetum*), the most scenic forest type of Białowieża and a typical association of deciduous forest in Central Europe, usually occupies rich brown or grey-brown podzolic soils. Oak, *Quercus robur* and lime, *Tilia cordata*, accompanied by spruce, *Picea abies* and maple, *Acer platanoides*, form the uppermost canopy layer. The second and the third layers are made by hornbeam, *Carpinus betulus* and lime (Fig. 1.8). Canopy coverage in full summer reaches 70–90%. Shrubs (mainly hazel, *Corylus avellana*) are sparse, but in many places regeneration of hornbeam, lime and maple adds to the shrub layer. Mass flowering of ground geophytes (e.g. *Anemone nemorosa, Isopyrum thalictroides, Corydalis cava, Allium ursinum*) occurs in early spring. The herb layer does not include true evergreen species, so the forest floor is covered by lushy greenness from spring to early autumn only (Fig. 1.9).

Vigorous growth of oak-lime-hornbeam forests is manifested by the extraordinary dimensions and ages reached by trees. Oaks grow up to 40 m in height and 200 cm dbh (diameter at breast height), and are estimated to be over 500 years of age. Not uncommon are lime trees of similar dimensions, maples of 100 cm dbh and 35 m high and hornbeam trees reaching 100 cm dbh and 30 m in height. However, the 'skyscrapers' of these forests are spruces with a maximum recorded height of 55 m (Faliński 1986).

Mixed spruce-oak-pine forests (*Pino-Quercetum* and *Querco-Piceetum*) grow on cryptopodzolic and podzolic soils. The main canopy is composed of pine, *Pinus silvestris*, spruce, oak, aspen, *Populus tremula* and birch, *Betula verrucosa*. In lower layers, spruce and deciduous trees (even hornbeam and lime) occur. Regeneration is usually dominated by spruce and hornbeam. The herb layer, in addition to deciduous species, contains many evergreens such as *Vaccinium myrtillus*, *V. vitis-idaea*, *Pirola chlorantha* and *P. minor*, as well as mosses, ferns and club-mosses. Mixed coniferous forests are in many

Fig. 1.8. Pristine oak-lime-hornbeam forest in Białowieża National Park in summer. The largest trees are oaks and spruces, the second layer is composed of lime and spruce trees. Abundant regeneration of young hornbeam and lime in the background. Herbaceous vegetation on the forest floor is rather sparse. Photo: W. Jędrzejewski

Fig. 1.9. Autumn (late September/early October) in the oak-lime-hornbeam forest, Białowieża National Park. Herbaceous vegetation on the forest floor is nearly gone and the ground is covered with fallen leaves. Photo: W. Jędrzejewski

respects transitional between deciduous and pure coniferous forests and most variable in response to climatic changes. In recent decades, an increase of deciduous trees in the undergrowth of mixed forests has been observed.

Pure coniferous forests grow on the poorest soils. Pine forests cover oligotrophic places from dry sand (*Peucedano-Pinetum*) to wetter sites (*Vaccinio uliginosi-Pinetum*) to peatbog (*Sphagnetum medio-rubelli pinetosum*). In dry forests growing on loose outwash sands, pine is the main component of the stand, but spruce competes for dominance (Fig. 1.10). Birches *Betula pubescens* and *B. verrucosa* are admixtures to such forests. In the past, forest fires promoted regeneration of pine, but in the 20th century, when every measure is undertaken to prevent fires, pine regenerates poorly and is being replaced by spruce. Forest floor vegetation is mainly composed of dwarf evergreen shrubs (e.g. *Vaccinium myrtillus, Calluna vulgaris, Cytisus ratisbonensis*), ferns and club-mosses. In these coniferous forests, pine trees reach spectacularly straight postures. They can be over 40 m high and up to 150 cm dbh. Bog pine forests grow on high peats covered by carpets of *Sphagnum* mosses and a dense layer of *Ledum palustre, Oxycoccus quadripetalus*, and *Andromeda polifolia*. In very wet places, the loose stands of dwarf, umbrella-like pine trees give way to open moors.

Fig. 1.10. Pristine coniferous forest (pine *Pinus silvestris* and spruce *Picea abies*) in Sierganowo backwood, Białowieża National Park. Photo: S. Wąsik

Black alder (*Alnus glutinosa*) is the main component of associations shaped by water: bog alderwoods and streamside forests. Bog alderwoods (*Carici elongatae-Alnetum*) occur at the peripheries of river and stream valleys and in depressions of the denudation plains. High groundwater level and its stagnation over a large part of the year are characteristic of such forests. The streamside forests (*Circaeo-Alnetum* dominated by alder and ash *Fraxinus excelsior*, and *Ficario-Ulmetum* dominated by ash, alder, and elm, *Ulmus scabra*) are situated along running waters on very fertile soils which more or less sporadically are inudated with flood water. Bog alderwoods and streamside forests usually occur near each other.

Bog alderwoods have admixtures of birch, *B. pubescens* and spruce. In the shrub layer, alder buckthorn, *Frangula alnus*, rowan, *Sorbus aucuparia*, willows, *Salix cinerea* and *S. aurita* together with regeneration of alder and spruce occur. The forest floor is distinctly structured with hollows and hummocks (Fig. 1.11). The herb layer of the hollows is greatly influenced by water which stagnates from late autumn, forms an ice lid in winter and remains until the following early summer. Many aquatic plants grow there (e.g. *Carex elongata, C. vesicaria, Solanum dulcamara, Calamagrostis canescens, Iris pseudoacorus*). The hummocks around tree bases are covered with plants typical of drier associations (e.g. *Oxalis acetosella, Vaccinium myrtillus, Majanthemum bifolium*).

The streamside alder-ash forests contain admixtures of spruce, elm and bird cherry *Prunus padus*. The shrub layer, in addition to ash, alder and elm regeneration, bird cherry, and hazel, includes the spindle tree, *Euonymus*

Fig. 1.11. Young bog alderwood of secondary origin (regenerated from both seedlings and vegetative offshoots of stumps) in the north of Browsk Forest District, Polish part of BPF. Note the hollow-and-hummock structure, typical of this forest association. Hollows, filled with stagnated water, harbour water-tolerant and aquatic plants, whereas hummocks, formed around tree bases, are overgrown with plants typical of drier habitats. The tree stand is dominated by black alder, with admixtures of young spruce and ash. Photo: W. Jędrzejewski

Fig. 1.12. Riverside ash-alder forest of natural origin along Orłówka brook, Białowieża National Park. The tree stand is composed of ash, with admixtures of black alder and bird cherry *Prunus padus*. The shrub layer is dominated by hazel *Corylus avellana*. Rich and wet soil supports abundant and tall herbaceous vegetation. Photo: S. Wąsik

europaea, and wild currants, *Ribes rubrum* and *R. nigrum*. The herb layer is dominated by perennial plants (*Filipendula ulmaria*, *Cirsium oleraceum*, *Chaerophyllum hirsutum*, *Urtica dioica*), which are over 1 m tall in midsummer (Fig. 1.12). In this community, ash trees reach over 40 m in height and 130 cm dbh, and elm trees over 35 m in height and 150 cm dbh. Elm, however, suffers from Dutch elm disease which, in recent decades, has decimated old elms in Białowieża Forest.

As long as five centuries ago, the flood-plain valleys of the rivers Narew, Narewka, Łutownia, Hwoźna, and Leśna were used as meadows by local people (Fig. 1.13). The natural open sedge marshes had been broadened at the expense of riverside forests. Those meadows, from 100 to over 1000 m wide, were scythed, utilised as pastures for cattle, and managed by the forest administration especially to support the population of European bison. During the last three decades, utilisation of the riverside meadows in the Polish part of BPF have been gradually abandoned, and nowadays alder forest and willow scrub have been slowly encroaching into river valleys. The floral composition of those marshy open habitats is characterised by a great diversity of sedges, grasses and tall forbs such as *Cirsium rivulare*, *C. palustre*, *Filipendu-*

Fig. 1.13. Sedge meadows in the valley of Narewka river (Belarussian part of BPF) in early winter. After a few centuries of having been scythed and used as pastures, the majority of such meadows are now abandoned, and secondary succession (encroachment of willow, black alders, and birches) takes place. Photo: W. Jędrzejewski

la ulmaria, Polygonum bistorta, Valeriana officinalis, Lythrum salicaria, and *Polemonium coeruleum*, which give the meadows a colourful appearance when in full blossom. In the Belarussian part of the forest, most of the riverside meadows are still in use for hay and grazing by domestic animals (cattle, sheep, and geese).

More detailed information on plant associations of BPF (including the rare and relic ones) is given by Faliński (1986), Sokołowski (1993), and Kwiatkowski (1994). Thorough description of secondary succession on abondoned riverside meadows is provided by Falińska (1991).

Over half (62%) of the BPF area is covered by coniferous and mixed forests dominated by pine and spruce. They are widely distributed on the whole terrain, but the largest continuous tracts of such forests are located in the south-eastern, south-western and northern regions (Fig. 1.14). Wet alder-ash forests cover, in total, 18% of the area and their distribution is shaped by rivers. Oak-lime-hornbeam forests grow on one-tenth of Białowieża Forest's area; they largely occur in its central and western regions.

The contemporary species composition of the forest is a result of both the potential of its soil and climatic conditions and the history of exploitation. In the strict reserve of Białowieża National Park, the actual and the potential composition of tree stands converge (Fig. 1.15). Forests dominated by oak cover 20% of the Park's area, and all deciduous forests dominate over 73% of the area. In the remaining forests of the Polish part of BPF, a marked decline in oak-hornbeam forests occurred as a result of selective cutting of oak-dominated stands and widespread replantation with pine and spruce that began in 1921, and – in major part – is continued today. Also, the shares of pioneering species, birch and aspen that spontaneously encroached onto the clearcuts, are larger than in the natural forests. In the Belarussian part of

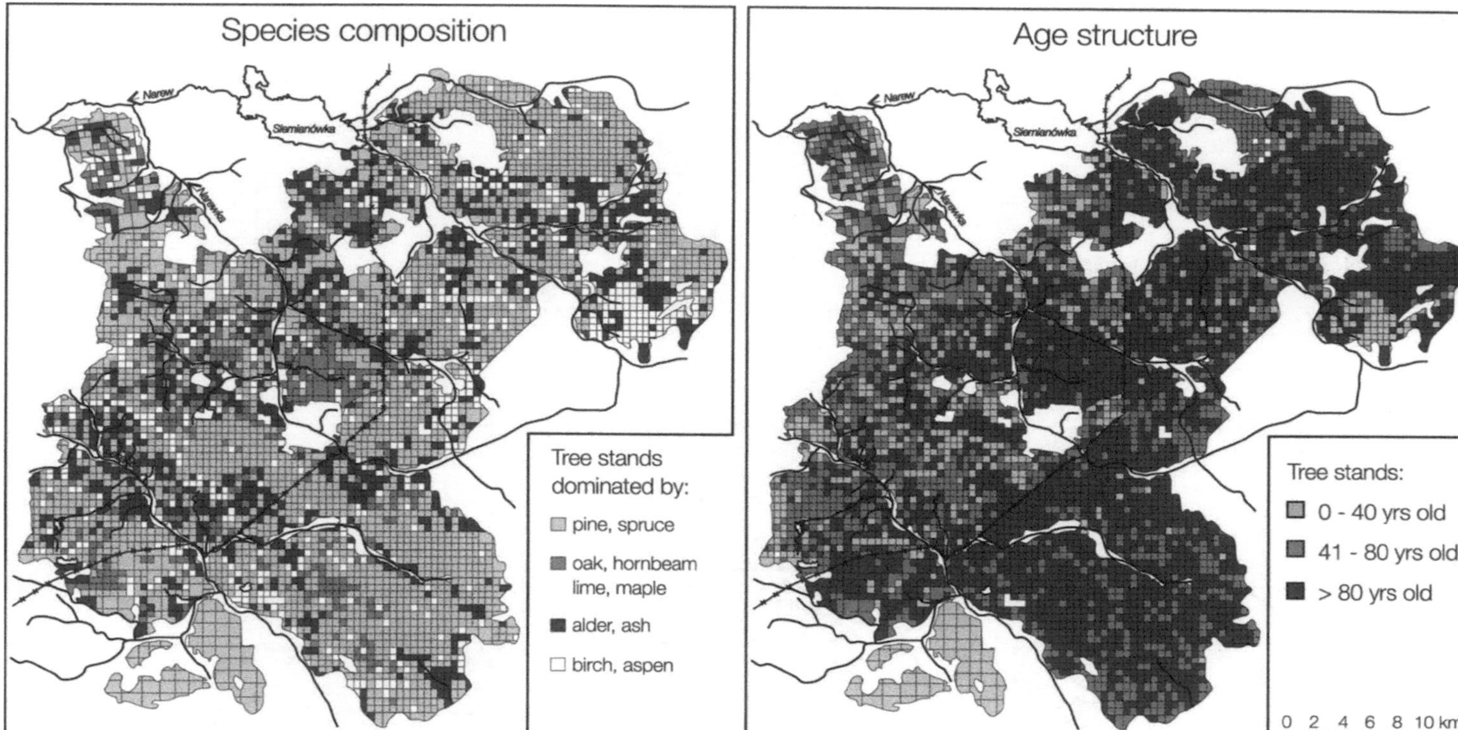

Fig. 1.14. Schematic map of the spatial distribution of tree stands dominated by various species of trees *left panel* and age classes of tree stands *right panel*. Data from detailed Forest Inventory and Management Maps were simplified by showing the dominating stands in 0.28-km^2 quadrats, i.e. one quarter of each forest compartment

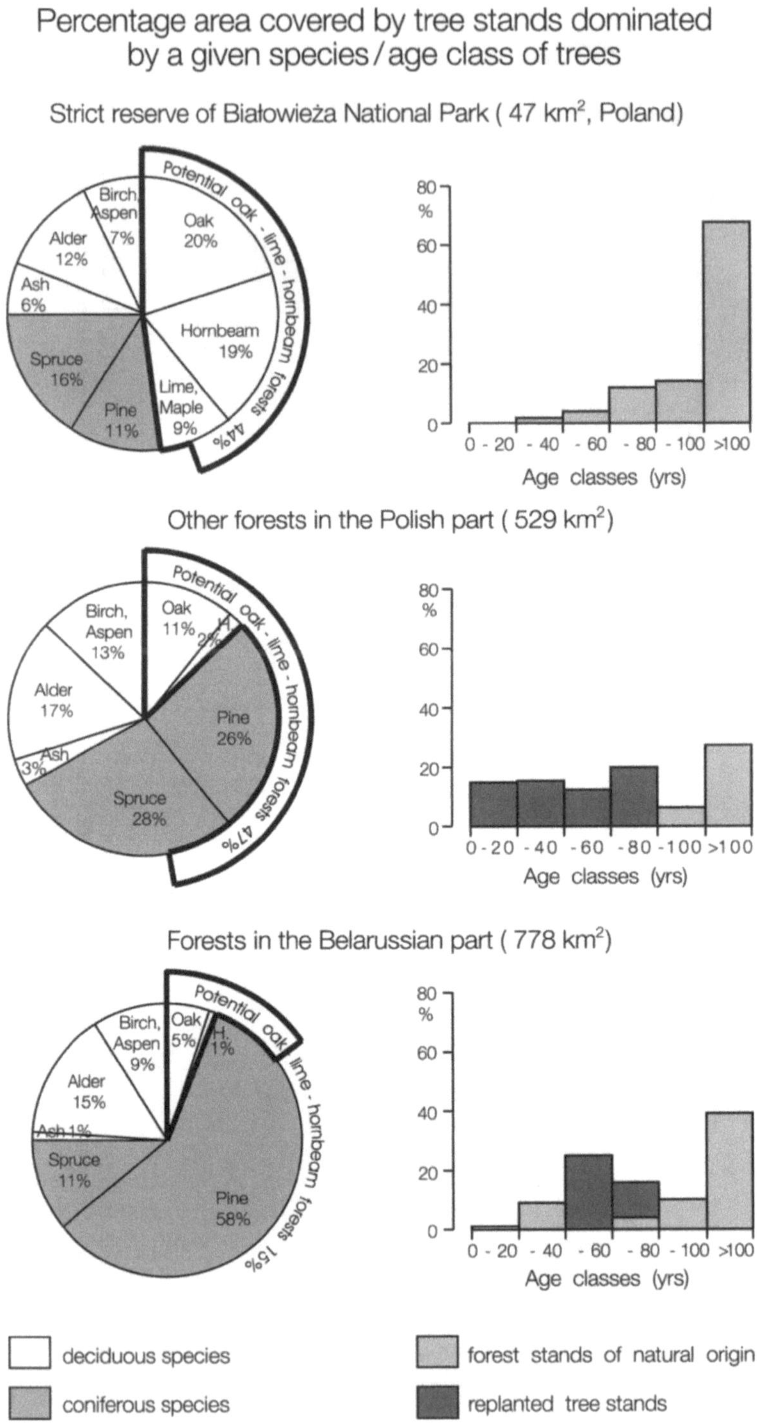
Percentage area covered by tree stands dominated by a given species / age class of trees
Strict reserve of Białowieża National Park (47 km², Poland)
Potential oak - lime - hornbeam forests 44%
Birch, Aspen 7%
Alder 12%
Ash 6%
Oak 20%
Hornbeam 19%
Lime, Maple 9%
Spruce 16%
Pine 11%
80 %
60
40
20
0
0 - 20 - 40 - 60 - 80 - 100 >100
Age classes (yrs)
Other forests in the Polish part (529 km²)
Potential oak - lime - hornbeam forests 47%
Birch, Aspen 13%
Oak 11%
H. 2%
Alder 17%
Ash 3%
Pine 26%
Spruce 28%
Age classes (yrs)
Forests in the Belarussian part (778 km²)
Potential oak - lime - hornbeam forests 15%
Birch, Aspen 9%
Oak 5%
H. 1%
Alder 15%
Ash 1%
Spruce 11%
Pine 58%
Age classes (yrs)
deciduous species
coniferous species
forest stands of natural origin
replanted tree stands

Fig. 1.16. Secondary forests in the south of the Belarussian part of BPF. Pine stands (the uppermost tree layer) were planted between 1921 and 1939 (robbery exploitation in 1915–1920 and 1940–1941 were not followed by replantation). The second layer and undergrowth is composed of self-sown spruces and oaks. Photo: W. Jędrzejewski

BPF, mixed coniferous forests dominate (Figs. 1.15 and 1.16), and the 25-year period of exploitation (1915-1941) caused decline in deciduous forests and an increase in the proportion of secondary pine stands.

In the whole wooded terrain of Białowieża (1354 km^2), old growths of natural origin (i.e. not planted by humans) cover about 590 km^2 (44%). The largest uninterrupted patch of such forests is located centrally and embraces the strict reserve of the Polish national park and the adjoining strict reserve of the Belarussian state national park (Fig. 1.14). Other fairly big remnants of natural old growths persisted in the north-eastern and south-western corners of the Belarussian part of BPF. The Polish part of BPF contains the least portion of old natural stands (30% of the area outside the strict reserve of Białowieża National Park). Worse still, mature forests survive there in small isolated patches and many of them have already suffered from plucking out the biggest trees for timber.

Old growth pristine forests are complex and intricate habitats. At first sight, they inspire respect for the large number of huge, old trees. But in fact, old trees are greatly outnumbered by young ones. In Białowieża National Park, on each hectare, on average 12049 'youngsters' (seedlings to small trees 16 cm dbh) grow and wait their turn (and luck!) to enter the canopy layer (Miścicki 1996). This young generation of the forest is 90% dominated by deciduous trees. Another inherent feature of pristine forests is the presence of dead wood. In addition to an average crop of 410 m^3/ha timber,

---

Fig. 1.15. Species and age structure of forests in the three parts of BPF differing in management rules: strict reserve of Białowieża National Park (pristine old growths), other forests in the Polish part (exploited for timber with varying intensity since 1915), and forests in the Belarussian part (exploited for timber in 1915-1941, and protected by varying regimes afterwards). Data from Polish and Belarussian Forest Inventory and Management Plans forests in the National Park contain 52-93 m^3/ha of fallen wood. In the exploited forests of the Polish part of BPF, the mean crop of timber is 268 m^3/ha and that of dead fallen wood 2-10 m^3/ha (Kirby et al. 1991, Jędrzejewski and Jędrzejewska 1995).

Making up half of all forest's area are stands of secondary origin. They were planted by foresters after the natural old growths had been cut. Some secondary stands grew from spontaneous generative or vegetative regeneration of trees on clear-cuts. In the Belarussian part of BPF, the majority of the secondary forests are 50–80 years old, since they originated in the period of massive exploitation in 1915–1941. In the Polish part, where timber exploitation has been continued, the secondary stands are from 1 to 80 years old (Fig. 1.15). Here, the principle of the economic management of the forests has been to cut natural pristine old growths and substitute them with plantations of trees (Fig. 1.17). After 1975, the process of demise of the remnant pristine stands has slowed, but the goals of forestry management remain unchanged. Replanted stands are usually composed of one species (rarely of two or three species) of trees. Little enriched by spontaneous regeneration from neighbouring stands, they grow as even-aged monocultures of regularly spaced trees (Fig. 1.18).

To herbivorous vertebrates (rodents, ungulates, birds) temperate forests provide three main sources of food: (1) ground vegetation of the forest floor, (2) browse, i.e. current year increments of shrubs and tree saplings from the ground level to about 2 m in height, and (3) seeds of trees.

The biomass of ground vegetation differs enormously among habitats. In deciduous forests, where sunlight, retained by dense canopies of multilayered tree stands, is a limiting resource, the midsummer standing crop of all plants on the forest floor is from 430 kg to 1075 kg dry weight/ha (Faliński 1973,

Fig. 1.17. Clear-cutting in the pristine old growths of mixed coniferous forest in Starzyna Forest District, Polish part of BPF in the winter of 1993/1994. Short rollers of poor quality wood are sometimes transported by horse carriages, logs are taken by motor transport. Branches of logged trees are usually burnt on clear-cuts. In the background are young secondary plantations of pine and spruce. Photo: B. Jędrzejewska

Fig. 1.18. Thicket of planted pine (20–30 years old) in the Browsk Forest District, Polish part of BPF. Self-sown birches, spruces, and oaks have made minor additions to this one-species, even-aged plantation of regularly spaced pine trees. Photo: W. Jędrzejewski

Aulak 1976). Since the herb layer is dominated by 'soft' vascular plants, the vegetation biomass in winter is only about 10% of that in summer (Fig. 1.19). Forest floor vegetation in coniferous (pine-spruce) forests is dominated by dwarf shrubs, evergreen plants and mosses. Therefore, the biomass accumulated from previous years exceeds that of the current year. In midsummer, the standing crop of alive ground vegetation is usually 2000–3000 kg dry weight/ha, and the seasonal changes are not so dramatic as those in deciduous forests (Fig. 1.19). Unmown riverside meadows with tall forbs, sedges and grasses are uncomparably richer in terms of biomass of ground vegetation. The summer standing crop of plants is at least 4000 kg dry weight/ha (Fig. 1.19). Seasonal variation in abundance of green vegetation is pronounced, but the winter biomass of vegetation in meadows is still many times greater than that in deciduous forests in the same season. It is the biomass of tree leaves of 2000 to 5600 kg/ha (review in Rohrig 1991) that composes the greater part of the total green mass in temperate deciduous forests. However, because of the short seasonal availability of leaves, the only arboreal folivores relying on them in this biome are insects.

Browse supply varies less among forest types than among tree stands of various ages. In autumn in the young plantations and thickets of Białowieża

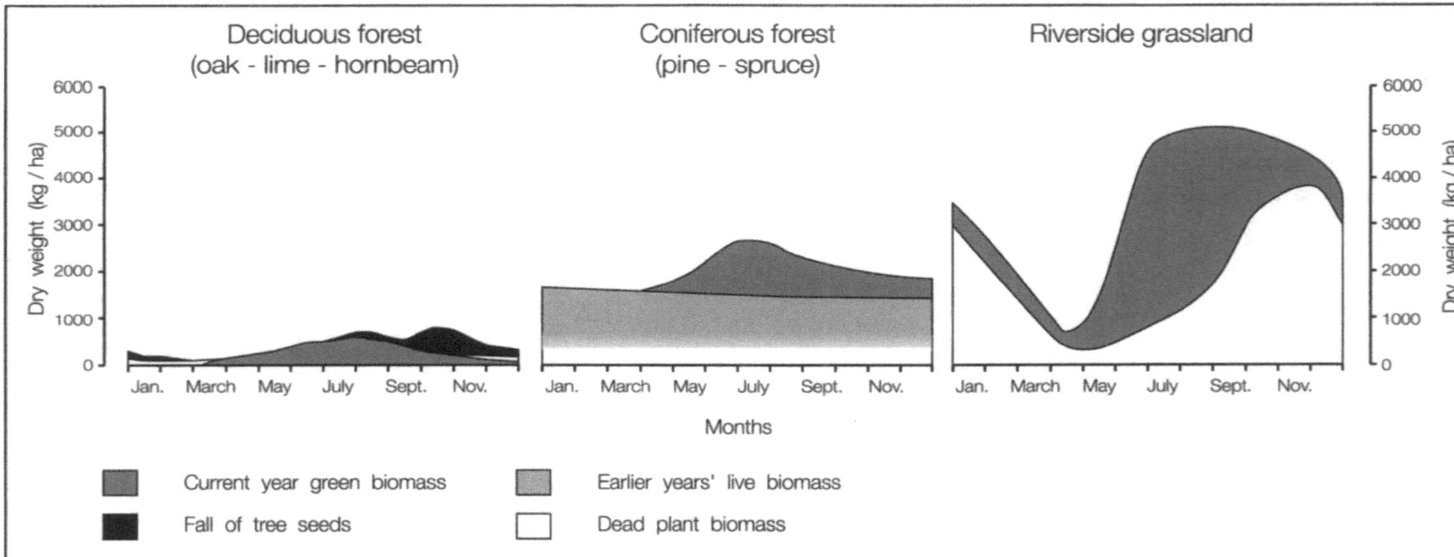

Fig. 1.19. Seasonal changes in the standing crop of biomass of ground vegetation in the main habitats of Białowieża Forest. Data from: Dąbrowski (1953), Karpiński (1954), Jankowska (1967), Falińska (1971), Traczyk et al. (1973), and Aulak (1976)

Forest, the total biomass of the current year's increment of woody plants from ground level to 2 m high was from 1875 to 7960 kg dry weight/ha (Borowski and Dzięciołowski 1980). In pole-sized and mature stands, 314 to 744 kg of browse/ha were recorded.

The oak-lime-hornbeam forests abound in tree seeds that are excellent food to many herbivorous mammals and birds. In 2 years of mast, the crop was 953 and 1922 kg/ha, and in other years, from 0 to 856 kg, on average 329 kg/ha (review in Pucek et al. 1993). In ash-alder forests, the seed crop averaged 75 kg/ha (range 20–200 kg; Aulak 1975), and in pine-spruce forests, 16 kg/ha (range 0–150 kg; Utenkova and Strelkov 1971; Tolkach and Ramlav 1973). In mixed forests, the biomass of seeds depends mainly on the share of mature oak, which can be a significant admixture to pine-spruce stands.

Seed crops vary from year to year. The warmth of summer determines the intensity of bud formation in oak, hornbeam and maple and, thus, governs their multiannual rhythm of fructification (Fig. 1.20). In 1958–1993, oak, hornbeam and maple produced heavy seed crops in synchrony at 6–9-year intervals. The mean daily temperature in June-July in summers preceding mast years (i.e. when trees formed flower buds) was on average 1 °C higher than in all other years (18.3 °C versus 17.3 °C, Pucek et al. 1993). The fructification pattern of ash diverged from that of oak, hornbeam, and maple. Only in two mast years (1967 and 1982) were ash trees heavy with seeds, too. Spruce, a boreal species, differed from deciduous trees in its pattern of cone production (Fig. 1.20).

Two other types of habitats deserve short descriptions: water bodies and human settlements. Białowieża Primeval Forest used to be rich in water. The rivers in its northern part are tributaries of the Narew River, and those in the southern part are tributaries of the Bug River. Both the Narew and the Bug are affluents of the Vistula River (Baltic sea basin). In the flat terrain of the forest, all rivers flow slowly, having an average slope between 0.35 m/km (Narewka River) and 0.64 m/km (Leśna). They are fed by surface waters and are rather poor in oxygen, typical of rivers depositing bog iron stone. Formerly, all medium-sized rivers of the forest meandered with numerous arms and old waterbeds. By the 18th century, some amelioration and drainage work had started (Hedemann 1939). Today, unnaturally straight fragments of Narewka, Hwoźna, Łutownia, Perevoloka, Solomenka and Kolonna show the past efforts to make them navigable for timber floating or to drain their marshy valleys for meadows. Nonetheless, some rivers (Leśna, Narew, Braszcza, Rudavka) remained largely natural in their course through the forest. The whole net of rivers, from very small ones (1–5 m wide and <1 m deep) to the biggest ones (11–15 m wide, 1.5–3 m deep) can be estimated at 225 km in length.

For centuries, an inseparable part of Białowieża Primeval Forest was its open moors: deep marshes, inudated with water, hardly accessible to people. The largest moor, Dziki Nikor, extended over 25 km^2 in the eastern part of BPF. In the 1950s, all open marshes in the Belarussian part of the forest (Dziki Nikor included) were drained by a network of canals and ditches, and

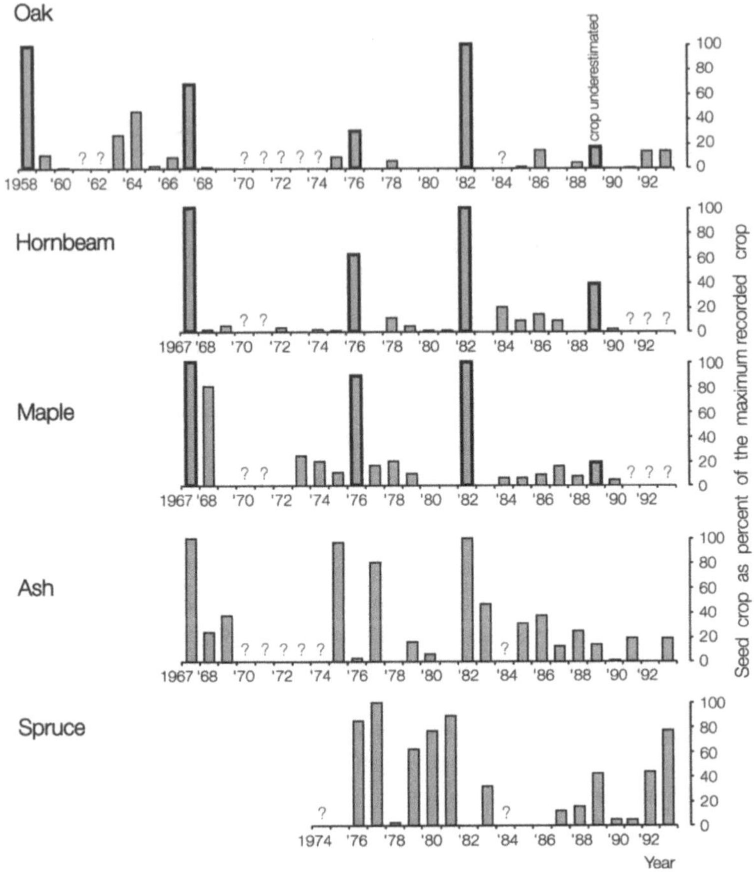

Fig. 1.20. Multiannual rhythm of seed production by five species of trees in BPF. Thick edges on *bars* denote years with synchronous masting of oak, hornbeam, and maple. To standardise the scale, the maximum recorded crop in each of the sources of data was taken as 100, and crops in other years were shown as percent of the maximum crop. *?* no data. From: Pucek et al. (1993), modified; and Forest Inventory and Management Plans of Białowieża Forest District

turned into meadows. Since these marshes fed the main rivers of Białowieża's woodland, soon the rivers shrunk, too. It would not be amiss if we mention here that some of the studies on predators synthesised in this book were done before the devastation of these unique moor ecosystems.

Along with the decades, indeed centuries of drainage works, man has created water bodies that never existed in the forest before: artificial lakes. The first ponds were constructed in Białowieża village on the Narewka River as early as the 16th century. In the 1930s, a reservoir was built on Perebel River (by daming it up) for storing timber. In the 1960s, daming up of Perevoloka

River in the Belarussian part gave rise to a large Lyackoe lake. Finally, in the 1980s, a dam built on Narew River at the northern outskirts of Białowieża Forest has created a lake (Siemianówka) of 32.5 km^2.

On its western and southern sides, and less so at other peripheries, Białowieża Forest adjoins extensive farmlands. Within the limits of the forest, however, very few glades have been cleared. The largest clearings, named after the villages built on them, are Masiewo (27 km^2), Dobrovola (19 km^2), and Białowieża (13 km^2) Glades (Fig. 1.2). They originated 200 to 400 years ago from small settlements. In all of these glades, villages with predominantly traditional wooden houses and barns are surrounded by ploughed fields and pastures. Worth mentioning is the fact that scientific stations, the bases for zoologists exploring the forest, have been located in Białowieża village (the heart of the forest, Poland), and in Kamenyuki (southern peripheries of the forest, Belarus).

2 Abundance of Prey

... over marshes
Where the moose echoes in a moon of acid frosts
And lynxes walk in springtime onto the drying holms.
Czesław Miłosz, *King Popiel*, 1958

2.1 Introduction

Predatory mammals and birds are on the top of a trophic pyramid, so the best way to understand their functioning is to go from bottom to top. In this chapter we describe the main groups of vertebrate prey (ungulates, rodents, shrews and moles, birds, amphibians) as well as mention those of secondary importance (squirrels, hare, beavers, bats, and reptiles). Both the scope and the order of presenting the various groups of animals are determined by their significance to the predator community. We attempted the same approach for all groups of prey. Wherever data were available, we analysed: (1) species structure of the community, body masses and reproductive parameters of animals, (2) spatial variation in abundance, (3) seasonal and multiannual population dynamics, and (4) mean level of densities (number of animals per unit area) in the pre-breeding (spring) and post-breeding (summer or autumn) seasons. As the first step towards explaining the role of predation in the life of prey, we quantified the influences of other factors, such as food and weather, on numbers and population dynamics of those animals. Finally, data on abundance of all prey are summarised. The purpose of pooling the numbers and biomasses of animals as different as deer, mice, tits, and frogs together was to compare their availability and predictability from the predators' perspective.

2.2 Methodical Outline

Data on animal abundance presented in this chapter were collected by methods that fall into two broad categories: counts of all animals living in a given area (density) and counts of animals recorded per standardised counting effort, not related to unit area inhabited by them (indices of abundance). Below we briefly characterise methods applied in our studies in 1985–1996 as well as other researches conducted in Białowieża Forest, the results of which are synthesised in this book.

Ungulates, the very reason for deliberate preservation of the forest since the 15th century, have been annually surveyed by game wardens for many decades, indeed for at least 200 years in the cases of bison and moose (Jędrzejewska et al. 1997). The basic method of survey was snow tracking after the new snow fall. Surveyors walked the fairly dense net of forest paths and compartment lines and counted the tracks of animals and noted the direction of their movement. The difference between 'in' and 'out' tracks gave the estimate of the number of animals that stayed in each compartment. Until now this has been the most frequent method of ungulate censusing by game wardens in the Polish and Belarussian parts of Białowieża Forest. To improve detectability of animals, Belarussian wardens snow track ungulates on two consecutive days, with obliteration of all recorded tracks during the first day of the survey (e.g. Romanov and Smoktunovich 1986).

Since the 1950s, the accuracy of snow tracking has been tested several times by comparing concurrent snow tracking with driving censuses, i.e. with counts of the number of ungulates driven out of the surveyed compartment by a wide 'battue' line moving through a censused area. Such tests conducted by Sablina (1955) in the Belarussian part of BPF showed that snow tracking on a dense grid (0.5×0.5 km) yielded the number of animals very close to the absolute. Censuses done on a looser grid (1×1 km) underestimated ungulate numbers. The smaller the ungulate, the more were they underestimated due to unrecorded individuals that stayed within surveyed quadrats. Studies by Sablina (1955) and later censuses in the Polish part of BPF (Miłkowski 1969; Pucek et al. 1975; Jędrzejewska et al. 1994, 1997) made it possible to work out correction factors for snow tracking indices of ungulate abundance. Densities of wild boar and moose appear to be on average twice as high, and those of roe deer three times higher than revealed by snow tracking. In annual censuses of European bison, snow tracking was always combined with visual observations and direct counts of herds (Krasiński et al. 1994).

Driving census is the most accurate method of estimating the densities of ungulates (Aulak and Goszczyński 1986) but if its results are to represent the whole forest and not only a single censused forest compartment, it must be done on a large enough sample of randomly chosen areas, covering at least 10% of the area in question and having the species/age structure of tree stands representative for the entire forest (Pucek et al. 1975). In practice, driving censuses conducted in the Polish part of BPF in 1969–1973, 1991–1993 and 1995 covered from 30 to nearly 70 km^2 (Pucek et al. 1975; Kossak 1995; Jędrzejewska et al. 1994), and each time it was a true levy in numbers of forestry personnel, game wardens, scientists, and students of the local Forestry High School. Driving censuses of ungulates were performed in winter, and only once (on a small area in Białowieża National Park) in summer (Jędrzejewska et al. 1994).

Difficult to census, ungulates are nonetheless relatively easy to observe. Records of ungulates seen in the forest, with notes of habitat, herd size, age and sex of animals, have collected over the years and provide good information on many ecological parameters of their populations (Lebedeva 1956; Krasiński 1978; Jędrzejewski et al. 1992a).

In Białowieża Forest, small mammals, i.e. rodents and shrews, have become a target of researchers since the late 1940s (Folitarek 1948; Borowski and Dehnel 1952; Aulak 1970). Rodents and shrews were first captured in snap traps and/or cylinders, arranged in the same spatial pattern from one trapping series to another, and operated with the same intensity to obtain comparable indices of abundance of Micromammalia between various habitats, seasons and years. These indices (number of animals captured per unit trapping effort, usually per 100 trap nights) were, however, not comparable among studies done by various researchers.

It was only during the International Biological Programme (IBP), to which studies in the Polish part of BPF contributed much, that methods of estimating densities of small mammals (rodents and shrews) were worked out and tested (Aulak 1967; Pucek 1969; Pucek and Olszewski 1971). The methods included both snap trapping and live trapping with individual marking by toe-clipping. Trapping was usually done on 5.76 ha (standard minimum).

Studies of Micromammalia often faced the problem of proper spatial and temporal scales. Trapping on grids to obtain density estimates is time consuming. Moreover, rodents and shrews are fairly numerous, so statistically sufficient numbers of animals can be captured on an area of a few hectares. No wonder that the majority of studies estimating densities of rodents were done on one study site and within a fairly short period of 1–3 (maximally 7) years (Gębczyńska 1966; Wołk and Kozłowski 1989; Wójcik 1993; Jędrzejewski et al. 1995). Long-term and/or large-scale censuses of rodents were, with no exception, done by estimating the indices of abundance only. Most of the data on rodents and shrews presented in this book come from two trapping series: one conducted since 1959 on a grid in oak-lime-hornbeam forest in Białowieża National Park (Pucek et al. 1993; Z. Pucek, unpubl. data), and another performed since 1986 on 14 stations (seven in forests and seven in riverside marshes) stretched over 6 km in BNP (Jędrzejewski et al. 1994a; W. Jędrzejewski, unpubl. data).

In most studies, Micromammalia were trapped two or three times per year: in spring (April), summer (July/August) and autumn (September/October). This procedure covered the breeding season of rodents and shrews, and recorded their seasonally lowest (in spring) and highest (in summer or autumn) numbers.

Methods of counting forest birds, based on audible and visual cues, developed in an entirely different way to those used for small mammals. The earliest regular ornithological surveys in Białowieża Forest were qualitative faunistic records of bird communities (Gavrin 1953) or regular standardised counts of seen and heard birds in order to obtain relative indices of their abundance (Karpiński 1954; Gavrin 1969). A real turning point in bird ecology was the mapping method based on recognition of bird songs in the breeding season. In the Polish part of BPF, the majority of censuses were done by the mapping method (Tomiałojć 1980). Censuses of birds singing in their territories are conducted from dawn to late morning (sometimes at dusk, too), and localisation of singing birds are mapped. As an auxilliary technique for correct differentiation of breeding territories, attempts are made to map as many concurrently singing neighbours as possible and to find nests. Counts are repeated 8–10 times in April–June. The most impressive long-term and large-scale ornithological study based on this method has been conducted by L. Tomiałojć and co-workers on several plots (24–33 ha each) in the pristine forests of Białowieża National Park since 1975 (Tomiałojć et al. 1984; Tomiałojć and Wesołowski 1990, 1996). Other researchers contributed data from uncommon forest types (Lewartowski and Wołk 1983; Piotrowska and Wołk 1983). Moreover, Pugacewicz (1997) conducted a large-scale survey aimed at evaluating the species composition and densities of avian fauna in the whole Polish part of Białowieża Primeval Forest, including its non-forest habitats.

Bird censuses conducted in the Belarussian part of BPF (e.g. Dackevich and Popenko 1987) were done on transects. According to the method proposed by Naumov (1965), walking observers recorded all birds they heard and then recalculated the obtained index of numbers into densities by applying the specific estimates of audibility distances of songs to each species (from the list in Naumov 1965). How accurate these estimates of densities were, remains unknown. Nearly all the above listed censuses of birds in BPF were done in spring. Data on bird numbers in summer (seasonally the highest abundance), autumn, and winter (seasonally the lowest numbers) are very rare.

An essential technique for studying birds' breeding success, reproductive parameters and production of young is finding nests, and then counting eggs, hatchlings, and fledglings during a few consecutive visits. Several species of birds were investigated that way in Białowieża Primeval Forest (e.g. Piotrowska and Wesołowski 1989; Wesołowski and Stawarczyk 1991).

Amphibians were much less studied in Białowieża Forest than the above described groups of vertebrates. Large-scale transect counts were performed in the Belarussian and Polish parts of BPF five times between 1955 and 1995 (Bannikov and Belova 1956; Bakharev 1986; Pikulik et al. 1987, and unpubl. data). The methodology was elaborated by Dinesman and Kaleckaya (1952). In midsummer (June–August), when terrestrial amphibians are well distributed throughout forest ecosystems but young froglets have not yet hatched, an observer walks the straight-line transects in various habitats and counts amphibians in a 1-m belt. The censuses are conducted during humid weather in early morning and late afternoon, i.e. during good amphibian activity. The numbers of recorded specimens are then recalculated into densities. Obviously, they give underestimated measures of amphibian abundance, nonetheless, due to constant methodology, the results are well comparable among habitats and years. Furthermore, we censused terrestrial amphibians on several 0.9 ha study plots in Białowieża National Park in 1992–1994, a few times from April till October. We applied the method based on live capturing and removal, elaborated during IBP by Głowaciński and Witkowski (1970). So far, no data are available on distribution and abundance of hibernating and spawning amphibians.

Despite a plethora of inventive techniques and devices, strenuous fieldwork, and careful analysis of data, all scientists cited above (us included) acknowledge that no method of count is perfect. Even the best techniques, critically tested and carefully applied, usually more or less underestimate the real numbers. Non-territorial silent birds, trap-shy rodents, nonactive amphibians and reptiles are present in all censused populations, and many of the already cited works attempted to learn how big that 'hidden' fraction is. Worse still, errors in counting one taxonomic group of animals may differ by an order of magnitude from errors in censusing another group. Yet inevitable weak points of the data should not forbear one from trying to understand the whole picture from the available pieces.

2.3 Ungulates – A Royal Feast Served on Too High Tables

No doubt the very soul of Europe's primeval forests is their large hooved mammals. It is enough to spend a night in the woods in September, during

Fig. 2.1. European bison *Bison bonasus*, a male. Photo: S. Wąsik

Table 2.1. Densities, percentage community structure (in 1991–1995) and biological features of native ungulates of Białowieża Primeval Forest: European bison (*Bison bonasus*), moose (*Alces alces*), red deer (*Cervus elaphus*), roe deer (*Capreolus capreolus*), and wild boar (*Sus scrofa*)

Parameter	Bison	Moose	Red deer	Roe deer	Wild boar	Total
Mean body mass of adult (kg)	400	200	100	20	80	
Mean body mass of juvenile (kg)	100	100	40	11	20	
Litter size (*N* juv)	1–2	1–2	1–2	1–4	2–13	
Annual breeding rate (*N* juv born/adult)	0.17	0.40	0.43	0.66	0.93	
Feeding ecology:						
grazing (%)	90	10	40	80	–	
browsing (%)	10	90	60	20	–	
rooting (%)	–	–	–	–	100	
Mean density (N/km^2)						
in late winter	0.46	0.16	4.61	3.80	3.01	12.04
in summer	0.54	0.22	6.53	6.35	5.23	18.87
Percentage of ungulate numbers	3	1	36	33	27	100
Crude biomass (kg/km^2)						
in late winter	184	32	461	76	241	994
in summer	192	38	538	104	285	1157
Percentage of total biomass	17	3	46	9	25	100

Sources of data on body masses (adults, approximate mean for the population in winter; juv, half-year olds) and reproduction rates: Sablina (1955), Lebedeva (1956), Dzięciołowski (1969), Miłkowski (1970, and unpubl.), Pucek and Sych (1984), Krasińska (1988, and pers. comm.), Pielowski (1988), Fruziński (1992), Dzięciołowski and Pielowski (1993), and authors' unpublished data. Feeding ecology (browsing, feeding on twigs of shrubs and small trees; grazing, feeding on ground vegetation: herbaceous vegetation and dwarf shrubs; rooting, getting food by moving the humus and upper layer of soil): Dzięciołowski (1970a, b), Kozlo (1975), Morow (1976), Gębczyńska (1980), Gębczyńska et al. (1991).

the red deer rut, when the vapour of mud and rot blends with musk scent, and chilly air carries the roars of bulls. Yet an unrivaled feature of Białowieża is the chance to encounter the herd of European bison: the majestic, as if leisurely mannered, animals of ancient times (Fig. 2.1).

Large ungulates are either exclusively herbivorous (bison, moose, deer), or predominantly such (wild boar). For the study of predation, the following characteristics of ungulates are essential. First and foremost, the crude biomass of ungulates per unit area exceeds by an order of magnitude that of any other group of herbivores (rodents, hare) or insectivores (birds, amphibians, shrews). Secondly, although the standing crop of ungulate biomass can be high, their productivity is low due to slow reproduction rates. Finally, despite their abundance, ungulates are available as food to those predators only, which are large enough to seize them. To dozens of other species, alive ungulates are inaccessible as prey and become food only as carcasses.

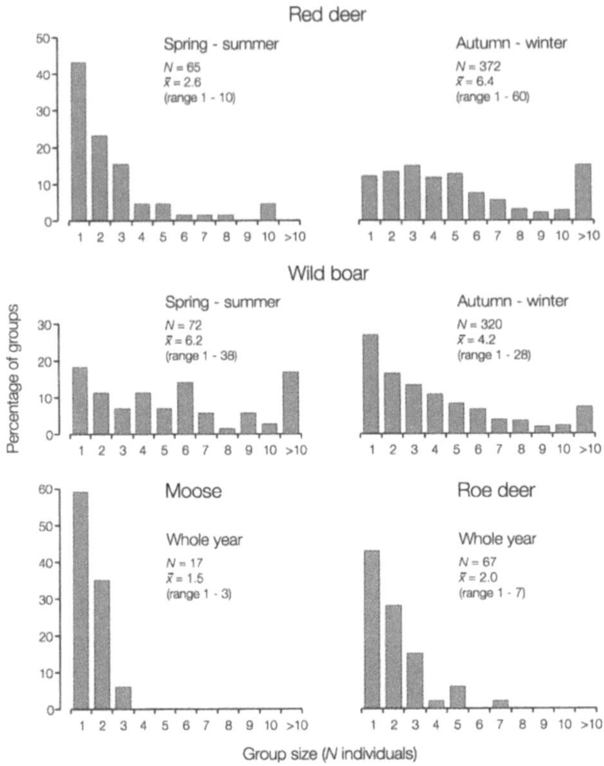

Fig. 2.2. Group size (frequency distribution and mean values) of ungulates as recorded by visual observations in the Polish part of Białowieża Primeval Forest in 1986–1995. *N* number of groups seen

Białowieża's community of ungulates consists of five species: European bison, moose, red deer, roe deer, and wild boar (Table 2.1). In 1991–1995, the red deer dominated (on average, 36% by numbers and 46% by biomass). Roe deer and wild boar were also very common, whereas bison and moose were rare. Mean densities of all ungulates in the Polish part of BPF were high; an average square kilometer of the forest harboured 12 ungulates in winter and 19 in summer (about 990 and 1160 kg of crude biomass, respectively) (Table 2.1).

Visual observations gathered in 1985–1995 showed that moose and roe deer lived in the smallest groups (Fig. 2.2). Many individuals (predominantly adult males) lead solitary lives, and the most frequently seen group is a fe-

Fig. 2.3. Schematic maps of spatial variation in densities of ungulates in various forest districts of BPF. Data from the late winter census in 1991 (driving census in the Polish part and double-count snow tracking on a dense grid of transects in the Belarussian part; Jędrzejewska et al. 1994; Letopis Prirody of the State National Park Belovezhskaya Pushcha)

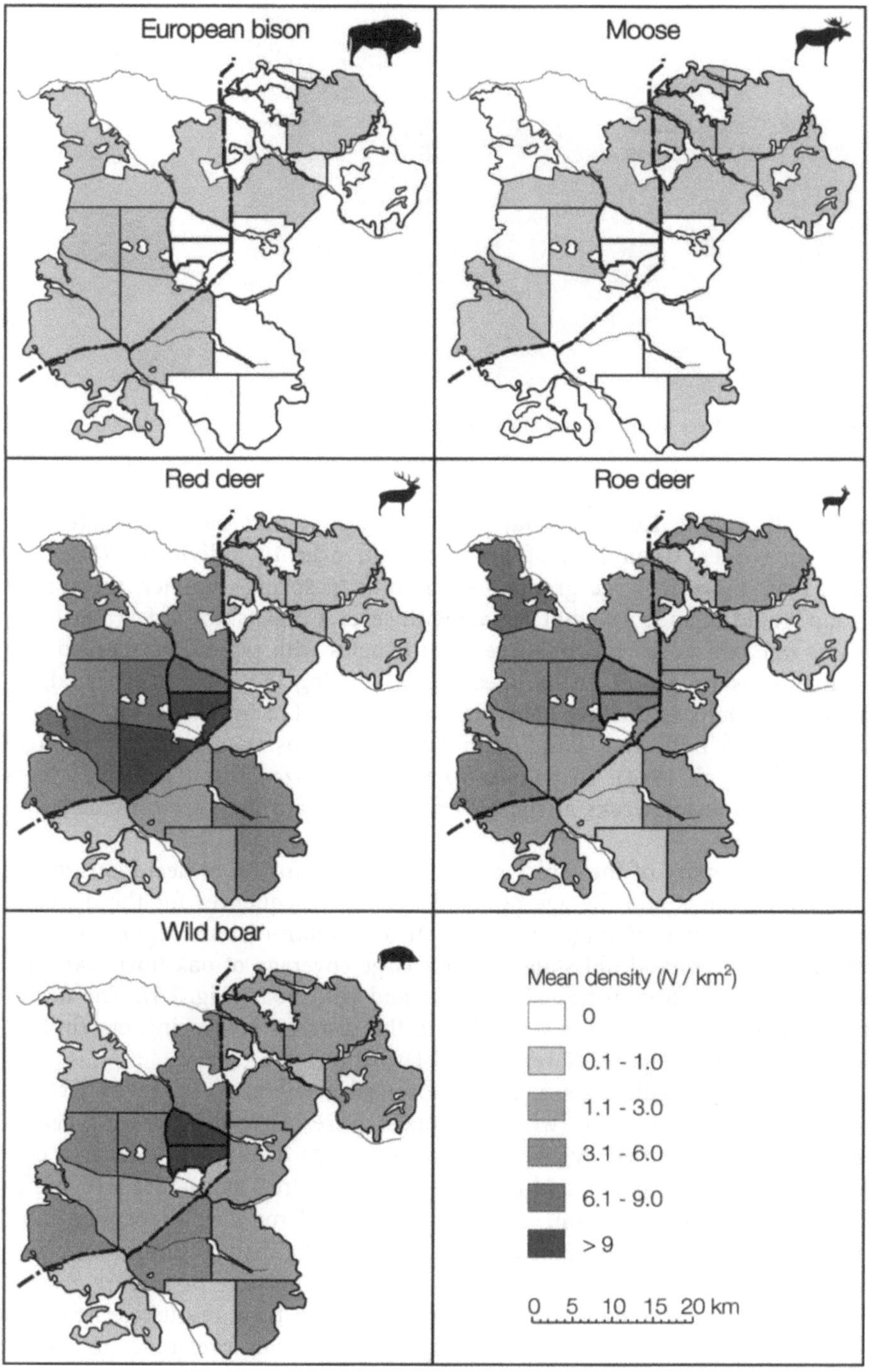
European bison
Moose
Red deer
Roe deer
Wild boar
Mean density (N / km²)
0
0.1 - 1.0
1.1 - 3.0
3.1 - 6.0
6.1 - 9.0
> 9
0 5 10 15 20 km

Fig. 2.4. Moose *Alces alces*, a young male. Photo: S. Wąsik

male with young. Red deer stay in larger groups and they aggregate even more in winter, when 15% of herds contain over 10 individuals (Fig. 2.2). Wild boar move in large groups year round. In spring–summer, the typical group would include an adult female with several piglets and a few yearlings. It was not rare to see two or three adult females with piglets, all merged into a large 'nursery' group. In European bison, during the snow-free period, females, subadults and calves move in large groups (12 inviduals, on average), whereas adult bulls stay alone or in small coalitions (two bulls, on average) (Krasiński et al. 1994). In winter, up to a few dozen bison gather in large herds and stay for weeks at the feeding racks where they are provided with winter fodder.

Spatial diversity of the forest, above all the proportion of deciduous versus coniferous stands and the age of stands, have consequences for the distribution and densities of ungulates. Central and south-eastern regions of BPF, which have preserved old stands with a large coverage of oak-hornbeam forests, harbour high densities of red deer and wild boar (Fig. 2.3). The occurrence of moose (Fig. 2.4) is restricted to the districts embracing marshy valleys of the Leśna, Narew, Narewka, and Hwoźna rivers.

Variation in ungulate abundance (expressed in biomass) in eleven sections of the Polish part of BPF (where ungulates were counted by driving census in 1991) was explained by two factors: percentage of the censused area covered by deciduous forests and percentage area covered by mature tree stands (Fig. 2.5). Herbivores (bison, moose, red deer, and roe deer) depended more on the proportion of deciduous forests, whereas the distribution of an omnivore (wild boar) was shaped largely by the proportion of mature forests. Such pattern of abundance is caused primarily by food resources. Red deer food is made up of tree and shrub twigs, with a strong preference for deciduous species such as hornbeam, oak, hazel, and maple (Dzięciołowski 1970 a). Also, moose and bison are browsers, although bison consumes large

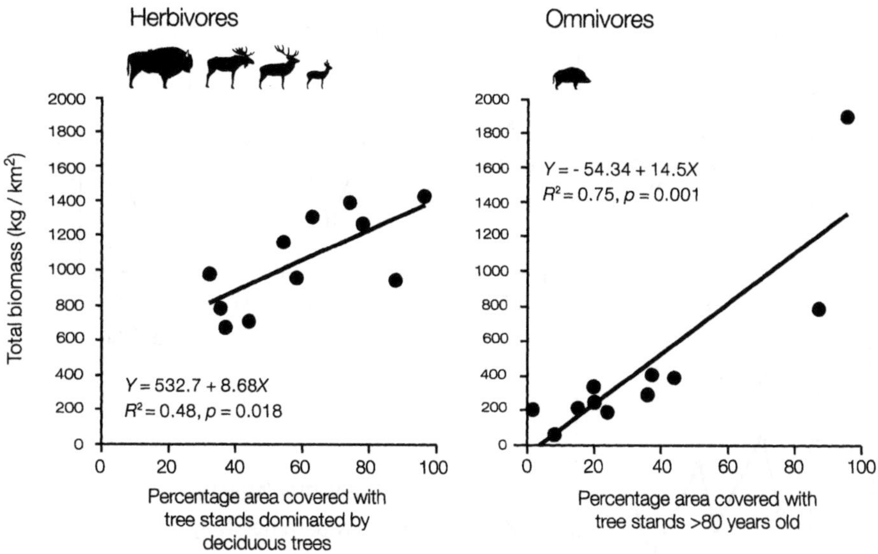

Fig. 2.5. Total crude biomass of ungulates (herbivores: European bison, moose, red and roe deer; omnivores: wild boar) in the eleven sections of the Polish part of BPF censused in 1991 (mapped in Fig. 2.3) in relation to percentage area covered by deciduous stands and mature stands in a given section. Source: Jędrzejewska et al. (1994), modified

quantities of forbs and grasses, too (Table 2.1). Roe deer and bison forage mainly on herb layer plants, but they also browse in winter. The supply of deciduous browse is much higher in all age classes of deciduous forests compared with parallel age classes of coniferous forests (Borowski and Dzięciołowski 1980).

The wild boar relies mainly on acorns (oak seeds), bulbs and roots of forest plants, and soil invertebrates (Jezierski and Myrcha 1975; Kozlo 1975). Mature deciduous forests with oaks produce large quantities of acorns (see Chapter 1), whereas coniferous and young tree stands do not yield any seeds edible for wild boar. Soil invertebrates are most abundant in the oak-hornbeam forests and alderwoods (>20 g/m^2) and least abundant in the pine-spruce forests (7–12 g/m^2) (Kozlo 1975).

Ungulates exhibit weak seasonal variations in numbers compared with small-sized herbivores, which produce >1 litter per year. Seasonal changes in ungulate densities in the Polish part of BPF in 1990–1995 are shown in Fig. 2.6. An adult female bison, moose or red deer usually gives birth to one calf, twins may occur in all species. Litters of one to four fawns (two on average) occur in roe deer (Table 2.1). Thus, the red deer (Fig. 2.7) with roughly one calf per breeding female in early summer (Fig. 2.8) is typical for the ruminants' reproductive capacity. The wild boar differs from other ungulates in this respect. An adult sow is capable of having from 2 to as many as 13 piglets in a litter (Pucek and Sych 1984). In spring, after a good crop of acorns, a mean number of seven to eight piglets per breeding female was recorded.

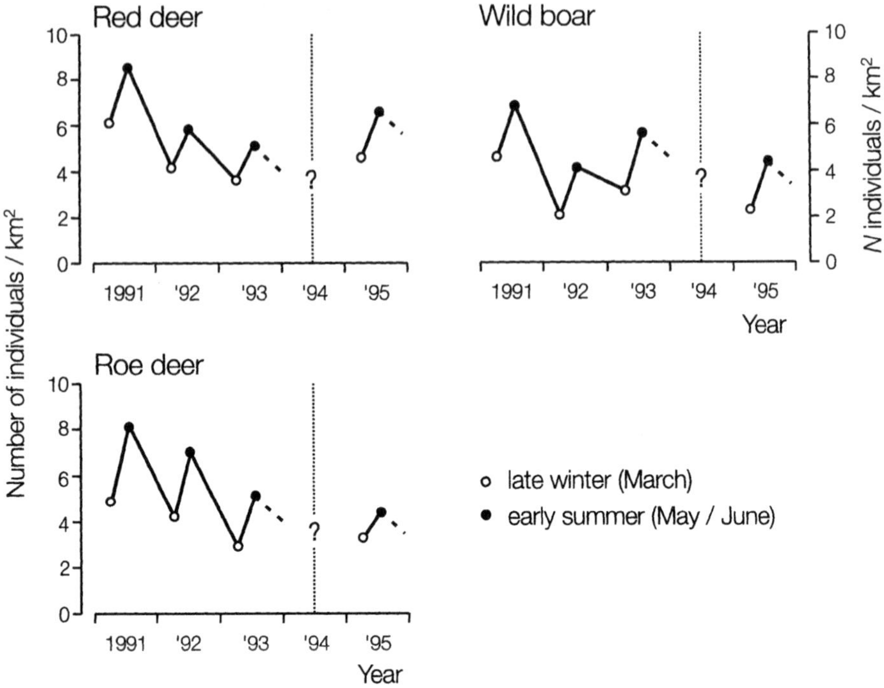

Fig. 2.6. Seasonal variation in mean densities of red deer, roe deer, and wild boar in the Polish part of BPF in 1991–1995. Late winter densities estimated by driving censuses, summer densities calculated based on reproductive parameters for each species. *Dotted line* separates data obtained by somewhat different methods, described in Jędrzejewska et al. (1994) and Kossak (1995)

Fig. 2.7. Red deer *Cervus elaphus*, an adult male guarding several does during a rut in September–October. Photo: S. Wąsik

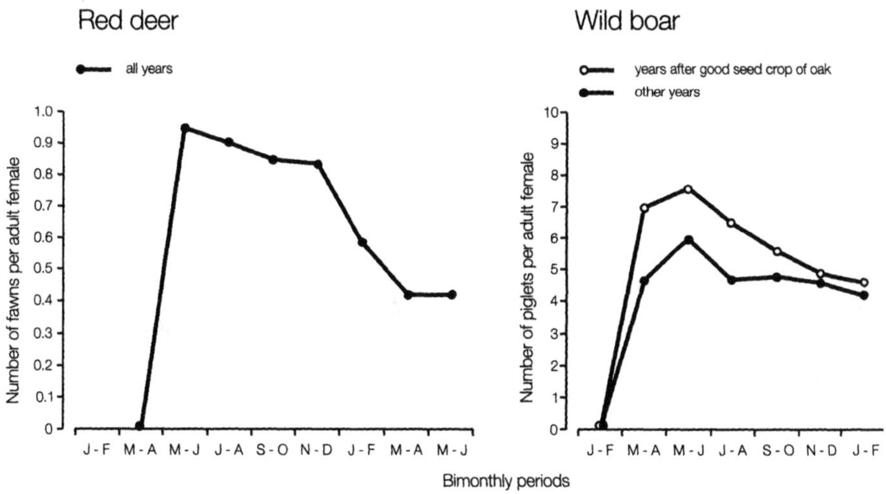

Fig. 2.8. Bimonthly changes in the mean number of juveniles in red deer (own data) and wild boar (after Lebedeva 1956) per adult female in BPF. Note the different *Y*-scales in the two species

In other years, sows led five to six young, on average (Fig. 2.8). Therefore, wild boar have potential for larger seasonal variations in numbers than deer, moose, or bison. In all species of ungulates, numbers decline from summer till the following spring, mainly due to high mortality of juveniles (Figs. 2.6 and 2.8).

In 1991–1995, year to year variations in ungulate numbers were not large. But this was not always so during the last two centuries (Jędrzejewska et al. 1997). Protected royal game in the times of peace and the poachers' target in the times of wars, ungulates experienced a series of ups and downs, including temporal extinctions (Fig. 2.9) caused either by man (bison and moose during World War I) or cooling of climate (red deer before 1800). All three species have returned to Białowieża: moose by spontaneous recolonisation from its eastern refuges (in 1946), red deer and bison by captive breeding and release to the wild (deer in 1890 and bison in 1952; Karcov 1903; Krasiński 1967). Also, for a short period (1890–1920) a non-native introduced fallow deer lived in the forest (Fig. 2.9).

Winter densities of all species (averaged for the whole BPF) varied from <2 individuals/km^2 (65 kg crude biomass/km^2) due to overexploitation by man (after World War I) to 14.4 ungulates/km^2 (1180 kg/km^2) due to 'overprotection' (e.g. 1890–1915, 1965–1990) based on small hunting harvests, supplementary winter feeding, and extermination of large predators (Fig. 2.10).

In the past few centuries, in addition to wild ungulates, the forest was utilised by large herds of domestic cattle in snow-free periods (Jędrzejewska et al. 1997). Since the second half of the 19th century, the impact of cattle became damaging for forest regeneration. In 1909, the maximum number of

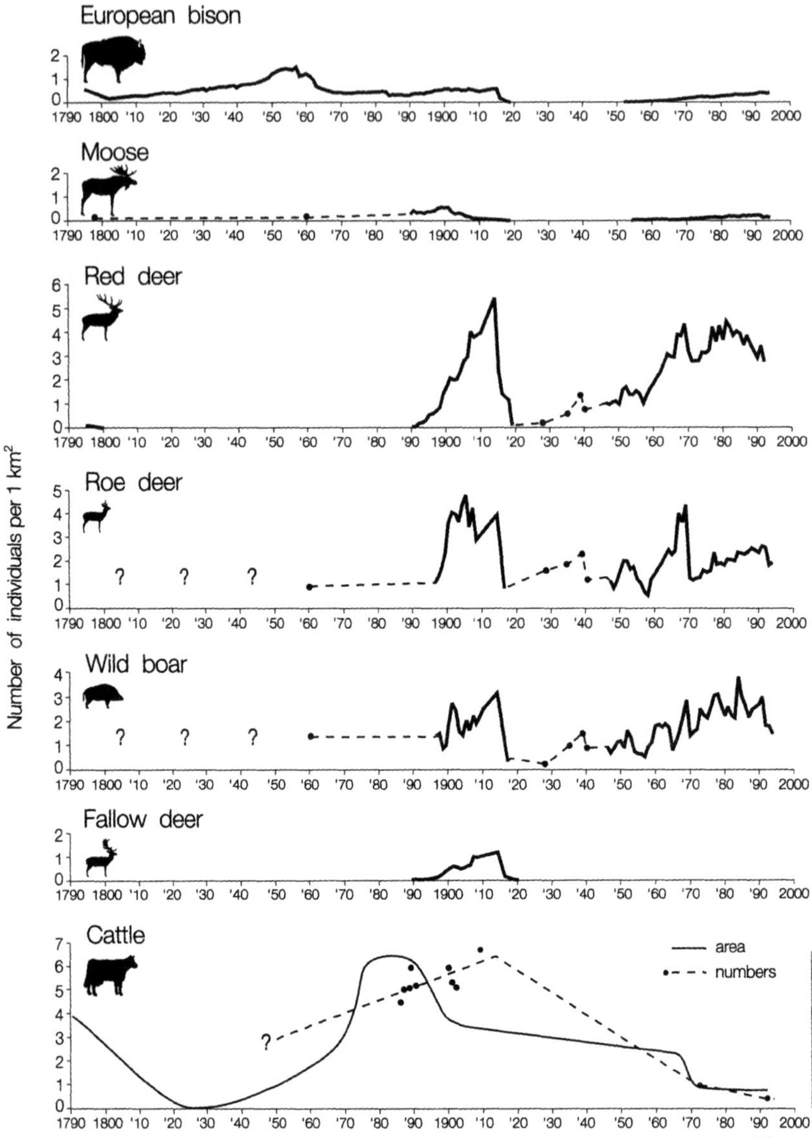

Fig. 2.9. Multiannual variation in winter densities of wild ungulates and summer estimates of number and area grazed by cattle in BPF in 1798–1993. *Thick lines* and *solid points* data based on censuses, *broken lines* interpolations. Jędrzejewska et al. (1997)

6.7 cattle/km^2 grazed in the forest (Figs. 2.9 and 2.10). It was only in the early 1970s that cattle grazing was forbidden by State Forestry Administration in the Polish part of BPF. In the Belarussian part, some 500–600 cattle still utilise riverside meadows, larger glades and openings in the forest.

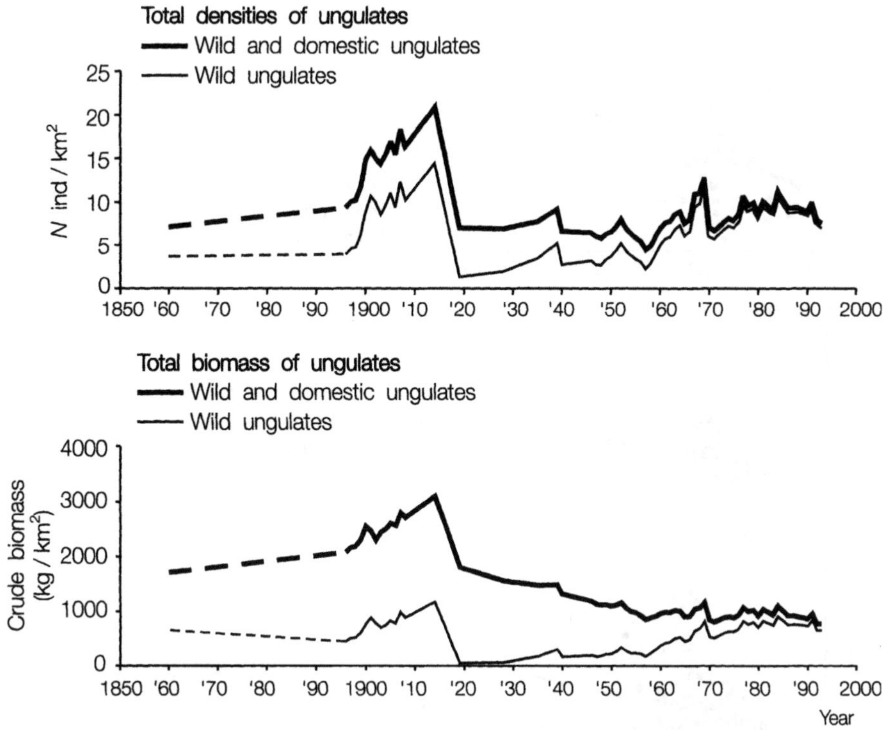

Fig. 2.10. Total densities *upper graph* and crude biomass *lower graph* of ungulates in BPF in 1860–1993 (summarised data from Fig. 2.9). Jędrzejewska et al. (1997)

Nowadays, in this respect, Białowieża Forest enjoys more natural functioning than during the last 200 years.

To explain the multiannual variation in ungulate populations, we need to show (1) factors that determine the level of densities, and (2) factors that influence their year-to-year changes in numbers and increase rates. Figure 2.5 indicates that food availability determines the present spatial distribution of ungulates. Also, in the long term, the upper level of densities seems to be set by food supply (mainly deciduous browse, acorn production, and availability of ground vegetation). Direct measures of changes in food supply are available for the acorn crop only. Nonetheless, climatic variables can be used as indirect measures of food supply and feeding conditions for ungulates. In the long run, warming of climate leads to better regeneration of deciduous trees and shrubs, and thus the expansion of nemoral elements. Cooling of climate favours coniferous trees and other boreal elements of the forest. Equally important is the effect of deep snow and frozen soil on availability of forage. Furthermore, an important role of food in the life of ungulate populations can be shown indirectly. The decrease in ungulate growth rates with growing densities of these animals is interpreted as evidence for food competition (e.g. Messier 1991).

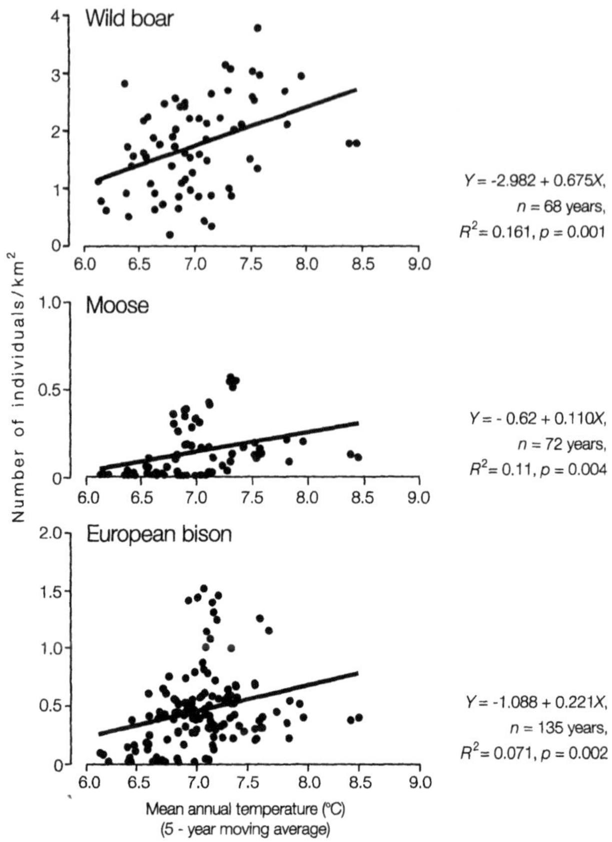

Fig. 2.11. Densities of wild boar, moose, and European bison in relation to the mean annual temperature (5-year moving average). Each point denotes 1 year. Jędrzejewska et al. (1997)

A significant positive relationship between population numbers and annual temperature was observed in wild boar, bison, and moose (Fig. 2.11). A positive, though weak (non-significant), effect was recorded also in red and roe deer (Fig. 2.12). In the latter two species, the presence and numbers of large predators (wolves and lynxes) appeared more important than anything else, and this will be discussed in Chapter 5.

In the two largest species, the bison and the moose, which have the highest food demands, the rate of population increase from year to year was shaped by intra- and interspecific competition for food resources. At high numbers of their own species or large biomasses of all other ungulates, bison and moose did not increase or even slightly declined (Fig. 2.13). Negative relationships between population increase and own density were found in red and roe deer and wild boar. Each of these three species was, however, weakly if at all affected by competition with other ungulates (Fig 2.13). In the wild boar (Fig. 2.14), the rate of population increase was positively correlated to the acorn crop of the preceding autumn-winter season (Fig. 2.15) and nega-

Fig. 2.12. A female roe deer *Capreolus capreolus*. Photo: S. Wąsik

tively to snow cover. At a mean daily snow depth of over 15 cm (in November through March) boar population usually declined. For all years in 1948–1993, average snow cover was 10 cm, and during that period, despite large fluctuations caused by varying crops of acorn, the wild boar population showed a clear trend of increase. The rate of population increase of the two deer species was critically dependent upon the presence and numbers of wolves and lynxes (see Chap. 5).

Periods of wars, insurgencies, political and economic instability in human society always took their tolls from ungulate populations, which served as 'free' food for people. All ungulates suffered population declines during such years, but none was affected more seriously than European bison and moose (Fig. 2.16). It is worth emphasising that, until 1990, the legal hunting harvest of ungulates had rather a small impact on their numbers (Jędrzejewska et al. 1997). In recent decades the planned hunting harvest of game species, i.e. moose, red and roe deer and wild boar were fairly small (usually below 10–15% of population per annum) and affected neither the level of densities nor their changes from year to year. In fact, in the Belarussian part of BPF (the Game Hunting Preserve before 1991), dense and healthy populations of game

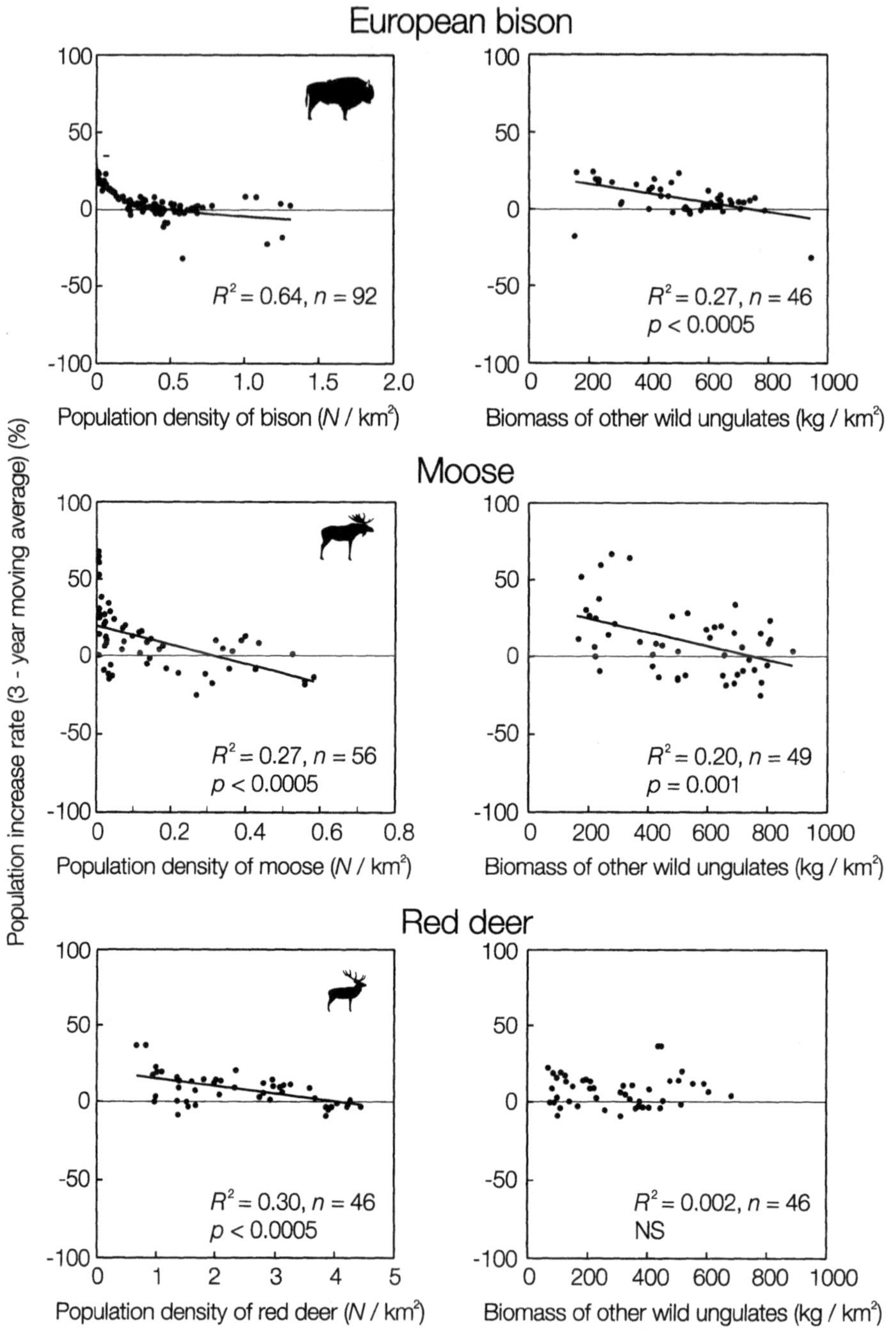

Fig. 2.13. Rate of annual population increase (3-year moving average) of ungulates in relation to density of own species *left panels* and biomass of other wild ungulates combined *right panels*. Each point denotes one year. Strong density-dependence in the time-series of bison, moose, red deer, roe deer, and wild boar numbers was also detected by randomization test (Pollard et al. 1987): $p = 0.001$ in each of the five species. Jędrzejewska et al. (1997)

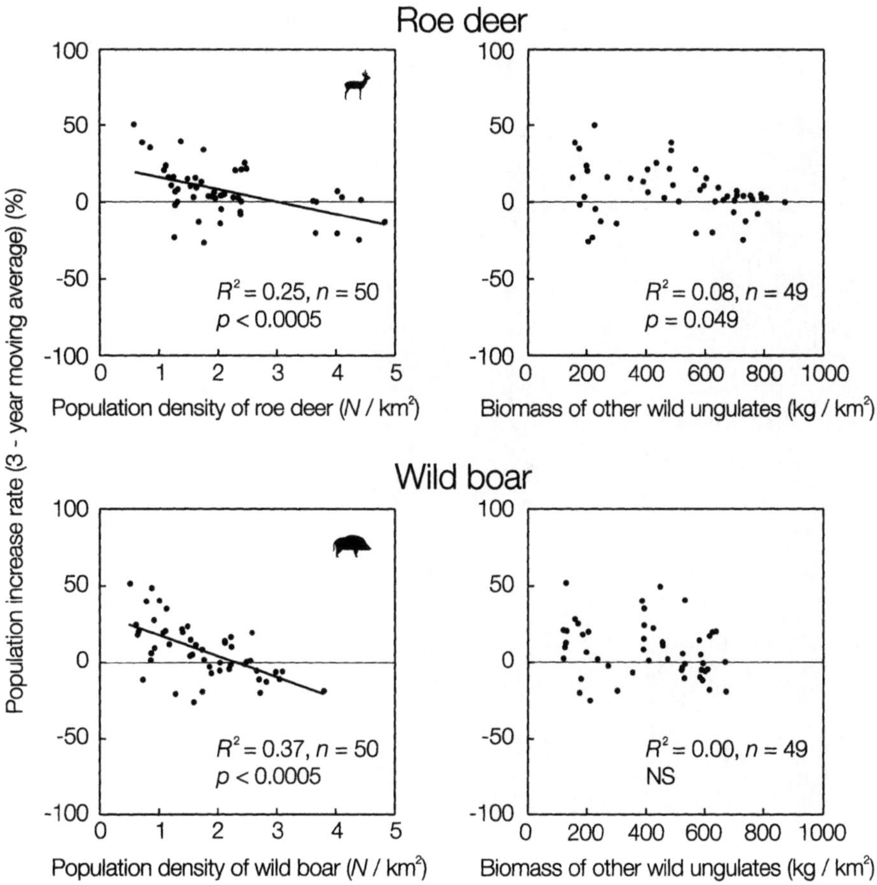

Fig. 2.13 (continued)

has been one of the management priorities. The disparity, so acute in the whole of Europe, between the goals of forestry and hunting management have only recently emerged in the Polish part of BPF. There, a growing area of replantations brought about a problem of damage caused by ungulates, a problem not existing in the natural forests. Since 1991, due to foresters' appeals, the annual hunting harvest of moose, roe deer, red deer, and wild boar increased two to three times compared with earlier years. More importantly, for the first time in the history of Białowieża Forest, the local forestry authorities explicitly stated that now the purpose of hunting is to reduce ungulates to very low numbers. European bison, a protected species in both Poland and Belarus, has so far been safe from such drastic decisions, but not from complaints by foresters.

Fig. 2.14. Wild boar *Sus scrofa*, an adult male. Photo: S. Wąsik

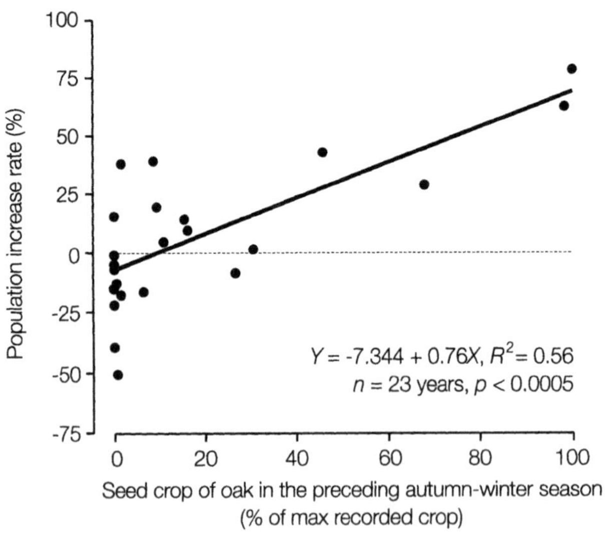

Fig. 2.15. Population increase rates of wild boar in 1960–1993 in relation to the crop of oak seeds in the preceding autumn-winter seasons. Each point denotes one year. Jędrzejewska et al. (1997)

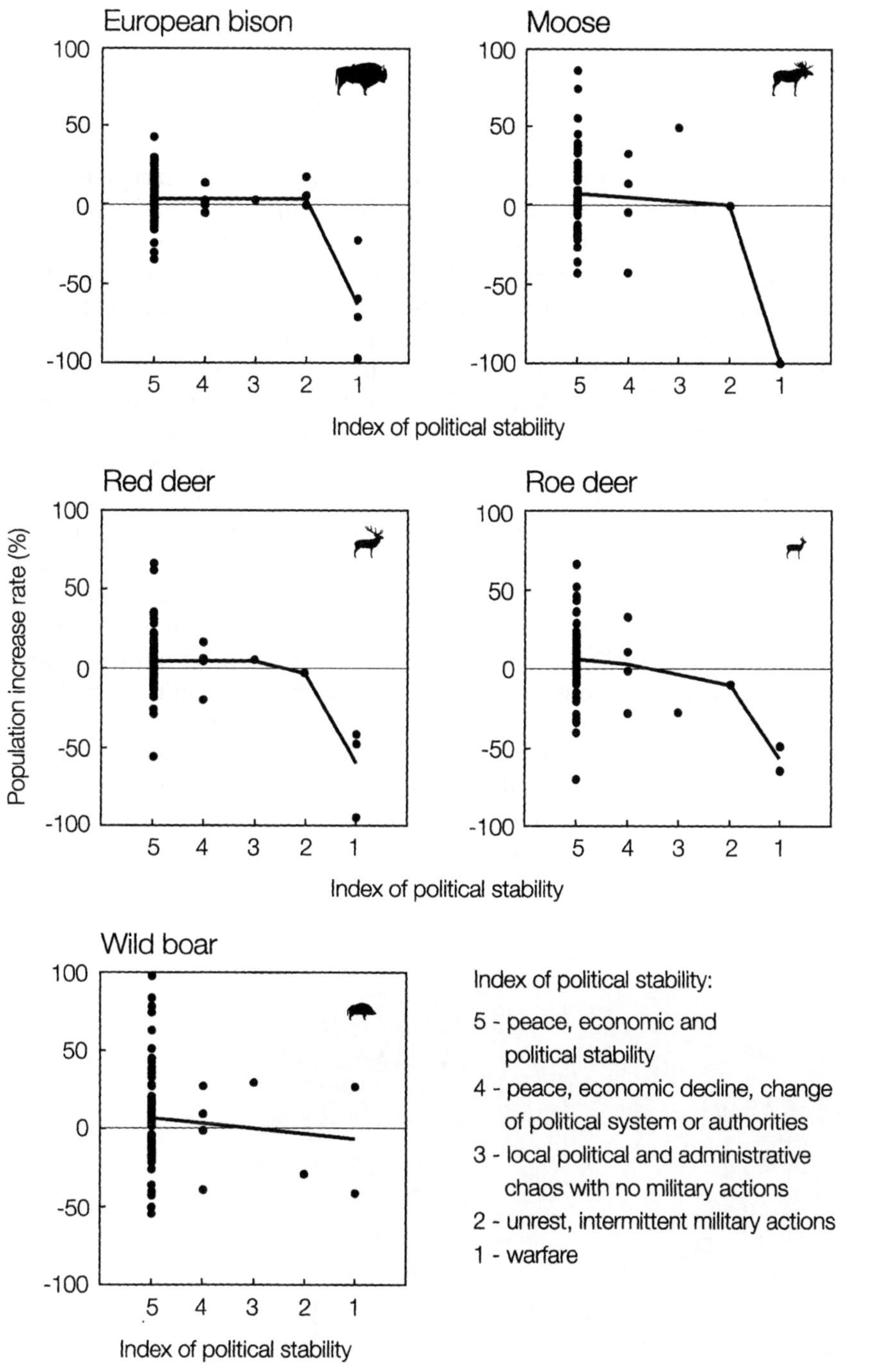

Fig. 2.16. Rates of annual population increase of ungulates in relation to an arbitrary index of political stability in the region of BPF in 1809–1993. Each point denotes 1 year. Lines join the mean values in each class of political stability. Jędrzejewska et al. (1997)

2.4 Rodents - Voles and Mice Everywhere

If we look at vertebrates of European temperate forests from the point of view of small and medium-sized predators, we will find three animal groups of particular interest as food: small terrestrial rodents, birds, and frogs. Of these three, rodents are the most predictable food resource. Except for the northern birch mouse, all species are active year round. They neither disappear for hibernation (as frogs do) nor migrate (as many birds do). In the natural habitats of Białowieża Primeval Forest, two communities of small terrestrial rodents can be distinguished: the forest community and the open marsh community. These two will be given detailed descriptions. Then, a few lines are devoted to arboreal rodents (squirrels and dormice) and beavers.

The community of forest rodents is dominated by the bank vole and the yellow-necked mouse, which together make up 79% of all rodents (Table 2.2). Their mutual proportions vary among forest types; mice prefer deciduous forests (especially oak-lime-hornbeam stands), whereas voles are more numerous in coniferous and mixed forests with richer vegetation on the forest floor (Aulak 1970). Bank voles (Fig. 2.17) feed on green plants and supplement their diet with seeds, fruit, and invertebrates (Gębczyńska 1976). Mice are granivorous (Fig. 2.18); they live on tree seeds but eat some green vegetation and invertebrates, too (Zemanek 1972; Gaiduk et al. 1985). The third most numerous forest rodent, the pine vole, contributes a mere 7% to the community. It feeds on roots and bulbs of herbs and green vegetation (Holišová 1965).

Open sedge and reed marshes in the river valleys are inhabited by voles in the genus *Microtus*, predominantly the root vole (65% of all rodents; Table 2.2). Common voles and striped field mice are quite numerous in drier grassy parts of the valleys, and bank voles occur in the forest-marshland ecotones and willow brushwoods. Voles in the genus *Microtus* are definitely her-

Fig. 2.17. Bank vole *Clethrionomys glareolus*, a herbivorous-granivorous rodent feeding largely on herbaceous vegetation of the forest floor and utilising tree seeds in years of their heavy crops. Photo: S. Wąsik

Table 2.2. Species structure of forest and open marsh communities of small terrestrial rodents inhabiting Białowieża Primeval Forest; data from 10-year trapping on seven plots in forests and seven plots located in the open habitats in the Narewka river valley, Białowieża National Park. Ranges of body masses after Kowalski et al. (1984). Mean body masses (calculated from trapped animals) in parentheses

Species	Body mass (g)	Percentage in forest community	Percentage in open marsh community
Bank vole (*Clethrionomys glareolus*)	10–38 (17)	53.1	13.3
Yellow-necked mouse (*Apodemus flavicollis*)	17–43 (31)	26.0	0.4
Pine vole (*Pitymys subterraneus*)	10–24 (17)	7.0	–
Northern birch mouse (*Sicista betulina*)	6–17 (8)	4.5	0.2
Wood mouse (*Apodemus sylvaticus*)	12–28	+	–
Harvest mouse (*Micromys minutus*)	5–10 (8)	1.2	0.9
Short-tailed vole (*Microtus agrestis*)	19–48 (23)	4.3	1.8
Root vole (*Microtus oeconomus*)	20–74 (26)	1.8	65.2
Common vole (*Microtus arvalis*)	18–37 (19)	1.1	10.6
Striped field mouse (*Apodemus agrarius*)	11–36 (17)	1.0	7.5
Water vole (*Arvicola terrestris*)	60–200	+	+
House mouse (*Mus musculus*)	6–25	–	+
Common rat (*Rattus norvegicus*)	166–455	–	+

Fig. 2.18. Yellow-necked mouse *Apodemus flavicollis* eating an acorn. Photo: S. Wąsik

bivorous; they consume green plants, their bulbs and roots, and bark of young woody saplings (Hansson 1971). Striped field mice live mostly on seeds of grasses and forbs (Holišová 1967). In both the forest and open marsh communities of rodents, the dominant species are those that also weigh most in the predators' diets. Thus, the fluctuations in the numbers or

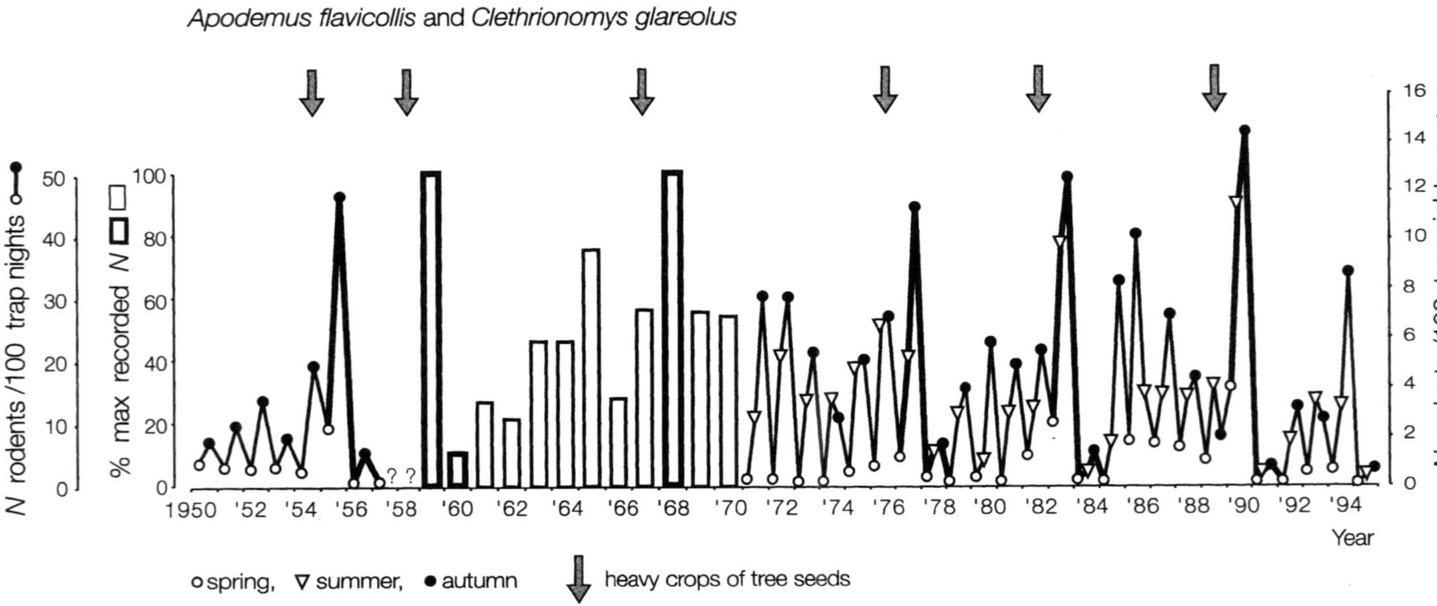

Fig. 2.19. Multiannual dynamics of forest rodents in Białowieża Primeval Forest (combined densities of yellow-necked mice and bank voles). *Arrows* heavy seed crops of oak and hornbeam trees. *Thick line* rodent outbreaks and crashes triggered by masts, *thin line* years of moderate abundance of rodents. Four series of trapping are presented: (1) 1950–1957 in the Belarussian part of BPF, rodents trapped in spring and autumn (after Golodushko, 1958, modified); (2) 1959–1964 and (3) 1965–1970 in Białowieża National Park (Polish part), the total number of rodents captured in each year is presented as a percentage of maximum numbers (recorded in 1959 and 1968, respectively) (after Pucek et al. 1993); (4) 1971–1995 in BNP, rodents trapped in spring, summer, and autumn (after Pucek et al. 1993, and Z. Pucek, unpubl.)

Fig. 2.20. In a year of good crop, the forest floor under large oak trees is covered with acorns. Photo: W. Jędrzejewski

availability of bank voles and yellow-necked mice in the forests, and root voles in the marshes are driving forces of a chain of phenomena in the predators' behaviour.

Small rodents reproduce very fast (Kowalski et al. 1984). A female bank vole, after a 20-day pregnancy, gives birth to a litter of five young, which leave their nest after 21 days, mature when 50 days old, and begin to reproduce. Meanwhile their mother, pregnant after post-partum oestrus, bears another litter. Conditions permitting, one pair of bank voles may give rise to 28 voles after 100 days, and 250 voles after 200 days! The reproductive potential of yellow-necked mice is the same and that of root voles even higher, 36 and 720 voles originating from one pair after 100 and 200 days, respectively. As homeothermic animals building well insulated underground or ground nests, rodents can reproduce during any season, severe winters with deep snow cover included.

However, in the temperate forests, scarce food resources in the cold season mean that, in most years, forest rodents reproduce from April until September/October. Spring numbers are low but grow until the autumn, and then decline throughout winter (Fig. 2.19). This pattern is caused by the seasonal course of food abundance. In deciduous forests, ground vegetation, composed of soft-tissue dicotyledons, grows in the shadow of multilayered canopies of trees and shrubs. It is moderately abundant in late spring and summer (430–1075 kg dry weight/ha, see Chap. 1) but it nearly vanishes in autumn and winter. Another important food to forest rodents are seeds of trees, particularly oak, hornbeam, maple, lime, and ash. Their fruiting pattern (described in Chap. 1) is characterised by spectacular masts that occur at 6–9-year intervals and poor or nil crops in other years (Fig. 2.20).

The multiannual pattern of population dynamics of forest rodents is characterised by 4–7 years of moderate densities and 2 years of outbreak and

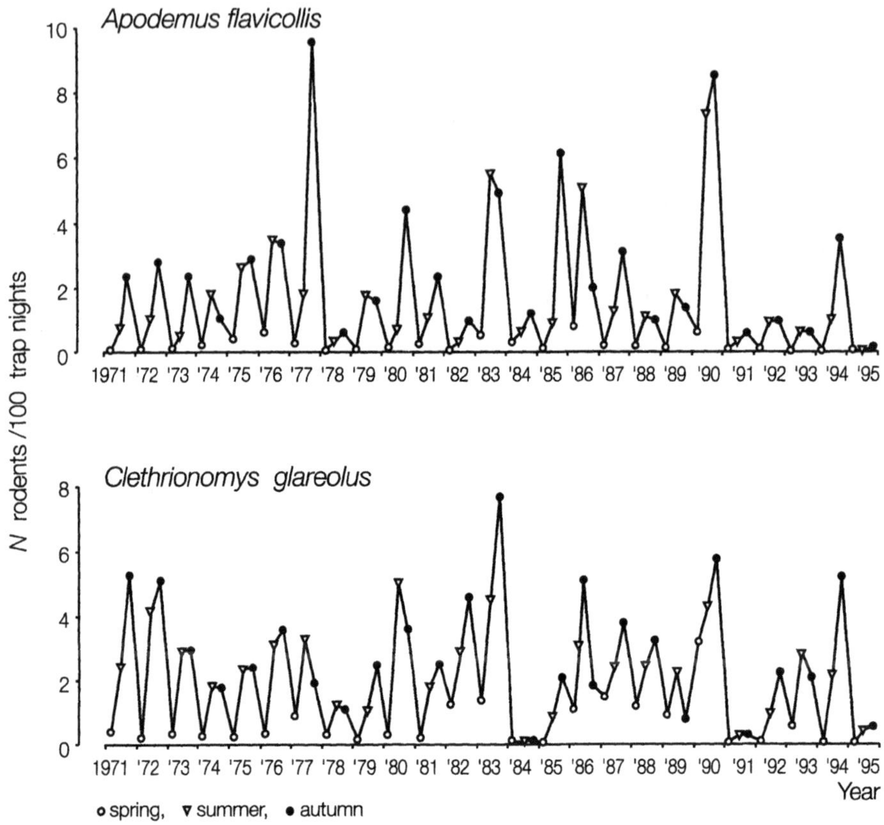

Fig. 2.21. Population dynamics of the yellow-necked mice and bank voles in 1971–1995 in Białowieża National Park. (After Pucek et al. 1993, and Z. Pucek, unpubl.)

crash triggered by heavy crops of tree seeds. Interestingly, rodents' response to masts is clearer at the community level than in each of the rodent species separately (Figs. 2.19 and 2.21). In winters following the mast, both yellow-necked mice and bank voles can reproduce. Due to breeding that lasts continually, although with varying intensity, for 1.5 years (e.g. from April 1989 until September 1990), rodents reach very high numbers (200–300 ind/ha) 1 year after masts. However, outbreaks last for 1 year only. They are followed by rapid winter decline and usually by a crash in rodent numbers the following year. After the crash, there are 4–7 years of moderate densities, when no winter breeding occurs and rodents exhibit seasonal fluctuations only.

The 21-year series (1971–1991) of data shown in Figs. 2.19 and 2.21 was analysed in detail (Pucek et al. 1993). Summer (July) densities of mice depended on the numbers of adults that survived winter, but outbreak and crash years did not fit the overall pattern (Fig. 2.22). In bank voles, summer densities depended on both numbers of overwintered adults and the temperatures recorded in June–July. The latter factor, indirectly reflecting the

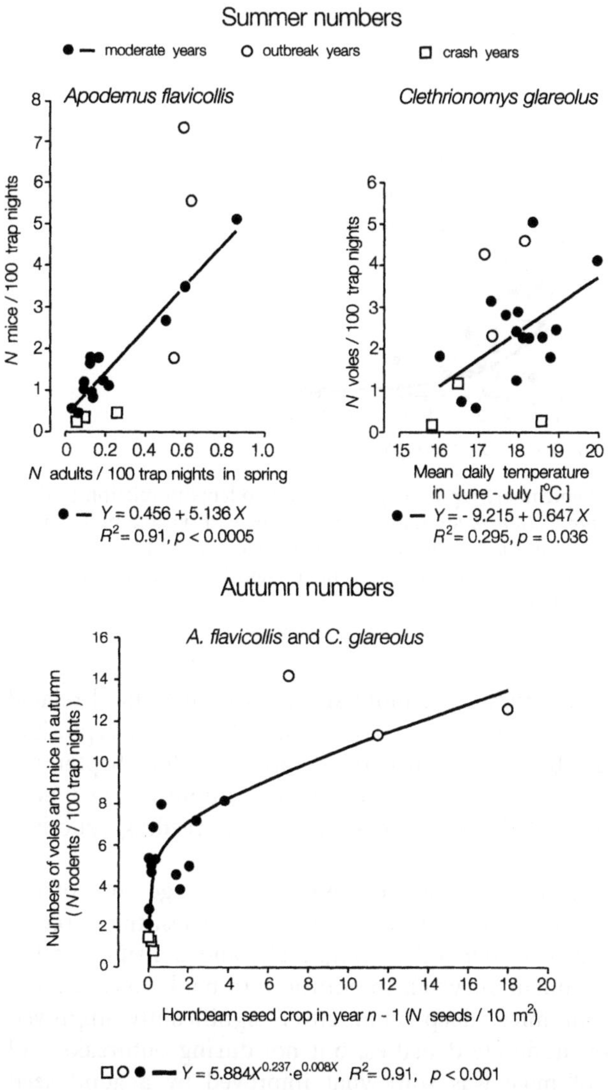

Fig. 2.22. Summer and autumn numbers of forest rodents in relation to factors affecting them (1971–1991). Each point denotes 1 year. *Upper graph* regressions calculated for moderate years. *Lower graph* regression line includes all years. Pucek et al. (1993)

abundance of food (green vegetation) was most important in years of moderate densities of voles. Autumn densities of forest rodents were influenced primarily by the seed crop of the previous year (Fig. 2.22).

Trapping in September–October coincided with the latest seasonal phase of rodent reproduction. At that time, breeding intensity of both voles and mice was inversely density dependent (Fig. 2.23). This shows that food shortage and competition for food might have been the prime factor governing

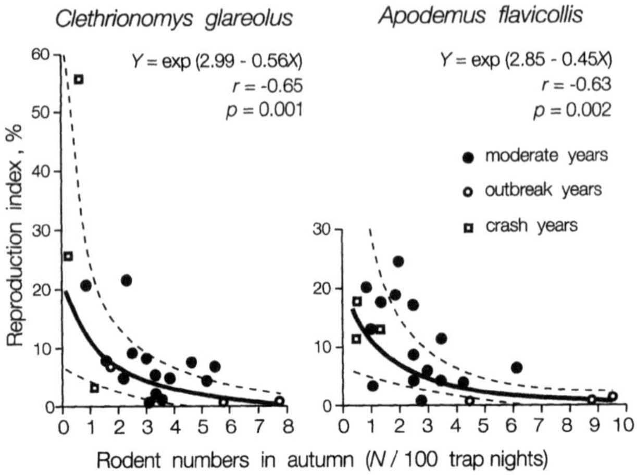

Fig. 2.23. Intensity of autumn (September) reproduction of forest rodents in relation to their density index (1971–1991). *Reproduction index* number of females with visible pregnancies and the youngest juveniles (<11 g in bank voles, and 14 g in yellow-necked mice) as percent of the total number of rodents captured in autumn. Regression lines include moderate, outbreak, and crash years. Pucek et al. (1993)

the cessation of breeding in autumn, the more so because other food-related variables contributed to explaining the total variation in autumn reproduction of rodents. Those variables were warm temperatures in July–September in the case of bank voles, and seed crop of the current autumn in mice; both of them were positively correlated with indices of autumn breeding (Pucek et al. 1993).

Winter mortality of forest rodents was very high: on average 86% of autumn numbers in mice and 77% in bank voles. In both species, the percentage winter mortality was density dependent (Fig. 2.24), and density was the decisive factor influencing rodent survival. In addition, in bank voles leading a more subnivean life than mice, deep snow cover significantly improved winter survival in years of moderate densities, but not during outbreaks and crashes. Winter survival of mice was somewhat improved by a good seed crop and warm ambient temperatures in the cold season (Pucek et al. 1993).

Based on regression between rodent density estimates and trapping indices (Jędrzejewski et al. 1995, 1996a), we calculated the abundance of forest rodents in Białowieża National Park in 1971–1993 (Fig. 2.25). Combined densities of forest rodents were, on average, 13.5 ind/ha (274 g/ha) in spring and increased sixfold, to 78 ind/ha (1673 g/ha), by autumn (Table 2.3). The largest overwinter declines in rodent abundance were about 30-fold. Abundance of rodents belonging to all other species dwelling in forests (species listed in Table 2.2) would be 10–20% of the two dominants.

Snow cover was an important factor modifying availability of rodents to predators. During snow tracking conducted in four winters of moderate densities of rodents (1985/1986–1988/1989) we recorded, on average, 8.2 rodent

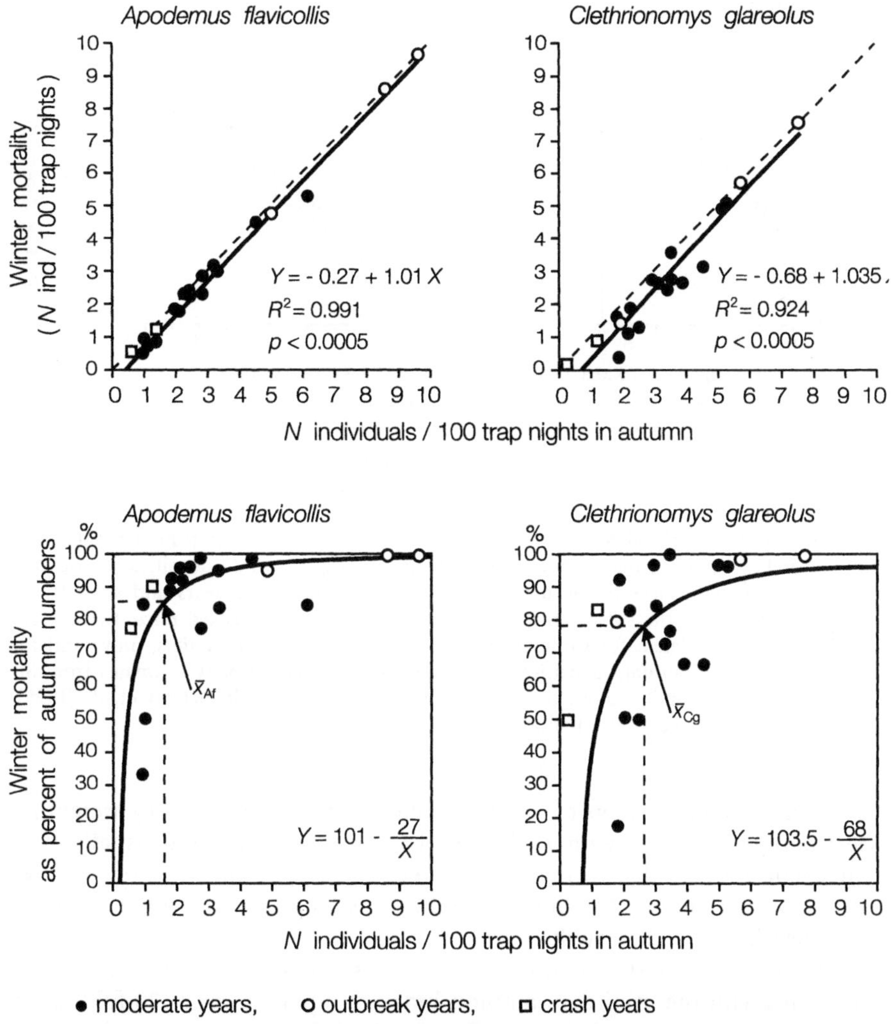

Fig. 2.24. Winter mortality of forest rodents in relation to autumn density index (1971–1991). *Upper graphs* winter mortality (calculated as the difference between the numbers trapped in autumn and numbers of overwintered adults in the following spring) plotted against autumn numbers of rodents. *Dotted line* $Y=X$ denotes the 'extinction line'. *Lower graphs* relationship between the percentage winter mortality (winter mortality expressed as percent of autumn numbers) and the autumn numbers. Empirical data points are overlaid on the expected curves calculated as the quotient of the two functions from upper graphs: the empirical regression line and the 'extinction line' in each species. $\bar{x}_{Af}$ and $\bar{x}_{Cg}$ mean percentage mortality of mice and voles, respectively. Pucek et al. (1993)

tracks per 1 km per day when snow was 1–10 cm deep, 5.3 tracks at snow depth 11–20 cm, and fewer than two tracks when snow cover exceeded 21 cm (Jędrzejewski and Jędrzejewska 1993).

To *Microtus* voles, sedge and tall forb marshes in the river valleys offer much more abundant food resources than those of the forests. In summer,

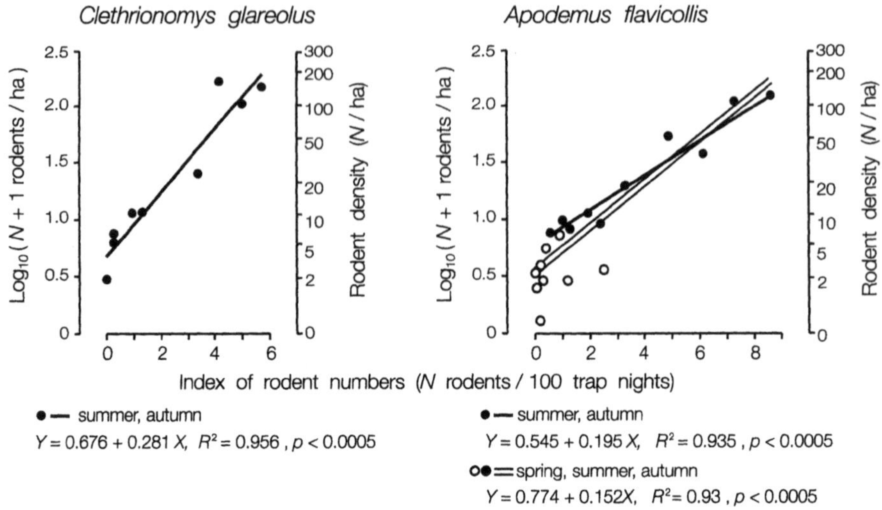

Fig. 2.25. Conversion of indices of rodent abundance (*N* rodents/100 trap nights; data from trapping in 1971–1991 in Białowieża National Park, see Fig. 2.21) into densities (*N* rodents/ha) by relating the indices to the density estimates from the same years and seasons, on study plots located in the same forest type of BNP. In the yellow-necked mice, two regression lines were calculated: for summer and autumn densities (*closed points*, only) and for spring, summer, and autumn densities (*open* and *closed points*). Density estimates from Sikorski (1989), Wołk and Kozłowski (1989), Wójcik (1993), and Jędrzejewski et al. (1995). From: Jędrzejewski et al. (1995, 1996a), modified

the standing biomass crop of live phytomass exceeds 4000 kg dry weight/ha (see Chap. 1). Habitats with such abundant ground vegetation usually harbour cyclic rodents due to the fact that even in winter the level of food resources is high enough to allow for winter breeding or good winter survival (Jędrzejewski and Jędrzejewska 1996).

In Białowieża Forest, microtine rodents in the open marshes exhibit 3–4-year cycles with the minimum autumn index of numbers about 20-fold lower than the maximum recorded index (Fig. 2.26). Unfortunately, no studies were done on the densities of *Microtus* voles in Białowieża. Buchalczyk and Pucek (1968) and Buchalczyk et al. (1970), who studied the root voles in an open grassy marsh in Augustów Forest (NE Poland), found that autumn densities varied from 65 to 756 voles/ha.

Arboreal rodents of Białowieża Primeval Forest include the fat dormouse *Glis glis*, the forest dormouse *Dryomys nitedula*, the common dormouse *Muscardinus avellanarius*, and the red squirrel *Sciurus vulgaris*. All dormice are rare as can be judged from their sporadic occurrence in the scats and pellets of predators as well as uncommon captures and findings by humans. The largest arboreal rodent, the red squirrel is fairly common. Squirrels are widely distributed in all mature forests of Białowieża, wherever large fructificating spruce trees occur. Squirrels attain seasonally highest numbers in July–August, when the spring-born cohort of young begins independent life. In 1985/1986–1991/

Table 2.3. Ranges of densities of forest rodents in deciduous forests of BNP in 1971–1995

Season	Range of densities in 1971–1995			Mean (SD) for 1971–1995
	Moderate	Outbreak	Crash	
	Density of bank voles (*N* ind/ha)			
Spring	4–12	7–116	4–5	10.4 (22.1)
Summer	5–127	39–90	4–9	28.8 (30.1)
Autumn	14–143	15–194 (700?)	4–8	53.3 (58.3)
	Density of yellow-necked mice (*N* ind/ha)			
Spring	3–4	5	3	3.1 (0.8)
Summer	5–35	10–76	5–6	13.2 (15.2)
Autumn	5–50	32–171	6–8	24.8 (38.6)
	Combined density of voles and mice (*N* ind/ha)			
Spring	6–15	12–121	7–8	13.5 (22.5)
Summer	10–162	49–146	10–15	42.0 (42.3)
Autumn	11–155	186–315 (730?)	11–15	78.0 (77.2)
	Standing crop of biomass of voles and mice (g/ha)			
Spring	150–290	270–2120	150–170	274 (389)
Summer	330–3230	980–3540	250–330	900 (912)
Autumn	260–2870	4300–7050	270–340	1673 (1726)

Data from trapping indices shown in Fig. 2.21 converted into densities using the regression equations in Fig. 2.25). Uncertain estimates (in parentheses with question mark) not included in mean densities. For calculating the biomass, mean body masses of 17 g in bank voles and 31 g in yellow-necked mice (data from trapping on seven plots in various forest types of BNP) were used.

1992, indices of winter abundance of squirrels were positively correlated with the crop of spruce cones in the previous year ($R^2=0.685$, $n=6$ years, $p<0.05$). Spruce seeds are the main food of squirrels. In winter, squirrels may be inactive for long periods if the weather is harsh. Snow-tracking data averaged for seven cold seasons in 1985/1986–1991/1992 showed the decline in squirrel activity from an average of 2 tracks km^{-1} day^{-1} at ambient temperatures of 0 to −5 °C to a mere 0.5 tracks km^{-1} day^{-1} when the temperature dropped to −20 °C.

The largest rodent of Europe, the beaver, *Castor fiber*, is a semiaquatic animal. In the winter of 1993/1994, beaver settlements were surveyed in Białowieża Forest (Sidorovich et al. 1996). The larger the river, the more abundant were beaver settlements. On the smallest rivers, which dry out in summer (Orłówka, Braszcza), beavers did not settle. Medium-sized rivers (e.g. Hwoźna, Perevoloka) harboured, on average, 2.4 settlements per 10 km of the river course, and large rivers (e.g. Leśna, Narew) 3.9 settlements/10 km. In total, about 60 settlements inhabited by 220 beavers were estimated to exist in the whole BPF, i.e. 15.2 beavers/100 km^2.

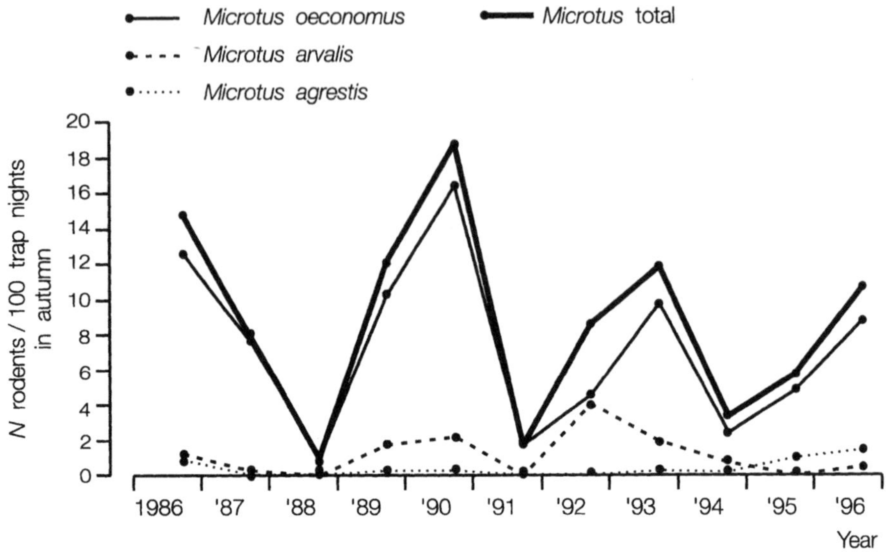

Fig. 2.26. Population dynamics of voles in the genus *Microtus* in the open marshy river valley, 1986–1996, as censused by autumn trapping. Data pooled for seven trapping plots located along the Narewka river, BNP

2.5 Brown Hare

A typical animal of open grasslands and farmlands, the brown hare *Lepus europaeus* is widespread but uncommon in Białowieża Primeval Forest. During snow tracking conducted in 1985/1986–1991/1992, we recorded on average 0.13 to 0.84 hare tracks km^{-1} day^{-1} on transects going through the pristine old growths of BNP, and 2.5–10 times higher indices of abundance (0.75–8.14 tracks km^{-1} day^{-1}) in the ecotone of forests and open marshes in the Narewka river valley or fallows in Białowieża Glade. In the winters of 1956/1957–1961/1962, the wardens of Białowieża National Park surveyed hares by snow tracking and estimated their densities. They recorded between 0.21 to 1.16 tracks km^{-1} day^{-1}, which corresponded to densities of 0.19 to 0.56 hares/km^2. Therefore, based on a total of 11 years of censuses we can reasonably assess that winter densities of brown hare in the woodland are usually below 0.5 ind/km^2, and in open marshes and fields within the forest they vary from 0.5 to 3 hares/km^2. These figures are remarkably low, compared not only with extensive farmlands on rich soils in western Poland (10–65 hares/km^2) but also with the average densities in agricultural landscapes in Poland (15 hares/km^2; Pielowski 1975).

Rarely though systematically, the arctic hare *Lepus timidus* is recorded in Białowieża Forest, the south-western periphery of its range (Polish Red Data Book of Animals, 1992).

Fig. 2.27. Common shrew *Sorex araneus*. Photo: P. Fabijański

2.6 Shrews and Moles

Tiny and voracious predators of invertebrates, the shrews (Fig. 2.27) are widespread in Białowieża Forest. It is not uncommon to see these nocturnal animals in late summer, as intact carcasses on a forest road or, in winter, as killed but uneaten prey on a fox trail. Besides five species of soricids, the order of Insectivora is represented by the eastern hedgehog (Fig. 2.28) and the mole (Table 2.4). All insectivores live both in the forests and open areas (natural and man-made). So far, no studies have been done on densities and abundance of hedgehogs in Białowieża Forest. In 1946–1994, a total of 190 moles were found in traps set for Micromammalia during many studies conducted by the Mammal Research Institute in the forests of Białowieża. Moreover, the collection included 146 moles that were found dead or deliberately captured in the gardens and fields of Białowieża Glade. Fairly big, though casual, the whole sample does not document multiannual variation in mole numbers. It does, however, approximate the seasonal changes in their abundance, both in terms of population size and availability to predators. Although most trapping efforts for small mammals occurred from April until October, moles were usually cap-

Table 2.4. Community of Insectivora inhabiting Białowieża Primeval Forest

Species	Body mass (g)	Percentage share in the Soricidae guild
Eastern hedgehog (*Erinaceus concolor*)	240–1230	
Mole (*Talpa europaea*)	70–120	
Common shrew (*Sorex araneus*)	5–16 (8)	72
Lesser shrew (*Sorex minutus*)	2–6 (3.5)	16
European water shrew (*Neomys fodiens*)	7–28 (14)	8
Mediterranean water shrew (*Neomys anomalus*)	6–20 (9.5)	3
Masked shrew (*Sorex caecutiens*)	3–8 (5)	1

Percentage shares in the community calculated for Soricidae only (data from trapping on seven plots in the BNP and seven plots in open marshes of the river valley). Body masses from Pucek (1984) and authors' trapping data. Mean body masses (calculated from trapped animals) in parentheses.

Fig. 2.28. Eastern hedgehog *Erinaceus concolor*. Photo: W. Jędrzejewski

tured in June–August, with a peak in July (Fig. 2.29). Affected by uneven trapping intensity in the cold and warm seasons, monthly distribution of trapped moles is nonetheless likely to reflect the availability of these subterraneous mammals to predators. The age structure of trapped moles allowed us to roughly reconstruct the seasonal changes in their population size. Female moles bear only one litter per year. The first juveniles were captured in June, and the peak of their numbers was recorded in July (Fig. 2.29). From midsummer till the following spring, numbers of moles declined.

Communities of Soricidae inhabiting forests and open sedge marshes in the river valleys do not differ much; the common shrew is dominant (72%, on average), followed by the lesser shrew (16%) (Table 2.4). Wet marshes

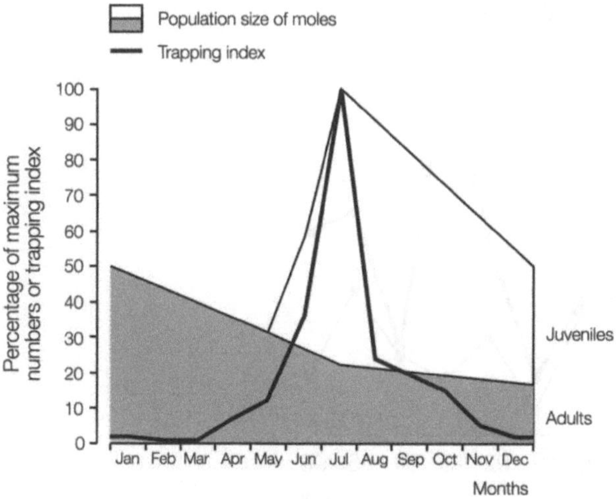

Fig. 2.29. Relative numbers of moles *Talpa europaea* trapped in 1946–1994 in Białowieża Primeval Forest ($N=336$). Total number of moles captured in July taken as 100%. Seasonal course of population dynamics reconstructed from age structure of the sample of trapped animals

close to rivers are the domain of water shrews, which dominate soricids trapped in such areas (L. Rychlik, pers. comm.). A 10-year trapping conducted by us on seven plots located in the mature forests of Białowieża National Park and seven plots in sedge marshes along the Narewka river showed that relative abundance of shrews in the open marsh was twice that of the forest (on average, 3.5 versus 1.7 shrews/100 trap nights; Fig. 2.30).

In spring (April), populations of all soricids consist of overwintered adults, and the reproductive season begins. In the common shrew, pregnancy lasts 25 days, and young are fully weaned when 22–25 days old. Until the end of the reproductive season (October), an adult female may give birth to three consecutive litters. The litter size decreases, from 6–9 in May, to 5–7 in July, and 3–6 in September (Z. Pucek et al., in prep). In some years, probably those with extra abundant food, a few young females of the spring cohort may attain sexual maturity (Pucek 1960). Unlike rodents, shrews do not breed in winter. The maximum reproductive capacity of the common shrew under conditions of unlimited food supply can be estimated at a total of 24 young produced by one pair of adults throughout spring, summer and early autumn. Compared with rodents, this is not an impressive figure.

Shrews are critically dependent on a constant supply of food. To stay alive, a shrew has to eat about 2000 invertebrates per day, that is one every 10 seconds of active foraging (Hanski 1984). European temperate forests offer two main food sources to shrews: (1) soil and forest floor invertebrates, which are extremely rich in species and available, though in varying quantities, year round, and (2) caterpillars and pupae of some tree-defoliating moths (mainly the win-

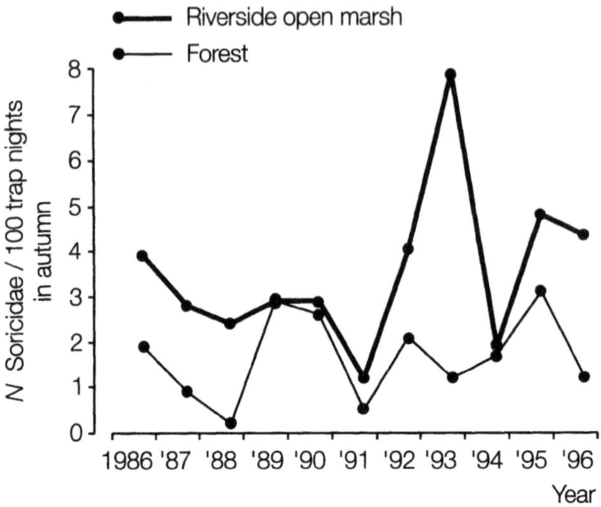

Fig. 2.30. Multiannual variation in autumn (October) abundance of shrews (Soricidae) in the forests and open habitats (marshy river valley) of Białowieża National Park; data pooled from seven trapping plots located in the forests and seven plots in the marshes

ter moth *Operophthera brumata*), which are available in summer and autumn, when caterpillars descend from canopies to pupate in soil. Forest floor invertebrates show a tenfold seasonal variation in abundance (with a minimum in January-February and a maximum in the summer) and, probably, a similar range in multiyear variation. Nonetheless, they form a dependable food to shrews in the most critical periods of winter and early spring. The seasonal course of shrew population dynamics follows the abundance of forest floor invertebrates (Fig. 2.31). Peak numbers of shrews are attained in summer, and then they usually decline until the following spring. In the common shrew, autumn decline was on average 60% of the summer index of abundance, and winter mortality took a mean of 50% of autumn numbers of shrews (Z. Pucek et al., in prep). A 3-year study by Borowski and Dehnel (1952) showed that between-year variation in numbers of the common shrews (especially juveniles) was related to the abundance of forest floor invertebrates (Fig. 2.32).

The supply of leaf-eating caterpillars is enormously variable. Eleven-year cycles of abundance were observed with peaks in 1982, and 1993 (Fig. 2.33). Entomological investigations conducted in Kampinos National Park (central Poland) indicated that, before 1982, an outbreak of winter moths occurred in 1968–1973 with a peak in 1971 (Kaczmarek and Wasilewski 1977, 1978). In Białowieża Forest, the lowest and the highest indices of caterpillar numbers varied 1500-fold! Caterpillars and pupae of winter moths are available to shrews from early summer to autumn, but not during the critical period of winter and early spring. Therefore, multiannual variation in the common shrew numbers was only loosely linked to fluctuations of winter moths (Fig. 2.33). However, numbers of caterpillars affected the breeding success of shrews. In the samples of common shrews trapped in July, the number of juveniles per

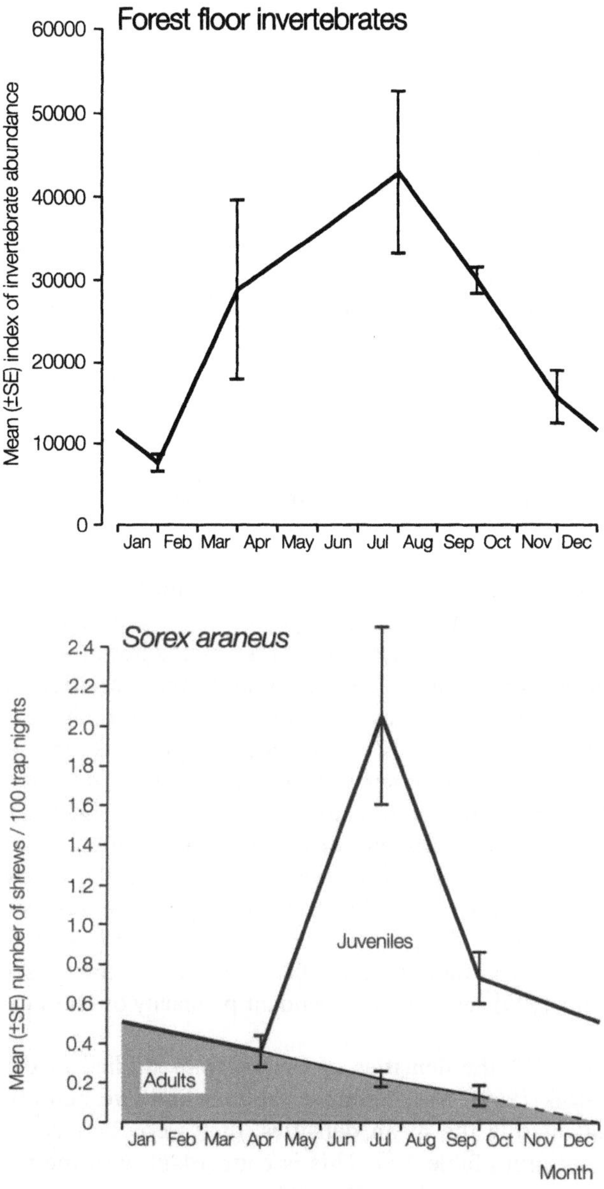

Fig. 2.31. Seasonal population dynamics of the common shrew *Sorex araneus* (data from 1971–1995) and forest floor invertebrates (mean bimonthly samples from year round trapping conducted in 1949–1950 by Borowski and Dehnel, 1952) in BNP. Z. Pucek et al. (in prep.)

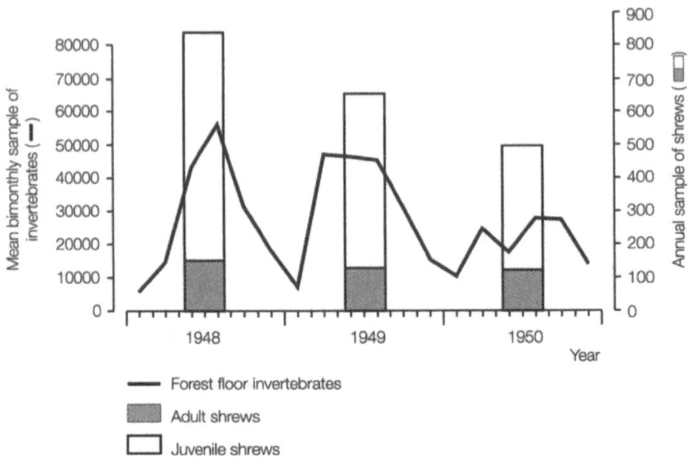

Fig. 2.32. Annual numbers of captured common shrews in relation to abundance of forest floor invertebrates in BNP. Data from: Borowski and Dehnel (1952)

adult varied from <1–8 (on average 4) in years with low and moderate abundance of caterpillars to between 2 and 23 (on average 10) in the pre-outbreak and outbreak years (Fig. 2.33). Obviously, not only winter moths but other invertebrates as well must have been responsible for the year-to-year variation in numbers and condition of shrews. Nonetheless, juveniles were discernibly heavier in years of good supply of caterpillars (Z. Pucek et al., in prep). Good nutrition seems to be a critical factor of shrew survival. The higher the body mass, the better was survival of common shrews from summer to autumn (Fig. 2.34). The same, though weak, relationship between body mass and percentage survival was found in the autumn–winter seasons (Fig. 2.34). Generally, shrew numbers were not affected by weather variables such as temperature, rain, or snow cover (Z. Pucek et al., in prep). These small predators holding large territories (400–2800 m^2 in the common shrew, 500–1950 m^2 in the lesser shrew; review in Churchfield 1990) seem to be dependent primarily on the supply of invertebrate prey.

In the pristine forests of BNP, the densities of shrews were studied in the late 1960s (Aulak 1967; Pucek 1969). Nine summer samples and two autumn ones were available for calculating the mean densities; they were 35 shrews/ha in summer and 14 in autumn (Table 2.5). This is concordant with the indices of seasonal abundance of soricids in 1971–1995: autumn numbers were, on average, 40% of those in summer. No data on spring (April) densities of shrews are available. A rough estimate based on indices of seasonal abundance (see Fig. 2.31) would be, on average, 7 shrews/ha.

Rare and small-sized in comparison with other vertebrates, such as rodents, birds, and frogs, the shrews are not per se attractive prey to raptors and carnivores. However, they *are* an interesting subject in the study of predation. Similar to rodents in their life habits, shrews are most often captured by rodent-oriented predators.

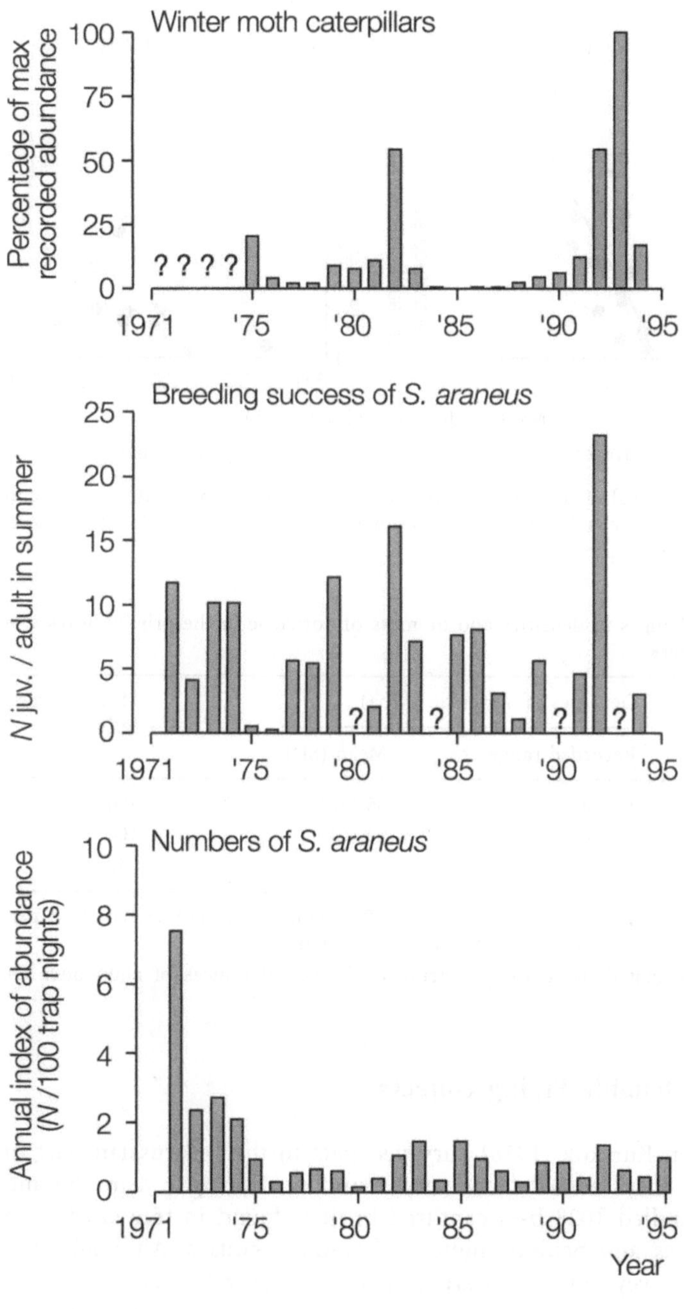

Fig. 2.33. Multiannual dynamics of common shrew abundance and breeding success in relation to fluctuations of the winter moth *Operophthera brumata* caterpillars in BNP. Annual index of shrew abundance = sum of spring, summer, and autumn trapping indices. Caterpillar abundance: Tomiałojć et al. (1984), Wesołowski and Stawarczyk (1991), Wesołowski and Tomiałojć (1997)

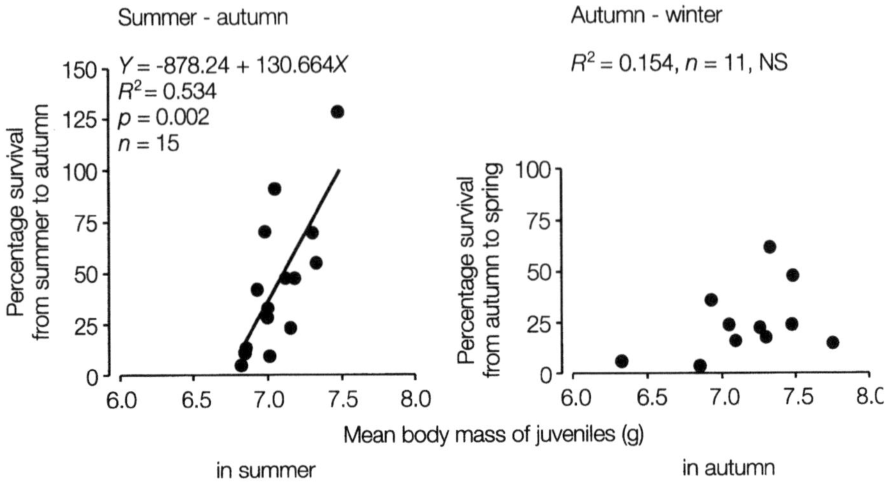

Fig. 2.34. Percentage survival of common shrews from summer to autumn and from autumn till spring in relation to mean body mass of juvenile shrews. Each point denotes 1 year. From: Z. Pucek et al. (in prep.)

Table 2.5. Seasonal changes in densities and biomass of Soricidae in the pristine forests of Białowieża National Park

Season	Densities of Soricidae (*N*/ha)		Mean biomass (g/ha)
	Recorded range	Mean (SD)	
Summer	11–83	35 (25)	260
Autumn	9–18	14 (6)	100
Spring	–	7[a]	60

Data on densities from Aulak (1967) and Pucek (1969). Biomass calculations based on percentage community structure and mean body masses (in Table 2.4).

[a] Spring densities estimated based on proportions of seasonal indices of abundance (in Fig. 2.31).

2.7 Bats – Uncatchable Flying Objects

In the 1950s, when Kurskov (1958) surveyed bats in the Belarussian part of Białowieża Forest, they seemed to be very abundant. During 2 years, he and his co-workers handled 1028 bats captured in nets, found in tree cavities in the woods, in attics and behind shutters of wooden houses. Although later mainly faunistic surveys on bats were conducted (Ruprecht 1976), it seems that nowadays bats are less numerous. Thirteen species of bats are reported from Białowieża Forest (Table 2.6). Most numerous are the common pipistrelle, common noctule, and parti-coloured bat. Kurskov (1958) investigated colonies of bats in the forests and villages, and found that forest colonies were dominated by pipistrelle and noctules (common noctule and lesser

Table 2.6. Community structure of bats in Białowieża Primeval Forest as surveyed by Kurskov (1958) and Ruprecht (1976)

Species	Body mass (g)	Percentage by numbers		
		Whole community	Forest colonies	Village colonies
Common pipistrelle (*Pipistrellus pipistrellus*)	6	38.4	39.2	37.7
Common noctule (*Nyctalus noctula*)	30	23.4	38.4	–
Lesser noctule (*Nyctalus leisleri*)	15	6.0	12.6	–
Common long-eared bat (*Plecotus auritus*)	8	4.7	9.8	–
Parti-coloured bat (*Vespertilio murinus*)	14	18.7	–	52.6
Whiskered bat (*Myotis mystacinus*)	4	2.4	–	5.1
Serotine bat (*Eptesicus serotinus*)	24	1.6	–	4.6
Daubenton's bat (*Myotis daubentoni*)	6	2.9	–	–
Northern bat (*Eptesicus nilssoni*)	10	1.0	–	–
Barbastelle (*Barbastella barbastella*)	10	0.5	–	–
Natterer's bat (*Myotis nattereri*)	8	0.3	–	–
Nathusius' pipistrelle (*Pipistrellus nathusii*)	7	0.1	–	–
Brandt's bat (*Myotis brandti*)	5	+	–	–

'Whole community' includes species found in colonies and captured in nets. Body masses after Kowalski and Ruprecht (1984).

noctule), whereas villages harboured predominantly populations of parti-coloured bats and pipistrelles (Table 2.6). No studies were done on densities of bats in BPF.

Hibernating or migrating away in the cold season and flying at dark during the warm season, bats are neither easy nor predictable prey for predators. No wonder that they are rare and adventitious prey to raptors and carnivores.

2.8 Birds – Endless Choice of (Expensive) Meals to Predators

Compared with a dozen species of small rodents, and a few species of amphibians, reptiles and ungulates, the birds of Białowieża, with 150 regularly breeding species, may seem an endless choice of prey to predators. Moreover, birds abundantly breed and forage in all habitats and all strata, from the ground to the crown level. With few exceptions, birds are diurnal, loud, and very mobile.

Three main habitat types of Białowieża Primeval Forest: woodlands, open marshes in the river valleys, and glades with villages and farmland, harbour different communities of birds. Woodlands are home to 83 breeding species. In spring, the pristine forests of Białowieża National Park are inhabited by an average of 11.3 birds/ha (total biomass 375 g). In the exploited forests, these figures are somewhat lower; 9.3 birds/ha and 323 g/ha (Table 2.7).

Table 2.7. Species structure of bird communities in the pristine forests of Białowieża National Park and managed (exploited) forests of the Polish part of Białowieża Primeval Forest

Species	Body mass (g)	Pristine forests of BNP			Exploited forests		
		Density N/km^2	Percent of number	Percent of biomass	Density N/km^2	Percent of number	Percent of biomass
Chaffinch (*Fringilla coelebs*)	25	245.68	**21.65**	**16.39**	169.70	**18.34**	**13.14**
Wood warbler (*Phylloscopus sibilatrix*)	10	120.68	**10.64**	3.22	69.10	**7.47**	2.14
Robin (*Erithacus rubecula*)	17	119.62	**10.54**	**5.43**	98.94	**10.69**	**5.21**
Collared flycatcher (*Ficedula albicollis*)	13	64.66	**5.70**	2.24	21.66	2.34	0.87
Song thrush (*Turdus philomelos*)	70	53.98	4.76	**10.08**	59.00	**6.38**	**12.79**
Hawfinch (*Coccothraustes coccothraustes*)	55	43.10	3.80	**6.33**	23.04	2.49	3.92
Blackcap (*Sylvia atricapilla*)	18	37.72	3.32	1.81	39.10	4.23	2.18
Goldcrest (*Regulus regulus*)	5	34.40	3.03	0.46	36.36	3.93	0.56
Wren (*Troglodytes troglodytes*)	8	34.26	3.02	0.73	19.34	2.09	0.48
Tree creeper (*Certhia familiaris*)	9	34.06	3.00	0.82	25.78	2.79	0.72
Blackbird (*Turdus merula*)	100	32.54	2.87	**8.68**	26.66	2.88	**8.25**
Chiffchaff (*Phylloscopus collybita*)	9	32.12	2.83	0.77	45.46	4.91	1.27
Great tit (*Parus major*)	18	31.26	2.75	1.50	32.42	3.50	1.81
Blue tit (*Parus caeruleus*)	11	28.44	2.51	0.83	21.22	2.29	0.72
Nuthatch (*Sitta europaea*)	24	21.98	1.94	1.41	11.52	1.24	0.86
Dunnock (*Prunella modularis*)	21	19.82	1.75	1.11	24.10	2.60	1.57
Pied flycatcher (*Ficedula hypoleuca*)	12	16.70	1.47	0.53	7.12	0.77	0.26
Marsh tit (*Parus palustris*)	11	16.16	1.42	0.50	10.38	1.12	0.37
Red-breasted flycatcher (*Ficedula parva*)	10	12.72	1.12	0.36	3.98	0.43	0.13
Middle spotted woodpecker (*Dendrocopos medius*)	60	9.80	0.86	1.57	3.30	0.36	0.61
Spotted flycatcher (*Muscicapa striata*)	15	8.90	0.78	0.37	6.00	0.65	0.29
Great spotted woodpecker (*Dendrocopos major*)	80	8.62	0.76	1.84	9.70	1.05	2.40
Willow warbler (*Phylloscopus trochilus*)	11	7.96	0.70	0.23	10.42	1.13	0.35
Crested tit (*Parus cristatus*)	11	7.76	0.68	0.23	15.54	1.68	0.53
Jay (*Garrulus glandarius*)	175	7.76	0.68	3.62	7.34	0.79	3.98

Table 2.7 (continued)

Species	Body mass (g)	Pristine forests of BNP			Exploited forests		
		Density N/km^2	Percent of number	Percent of biomass	Density N/km^2	Percent of number	Percent of biomass
Tree pipit (*Anthus trivialis*)	20	7.62	0.67	0.41	19.72	2.13	1.22
Garden warbler (*Sylvia borin*)	19	7.54	0.66	0.38	16.96	1.83	1.00
Wood pigeon (*Columba palumbus*)	500	7.02	0.62	**9.37**	4.48	0.48	**6.94**
Siskin (*Carduelis spinus*)	14	5.52	0.49	0.21	7.28	0.79	0.32
Long-tailed tit (*Aegithalos caudatus*)	9	4.28	0.38	0.10	4.66	0.50	0.13
Golden oriole (*Oriolus oriolus*)	73	4.14	0.36	0.81	4.10	0.44	0.93
Bullfinch (*Pyrrhula pyrrhula*)	25	4.14	0.36	0.28	5.38	0.58	0.42
Swift (*Apus apus*)	40	3.98	0.35	0.42	1.82	0.20	0.23
Hazel hen (*Bonasa bonasia*)	420	3.64	0.32	4.08	5.78	0.62	7.52
Lesser spotted woodpecker (*Dendrocopos minor*)	21	3.22	0.28	0.18	0.74	0.08	0.05
Three-toed woodpecker (*Picoides tridactylus*)	65	2.72	0.24	0.47	0.34	0.04	0.07
Starling (*Sturnus vulgaris*)	80	2.62	0.23	0.56	2.54	0.27	0.63
Coal tit (*Parus ater*)	10	2.44	0.22	0.07	4.62	0.50	0.14
River warbler (*Locustella fluviatilis*)	20	2.38	0.21	0.13	1.18	0.13	0.07
Woodcock (*Scolopax rusticola*)	275	2.16	0.19	1.59	1.96	0.21	1.67
Cuckoo (*Cuculus canorus*)	100	1.90	0.17	0.51	1.94	0.21	0.60
Stock dove (*Columba oenas*)	265	1.68	0.15	1.19	0.88	0.10	0.72
White-backed woodpecker (*Dendrocopos leucotos*)	80	1.68	0.15	0.36	0.60	0.06	0.15
Icterine warbler (*Hippolais icterina*)	14	1.60	0.14	0.06	1.86	0.20	0.08
Green sandpiper (*Tringa ochropus*)	90	1.56	0.14	0.37	1.58	0.17	0.44
Black woodpecker (*Dryocopus martius*)	310	1.56	0.14	1.29	1.08	0.12	1.04
Redwing (*Turdus iliacus*)	70	1.56	0.14	0.29	0.90	0.10	0.20
Scarlet rosefinch (*Carpodacus erythrinus*)	22	1.30	0.11	0.08	1.54	0.17	0.10
Firecrest (*Regulus ignicapillus*)	5	1.12	0.10	0.02	0.22	0.02	+

Table 2.7 (continued)

Species	Body mass (g)	Pristine forests of BNP			Exploited forests		
		Density N/km^2	Percent of number	Percent of biomass	Density N/km^2	Percent of number	Percent of biomass
Willow tit (*Parus montanus*)	11	0.84	0.07	0.02	3.84	0.41	0.13
Turtle dove (*Streptopelia turtur*)	150	0.80	0.07	0.32	1.92	0.21	0.89
Thrush nightingale (*Luscinia luscinia*)	25	0.78	0.07	0.05	1.30	0.14	0.10
Nutcracker (*Nucifraga caryocatactes*)	200	0.42	0.04	0.22	1.84	0.20	1.14
Mallard (*Anas platyrhynchos*)	1000	0.38	0.03	1.01	0.40	0.04	1.24
Common snipe (*Gallinago gallinago*)	100	0.38	0.03	0.10	0.72	0.08	0.22
Whitethroat (*Sylvia communis*)	14	0.34	0.03	0.01	0.42	0.05	0.02
Yellowhammer (*Emberiza citrinella*)	30	0.34	0.03	0.03	13.16	1.42	1.22
Raven (*Corvus corax*)	1000	0.26	0.02	0.69	0.36	0.04	1.11
Red-backed shrike (*Lanius collurio*)	30	0.22	0.02	0.02	4.10	0.44	0.38
Greenfinch (*Carduelis chloris*)	28	0.22	0.02	0.02	0.74	0.08	0.06
Goldfinch (*Carduelis carduelis*)	14	0.22	0.02	0.01	0.34	0.04	0.01
Wryneck (*Jynx torquilla*)	37	0.22	0.02	0.02	0.78	0.08	0.09
Redstart (*Phoenicurus phoenicurus*)	16	0.22	0.02	0.01	0.16	0.02	0.01
Black stork (*Ciconia nigra*)	3000	0.18	0.02	1.44	0.04	+	0.37
Crane (*Grus grus*)	5150	0.12	0.01	1.65	0.12	0.01	1.91
Mistle thrush (*Turdus viscivorus*)	115	0.12	0.01	0.04	0.72	0.08	0.26
Crossbill (*Loxia curvirostra*)	40	0.12	0.01	0.01	0.18	0.02	0.02
Greenish warbler (*Phylloscopus trochiloides*)	8	0.12	0.01	+	-	-	-
White wagtail (*Motacilla alba*)	24	0.08	0.01	0.01	0.68	0.07	0.05
Lesser whitethroat (*Sylvia curruca*)	12	0.08	0.01	+	1.36	0.15	0.05
Grey-headed woodpecker (*Picus canus*)	120	0.06	0.01	0.02	0.18	0.02	0.07
Hoopoe (*Upupa epops*)	60	0.04	+	0.01	0.04	+	0.01

Table 2.7 (continued)

Species	Body mass (g)	Pristine forests of BNP			Exploited forests		
		Density N/km^2	Percent of number	Percent of biomass	Density N/km^2	Percent of number	Percent of biomass
Green woodpecker (*Picus viridis*)	200	0.02	+	0.01	0.14	0.02	0.09
Kingfisher (*Alcedo atthis*)	40	0.02	+	+	–	–	–
Nightjar (*Caprimulgus europaeus*)	63	–	–	–	1.14	0.12	0.22
Wood lark (*Lullula arborea*)	28	–	–	–	0.54	0.06	0.05
Marsh warbler (*Acrocephalus palustris*)	12	–	–	–	0.22	0.02	0.01
Corncrake (*Crex crex*)	160	–	–	–	0.18	0.02	0.09
Whinchat (*Saxicola rubetra*)	18	–	–	–	0.12	0.01	0.01
Teal (*Anas crecca*)	300	–	–	–	0.10	0.01	0.09
Barred warbler (*Sylvia nisoria*)	24	–	–	–	0.10	0.01	0.01
Bluethroat (*Luscinia svecica*)	18	–	–	–	0.06	0.01	+
Serin (*Serinus serinus*)	13	–	–	–	0.02	+	+
Total density (*N* birds/km^2) and biomass (g/km^2)		1134.68		37476	925.36		32296

Data compiled from censuses by Tomiałojć et al. (1984), Tomiałojć and Wesołowski (1994; pristine forests) and Pugacewicz (1997; exploited and pristine forests). Boldface type indicates the dominant species.

An unchallenged dominator of forest birds is the chaffinch (about 20% of all birds), followed by wood warbler, robin, song thrush, chiffchaff, collared flycatcher, and blackcap. Most of the common birds are very small. Thus, in terms of biomass, the forest bird community is dominated by the chaffinch (13–16% of total biomass), song thrush (10–13%), blackbird, and woodpigeon (Table 2.7). Sixty percent of species dwelling in the forests (including all dominant species mentioned above) are migratory and come here only to breed.

Year-round residents comprise 30 species (Fig. 2.35 and 2.36). Their mean density in spring was 2–3 birds/ha (about 100 g biomass/ha) (Fig. 2.37). This is about 25% of spring numbers of birds, and 27–30% of their total biomass. A similar estimate of winter abundance of birds was given by Karpiński (1954), who conducted year-round counts of relative numbers of birds in nine habitats of Białowieża National Park in 1948 and 1949. Seasonally low-

Fig. 2.35. Blue tit *Parus caeruleus* with a beak full of invertebrates, going to feed the nestlings. Photo: S. Wąsik

est numbers of birds were recorded in February and constituted 32% of April numbers. Moreover, in 1982, Dackevich and Popenko (1987) estimated the densities of birds in forests and open terrains on the southern peripheries of the Belarussian part of BPF in winter and late spring. In January, they recorded from 0.43 to 1.91 birds/ha (mean 1.33 birds/ha), which was, on average, 20% of the spring densities of birds in these forests.

Forest communities of birds dwelling in wet alder-ash forests, oak-lime-hornbeam forests, and coniferous pine-spruce forests show some distinct features. The highest densities occur in both types of deciduous forests, whereas the number of birds inhabiting coniferous stands is only half that of deciduous stands (Table 2.8). In the species composition of breeding and wintering birds, coniferous forests markedly differ from deciduous forests. Noteworthy is the fact that in the coniferous forests of the Belarussian part of BPF, a declining population of capercaillie *Tetrao urogallus* still survives (Dackevich and Vakula 1980).

In the Polish part of BPF, open marshes along the Leśna, Narewka, and Hwoźna rivers harbour distinctly fewer birds (61 species, mean density 4.75 birds/ha) than forests, but the total biomass of birds is not lower because many birds living near rivers are medium-sized and large (Table 2.9). In these riverside unmown marshes with willow shrubs and single alder trees, the community of birds is dominated by scarlet rosefinch and a few species of warblers (sedge, marsh, and river warbler), invisible among the tall reeds but conspicuous due to their restless vibrating songs. In terms of biomass, this 'crowd' of small birds cannot compare with the fewer but larger birds: mallards (45% of total biomass), swans (9%), cranes, and corncrakes (Table 2.9).

Fig. 2.36. Black woodpecker *Dryocopus martius* feeding a fledgling. Black woodpeckers are true forest residents. Photo: S. Wąsik

The marshland bird community shows the strongest seasonal variation (Fig. 2.37). Year-round residents comprise only seven species. Their spring density is on average 0.13 ind/ha (3% of all breeding birds). In severe winters, when rivers freeze, the marsh becomes an inhospitable ice and snow desert with neither suitable refuge nor food for birds other than yellowhammers and finches that forage on seeds of tall forbs. Dackevich and Popenko (1987), who counted birds in such habitats in winter, found on average 0.08 birds/ha, i.e. merely 4% of the densities recorded there in spring. However, besides resident birds, migratory species such as siskin, redpoll, *Carduelis flammea*, and goldfinch, which feed on seeds of alder trees, can be fairly common along forest rivers in winter. Their relative densities, estimated in January 1984–1993 on several small rivers in eastern Poland, were 0.6–19.3 siskins, 0–5 redpolls, and 0.04–0.9 goldfinches per kilometre of the river course (M. Jędrzejewski 1993). Siskins and redpolls are highly invasive species, with numbers extremely vari-

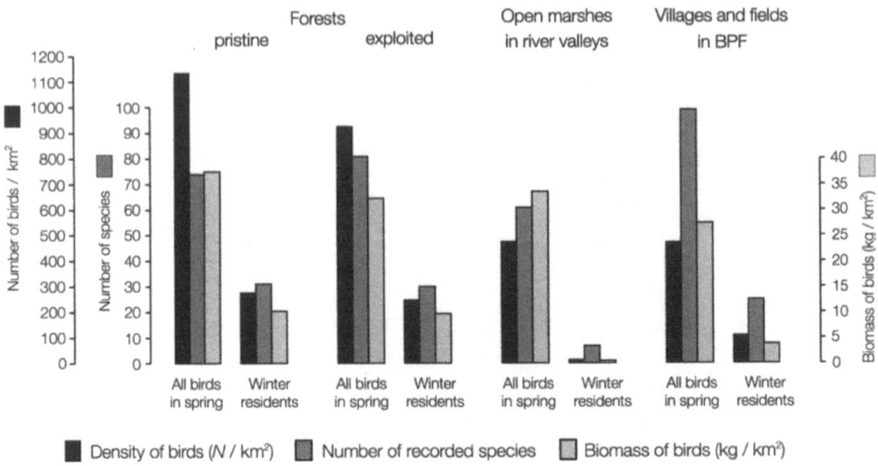

Fig. 2.37. Comparison of bird abundance in the forests (pristine forests of Białowieża National Park and exploited forests), open marshes in the valleys of forest rivers, and settlement glades located in the Polish part of Białowieża Forest. In each habitat type, a spring community of birds (all breeding species) and a winter one (local residents only, with their spring densities) are shown. Data from: Tomiałojć et al. (1984), Tomiałojć and Wesołowski (1994, 1996) and Pugacewicz (1997)

able from year to year and place to place. In Dackevich and Popenko's (1987) survey, redpolls occurred on one of the study plots in numbers (1.8 birds/ha) over 20 times those of wintering residents.

Unlike the forest community of birds, which undergoes fluctuations from year to year but does not show directional changes in species structure, birds inhabiting river valleys are still in the process of succession triggered by abandonment of utilisation of riverside meadows by local people in recent decades. No long-term census followed the quantitative and qualitative changes in the bird communities. However, the study by Golodushko and Danilyuk (1961), conducted in 1958–1959 in the mown meadows along the Leśna River, showed that, in the total density of 5–6 birds/ha, the sky lark and meadow pipit were most numerous, followed by whinchat, red-backed shrike, white wagtail, and yellow wagtail. Of these species, only the red-backed shrike and whinchat are still very numerous. Ornithologists working in BPF for years have noticed the decline in the once very common lapwings in riverside meadows (E. Pugacewicz, pers. comm.). Also, the high numbers and dominance of the scarlet rosefinch are the result of its fast 'invasion' in recent years (E. Pugacewicz, pers. comm.).

Several villages (surrounded by orchards, farmlands, fallowland, meadows) located in small and large glades within the Białowieża Forest are habitats of 99 species of breeding birds (many of them of forest origin) (Table 2.10). House martin, sky lark, house sparrow, starling, yellowhammer, and swallow dominate by numbers, whereas white storks, inhabitants of many barn roofs, make as much as 32% of bird biomass. In winter, quite a few birds stay in

Table 2.8. Bird densities and dominant species (percentages of numbers of all birds) in the three main types of forest in Białowieża National Park

Parameter and species	Wet alder-ash forest	Oak-lime-hornbeam forest	Pine-spruce forest
Spring density of breeding birds (*N*/ha)	16.7	14.6	8.4
Minimum density of birds in winter (*N*/ha)	4.1	4.1	2.2
Dominant species in spring (%)			
Fringilla coelebs	19.7	24.5	28.5
Erithacus rubecula	9.0	9.5	11.7
Phylloscopus collybita	6.3		
Sylvia atricapilla	5.4		
Ficedula albicollis	5.0	7.3	
Phylloscopus sibilatrix		6.3	12.8
Regulus regulus			6.6
Turdus philomelos		5.8	5.5
Dominant species in winter (%)			
Sitta europaea	16.7	11.5	
Parus major	16.1	15.8	
Coccothraustes coccothraustes		15.8	
Parus caeruleus	12.8	11.5	
Certhia familiaris	12.2	11.5	13.5
Regulus regulus	8.9		26.9
Parus cristatus			13.5
Parus ater			12.5
Pyrrhula pyrrhula			6.7

Calculated from Tomiałojć et al. (1984), Tomiałojć and Wesołowski (1994, 1996). Minimum winter density estimated from spring density of resident species.

villages and surrounding farmlands (Fig. 2.37), most of them are house sparrows and tits. The censuses made in villages included wild birds, only. However, to predators, which do not share ornithologists' prejudice against recognising poultry as birds, domestic hens contribute high numbers and biomass to the village community of birds in both summer and winter.

Most important for this book is the community of birds inhabiting woodland, especially the pristine mature forests. Of all birds recorded in spring in the old growth forests of Białowieża National Park about half (by numbers of individuals) are short-distance migrants (wintering in south-west Europe), one quarter are winter residents, and one quarter are tropical migrants (wintering south of Sahara; Fig. 2.38). In terms of biomass, short-distance migrants prevail (63%), whereas tropical migrants, mostly small birds, contribute a mere 11% to the community. The majority of forest birds are insectivores, with three main adaptations for collecting insects: from tree bark (only residents such as woodpeckers, nuthatches, tree creepers), from the

Table 2.9. Species structure of bird community in the open marshes in river valleys of the Polish part of BPF

Species	Body mass (g)	Density *N*/km^2	Percent of number	Percent of biomass
Scarlet rosefinch (*Carpodacus erythrinus*)	22	52.70	**11.10**	3.44
Sedge warbler (*Acrocephalus schoenobaenus*)	12	52.46	**11.04**	1.94
Marsh warbler (*Acrocephalus palustris*)	12	47.30	**9.96**	1.75
River warbler (*Locustella fluviatilis*)	20	43.90	**9.24**	2.60
Reed bunting (*Emberiza schoeniclus*)	20	37.62	**7.92**	2.23
Whitethroat (*Sylvia communis*)	14	24.66	**5.19**	1.02
Red-backed shrike (*Lanius collurio*)	30	18.74	3.95	1.67
Savi's warbler (*Locustella luscinioides*)	17	17.86	3.76	0.90
Whinchat (*Saxicola rubetra*)	18	16.86	3.55	0.90
Mallard (*Anas platyrhynchos*)	1000	15.34	3.23	**45.50**
Tree pipit (*Anthus trivialis*)	20	14.16	2.98	0.84
Thrush nightingale (*Luscinia luscinia*)	25	13.96	2.94	1.04
Barred warbler (*Sylvia nisoria*)	24	13.34	2.81	0.95
Common snipe (*Gallinago gallinago*)	100	10.70	2.25	3.17
Yellowhammer (*Emberiza citrinella*)	30	10.06	2.12	0.90
Corncrake (*Crex crex*)	160	9.44	1.99	4.48
Grasshopper warbler (*Locustella naevia*)	14	9.44	1.99	0.39
Reed warbler (*Acrocephalus scirpaceus*)	12	9.18	1.93	0.34
Great reed warbler (*Acrocephalus arundinaceus*)	32	6.28	1.32	0.60
Water rail (*Rallus aquaticus*)	140	5.72	1.20	2.38
Garden warbler (*Sylvia borin*)	19	5.72	1.20	0.32
Spotted crake (*Porzana porzana*)	80	5.40	1.14	1.28
Meadow pipit (*Anthus pratensis*)	19	4.16	0.88	0.23
White wagtail (*Motacilla alba*)	24	4.02	0.85	0.29
Willow warbler (*Phylloscopus trochilus*)	11	2.90	0.61	0.09
Chaffinch (*Fringilla coelebs*)	25	2.58	0.54	0.19
Lapwing (*Vanellus vanellus*)	200	1.88	0.40	1.12
Moorhen (*Gallinula chloropus*)	300	1.50	0.32	1.33
House martin (*Delichon urbica*)	19	1.50	0.32	0.08
Greenfinch (*Carduelis chloris*)	28	1.50	0.32	0.12
Cuckoo (*Cuculus canorus*)	100	1.44	0.30	0.43
Skylark (*Alauda arvensis*)	37	1.44	0.30	0.16
Chiffchaff (*Phylloscopus collybita*)	9	1.06	0.22	0.03
Teal (*Anas crecca*)	300	1.00	0.21	0.89
Blackbird (*Turdus merula*)	100	1.00	0.21	0.30
Dunnock (*Prunella modularis*)	21	0.94	0.20	0.06
Little crake (*Porzana parva*)	55	0.76	0.16	0.12
Icterine warbler (*Hippolais icterina*)	14	0.76	0.16	0.03
Magpie (*Pica pica*)	200	0.62	0.13	0.37
Swallow (*Hirundo rustica*)	18	0.56	0.12	0.03

Table 2.9 (continued)

Species	Body mass (g)	Density N/km^2	Percent of number	Percent of biomass
Song thrush (*Turdus philomelos*)	70	0.50	0.11	0.10
Great snipe (*Gallinago media*)	155	0.44	0.09	0.20
Linnet (*Acanthis cannabina*)	18	0.38	0.08	0.02
Crane (*Grus grus*)	5150	0.32	0.07	4.89
Kingfisher (*Alcedo atthis*)	40	0.32	0.07	0.04
Mute swan (*Cygnus olor*)	12000	0.26	0.05	**9.25**
Garganey (*Anas querquedula*)	435	0.26	0.05	0.34
Bluethroat (*Luscinia svecica*)	18	0.26	0.05	0.01
Yellow wagtail (*Motacilla flava*)	17	0.26	0.05	0.01
Blackcap (*Sylvia atricapilla*)	18	0.26	0.05	0.01
Goldeneye (*Bucephala clangula*)	780	0.18	0.04	0.42
Wryneck (*Jynx torquilla*)	37	0.18	0.04	0.02
White-backed woodpecker (*Dendrocopos leucotos*)	100	0.18	0.04	0.05
Redshank (*Tringa totanus*)	130	0.12	0.03	0.05
Lesser spotted woodpecker (*Dendrocopos minor*)	21	0.12	0.03	0.01
Lesser whitethroat (*Sylvia curruca*)	12	0.12	0.03	+
Penduline tit (*Remiz pendulinus*)	10	0.12	0.03	+
Starling (*Sturnus vulgaris*)	80	0.12	0.03	0.03
Hawfinch (*Coccothraustes coccothraustes*)	55	0.12	0.03	0.02
Total density (N birds/km^2) and biomass (g/km^2)		474.98		33717

Data from Pugacewicz (1997). Dominant species are in boldface type.

ground (e.g. blackbird and song thrush, Fig. 2.39), and high in the canopies, from leaves and needles of trees and in air (e.g. chaffinch, warblers; Fig. 2.40). The occurrence of a particular type of insectivory differs among birds of various migratory habits (Fig. 2.38). Vegetarians (both granivorous birds such as the bullfinch, siskin, and herbivorous birds, e.g. hazel hen) occurred in small numbers in all groups except for tropical migrants.

Accustomed to European monoculture-type plantations of trees, we can hardly imagine the intricacy of the structure of the pristine temperate forest. Multilayered canopy, stretching from 5 to 50 m high, shrubs, forbs, standing and lying snags of dead trees, root plates of windfallen trees, all offer endless possibilities for bird nests. In fact, all layers from the ground to canopies serve as nesting sites, yet the most of bird nests are placed in lower layers of tree crowns: over 250 breeding pairs/km^2 (Fig. 2.41). The real numbers of birds are still higher there, because many ground-, shrub-, and hole-nesters forage in the canopy layer. Such vertical distribution of forest birds suggests that a successful predator of birds should be able to fly or at least skillfully climb trees.

Table 2.10. Densities and dominance structure of bird community in the settlement glades and villages located in the Polish part of BPF

Species	Body mass (g)	Density N/km^2	Percent of number	Percent of biomass
House martin (*Delichon urbica*)	19	51.94	**11.04**	3.60
Sky lark (*Alauda arvensis*)	37	42.18	**8.97**	**5.69**
House sparrow (*Passer domesticus*)	32	37.10	**7.89**	4.33
Starling (*Sturnus vulgaris*)	80	29.22	**6.21**	**8.53**
Yellowhammer (*Emberiza citrinella*)	30	27.20	**5.78**	2.98
Swallow (*Hirundo rustica*)	18	23.90	**5.08**	1.57
Whinchat (*Saxicola rubetra*)	18	20.16	4.29	1.32
White wagtail (*Motacilla alba*)	24	15.08	3.21	1.32
Scarlet rosefinch (*Carpodacus erythrinus*)	22	11.86	2.52	0.95
Chaffinch (*Fringilla coelebs*)	25	11.60	2.47	1.06
Whitethroat (*Sylvia communis*)	14	10.64	2.26	0.54
Greenfinch (*Carduelis chloris*)	28	10.58	2.25	1.08
Marsh warbler (*Acrocephalus palustris*)	12	9.34	1.99	0.43
Red-backed shrike (*Lanius collurio*)	30	8.58	1.82	0.94
Tree pipit (*Anthus trivialis*)	20	8.38	1.78	0.61
Meadow pipit (*Anthus pratensis*)	19	8.24	1.75	0.57
Linnet (*Acanthis cannabina*)	18	7.60	1.62	0.50
Great tit (*Parus major*)	18	7.02	1.49	0.46
Robin (*Erithacus rubecula*)	17	6.60	1.40	0.41
Willow warbler (*Phylloscopus trochilus*)	11	6.60	1.40	0.26
Goldfinch (*Carduelis carduelis*)	14	6.42	1.37	0.33
Tree sparrow (*Passer montanus*)	23	6.36	1.35	0.53
Barred warbler (*Sylvia nisoria*)	24	5.38	1.14	0.47
Spotted flycatcher (*Muscicapa striata*)	16	5.26	1.12	0.30
Icterine warbler (*Hippolais icterina*)	14	5.00	1.06	0.26
Blue tit (*Parus caeruleus*)	11	4.80	1.02	0.19
White stork (*Ciconia ciconia*)	3510	2.48	0.53	**31.75**
63 Other species in total	–	80.76	17.19	70.98
Total density (*N* birds/km^2) or biomass (g/km^2)		470.28		33717

Data from Pugacewicz (1997). Dominant species are in boldface type.

In 1976–1994, mean numbers of all breeding birds in spring varied from 9.8 ind/ha in 1978 to 16.3 ind/ha in 1993, i.e. 1.7-fold. The two most important factors affecting spring numbers of forest birds were snow cover in the preceding winter and abundance of leaf-eating caterpillars in the preceding summer (Fig. 2.42). From 1975 till 1994, outbreaks of caterpillars (mainly the winter moths *Operophthera brumata* feeding on hornbeam, maple, and lime leaves) occurred at 11-year intervals. The peaks (in 1982 and 1993) were preceded by a few years of exponentially increasing abundance of the moth. During a

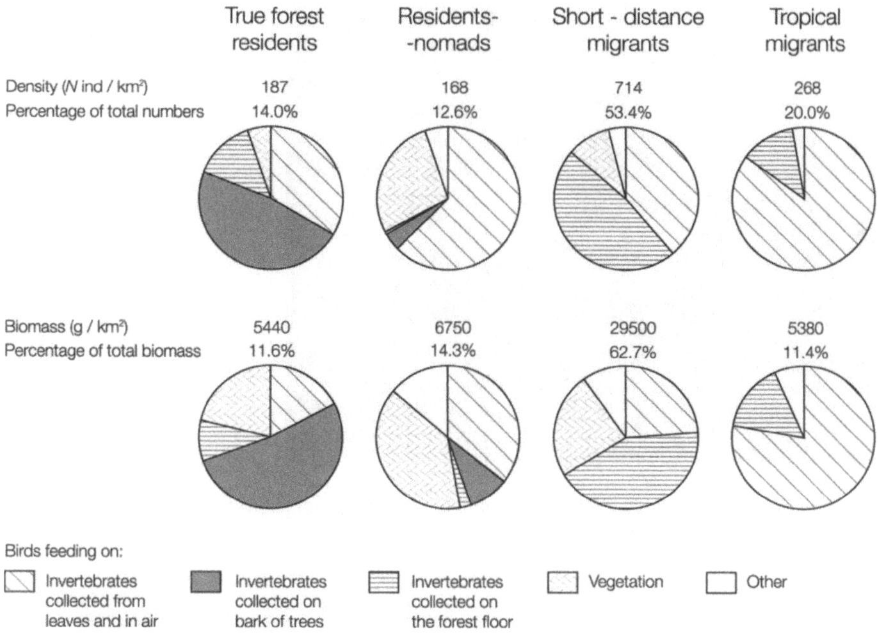

Fig. 2.38. Structure of the forest bird community in respect of migratory habits and feeding ecology. Density and biomass censused in spring (Białowieża National Park, mean for seven plots in 1976–1994; from Tomiałojć et al. 1984; Tomiałojć and Wesołowski 1994, 1996). *Circles* are birds in each migration group expressed as 100% to show contributions of various feeding groups

Fig. 2.39. Song thrush *Turdus philomelos* (a short-distance migrant) at the nest.
Photo: S. Wąsik

Fig. 2.40. Wood warbler *Phylloscopus sibilatrix* (a tropical migrant) bringing a caterpillar to nestlings. Photo: P. Fabijański

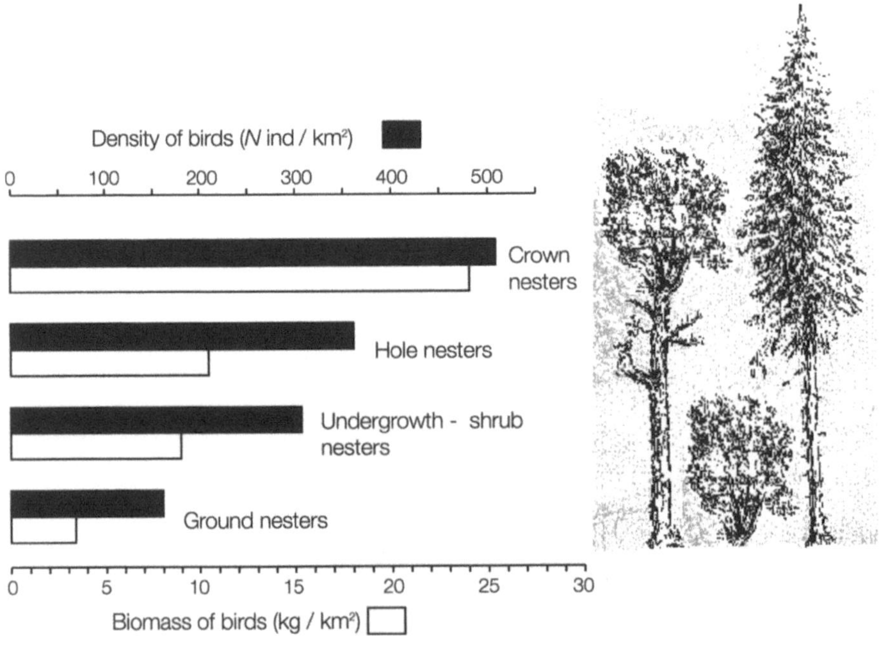

Fig. 2.41. Vertical distribution of spring numbers and biomass of birds according to their nesting habits in the pristine forests of Białowieża National Park. Data averaged for 1976–1994 from Tomiałojć et al. (1984); Tomiałojć and Wesołowski (1994, 1996)

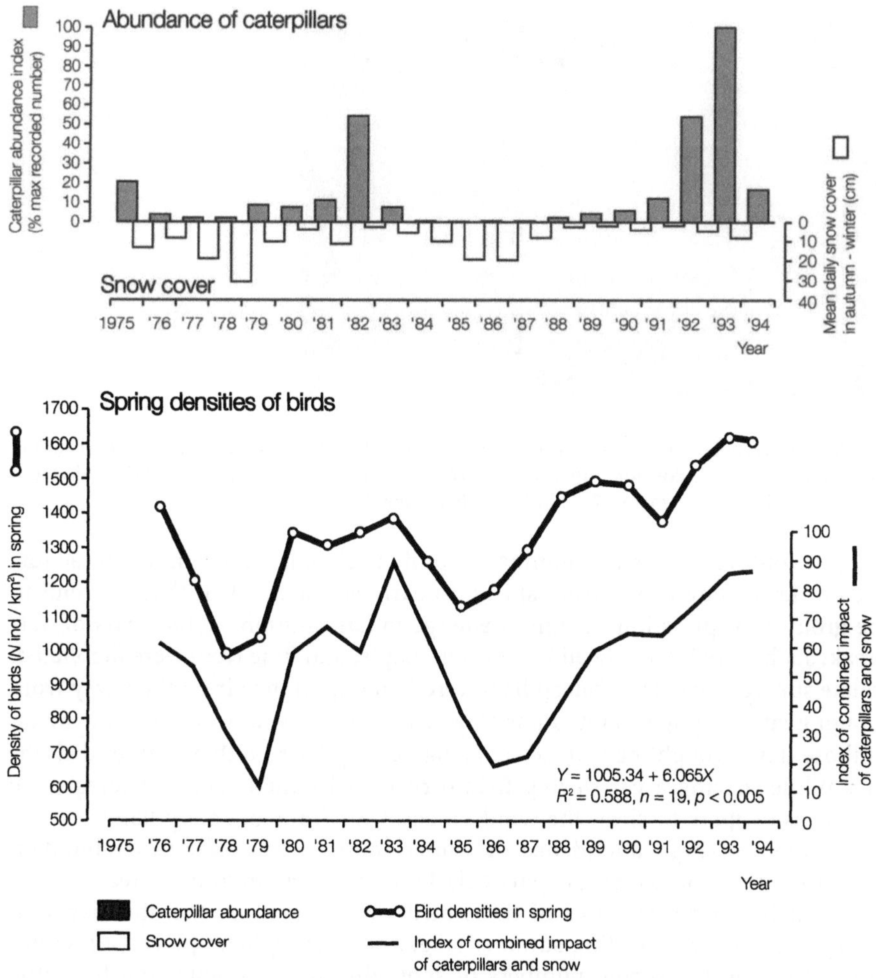

Fig. 2.42. Multi-year variation in spring densities of birds breeding in Białowieża National Park in relation to abundance of tree-defoliating caterpillars (mainly the winter moth *Operophthera brumata*) in the preceding summer and snow cover in the preceding winter. *Upper graph* snow cover and index of caterpillar abundance. Maximum abundance of caterpillars (on average 44.2 larvae per standard twig of hornbeam, recorded in spring-summer 1993) was taken as 100%. *Lower graph* spring densities of forest birds (averaged from seven plots) in relation to the index of snow and caterpillars combined. To calculate that index, variation in snow cover was also expressed on a 0–100 scale (maximum recorded snow as 0, and minimum as 100) and the arithmetic means of both indices (snow in the preceding winter and caterpillars in the preceding summer) were calculated and related to bird densities in spring. Data on caterpillar abundance from: Wesołowski and Tomiałojć (1997). Data on bird densities from: Tomiałojć et al. (1984), Tomiałojć and Wesołowski (1994, 1996)

Fig. 2.43. Caterpillars of winter moth *Operophthera brumata* in the stage of descending from canopies to the soil for pupation; outbreak of winter moth, mid-May 1993, Białowieża National Park, compartment 371. Photo: W. Jędrzejewski

few post-outbreak years, its numbers rapidly declined. Winter moth larvae forage on developing leaves from spring to early summer, when they descend to the ground for pupation. Imagines emerge in late autumn. In the years of outbreak, such as 1993, nearly all hornbeam, maple, and lime trees were defoliated. In late spring, miriads of caterpillars were hanging in the air on their way from foraging sites to pupation habitats (Fig. 2.43). People who worked in the forest in those days brought back dozens of small caterpillars on their hair and clothing. While driving a car during fieldwork in oak-lime-hornbeam forests, we had to use wipers to clean the windscreen of the 'rain' of caterpillars!

Summers of high abundance of winter moth larvae offer superabundant food to insectivorous birds, particularly to tropical migrants. Year-to-year variation in spring densities of these birds was shaped by the abundance of caterpillars in the preceding summer (Fig. 2.44). Caterpillars played a less important role in the spring numbers of short-distance migrants, and had little effect on densities of residents. Instead, birds wintering in Europe, and particularly so the local residents, were heavily affected by the depth of snow cover (Fig. 2.44). Many of these birds forage by collecting invertebrates on tree bark and on the ground. Thick, frozen snow cover and snow formations on trees, the essence of winter beauty to humans, mean hard times and starvation to those birds. An index that combined the effects of both factors (snow depth and caterpillar abundance) explained 59% of year to year variation in the spring numbers of all birds in the old growth forests of Białowieża National Park (Fig. 2.42).

Densities of migratory birds depended on the *previous* year's abundance of caterpillars. Thus, caterpillars as food must have primarily shaped the production of fledglings which, in the following year, return to their natal region to breed. Indeed, available data on the collared flycatcher, wood warbler, and chiffchaff feeding on caterpillars in tree canopies showed that their breeding success was higher with more abundant food (Fig. 2.45). The per-

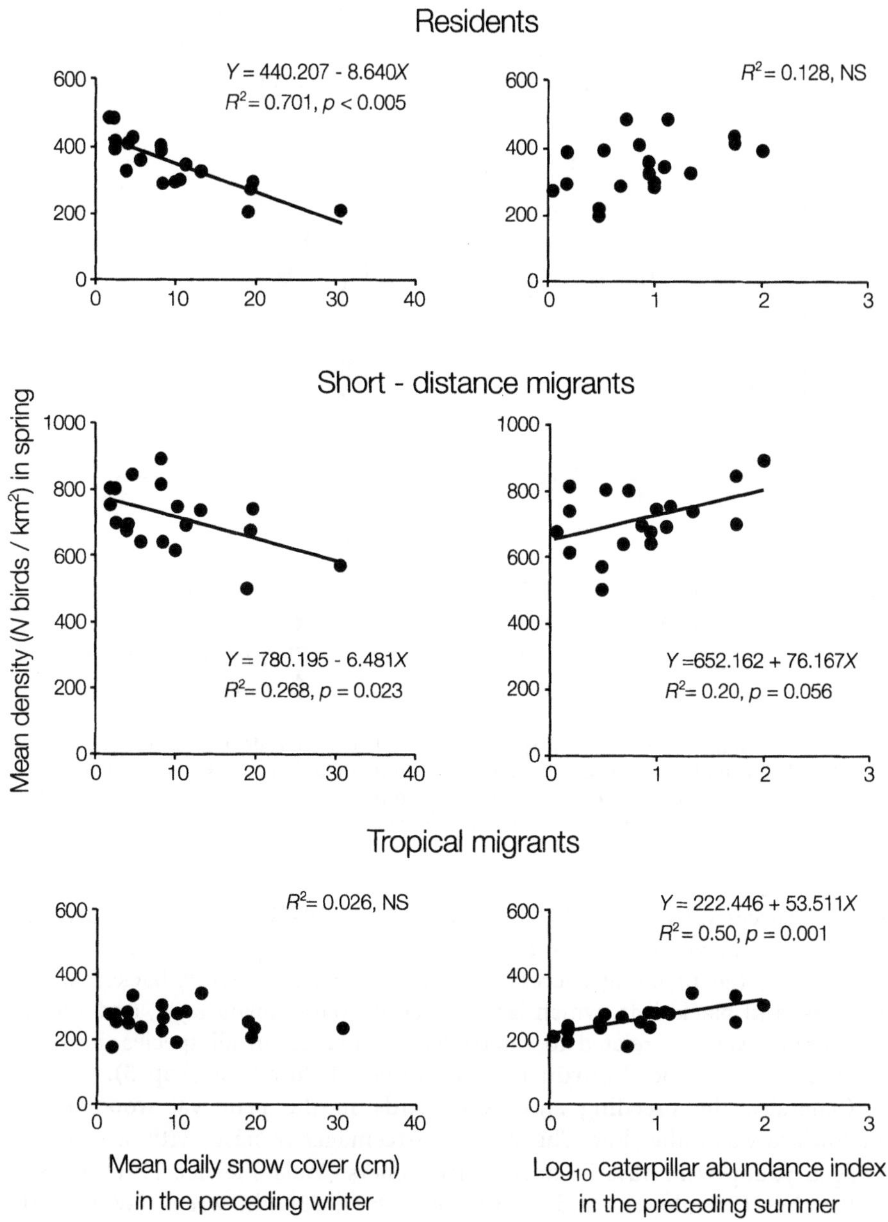

Fig. 2.44. Spring densities of birds from various migration groups in BNP (data for 1976–1994 from Tomiałojć et al. 1984; Tomiałojć and Wesołowski 1994, 1996) in relation to snow cover in the preceding winter and caterpillar abundance of the preceding summer. Each point denotes 1 year

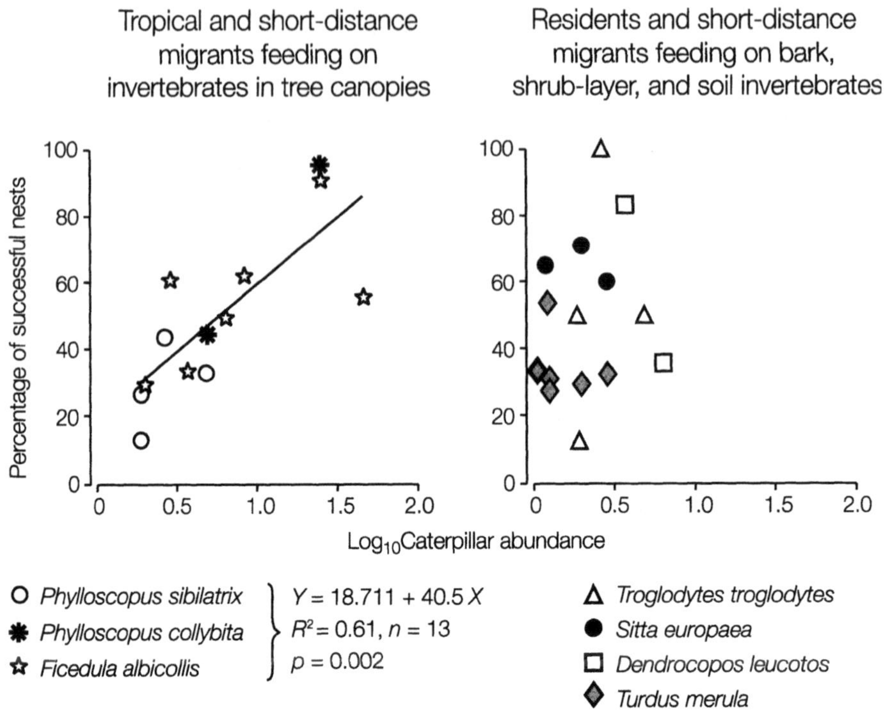

Fig. 2.45. Breeding success of seven species of birds (each point denotes 1 year) in relation to caterpillar abundance during the breeding season. Data from: Wesołowski (1983, 1985, 1995), Piotrowska and Wesołowski (1989), Wesołowski and Stawarczyk (1991), Walankiewicz (1991, and unpubl. data), and Tomiałojć (1994)

cent of successful nests varied from 10–30% to 50–90%, when tree-defoliating larvae ranged from about 2 to 44 per standard twig of hornbeam. At the same time, the breeding success of wrens, nuthatches, white-backed woodpeckers, and blackbirds, which largely feed on bark, shrub layer, and soil invertebrates was not related to caterpillar abundance. In all species, however, predation was a crucial proximate cause of nest failure (see Chap. 5).

Generally, the breeding success of birds in the primeval woodland of Białowieża was rather low. The average percentages of nests with at least one fledged young were 42% in collared flycatchers (Walankiewicz 1991), 65% in nuthatches (Wesołowski and Stawarczyk 1991), 53% in wrens (Wesołowski 1983), 33% in blackbirds (Tomiałojć 1993), 44% in wood warblers (Piotrowska and Wesołowski 1989), and 59% in white-backed woodpeckers (Wesołowski 1995). Again, in all cases, predation was reported as the most important proximate factor of losses.

Seasonal changes in bird densities in the temperate forests are not easy to assess. Spring censuses, done by the mapping method, recorded only the territorial birds (mean density 13.4 ind/ha). As shown experimentally for tits and wrens, non-territorial, non-breeding birds may be fairly numerous

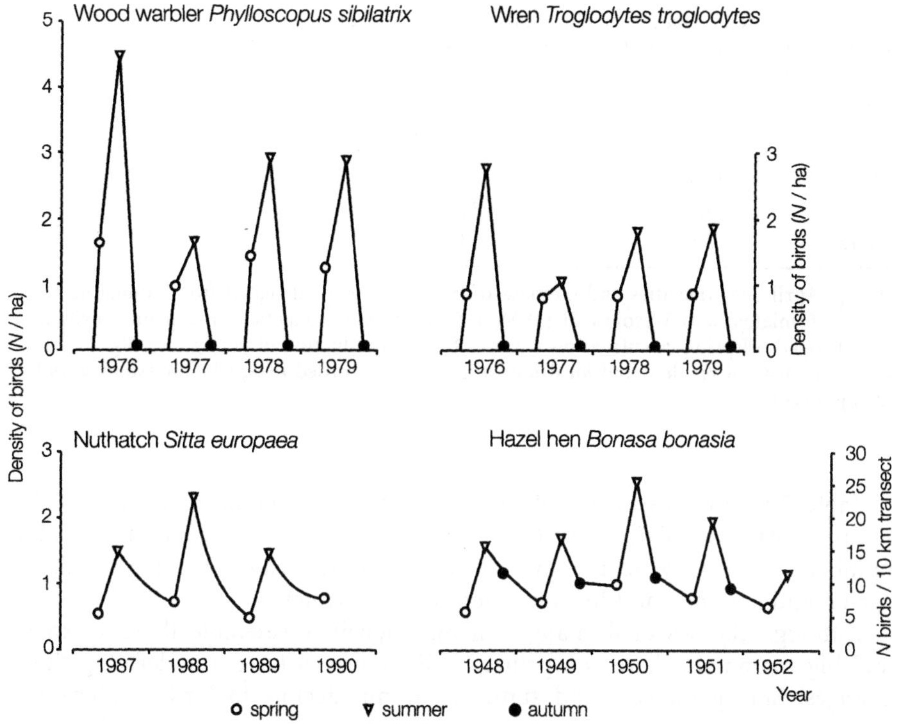

Fig. 2.46. Seasonal and year to year changes in densities of four species of birds in BPF. Data from: Gavrin (1969), Wesołowski (1983, 1985), and Wesołowski and Stawarczyk (1991)

(Wesołowski 1981; Wesołowski et al. 1987). Summer (post-breeding) numbers or indices of abundance have so far been estimated in four species during 16 years in total (Fig. 2.46). In these species that represented all ecological groups of migrants and residents, the post-breeding numbers of birds constituted from 131 to 350% of spring numbers of territorial birds (on average 239%, SD 63). In autumn, all local migratory birds have left their breeding habitats, but numerous and often large flocks of migrants from the north and north-east temporarily stop in Białowieża Forest on their way to wintering quarters. During migrations, the abundance of passing birds may change overnight from scarcely any birds to large numbers of warblers, redwings, wood pigeons, and – as winter comes – waxwings and siskins.

In winter, the density of birds is affected by weather conditions. In 1984–1993, the abundance of forest birds (great tit, wren, blue tit) and non-forest ones (yellowhammers, tree sparrows) wintering in river valleys in eastern Poland were strongly positively correlated with temperature (M. Jędrzejewski 1993). In contrast, the numbers of northern migrants were negatively related to temperature; censuses conducted in January recorded more redpolls and bramblings, *Fringilla montifringilla* during harsh winters than during mild ones.

Table 2.11. Seasonal changes in the mean densities and biomass of birds in the pristine forests of Białowieża National Park

Season	Mean density (*N* birds/ha)	Mean biomass (g/ha)
Spring	13.4	470
Summer	32.6	1140
Winter	3.5	120

Spring, April–May densities and biomass of breeding birds calculated from Tomiałojć et al. (1984), Tomiałojć and Wesołowski (1994, 1996). Summer, densities of breeding birds and their fledged offspring calculated on a basis of mean production of young (see Fig. 2.46). Winter, density of resident and local nomadic species recorded in April–May (sources as for spring densities).

Table 2.11 gives estimates of mean densities and biomass of birds in the pristine forests of BNP. These figures should be treated as minimum reliable numbers, since they do not include non-territorial (non-breeding) birds in spring and summer, and incoming migrants in winter.

Although the seasonal changes in bird densities resemble those in small and big mammals, there is a principal difference in factors underlying these changes. Spring increase and summer-autumn decline in bird numbers are largely due to migrations, in other words, due to their repeated short- or long-distance passages between different environments for breeding and survival. Indeed, their ability to migrate and vulnerable nesting stage seem the principal ecological features of birds creating foundations for their coexistence with predators.

2.9 Reptiles – Rare Animals of Unpredictable Activity

In temperate deciduous and mixed forests, a reptile is an unusual sight. Indeed, a lizard or a snake is easier to spot on a sun-baked clear-cut, sandy road, or wet meadow by a river than in the forest proper. The seven species of reptiles dwelling in Białowieża Forest are predaceous heterotherms. They are inactive in the cold season (October–mid April) and rather unevenly distributed. One can walk kilometres without seeing a single reptile while in another place count several grass snakes along a dozen metres of the river bank.

To show a general picture of the reptile community of Białowieża Forest, we averaged the results of five large-scale censuses conducted between 1922 and 1994 (Table 2.12). Viviparous lizards and grass snakes (Fig. 2.47), both preferring wet and boggy meadows and forests, strongly dominate in the total numbers of reptiles. However, in terms of biomass, grass snakes and slow-worms are the most important species. Data on year-to-year variation of reptiles are very scarce. Midsummer counts done on a few dozen kilometres were conducted in 4 years, only. In 1955, the mean number of reptiles

Table 2.12. Species structure and dominance in the reptile community in Białowieża Primeval Forest

Species	Body mass (g)	Mean percentage of numbers	Mean percentage of total biomass
Viviparous lizard (*Lacerta vivipara*)	2–6 (5)	52	6.5
Grass snake (*Natrix natrix*)	75–260 (85)	33	82.4
Slowworm (*Anguis fragilis*)	15–31 (30)	11	8.2
Viper (*Vipera berus*)	21–162 (35)	3	2.6
Sand lizard (*Lacerta agilis*)	7–25 (12)	1	0.3
Smooth snake (*Coronella austriaca*)	?	+	+
Swamp turtle (*Emys orbicularis*)	?	+	+

Data averaged from five large-scale censuses conducted in 1922 (entire BPF, Koźmiński 1923), 1955 (eastern part, Bannikov and Belova 1956), 1957 (eastern part, Golodushko 1961a), 1981–1983 (eastern part, Bakharev 1986), and 1994 (western part of BPF, M.M. Pikulik and co-workers, unpubl. data). Records of *Coronella austriaca* after Bakharev (1986) and Pikulik et al. (1988), and that of *Emys orbicularis* after Ruprecht (1989). Body masses (mean values in parentheses) after Pikulik et al. (1988).

Fig. 2.47. Grass snake *Natrix natrix*. Photo: S. Wąsik

recorded per kilometre of transect was 2.8; in 1957, 0.4; in 1981, 5.9; and in 1994, 2.6 (Bannikov and Belova 1956; Golodushko 1961a; Bakharev 1986; M. M. Pikulik and co-workers, unpubl. data). Thus, at least 15-fold changes in availability of reptiles may occur.

It must be emphasised that day-to-day, season-to-season and year-to-year variations in reptile abundance involve changes in both numbers (mortality and natality) and activity. All reptiles are relatively long-lived and slowly reproducing animals (one litter per year with 4–30 offspring; Pikulik et al. 1988). During unfavourable abiotic or trophic conditions, they can remain inactive for long periods.

2.10 Frogs - Plentiful in Rainy Summers, None in Winters

With recent drying of the climate and large-scale draining, choruses of frogs resound no longer in many parts of Europe, its forests included. In Białowieża Primeval Forest, amphibians are still fairly widespread. In late April, shallow flood waters of brooks, old river beds, and forest ponds sound the croaks of frogs and toads, which are there for mating and spawning. The common frog, a typical forest species (Fig. 2.48), is most numerous (Table 2.13). All other species are much rarer. They are either distributed patchily or they are ubiquitous but very rare.

Amphibians are obligatorily aquatic during mating and spawning, then concentrating in and near stagnating and running waters. After spawning, adults of all species except for green frogs (*Rana esculenta*) disperse far from waters. Also, their juveniles leave water and move on land as soon as they metamorphose in late summer or early autumn. All amphibians hibernate in winter so they usually 'disappear' by October.

Amphibians are very prolific animals. A female common frog deposits 840–4000 eggs, and a female moor frog from 220 to 2700 eggs (review in Pikulik 1985). However, the mortality of embryonic and larval amphibians is very high. In various inland waters of Belarus, from 1 to 3.9% of common frog eggs gave rise to froglets emerging on land (Pikulik 1985). Beattie et al. (1993) reported a similar figure (2.1%) in the common frog population of northern England. Thus, an adult female common frog would produce from 8 to 156 froglets/year. Maturation time for frogs is long. Common frogs begin to breed in the fourth year of their life, and moor frogs in the third year (Juszczyk 1987). Potential longevity of frogs is several years and that of toads may be over 20 years (Pikulik 1985; Juszczyk 1987).

In summer, when the spawning season of the most numerous species is over, amphibians are widely, but not evenly distributed in various habitats of

Fig. 2.48. Common frog *Rana temporaria* leading a terrestrial life in forests during summer. Photo: K. Zub

Table 2.13. Species structure of the amphibian community in Białowieża Forest

Species	Body mass (g)	Percentage share in the community	
		Eastern part	Western part
Common frog (*Rana temporaria*)	1–44 (13, 5)	69.4	81.4
Edible frog (*Rana esculenta*)[a]	1–28	19.9	8.6
Moor frog (*Rana arvalis*)	1–29 (10, 8)	2.9	6.3
Common toad (*Bufo bufo*)	1–200 (9, 2)	5.4	3.7
Fire-bellied toad (*Bombina bombina*)	1–10	0.4	+
Green toad (*Bufo viridis*)	1–51	+	+
Natterjack (*Bufo calamita*)	1–22	+	+
Common spadefoot (*Pelobates fuscus*)	1–20	+	+
Tree frog (*Hyla arborea*)[b]	1–8	1.7	+

Eastern part: Bannikov and Belova (1956), census in 1955, western part: M.M. Pikulik and co-workers (unpubl. data), census in 1995 and observations of rare species in 1985–1995. The smooth newt *Triturus vulgaris* and the crested newt *T. cristatus* occurred but were not surveyed. Body masses after Pikulik (1985) and own data. In parentheses, mean body mass in summer (first number: overwintered individuals) and autumn (second number: overwintered and juvenile amphibians); data from trapping in BNP in 1992–1994 (M. Brzeziński, B. Jędrzejewska, W. Jędrzejewski, unpubl. data).

[a] Included *Rana lessonae* and *R. esculenta* (Bakharev 1986).

[b] Censuses of terrestrial amphibians underestimate the share of this arboreal species in the community.

Białowieża Forest (Fig. 2.49). Medium-sized and small forest brooks and wet alder-ash forests growing along them are most densely populated by frogs. The drier the forest, the fewer amphibians live in it, and the higher is the relative contribution of toads to the community (up to 19% in the mixed coniferous forests). Bigger rivers with open marshes in their valleys harbour relatively few amphibians in summer, and the dominant role is played by green frogs (up to 77% of all amphibians in Leśna river valley) (Fig. 2.49). The moor frog was rare in all censused habitats, but in fact, the preferred biotope of this species was not surveyed. Moor frogs can be very numerous in meadows, sedge marshes with stagnant waters and low moors along the Narew River (from 300 to 1800 ind/ha, Pikulik et al. 1987).

Over the last four decades, large-scale summer censuses of amphibians in Białowieża Forest were conducted seven times (Fig. 2.50). The summer index of abundance of overwintered amphibians varied from 0.6 to 24 ind/km of transect. The prime factor responsible for that variation was the amount of rainfall in April-June of the previous year, i.e. the year when most of the censused amphibians were born. Heterothermic animals very susceptible to water shortage, amphibians appeared critically dependent on water supply during the 3 months of embryonic and tadpole stages. Older individuals are

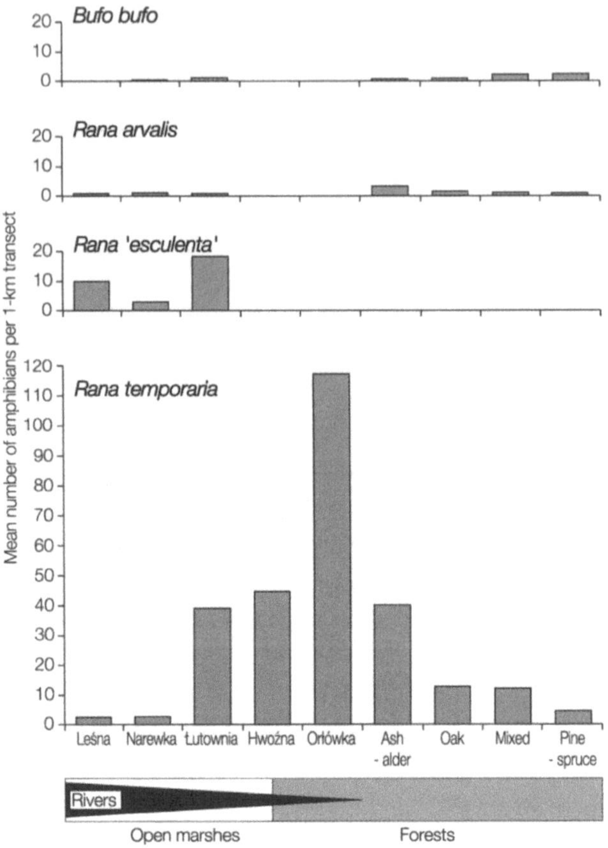

Fig. 2.49. Habitat distribution of terrestrial amphibians in Białowieża Primeval Forest, as censused in summer 1995 (M.M. Pikulik and co-workers, unpubl. data). Habitats arranged in a gradient of decreasing humidity (from marshes and bog forests to dry coniferous forests). Rivers are arranged from the biggest to the smallest ones

susceptible to drought, too, but they can protect themselves by changes in activity and microhabitat selection. Moreover, amphibians can survive long periods without food.

We studied the seasonal dynamics of amphibians in the forests of Białowieża National Park from summer 1992 until spring 1994. Unfortunately, this was the driest 3-year period in the last 50 years (see Fig. 1.4 in Chap. 1) and frogs were very scarce. In the second half of April, when all adult males and females gather in spawning waters, only overwintered subadults not yet reproducing begin to disperse into the forests. In late spring and summer, adults are also captured in forests (Fig. 2.51). At first, most frogs are found in the riverside ash-alder forests, but as summer proceeds, frogs begin to appear in oak-lime-hornbeam forests, too. In late July, newly metamorphosed young spread in the forests (predominantly the wet ash-alder stands). The peak abundance of the juvenile co-

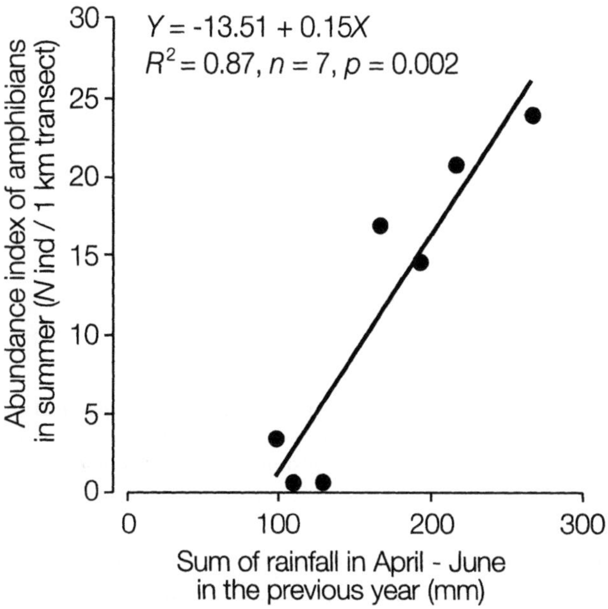

Fig. 2.50. Summer abundance index of overwintered amphibians in BPF in relation to sum of rainfall in April–June of the previous year. Data from: Bannikov and Belova (1956), Bakharev (1986), Pikulik et al. (1987), M. Brzeziński, B. Jędrzejewska, W. Jędrzejewski, and M.M. Pikulik (unpubl. data)

hort is recorded in September. Both numbers and biomass of amphibians are seasonally highest then (Fig. 2.51). In 1992–1993, from summer till early autumn, numbers of amphibians increased five times but biomass only two times, since in September most individuals were 1–3-g froglets. By the second half of October hardly any amphibians are seen in the forests. Those hibernating on land (common toad, moor frog) have gone to their winter retreats, and common frogs have migrated to forest brooks and rivers to immerse themselves in their muddy bottoms and banks. During severe winters, the only signs of amphibian existence from November till March are the leftovers from polecat, otter or mink meals on river banks. But in mild winters with spells of warm weather, it is not unusual to see an awakened frog or the track of its leaping on thawing snow by brooks and rivers. In 1992/1993 and 1993/1994, winter mortality of the common frog was about 80% of autumn numbers, with the juvenile cohort suffering most.

During 1985–1995, censuses of amphibians were done in BPF over 5 years, including dry and rainy ones. In wet summers, densities varied from 0–50 ind/ha in coniferous forests to 1000–2000 ind/ha in riverside sedge marshes and alderwoods (Pikulik et al. 1987). In very dry summers, from 0 in coniferous forests to 220 ind/ha in alder-ash forests were recorded (M. Brzeziński, B. Jędrzejewska, W. Jędrzejewski, unpubl. data). Thus, the largest multiannual variation in amphibian density occurred in good, water-rich habitats. The

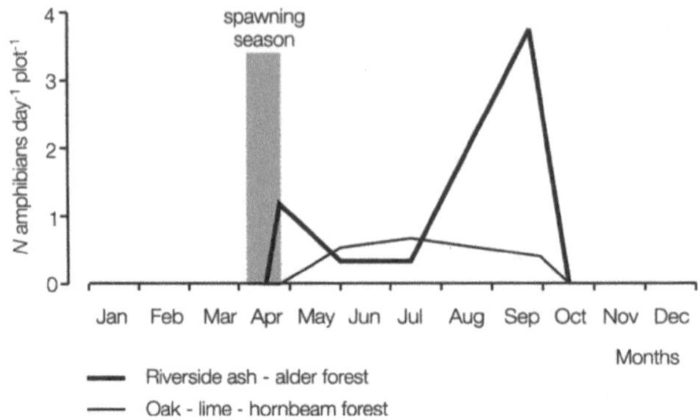

Fig. 2.51. Seasonal variation in amphibian abundance (predominantly *Rana temporaria*) in two types of forests; data averaged from four to eight plots (0.09 ha each) surveyed in 1992–1994 in Białowieża National Park (M. Brzeziński, B. Jędrzejewska, W. Jędrzejewski, unpubl. data)

average densities estimated for the whole forest varied from 15 to 240 amphibians/ha in 1985–1995 (Pikulik et al. 1987; M.M. Pikulik, unpubl. data; M. Brzeziński, B. Jędrzejewska, W. Jędrzejewski, unpubl. data). The mean summer density for all these years was 100 ind/ha, mean biomass about 1300 g/ha. Information on autumn densities, though coming only from years of drought, indicates a fivefold increase in numbers and a twofold increase in biomass from summer until autumn. Undoubtedly, to raptors and carnivores, amphibians are very unpredictable prey. Semiaquatic predatory mammals (polecat, otter, and mink) largely rely on frogs but they too have developed adaptations for hunting other prey, just in case.

2.11 Vertebrate Herbivores and Insectivores of Temperate Forests – Characteristics of Prey Resources for Carnivores and Raptors

Carnivores and raptors of the temperate lowland forests of Europe coexist with nearly 200 species of other vertebrates (fish and aquatic amphibians excluded) that are their potential prey. Let us ponder on the essential features of these species as a whole in order to determine the factors that have influenced the structure of a local predator community. First, one has to realise, that in various biomes of the Palaearctic region, from the Turkmen deserts and dry Mediterranean scrubland to the northern tundra, seven major groups of animals support diverse communities of carnivores and raptors preying on them. These groups are: (1) invertebrates, (2) fish, (3) amphibians, (4) reptiles, (5) birds, (6) small and medium-sized mammals, and (7) ungulates. It is their varying relative contributions to the total numbers and biomass of potential prey that plays a decisive role in shaping predator communities in each biome.

In the temperate deciduous and mixed forests of Europe, as exemplified by the pristine woodland of Białowieża Primeval Forest, the terrestrial fauna of potential vertebrate prey is dominated by homeotherms: birds (68% of species), mammals (23%), with a few cold-blooded animals (5% amphibians, and 4% reptiles). Of all non-aquatic species of potential vertebrate prey, 16% are tiny creatures with an adult body mass below 10 g, and as many as 58% of species are small (11–100 g) (Fig. 2.52). A mere 19% falls into the weight class 0.1–1 kg, and very few (7%) non-predatory species are larger (brown hare, beaver, ungulates). The same frequency distribution of various sized prey animals was calculated for a number of individuals, taking spring densities as a basis. Uncommon species with no reliable density estimates (reptiles, bats, moles) were omitted. Of over 13200 terrestrial vertebrates living on an average square kilometre of Białowieża Primeval Forest, about three quarters (76%) are amphibians, 15% are mammals, and 9% birds. Again, in all these taxonomic groups, the largest number of individuals (77%) falls in the weight class 11–100 g (Fig. 2.52). A further 21% of individuals are diminutive (body mass <10 g), and only 2% of individuals of non-predatory animals are bigger than 100 g.

However, the few largest herbivores (ungulates) make up 83% of the crude biomass (Fig. 2.52) of all potential vertebrate prey. Next, a substantial amount of biomass (13%) comes as small animals (body masses 11–100 g). Although during the study (1985–1996), the densities of ungulates were fairly high, this striking predominance is not aberrant. Ungulates would have dominated in terms of biomass even at the low densities recorded in the 1940s and 1950s. The unrivaled importance of ungulates in the total crude biomass of vertebrate prey seems typical for the pristine woodlands of the temperate zone of Europe.

Another indicative feature of the temperate forests' prey assemblage is the fact that the most abundant (in terms of biomass) prey resources occur in two very distant size classes: 11–100 g and 10–1000 kg. There are very few medium-sized, non-predatory mammals and birds of body masses between 1 and 10 kg. The few brown hares and beavers are essentially inhabitants of open or aquatic habitats that occur marginally in vast woodlands. Among birds, only a few rare species, such as the black stork, capercaillie, crane, and ducks fall into this size category. This is in striking contrast to large open habitats of the temperate zone (grasslands, farmlands, marshes, seashores and lakes), where medium-sized mammals (hares, rabbits, sousliks etc.) and birds (waterfowl, bustards, gulls) are widespread and numerous. One might expect that a consequence of such polarisation of sizes in potential prey would be a division of carnivores and raptors into predators of small prey and predators of large prey, with very little overlap between the two groups.

Besides numbers and biomass, two other features of prey animals are critical for their predators: productivity (i.e. resource renewal rate from the predators' point of view) and range of temporal and spatial variation in abundance (i.e. predictability). Productivity parameters of the six most important

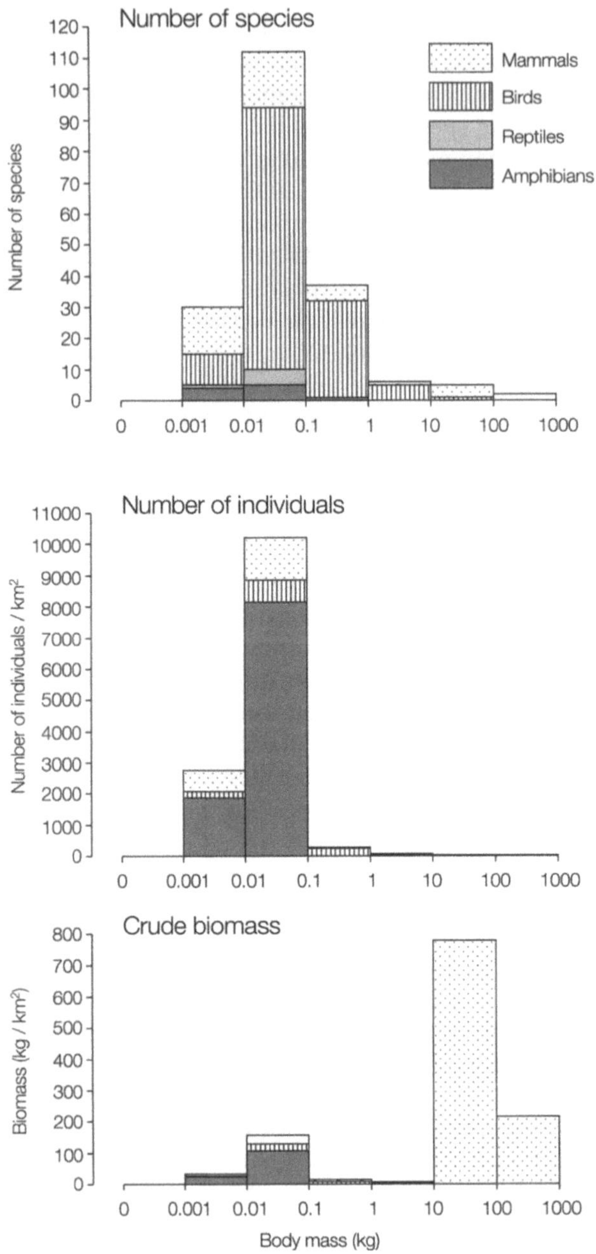

Fig. 2.52. Numbers and diversity of vertebrate animals (prey to carnivores and raptors) inhabiting Białowieża Primeval Forest. *Upper graph* frequency distribution of species in classes of body mass ($N=192$ species), *middle graph* frequency distribution of individual prey in classes of body mass (based on mean densities in spring, $N=13220$ individuals; bats, reptiles, and moles were omitted because of lack of data on densities). *Lower graph* distribution of the amount of crude biomass of animals of various body masses (mean standing crop of biomass in spring, in total 1196 kg; groups of vertebrates as in the middle graph)

groups of vertebrate prey were described in detail in Sections 2.3–2.10. Here, it suffices to compare two essential features: the biggest possible number of young produced during 200 days of the breeding season by a pair of adults, and the magnitude of actual population increase from spring to summer/autumn (Table 2.14). Both parameters unequivocally point to small rodents as the most productive animals. Only can amphibians compete with them in this respect, however, there is a qualitative difference between these two groups of vertebrates. Frogs can produce enormous numbers and biomass of young, but all in a single outflow per year (in late summer/early autumn) which soon completely disappear for winter hibernation. In contrast, rodents' production of young lasts continually from spring till autumn (and sometimes throughout winter) due to the fact that adults bear a few litters in sequence and young of the spring cohort regularly reproduce in the same summer. This makes rodents more attractive than amphibians as a quickly renewable food resource to carnivores and raptors.

Predictability or dependability of prey is based on spatial and temporal stability of their densities. In four groups of vertebrates, we compared (1) variation in densities among habitats, (2) multiannual variation in density, and

Table 2.14. Comparison of productivity of the main groups of vertebrate prey in Białowieża Primeval Forest, as approximation of their renewal rate to predators

Group of vertebrates	Max *N* juveniles produced by a pair of adults in 200 days (spring-autumn)	Ratio of mean post-breeding to mean pre-breeding density
Voles and mice	250	5.8
Amphibians	160	5.0
Shrews	30	5.0
Moles	5–10	3.4
Birds	5–15	2.4
Ungulates	1–13	1.6

Table 2.15. Predictability of major groups of vertebrate prey to their predators in the temperate woodland of BPF, expressed as the mean magnitude of spatial and temporal variation in densities. Winter decline of the number of species results from migration (birds) and hibernation (mammals and amphibians)

Source of variation	Birds	Ungulates	Small rodents	Amphibians
Density variation among habitats (*n*-fold)	2	4	5	10–100
Seasonal variation in mean biomass (*n*-fold)	9.5	1.2	6	Great
Multiannual density variation (*n*-fold)	2	10	30	100
Percentage of species active or present in winter	35%	100%	92%	0%

Table 2.16. Seasonal changes in the estimated standing crop of biomass of the main small prey in the forests, and percentage contributions by each group to the total biomass

Group of prey	Percentage contribution to the total standing crop of biomass			
	Spring	Summer	Autumn	Winter
Forest rodents	12.9	25.0	34.5	66.7 (21.1)
Birds	22.3	31.7	9.7	20.0 (6.3)
Amphibians	61.9	36.1	53.7	0 (68.4)
Shrews	2.8	7.2	2.1	13.3 (4.2)
Total biomass (kg/km^2)	210	360	484	60 (190)

Numbers in parentheses include the biomass of hibernating amphibians (available to only a few carnivores).

(3) variation between seasons. Furthermore, in each group, percentage of species active or present in winter as compared with spring–summer is given (Table 2.15). Birds and ungulates appear definitely the most predictable and dependable food resources to predators. Conceivably, these two groups should give rise to greater specialisation of their predators than any other groups of prey.

Finally, the conspicuous feature of prey abundance in the temperate forests of Europe is its seasonal variation. It is most pronounced in small prey (forest rodents, birds, shrews, and amphibians), the combined biomass of which varies 2.5-fold from autumn till late winter, or even eightfold, if hibernating amphibians (not available to most predators) are excluded from total winter biomass (Table 2.16). In spring and summer, amphibians and birds dominate in the pooled biomass of small prey, whereas in autumn and winter, amphibians and rodents contribute most. Seasonal variation in numbers of ungulates is less pronounced than that of small vertebrates (see Table 2.15), and their share in the total biomass of all prey varies fairly little, from 71% in autumn to 84% in winter.

Small and large, highly productive and slowly reproducing, of stable and changing numbers, the vertebrate herbivores and insectivores of the temperate forests offer a variety of potential foods to carnivores and raptors. In the following chapters we will show how almost forty species of predators make deft use of these resources.

3 Numbers, Distribution and Home Ranges of Predators

On the ruins of their homes grows a young forest.
Wolves are returning and a bear sleeps secure in a raspberry thicket.
Czesław Miłosz, *Notes. In Reverse*, 1978

3.1 Introduction

The following pages describe all the carnivores and raptors that live in Białowieża Primeval Forest. The first and foremost aim of these brief 'portraits' is to show population sizes and densities of predators. The question of how many wolves, goshawks, lesser spotted eagles, and martens inhabit the forest is interesting in itself, particularly in the case of rare and endangered species. However, from an ecological standpoint, it is just the first step towards asking 'Why that many or that few?' Although in some species the answer would be past or current persecution by humans, the overwhelming majority of predators of BPF enjoy relatively undisturbed living conditions, and their densities reflect the potential of pristine temperate forests, where large trees or cavities for nest placement and suitable locations for dens and burrows are excessively abundant, and prey densities are shaped by natural factors.

In pursuing the explanation of observed densities, we analysed two aspects: (1) the spatial distribution of predators, and (2) temporal variation in their numbers. Carnivores and raptors are known for being among the most fiercely territorial animals. We noted the size of home ranges occupied by individuals or social groups, and how densely these home ranges were packed in space. Moreover, we analysed whether predators showed preferences for any particular types of habitat.

It seems that the abundance and availability of food resources are vital for shaping the spatial distribution and densities of carnivores and raptors. This was manifest in a whole array of species, from examples as clear as river otter living exclusively on rivers, in higher densities on larger rivers than on smaller ones, to cases less obvious, such as polecats, for instance, which attain higher densities in habitats where brown frogs are more abundant.

The second aspect of the question 'Why such numbers of predators?' is temporal variation in densities. Definitely, seasonal migrations of raptors, varying reproductive success of all predators and varying mortality rates are driven by changes in abundance and availability of prey. Predators' reaction to growing prey numbers by the increase in their own numbers was called numerical response by Solomon (1949), who first defined it as one of the two essential elements of predators' impact on prey (the second one being functional or dietary response). Since that time, numerical response has received much interest from ecologists studying vertebrate predators and prey.

Two aspects have been particularly considered. One is the biological mechanism of that response, i.e. changes in reproduction and mortality rates of resident predators versus changes in immigration and emigration rates. It seems well documented that the first reaction is more common among carnivores and resident (non-migratory) raptors, whereas the second is more typical of migratory raptors, which can quickly find the regions of temporarily high abundance of prey (Kadochnikov 1953; Gibet 1960; Korpimäki 1994). Another aspect of predators' numerical response is its promptitude. Do predators react immediately or do they lag behind the fluctuations of their prey? If they respond with a time delay, how long is it and what is it determined by? The time lag, if any, has profound consequences to the predators' impact on prey populations. Mathematical modelling has shown that delayed numerical response leads to heavy predation at low numbers of prey and thus has a destabilising effect on prey populations which leads to cycles of prey and predators (May 1981; Hanski et al. 1991).

However, when talking about numerical responses, one must admit that most of the empirical studies and theoretical modelling in vertebrate predator and prey focused on rodent-predator relationships (e.g. Hanski et al. 1993; Korpimäki 1994) and ungulate-wolf relationships (e.g. Peterson and Page 1983; Messier 1991, 1995a, b). This book is not an exception to the rule. The most comprehensive information was available on density variation in predators of rodents and ungulates. However, wherever possible, we analysed the changes in numbers of predators feeding on non-mammalian prey. No doubt amphibian prey-predator and bird prey-predator relationships differ from those of mammalian prey-predator systems. One need only think of the great seasonal changes in both frog and bird numbers and the large variation in their availability to predators, resulting not only from migrations and fluctuations in numbers but from changing abiotic conditions as well.

In the synthesis of the presentation of all species of Białowieża's predators, we sketch the traits of the whole community of carnivores and raptors of Białowieża Forest. Quantitative presentation of dominants, as well as the species barely present, is followed by an attempt to explain what features of dominants made them so successful in the biome of temperate forests.

3.2 An Outline of Methods Used to Survey Predators

Fifty years have elapsed since V.F. Gavrin started the first ecological investigation of predators in the Belarussian part of Białowieża Forest (e.g. Gavrin 1953; Gavrin and Donaurov 1954). His studies on raptors were continued by B.Z. Golodushko in the 1950s (e.g. Golodushko 1961a, b, c) and V.A. Dackevich in the 1970–1980s (e.g. Dackevich et al. 1985). Meanwhile, the Belarussian part of the forest witnessed severe, indeed consequently disastrous, persecution of carnivores and raptors in the 1960s (Korochkina et al. 1980).

In the Polish part of BPF, predators became the target of scientific investigations in the late 1970s. Censuses of owls and diurnal raptors were fully developed by ornithologists of the North Podlasie Society for Bird Protection in the 1980s (e.g. Domaszewicz 1993; Pugacewicz 1996). In 1985, we began the study of the entire community of carnivores and raptors inhabiting the most pristine part of the forest – Białowieża National Park (e.g. Jędrzejewski et al. 1989; Jędrzejewski et al. 1992b; Jędrzejewski and Jędrzejewska 1993). The two streams of information, the Polish and the Belarussian, remained isolated during all the years of communist regime. It was only in the early 1990s that exchange of data, experience and cooperative research started (e.g. Jędrzejewska et al. 1996; Jędrzejewski et al. 1996b; Sidorovich et al. 1996).

Shy, elusive, and rare forest carnivores and raptors have always been very difficult to census, despite the obvious progress in research techniques in the last 50 years. The area covered by such censuses must reckon from several to hundreds of square kilometres, and the time span should be from several years to decades. The earliest attempts at censusing carnivores were based on snow tracking (Fig. 3.1). This simple way of recording an animal's presence, necessarily requiring the researcher to have intimate knowledge of animal tracks, has developed into several methods widely practised in the temperate, boreal and arctic zones of the former Soviet Union (e.g. Nasimovich 1948;

Fig. 3.1. Tracks of lynx *Lynx lynx* family (female and two kittens) in the snow in an alder wood (Białowieża State Forestry District, BPF). Root plates of fallen trees are lynx's favoured places for urine marking. Photo: W. Jędrzejewski

Formozov 1952; Zheltukhin 1984), and central and northern Europe (e.g. Haglund 1968; Goszczyński 1989; Oksanen et al. 1992).

In Białowieża Forest, wolves and lynxes were surveyed on a grid of forest compartment lines (1×1 km) by game wardens during the annual winter censuses of ungulates. Observers recorded the tracks crossing the lines, number of wolves in a pack or lynxes in a family, and the direction of their movement. Then all tracks were mapped, and by drawing the reconstructed routes of wolf and lynx movements, their daily rest sites were determined (i.e. forest compartments where a track went into but not out of it). Data were collected consistently over the last 50 years (Gavrin and Donaurov 1954; Jędrzejewska et al. 1996; Jędrzejewski et al. 1996b). Snow tracking along river banks, combined with measurements of tracks and identification of sex by signs of urine marking, was done to survey river otters, American mink, and polecats on the rivers in the Polish and Belarussian parts of BPF (Sidorovich et al. 1996).

Medium-sized carnivores were censused on the dense grid of transects (0.5×0.5 km and 0.25×0.5 km) covering 11.2 km^2 in Białowieża National Park (Jędrzejewski and Jędrzejewska 1993). During each winter (1985/1986–1990/1991), the transects were walked on the second and third days after a new snow fall (in total two to ten times each winter), and tracks of each species crossing the transects were noted. The number of recorded tracks was converted into the number of tracks per day per kilometre. Then, densities of red fox, pine marten, and polecat were estimated according to the method by Priklonsky (1965). In each species, the mean index (N tracks km^{-1} day^{-1}) for the entire study area for each winter was calculated. This index was then multiplied by a constant 1.57 to obtain the total length (L) of trail of a given predator species laid on an average 1 km^2 during 1 day. L divided by the average daily distance covered by a predator gives the number of animals present per square kilometre. The rationale of Priklonsky's method is that an animal moves along any curve within a circular home range. The constant 1.57 ($=1/2\pi$) is the average length of animal trail inside a square of side$=1$, provided that each side of the square is crossed once. Critical to the accuracy of Priklonsky's method is use of the correct measures of daily movement distances of carnivores. For pine marten and polecat, we applied our own data obtained by combination of snow tracking and radio-tracking (Brzeziński et al. 1992; Zalewski et al. 1995). Furthermore, snow tracking on the grid of transects and, later, mapping of territories, allowed direct estimation of the numbers of small carnivores, the weasel and the stoat (Jędrzejewski and Jędrzejewska 1993; Jędrzejewski et al. 1995). Yet another method based on mapping snow tracks was used to estimate winter home ranges and numbers of raccoon dogs (Bunevich 1983a).

In the three species of denning carnivores, the badger, the red fox, and the raccoon dog, a survey of dens occupied in spring and summer was conducted, along with observations done at some of them to count the number of cubs and adults (Bunevich 1983b, 1984). This is a fairly robust method of estimating badger and fox densities, but only an auxiliary method for censusing raccoon dogs, which show rather weak tenacity to the dens.

Live trapping and marking of individuals, a method widely used in studies of rodents and other small mammals, is not that useful for carnivores because of rather low recapture rates. Thus, live trapping was usually done in order to fit an animal with a radiocollar. However, weasels and polecats were successfully studied by live trapping (capture-mark-recapture) in order to learn their home ranges (polecats) and relative numbers (weasels) (Brzeziński et al. 1992; Jędrzejewski et al. 1995).

Radiotracking, conducted in the Polish part of BPF since 1990, has so far involved five species of forest carnivores: weasel, polecat, pine marten, lynx (Fig. 3.2), and wolf (Brzeziński et al. 1992; Jędrzejewski et al. 1995, 1996b; Zalewski et al. 1995; Okarma et al. 1998). In all species, besides routine localisations on a daily basis, we conducted numerous series of continuous (usually 5-day) radio-tracking sessions, recording an animal's location and activity every 15 or 30 min. Home ranges of radio-collared carnivores presented in this book are the minimum convex polygons with 100% of daily localisations.

Since each census method, the high-tech ones included, has its strong and weak points, we found that the best results were obtained when any carnivore species was studied by two or more methods (usually it was a combination of snow tracking and radio-tracking) so that results obtained by one method could be confirmed and complemented by those from another method (cf. Jędrzejewski et al. 1995; Zalewski et al. 1995).

When it comes to censusing diurnal birds of prey, no method serves better than binoculars plus hours and hours spent in the forest. Regular searches of raptors' nests conducted in early spring (before the development of leaves) and later observations of adult birds carrying prey to young was the basic method of surveying diurnal raptors (Gavrin 1953; Golodushko 1961b; Jędrzejewski et al. 1994b; Pugacewicz 1996; Dackevich 1997). Map-

Fig. 3.2. A radio-collared lynx 'Makary', an adult male; Białowieża National Park, winter 1992/1993. Photo: K. Schmidt

ping of all territorial pairs of raptors gave a basis for calculating their breeding densities. In most species, some nests were reached by climbing to check the number of eggs and/or young. Thanks to the principle of ornithological fascination ('the rarer the species, the more attention it attracts'), even the raptors ephemerally breeding in the forest have been fairly well documented (e.g. Pugacewicz 1993).

All species of owls were censused by mapping their hooting in autumn-early spring, when these birds are very active vocally to establish or confirm their territories (Dackevich et al. 1985; Domaszewicz 1993; Jędrzejewski et al. 1996 a). Surveyors usually stimulated the owls by broadcasting the calls of conspecifics from portable tape-recorders. In early summer, food cries of young owlets made it possible to estimate the owls' breeding success (Domaszewicz 1993).

The sizes of raptors' breeding ranges were measured by the mapping method in tawny owls and pygmy owls only (Domaszewicz 1993, 1997). In all other raptors, it was assessed based on the average nearest neighbour distance between occupied nests (diurnal raptors) or centers of vocal activity (Tengmalm's owl). Rough estimate as it is, the method is justified by the character of the habitat: a large continuous forest, where nearly all species of raptors are fairly evenly distributed.

Last, but not least, the annual harvest and persecution quotas of carnivores and raptors, documented by Game Departments of the Polish and Belarussian headquarters of the forest, provided an important source of information on abundance and status of these animals in BPF. For wolf and lynx, archival data on hunting harvests enabled us to reconstruct the course of their population dynamics in the last 130–150 years (Jędrzejewska et al. 1996; Jędrzejewski et al. 1996 b). We note in concluding that after several research programmes by Polish and Belarussian scientists, a diversity of methods applied, sophisticated equipment used, and a great number of people involved in the strenuous field work, still many questions about Białowieża's predators await future enthusiastic investigators.

3.3 Lynx (*Lynx lynx*)

The sight of large feline tracks on snow is just about the closest acquaintance with a lynx an observer can experience. In recent decades, information on lynxes was collected by carefully reporting occasional observations, winter snow tracking, and recently by radio-tracking. Still longer series of data were available from game inventories and hunting records.

During the last 125 years, human attitude towards lynxes has oscillated from severe persecution to protection. In consequence, the densities of lynxes varied from nearly zero (during periods of deliberate extermination in 1890–1915 and 1960–1968) to 6 ind/100 km^2 (Fig. 3.3). Since 1945, when the state border divided the forest, intense hunting and persecution done in the Belarussian part (in the 1960s) led to temporary eradication of the lynx in

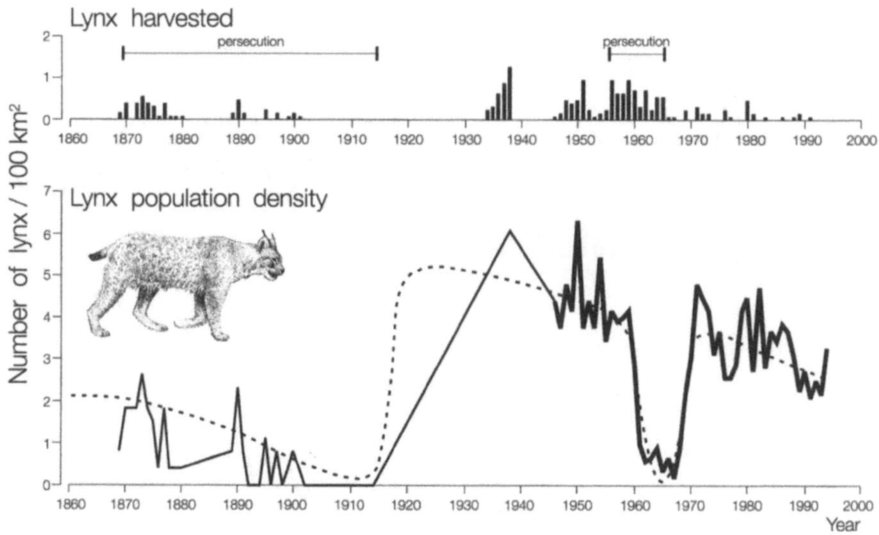

Fig. 3.3. Long-term dynamics of lynx density and hunting harvest in Białowieża Primeval Forest (1869–1994). Density in 1946–1994 *thick line* – snow tracking censuses. Density in 1869–1946 *thin line* reconstructed based on regression with the number shot. *Broken line* – smoothed population dynamics. Jędrzejewski et al. (1996b)

the whole forest (Fig. 3.4). Fortunately, the area was quickly recolonised because the local population is in contact (by dispersal and migrations) with lynxes in large forest tracts adjoining Białowieża from north-east, north-west, and south-west (Schmidt, in press).

In 1991/1992–1995/1996, while the lynx has been protected in both the Polish and Belarussian parts, from 3 to 5.2 lynxes/100 km^2 were recorded in winters (Table 3.1). In the winter population, adult males constituted 24–34%, adult reproducing females 19–23%, subadults and non-breeding adult females 11–26%, and dependent kittens 20–45%. The reproductive success of lynx population depended primarily on food supply (densities of roe deer). In 1991–1995, densities of roe deer were deliberately reduced by forestry personnel by 35–40% (Okarma et al. 1997; Jędrzejewski et al., in press). The lynx reproduction rate was highest in 1991/1992 (on average 2 kittens/mother) and lowest in 1995/1996 (1 kitten/mother) after a decline in roe deer density. Kittens are born in May and remain with their mothers until the mating season of the following year. Only mothers care for young. Data collected over a few decades showed that, on average, 3.3 kittens/mother are born, and 1.6 of them survive till independence (48.5%). Mortality is greatest during the first 100 days of kitten life. In 1991–1994, annual mortality of subadults and adults in a non-hunted population was on average 37%, almost three quarters of that figure being deaths in snares set by poachers for ungulates and hare. Other causes of mortality were rabies, traffic accidents, and possibly accidental poisoning (Jędrzejewski et al. 1996b).

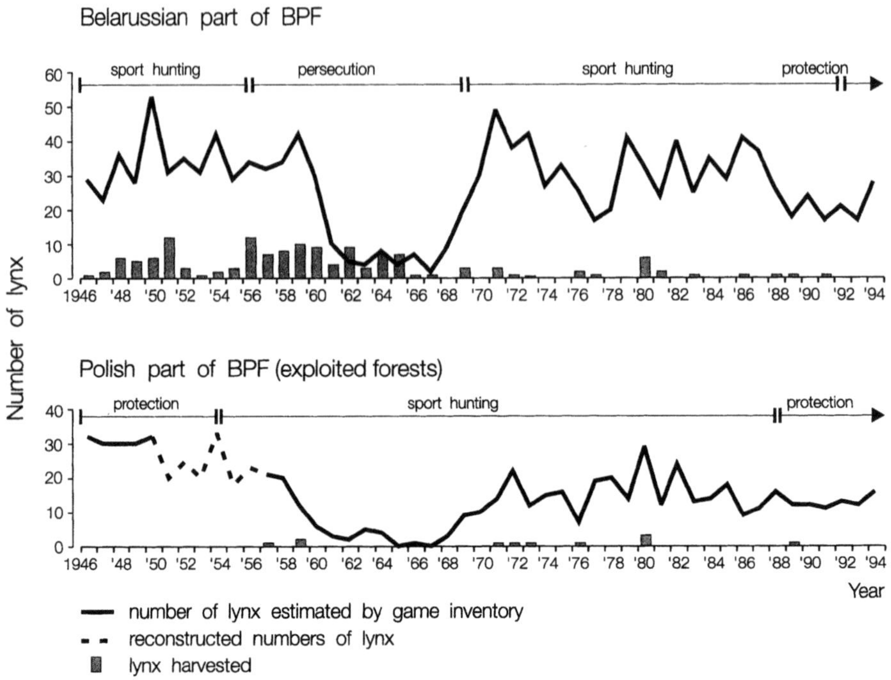

Fig. 3.4. Dynamics of lynx numbers (censused by snow tracking) and hunting harvest of lynx in the Belarussian and Polish parts of BPF in 1946–1994. Jędrzejewski et al. (1996b)

Table 3.1. Densities and population structure of the lynx in the Polish part of BPF as studied by radio-tracking and snow tracking, in the winters of 1991/1992–1993/1994 and 1995/1996. Data are from Jędrzejewski et al. (1996b) and Okarma et al. (1997)

Parameter	Autumn-winter season				Mean (SD)
	1991/1992	1992/1993	1993/1994	1995/1996	
Total density of lynxes (N/100 km^2)	5.1	4.9	5.2	3.0	4.55 (1.0)
Density of adult lynxes (N/100 km^2)	2.8	3.2	3.2	2.4	2.90 (0.4)
including: adult males	1.26	1.57	1.25	1.01	1.27 (0.2)
reproducing adult females	0.98	1.15	1.09	0.60	0.96 (0.2)
subadults and single adult females	0.56	0.48	0.86	0.79	0.67 (0.2)
Density of kittens (N/100 km^2)	2.3	1.7	2.0	0.6	1.65 (0.7)
Mean number of kittens per mother	2.0	1.4	1.8	1.0	1.6 (0.4)

Lynxes are territorial. The lifetime home ranges held by adult males covered from 190 to 343 km^2 (mean 248 km^2) and those of adult reproducing females from 127 to 147 km^2 (mean 133 km^2). Subadult males utilised, on average, an area of 199 km^2, and subadult females 85 km^2 (Schmidt et al. 1997). The size of the area utilised by lynxes varied throughout a year. In May–June, adult females, which nursed kittens, stayed within an area of 7–14 km^2 only. As the

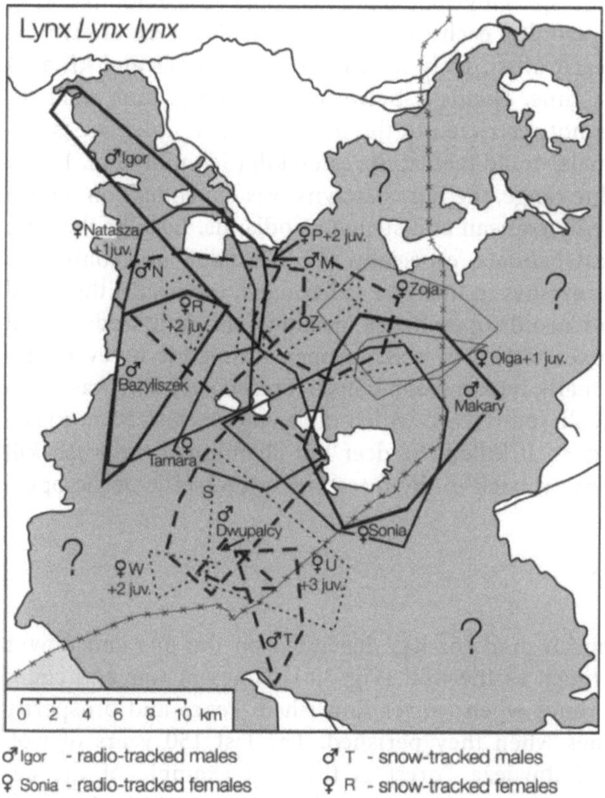

Fig. 3.5. Spatial distribution of lynxes in BPF in the autumn-winter season 1993/1994. *Solid lines* (and all names of lynxes except of Dwupalcy) denote radio-tracked lynxes. Minimum convex polygon home ranges are plotted for all of them, irrespective of the number of localisations. *Broken lines* (lynx Dwupalcy and lynxes denoted by letters) are non-collared individuals recorded by snow tracking. Polygons of 'ranges' embraced all tracks that were unambiguously prescribed to a given lynx. Jędrzejewski et al. (1996b)

kittens grew and were able to follow their mother, her range increased to 32–38 km^2 in July–August and 65 km^2 in September–October. In winter (January–February), family groups moved over the largest area of 61–90 km^2. Subadult females did not show such pulsation of territory size (Schmidt et al. 1997). Adult males held fairly stable ranges (on average 89–99 km^2) in June–November. In December–January, before the mating season, males nearly doubled their ranges (mean 175 km^2). In February–May, males' territories were still larger by one-third (on average 120 km^2) than in summer-autumn. Again, no such changes were observed in a subadult male. In total, mean home ranges of males were 143 km^2 in spring–summer and 165 km^2 in autumn–winter, and those of females 55 km^2 and 94 km^2, respectively (Schmidt et al. 1997).

Home ranges of lynxes overlap much intersexually (62% on average), but less so within each sex (30% among males, and 6% among females). Neighbouring mothers with kittens seem especially well separated (Fig. 3.5). Radiotracking revealed that adult males avoided each other; various males were always located >1 km away from each other. Only family groups (mother and kittens) stay together most of the time. Contacts between adult males and adult females were rare, although not restricted to the mating season only. Occasionally, adult females and a male would feed at the same kill (Schmidt et al. 1997).

In its original geographic range, the Eurasian lynx was dependent on forests from northern taiga to Mediterranean and steppe woodlands. Locally, the lynx also strongly prefers forest habitats, especially wet ash-alder and coniferous stands. It can utilise open marshes in the river valleys and, especially, the small clear-cuts in the forests, but avoids open fields, meadows and farmlands. Rabid or otherwise diseased lynxes fearlessly walk in open farmlands or even into village farmyards. But normally, lynxes seem to need forest cover. This was manifest in one of our radio-collared lynxes, which once entered just 50 m into a field while pursuing roe deer. It killed the deer but abandoned the fresh kill not being able to either drag it back to the forest or to feed on it in the open.

3.4 Wolf (*Canis lupus*)

No other species of European predator has depended on the ups and downs of human civilisation as much as the wolf (Fig. 3.6). Times of war and chaos in human societies were times when wolves flourished. Peace and prosperity in our history meant times when they perished. The last 150 years of wolf population dynamics in Białowieża Forest is the best example of the entangled history of wolves and man (Gavrin and Donaurov 1954; Bunevich 1988; Jędrzejewska et al. 1996). Wolf numbers ranged from nearly zero in 1882–1915 and 1960–1971 to very high in 1870–1880, 1925–1930, and 1945–

Fig. 3.6. Wolf *Canis lupus*, a subadult male. Photo: W. Jędrzejewski

1948 (Fig. 3.7). Maximums of 7.2 to 9.1 ind/100 km^2 were recorded. In recent decades, especially intense persecution of wolves took place in 1946–1960 in both the Polish and Belarussian parts of BPF (Fig. 3.8). Since the early 1970s, wolves have spontaneously recolonised Białowieża Forest from the east. In 1980–1993, winter snow tracking censuses recorded from 11 to 46 wolves in the whole BPF, on average 33, i.e. 2.6 wolves/100 km^2 (Jędrzejewska et al. 1996). In the Polish part of BPF, where on average 13 wolves were recorded every year, negligible hunting occurred in 1980–1989, and since 1989 wolves have been protected. In the Belarussian part, wolves have been treated as pests, and every year on average 16 wolves were shot out of the mean number of 20 individuals recorded (80%). Most of the killed wolves (70% of harvest) were shot in December, January, and February, during traditional snow tracking-driving hunts with 'fladry' (lines of flags, Fig. 3.9) (Jędrzejewska et al. 1996).

Wolves are social animals with family packs formed around the breeding pair and including young of the current and, possibly, previous year(s). In 1980–1993, although groups varied from lone wolves to 11 individuals in a group, four to five wolves was the most common pack size in late autumn. Throughout winter, mean pack size did not change much (3.6–4.2 ind) in the Polish part of the forest (wolves protected). It did so, however, in the Belarussian part: intense hunting meant that, on average, each original pack of four wolves, lost two members during the winter. In some cases, the survivors

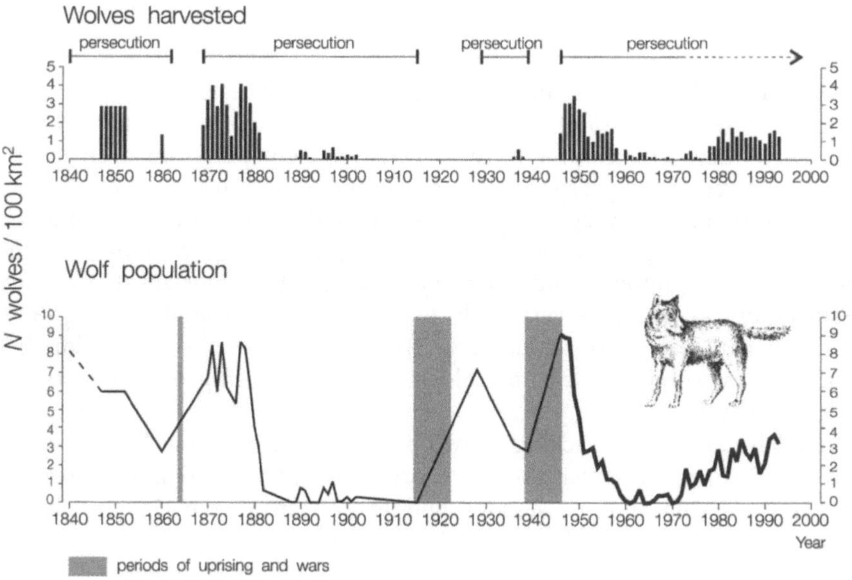

Fig. 3.7. Long-term dynamics of wolf density and hunting harvest in BPF (1847–1993). *Thick line* – density surveyed by snow tracking. *Thin line* – density reconstructed based on numbers shot. Jędrzejewska et al. (1996)

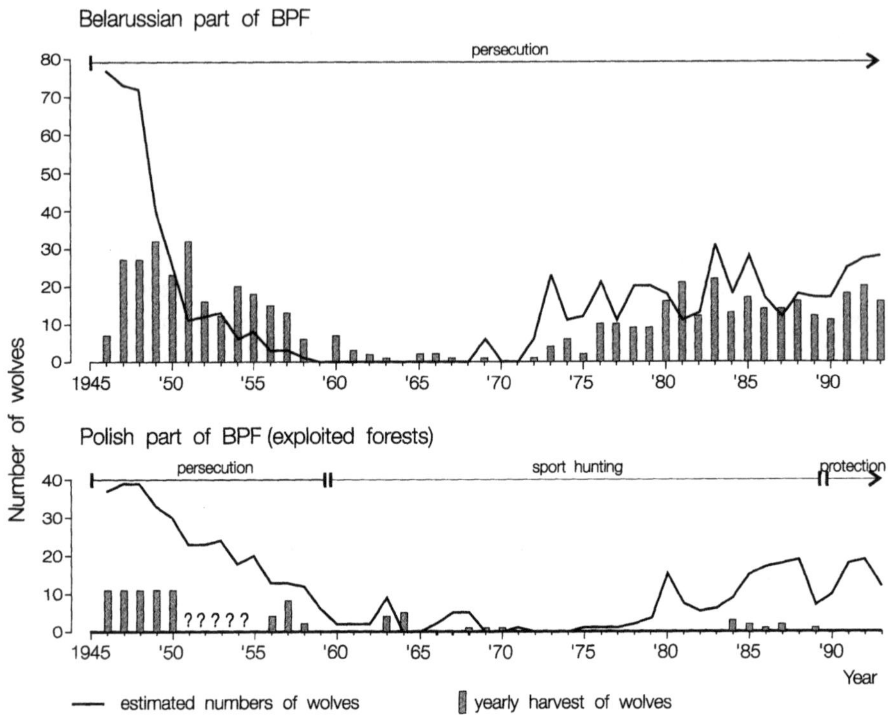

Fig. 3.8. Dynamics of wolf numbers (surveyed by snow tracking) and hunting harvest of wolves in the Belarussian and Polish parts of BPF in 1946–1993. Jędrzejewska et al. (1996)

broke apart and no longer formed a pack. By March each year, the typical pack was a pair and 14% of wolves roamed alone (Jędrzejewska et al. 1996).

In early May, females give birth to three to eight pups (on average 6.2) which have very poor prospects of survival due to both natural factors (food deficiency because of very small pack sizes) and human pressure (common tradition of taking pups for pets). Only 2.2 pups/litter survive more than 100 days, and 0.9 pup/litter would reach 1 year of age (85% mortality). Mortality of subadults and adults is very high, too, mainly due to legal hunting (93% of death) and less so poaching (5%). In consequence, the population has been constantly 'rejuvenated', the oldest individuals being only 5 years old (Jędrzejewska et al. 1996).

In 1994–1996, radio-tracking of two packs of wolves in the Polish part of BPF (Fig. 3.10) revealed that their territories covered 141–168 km^2 (mean 153 km^2) in the spring–summer seasons, and 99–271 km^2 (mean 175 km^2) in autumn–winter (Fig. 3.11). The area utilised by the two packs during the total period of radio-tracking covered 196 and 294 km^2. Home ranges of neighbouring packs overlapped by 8% only (Okarma et al. 1998).

Radiotracking done in 1994–1996, and snow tracking censuses conducted in 1985–1996 indicated a high year-to-year stability of wolves in the Polish

Fig. 3.9. A hunt with 'fladry' for wolves; Khvoinik Forest District, Belarussian part of BPF, winter 1993/1994. A pack of wolves, staying in its diurnal resting site, is surrounded (0.5–1 km^2) by ropes with loosely hanging red flags ('fladry'). Wolves, which are afraid of 'fladry' and will not cross them, are driven by a battue line of several people towards hunters waiting at the outlet opened in the 'fladry' line. Photo: W. Jędrzejewski

Fig. 3.10. A subadult female wolf 'Bystra', anaesthetised (after being captured in nets, see Okarma and Jędrzejewski 1997) and radio-collared; Hajnówka State Forestry District, BPF, December 1995. Photo: W. Jędrzejewski

part of BPF where wolves were protected (Jędrzejewski et al., in press). Usually, three packs (range 2–4) were recorded with little variable spatial distribution, partly determined by the high wire fence on the state border as well as the shape and size of the forest (Fig. 3.11). In 1985–1996, their average density in the Polish part of BPF was 2.3 ind/100 km^2 (Jędrzejewski et al., in press). In the Belarussian part, intense exploitation of wolves causes their social and spatial instability. Numbers of packs, their composition and location of territories varied from year to year. Data on wolf demography and densities showed that, in the 1980s–1990s, breeding of wolves in the Pol-

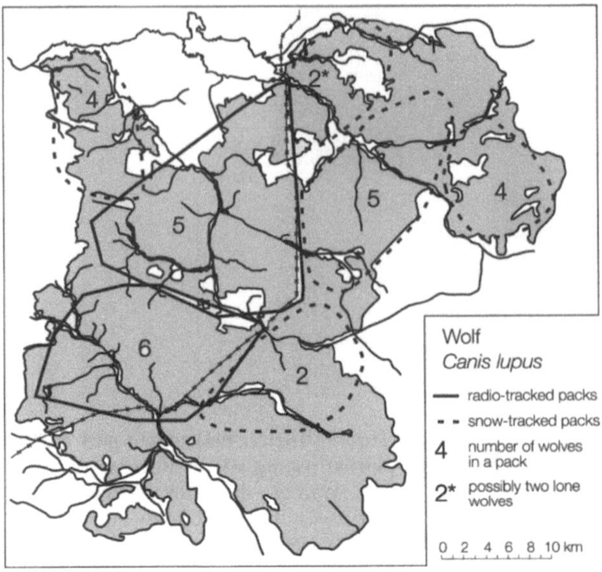

Fig. 3.11. Spatial distribution of wolf packs in BPF in the winter of 1995/1996. *Solid lines* denote home ranges (minimum convex polygon) of radio-tracked packs. *Broken lines* are 'ranges' of snow-tracked packs and embrace all tracks that were unambiguously prescribed to a given pack. Pack size and distribution in the Belarussian part reflects the situation after the hunting harvest in the cold season of 1995/1996. Okarma et al. (1998)

ish part of BPF gave a surplus of individuals (annually about four wolves) that had to leave this area. In contrast, in the Belarussian part, breeding could not compensate for mortality. Annually, on average six wolves must have immigrated there. The 'surplus' wolves from the Polish part could reinforce the heavily exploited wolves in the Belarussian part of BPF. In the whole forest, stable or even slightly increasing wolf numbers were due to constant immigration of wolves (mainly from the forests east of BPF) at a mean rate of 2 wolves/year (Jędrzejewska et al. 1996).

Białowieża's wolves did not have any distinct habitat preferences. All types of forests (with some positive selection of coniferous stands), open marshy river valleys, and - sporadically - even open farmlands are walked by wolves in search of prey and refuge. Wolves do venture into farmlands with small woodlots adjoining Białowieża Forest. In 1993, villagers of Dubiny checked their snares set for hare in patches of willow bushes and copses of young pine amongst fields, and found a large female wolf in one of the snares. However, the most pristine wolf habitats are dense mixed and coniferous old stands in the heart of the forest, not far from marshy valleys of brooks and rivers. In such places along the Hwoźna and Leśna rivers, we spent the most memorable nights, tending at a distance the radio-collared wolves and occasionally hearing their chorus howling.

3.5 Red Fox (*Vulpes vulpes*)

Active at dawn and even in full daylight while hunting voles in meadows by the forest, the red fox is an easy and rewarding target for observer-naturalists as well as hunters (Fig. 3.12). The status of fox, and in consequence, its numbers in the Białowieża Forest have varied greatly during the last century. Alleged pest of game, the fox underwent two periods of severe persecution: during 1889–1915 in the whole forest and in 1958–1970 in the Belarussian part of BPF. In 1885–1902, the years of most intense extermination, on average 24 foxes/100 km^2 were killed annually (Karcov 1903). Still heavier tolls were taken in the Belarussian part in 1960–1965: on average 55 foxes/100 km^2 per year (Fig. 3.13). In those years, foxes were not only shot but also de-

Fig. 3.12. Red fox *Vulpes vulpes*, an adult female. Photo: S. Wąsik

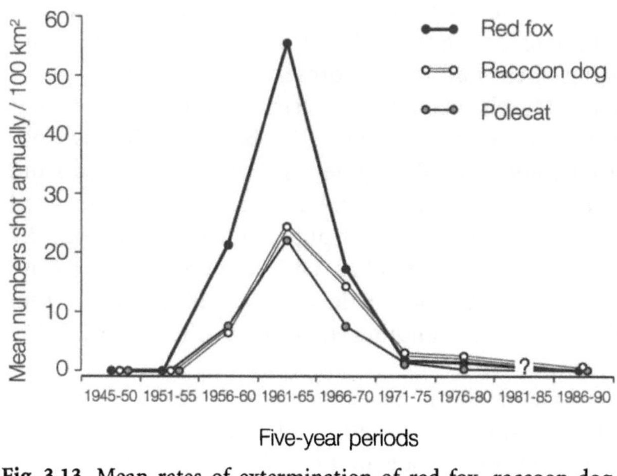

Fig. 3.13. Mean rates of extermination of red fox, raccoon dog, and polecat in the Belarussian part of BPF between 1945 and 1990. Data from: Korochkina et al. (1980) and Proekt organizacii i razvitiya lesnogo khozaistva GNPBP (1993)

stroyed by digging up their dens with young. This practise caused the destruction of many badger's dens as well, although officially badgers were not controlled. Besides the years of persecution, foxes were subject to sustainable hunting (and inevitably poaching) for their high quality pelts. Only in the strict reserve of Białowieża National Park (47.5 km^2) have they been protected since 1921.

Though full records of fox population size are not available, undoubtedly periods of predator control markedly reduced or even temporarily eradicated foxes from parts of the forest. However, it seems that they began recovering soon after the deliberate persecution ceased. For instance, in the Belarussian part of BPF, Gavrin and Krapivnyi (1968), based on snow tracking censuses of foxes done in the late winters of 1946–1954 (before persecution), estimated that their densities varied from 14.5 to 29.4 foxes/100 km^2 (on average 21.2 foxes/100 km^2). By 1982, about 10 years after the extermination ceased, Bunevich (1983 a) found 41 dens inhabited by foxes, and estimated that summer densities were 20.6 foxes/100 km^2. In the winter of 1981/1982, snow tracking census revealed, on average, 1.11 fox tracks km^{-1} day^{-1} (Bunevich 1983 a), which, recalculated into densities according to Priklonsky's (1965) method, yielded an estimate of 19 foxes/100 km^2. In 1985/1986–1988/1989, in Białowieża National Park (Polish part of BPF), based on the same snow tracking method, we found somewhat higher densities of 2.5–3.5 foxes per 10 km^2.

Generally, in the 1980s, 25 to 41 breeding dens of red foxes were found in the Belarussian part of BPF (A.N. Bunevich, unpubl. data) and three to five dens in Białowieża National Park (authors' unpubl. data). As shown for 1987 (Fig. 3.14), most dens are located along forest edges: near open farmland, meadows and pastures, or open sedge marshes along rivers. Foxes strongly prefer open habitats as foraging ground. Breeding dens, however, were placed in all types of forests except for wet ash-alder forest. The neareast neighbour distance between breeding dens varied from 1.4 to 10.7 km, on average 4.62 km (Fig. 3.14). Thus, there was one occupied breeding den per 16.8 km^2, with much of that area being open land by the forest.

After mating in February–March, vixens give birth in April. In late May, pups may be observed at the den entrance (Fig. 3.15). Based on 76 such observations collected between 1949 and 1981, Bunevich (1983 b) reported from one to seven pups (in one case 13, 10 alive and 3 dead) seen at the natal den. In most cases (30%) a litter consisted of 3 pups, and the mean was 3.2 pups/litter. Young disperse in August–September. Of five pups eared by Gavrin and Krapivnyi (1968) in June 1953, one male was shot 40 km away from the natal den in December 1953, and another male was killed just 2.5 km from the den in January 1954.

Besides human exploitation, epizootic diseases such as sarcoptic mange might have contributed to year-to-year variation of fox numbers. In BPF, such epizooty occurred in 1944–1947 (Gavrin and Krapivnyi 1968). A few carcasses of foxes dead from mange were found and 25% of shot foxes were infected. Rabid foxes have also been noted in Białowieża Forest.

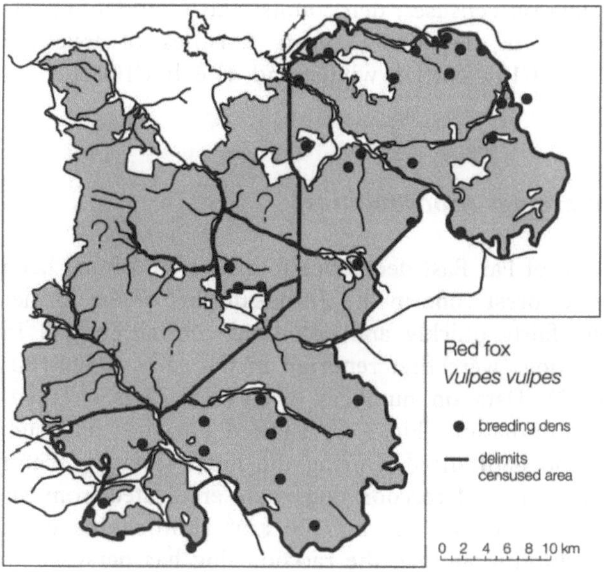

Fig. 3.14. Spatial distribution of breeding dens of red foxes in the whole Belarussian part of BPF and Białowieża National Park (Polish part) in 1987. Sources: A.N. Bunevich (unpubl. data) and authors' unpubl. data

Fig. 3.15. Red fox pup at the entrance of its natal den, Białowieża National Park, compartment 401, late spring 1995. Photo: S. Wąsik

Based on all available data on censuses, den counts, and reproductive parameters, we estimated that in the 1980s, mean densities of foxes in Białowieża Forest were 10 ind/100 km^2 in winter and 21.4 ind/100 km^2 in summer.

3.6 Raccoon Dog (*Nyctereutes procyonoides*)

The raccoon dog, a predator of Far East deciduous forests, after having been introduced to the temperate forest zone in the European part of the former Soviet Union, spread west fairly quickly and colonised central Europe. In Białowieża Forest, raccoon dogs were first reported in the early 1950s (Bunevich and Dackevich 1985). Data on numbers of raccoon dogs shot annually during predator extermination (Fig. 3.13) showed how fast and successfully this species became naturalised. During the most severe predator control in 1961–1965, on average 214 raccoon dogs were eradicated from the Belarussian part of BPF annually, i.e. 24.6 ind/100 km^2 (Korochkina et al. 1980). Nonetheless, a very prolific animal, the raccoon dog has persisted in Białowieża Forest (Fig. 3.16). Nowadays, it varies from nearly zero in bog pine forests (Bunevich and Dackevich 1985) to 5 ind/10 km^2 in rich deciduous old growths of Białowieża National Park (Jędrzejewski and Jędrzejewska 1993). In 1982, a winter census done in the Belarussian part of BPF yielded a density estimate of 16.6 ind/100 km^2 (Fig. 3.17).

Similar to badgers in their life habits, raccoon dogs are almost completely inactive from early December, when they begin their winter sleep, preferably in badger den systems (with or without the host wintering in it). Although raccoon dogs sometimes wake up and go out to forage in winter, they leave their winter 'beds' definitely in early February in mild winters and in early

Fig. 3.16. Raccoon dog *Nyctereutes procyonoides*. Photo: S. Wąsik

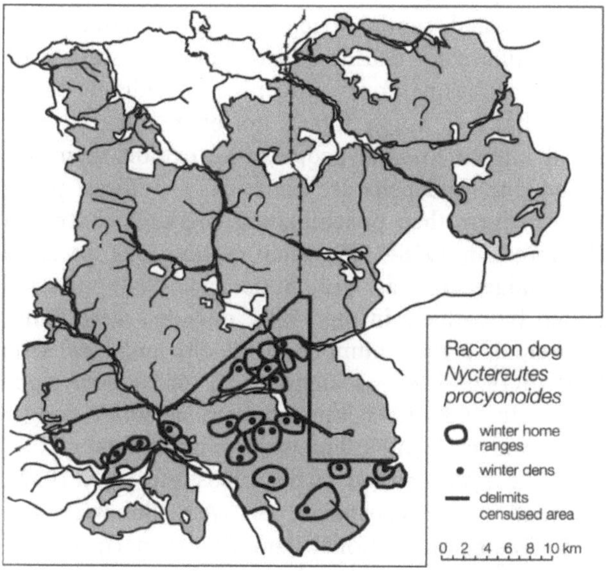

Fig. 3.17. Spatial distribution of winter dens and home ranges of raccoon dogs in the southern districts of the Belarussian part of BPF in 1981/1982. Source: Bunevich (1983 a)

March in more severe ones. Raccoon dogs do not dig dens but readily occupy dens of foxes and badgers as well as hollows in tree trunks, hide under the root plates of wind-fallen trees and other refuges. They live in pairs. In the cold season, a pair utilises a home range of 0.5 to 4 km^2 (Fig. 3.17). Summer ranges of raccoon dogs are unknown.

The reproductive potential of raccoon dogs is very high; from four to eight pups (on average 6.2) were seen in March through July (Bunevich 1983 a; unpubl. data). However, mortality of pups must be high because family groups leave a breeding den as soon as the pups can follow their mother (in June/July) and forage. Year-round, raccoon dogs particularly favour sedge marshes by rivers, wet alder-ash forests and rich oak-lime-hornbeam forests as their foraging quarters. The ecological success of raccoon dogs in a predator community of Białowieża has its dark side. This carnivore is very susceptible to rabies. Nearly every year rabid raccoon dogs, walking fearlessly in streets or fighting with chained dogs in farmyards, are killed in the village of Białowieża.

3.7 Badger (*Meles meles*)

After busy, intense foraging in October, badgers hide in their large deep dens and usually do not emerge until late March. Unlike foxes and raccoon dogs, which often shift dens from year to year or even from season to season, badgers (Fig. 3.18) usually overwinter, breed and live in the same den system for years. Carefully situated in elevated sandy strips among deciduous and mixed forests, many dens are utilised for decades. In Białowieża Forest, badgers strongly prefer oak-lime-hornbeam forests and mixed deciduous-coniferous stands. They inhabit neither wet, boggy forests nor open areas.

Most badger setts are well known to local people, who exploit them because badger fat (as folk medicine) and pelts are much desired 'forest products'. Yet more dangerous to badgers than poaching was predator control in the Belarussian part of the forest in 1958–1970, when digging out fox dens caused mass destruction of badger setts as well. Bunevich (1984) reported that of over 50 setts inhabited by badgers in the 1950s, only ten were found active in 1978–1979. Moreover, of a total number of 80 old and new setts mapped by Bunevich (1984) in 1978–1982, as many as 52 (65%) bore signs of earlier digging by humans. In 1980–1995, the numbers of active badger setts varied from 4 to 21, on average 12 (A.N. Bunevich, unpubl. data).

In 1987, a year typical for the whole period of the 1980s and early 1990s, 11 inhabited dens were recorded in the Belarussian part (A.N. Bunevich, unpubl. data), and two dens in Białowieża National Park (Polish part; unpubl. data), i.e. 1.4 active dens/100 km^2 (Fig. 3.19). Nearly all of them were located deep in the forest. The nearest neighbour distance among occupied setts varied from 2.1 to 13.5 km (mean 8.1 km). However, one must remember that due to former exploitation of badgers, even the suitable habitats were seriously undersaturated. Since 1995, the badger population in Białowieża Forest (both the Polish and Belarussian parts) has begun to increase (R. Kowalczyk, A.N. Bunevich, pers. comm.). In the strict reserve of Białowieża National Park (47.5 km^2), six active setts were found in 1996 (R. Kowalczyk, pers. comm.).

Fig. 3.18. Badger *Meles meles*, an adult male.
Photo: W. Jędrzejewski

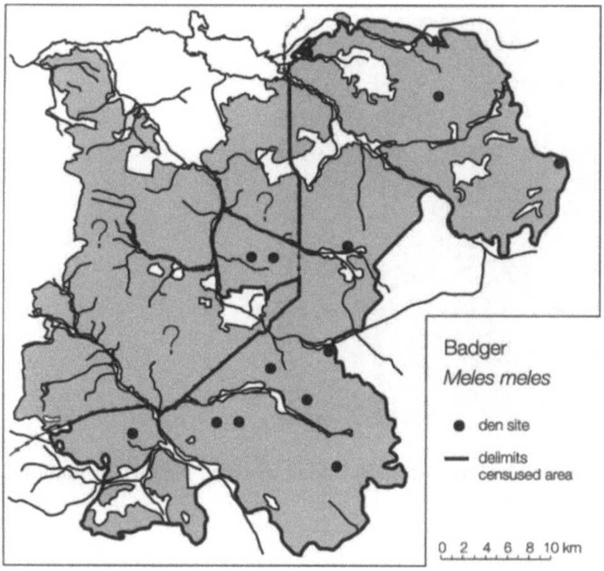

Fig. 3.19. Spatial distribution of dens inhabited by badgers in the whole Belarussian part of BPF and Białowieża National Park (Polish part) in 1987. Source: A.N. Bunevich (unpubl. data) and authors' unpubl. data

Badgers live solitarily, in pairs or larger family groups. So far, no data are available on numbers of individuals living in one den system in BPF. Śliwiński (1987), who studied badger population in the deciduous and mixed forests of Rogów (central Poland), found that from one to ten badgers lived in one sett in summer, on average 3.2 individuals per active sett. This number included a mean of 0.9 cubs/sett (range from 0 to 6). In ten litters observed by Śliwiński (1987) in 1984–1986, the number of cubs varied from one to six, most often two cubs (6 litters). Mean litter size was 2.3 cubs. Bunevich (1984) reported similar reproductive success (2.2 cubs/litter) from Białowieża Forest.

No data are so far available on badger densities in the exploited forests of the Polish part of BPF. Therefore, based on information from the Belarussian part and Białowieża National Park, and the above-cited demographic parameters, we estimated that in 1980–1995, the forest harboured at least 3.6 badgers/100 km^2 in winter and 5 badgers/100 km^2 in summer.

3.8 Pine Marten (*Martes martes*)

The Polish name for pine marten, *kuna leśna*, means forest marten, an appropriate characteristic of this predator in both its geographic range and local habitat preferences (Fig. 3.20). In central and eastern Europe, from Medi-

Fig. 3.20. Pine marten *Martes martes*, a radio-collared adult female; Białowieża National Park, June 1992. Photo: W. Jędrzejewski

eval times until the 17th century, marten pelts were a kind of currency in the economic exchange of goods and duties. In some countries, 'marten' became synonymous with 'money', as in Croatia, where contemporary currency is called *kuna*. Nowadays, a game species in both Poland and Belarus, the pine marten is a valuable although casual trophy of hunters and much desired target of poachers. In the past, pine martens in Białowieża underwent periods of extermination (with a peak in 1895–1902, when on average 1.2 martens/10 km^2 were killed annually; Karcov 1903) but in recent years, its official harvest in the forest has been rather small.

In autumn-winter, home ranges of pine martens studied by radio-tracking were from 1.4 to 2.7 km^2 (mean 1.9) in males and from 1.2 to 2 km^2 (mean 1.6) in females. In spring-summer, home ranges of males covered 1.9–3.5 km^2 (mean 2.7) and those of females 0.7–2.6 km^2 (mean 1.4) (Zalewski et al. 1995). Home ranges overlapped extensively between sexes but less so intrasexually (Fig. 3.21).

By snow tracking in the winters of 1985/1986–1990/1991, verified by live trapping and radio-tracking in 1990/1991 and 1991/1992, we estimated that the densities of pine marten in the pristine forests of Białowieża National Park ranged from 3.6 to 7.6 ind/10 km^2 (mean 5.7) (Jędrzejewski and Jędrzejewska 1993, Zalewski et al. 1995). Year-to-year variation in marten numbers were predominantly influenced by changes in forest rodent abundance, to which martens responded with a 1-year lag (Figs. 3.22 and 3.23). This carnivore has a pregnancy with delayed implantation, so mating in July–August results in young born in March–April of the following year. Thus, the highest densities of martens were observed 1 year after the rodent outbreak, and the lowest ones 1 year after the rodent crash. From one to four kittens in a litter (on average 2.6) were seen when they leave nests and fearlessly chase around, scream, and play together.

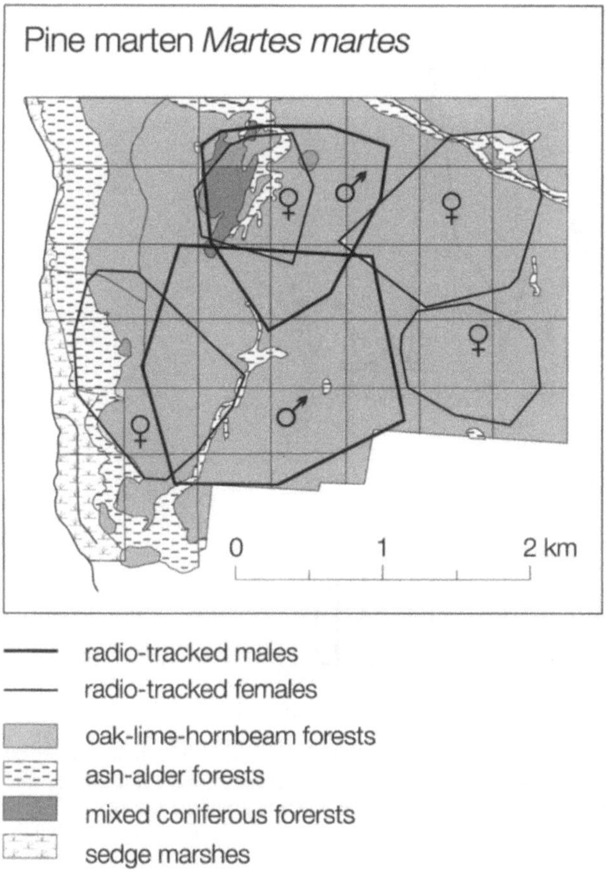

Fig. 3.21. Distribution of pine marten home ranges (minimum convex polygons) in the south-western part of Białowieża National Park in the autumn-winter of 1991/1992. Source: Zalewski et al. (1995), modified

Pine martens lead the most arboreal life of all European predatory mammals. Most of their daytime resting sites are cavities in old trees, squirrel nests, and thick branches in the canopies (Zalewski 1997). Białowieża's martens preferred coniferous forests and oak-lime-hornbeam forests, but they also readily penetrated narrow strips of sedge marshes along forest rivers. Strong avoidance of fields, meadows and any large open areas was typical of this predator.

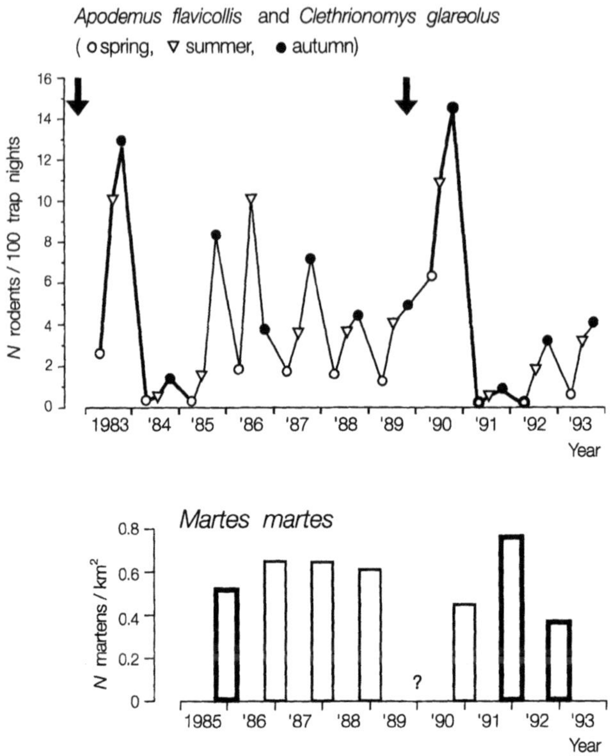

Fig. 3.22. Autumn-winter density indices of pine martens *lower panel* in relation to forest rodent abundance *upper panel* in BNP. *Arrows* denote heavy crops of oak, hornbeam, and maple seeds. *Thick lines* denote outbreaks and crashes of rodents after the heavy seed crops and marten numerical response to the extreme numbers of rodents. Zalewski et al. (1995)

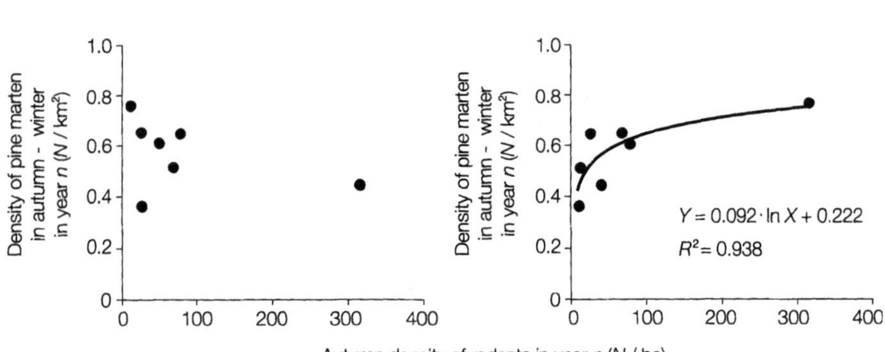

Fig. 3.23. Numerical response of pine martens (autumn-winter densities) to the variation in autumn numbers of forest rodents in BNP in the same year *left panel*, and with 1-year lag *right panel* in 1985/1986–1992/1993. Zalewski et al. (1995)

3.9 Stoat (*Mustela erminea*)

Open sedge marshes scattered with willow bushes by rivers are the preferred habitats of the stoat. In the contiguous forests of Białowieża National Park, on average 2 stoats/10 km^2 were found in the winters of 1985/1986–1988/1989 (Jędrzejewski and Jędrzejewska 1993). A winter survey done in 1993/1994 (Sidorovich et al. 1996) along the rivers revealed from 1.3 to 6.9 stoats per 10 km transect along the river banks (on average 5.1). Stoats were twice as common along rivers with open floodplains (5.6 ind/10 km) as along rivers with forested valleys (2.7 ind/10 km). Recalculated into the area of open valleys, the numbers of stoats recorded per 10 km transect would correspond to about 27 ind/10 km^2 in the open marsh habitat. This figure is most likely an overestimate, since stoats recorded in the narrow valleys could have penetrated the adjacent forests. Nonetheless, it shows that these predators are predominantly concentrated along the rivers.

3.10 Weasel (*Mustela nivalis*)

One of the rewarding experiences of working in Białowieża Forest is to spot the tiny, impeccably white weasel dogding on the mossy forest floor on a snowless winter day. However, despite the fact that weasels are the most numerous predatory mammals and are totally diurnal, it was only during the radio tracking sessions that we began to see these animals in the forest. The radio-collared weasels (Fig. 3.24) did not pay any attention to a person holding an antenna, and occasionally would even run straight to his or her feet in passionate pursuit of a rodent.

Fig. 3.24. Weasel *Mustela nivalis*, a radio-collared adult male; Białowieża National Park, autumn 1990. Photo: K. Zub

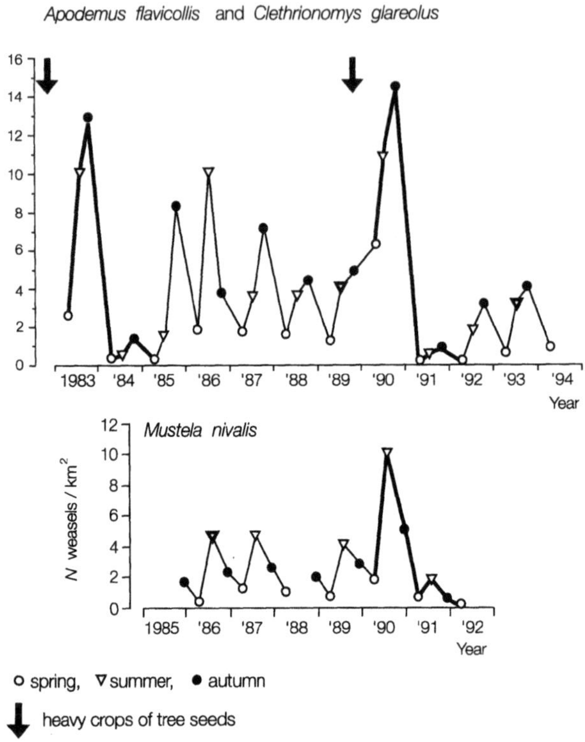

Fig. 3.25. Densities of weasels *lower panel* in relation to forest rodent abundance *upper panel* in BNP. *Thick lines* denote outbreaks and crashes of rodents after heavy seed crops and weasel numerical response to the extreme numbers of rodents. Jędrzejewski et al. (1995)

In Białowieża Forest, weasels can be found in all habitats, from pristine and exploited forests of all types, to natural sedge marshes, to farmlands and villages. Every aspect of weasel biology, particularly its small size and fast reproduction rate, is an adaptation to exploit rodents and to trace the changes in their abundance as quickly as possible. As shown by the study in Białowieża National Park in 1985/1986–1991 (Jędrzejewski et al. 1995), densities of weasels were lowest in spring: 0–1.91 ind/km^2, on average 0.88 (Fig. 3.25). Weasels reproduce from spring until early autumn. After 35 days of pregnancy, females give birth to four to eight young (on average 5.2), and may become pregnant again while still nursing their first litters (Jędrzejewska 1987). Thus, an adult female may bear two litters in one summer.

By July, weasels reached their seasonally highest densities, from 1.9 ind/km^2 during a crash of forest rodents, to 4.2–4.8 ind/km^2 at moderate rodent density, to as many as 10.2 ind/km^2 during an outbreak of rodents (Fig. 3.25). Increase of weasel population from spring to midsummer was positively related to the spring numbers of rodents. However, the summer peak of weasels did not last long. From July until April of the following year, their numbers declined due to mortality and dispersal of young. Winter numbers of weasels were less variable than summer ones, from 0.5 to 2.7 ind/km^2, on average 1.9 (Fig. 3.26).

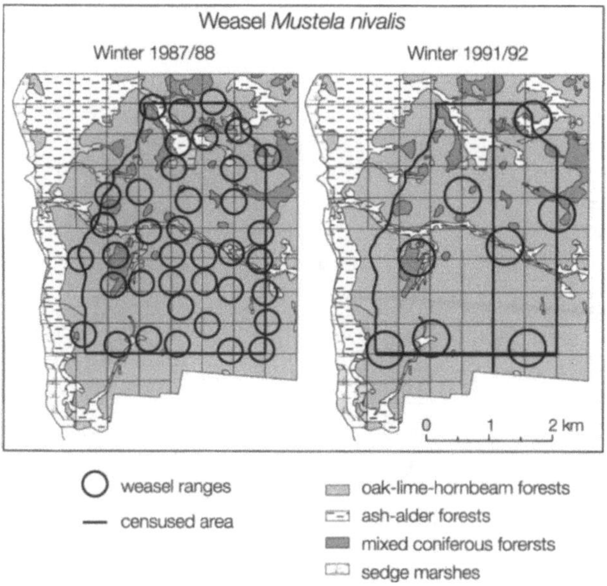

Fig. 3.26. Schematic winter 'territories' of weasels in BNP mapped by recording tracks in snow on an 11.2-km^2 grid of transects in the winters of 1987/1988 (moderate density of forest rodents) and 1991/1992 (crash of rodents). Diameter of 'territory' was 0.4 km in 1987/1988 and 0.6 km in 1991/1992. Jędrzejewski et al. (1995), modified

Weasels truly live fast and die young, mainly because apart from being predators they are prey, too. The mortality rate estimated by radio-tracking (according to the method of Heisey and Fuller 1985) was 87% during 6 months. Major causes of that mortality were predation (65%) and intraspecific combats (22%). We found weasel remains in scats or pellets of several species of predators (e.g. buzzard, lesser spotted eagle, tawny owl, pine marten, red fox) with a mean frequency of 0.5%.

Densities of weasels estimated in breeding and nonbreeding seasons were plotted against rodent densities. The numerical response of weasels to fluctuations in their prey population was always of logarithmic type (Fig. 3.27). Numbers of weasels were strongly correlated with those of rodents with no time delay. In spring and summer, weasels immediately responded to rodent increase by increasing, too. In autumn and winter, when rodents declined, weasel numbers declined as well.

Home range sizes of weasels, as determined by radio-tracking, are shaped by rodent abundance. Territories of males varied from 0.11–0.37 km^2 (mean 0.24 km^2) during a rodent outbreak (300 rodents/ha) to 1.17–2.16 km^2 (mean 1.7 km^2) during a year of rodent crash (10 rodents/ha) (Fig. 3.28). No data are available on females' home ranges. In a year of plentiful food resources, weasels tolerated much overlap between home ranges as well as the presence of transient (dispersing) weasels (Jędrzejewski et al. 1995).

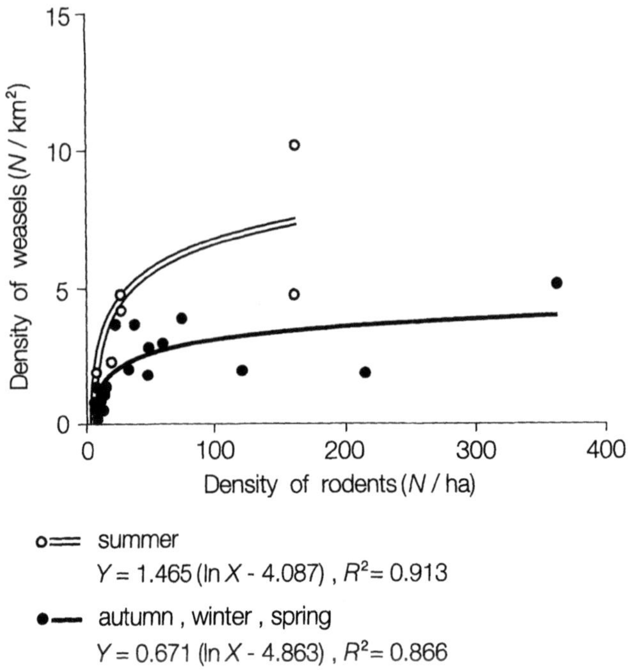

Fig. 3.27. Numerical response of weasels to variation in numbers of forest rodents in BNP in 1985/1986–1992. Jędrzejewski et al. (1995)

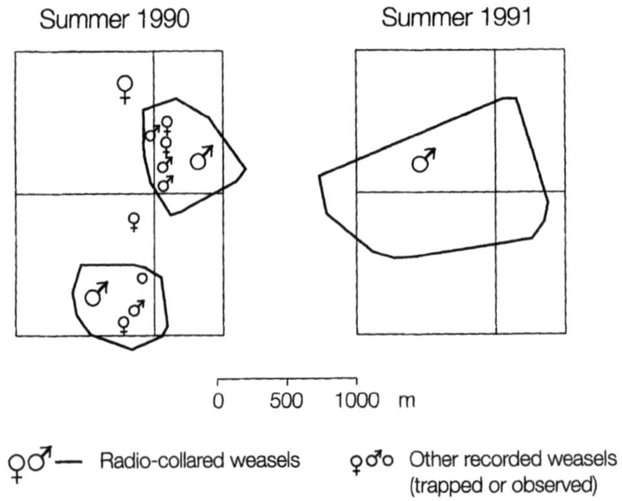

Fig. 3.28. Changes in home range size of male weasels (studied by radio-tracking) and total number of weasels recorded on the 2.1-km transect in BNP in summer (July–August) of 1990 (rodent outbreak) and 1991 (crash of rodents). Jędrzejewski et al. (1995)

3.11 Polecat (*Mustela putorius*)

In their pristine habitats of European temperate forests, polecats (Fig. 3.29) are semiaquatic predators, but of the three such predatory mustelids (the other two being otter and mink), it is the most 'terrestrial' one. Its preferred habitats are riverside ash-alder forests, forested valleys of the smallest forest brooks with shallow, slowly flowing waters, and bog alderwoods with stagnated water (Jędrzejewski et al. 1993 a). Polecats are known and blamed as poultry killers, but one must admit that large-scale drainage and land reclamation all over Europe deprived these animals of their primeval habitat and, in many regions, forced them to live under conditions, which used to be their secondary, 'last choice' habitats. In the past, polecats were traditionally alleged as damaging small forest game and were persecuted. In Białowieża Forest, polecats suffered most in 1895–1902 (when annually on average 1 polecat/10 km^2 was killed; Karcov 1903) and again in 1960–1965 in the Belarussian part of BPF (when 2.3 polecats/10 km^2 per year were eradicated; Fig. 3.13).

A snow tracking census conducted in the winter of 1993/1994 (Sidorovich et al. 1996) showed that polecats were most numerous along the smallest brooks in the alder-ash forests, on average 2.5 individuals per 10 km of water course (and up to 5 polecats/10 km on Orłówka brook in Białowieża National Park). Medium-sized rivers harboured on average 1.5 ind/10 km, and the biggest ones 0.9 ind/10 km (Fig. 3.30). Such a distribution of polecats was determined by the abundance of their vital food resource, the common frog (Fig. 3.31). Polecats also live in bog alderwoods and wet ash-alder forests, which are the best of all forest types for brown frogs (see Chap. 2). In 1985/1986–1990/1991, winter densities of polecats in Białowieża National Park varied from 3.5 to 6.3 ind/10 km^2 (mean 4.4 ind/10 km^2).

Fig. 3.29. Polecat *Mustela putorius*, an adult male. Photo: S. Wąsik

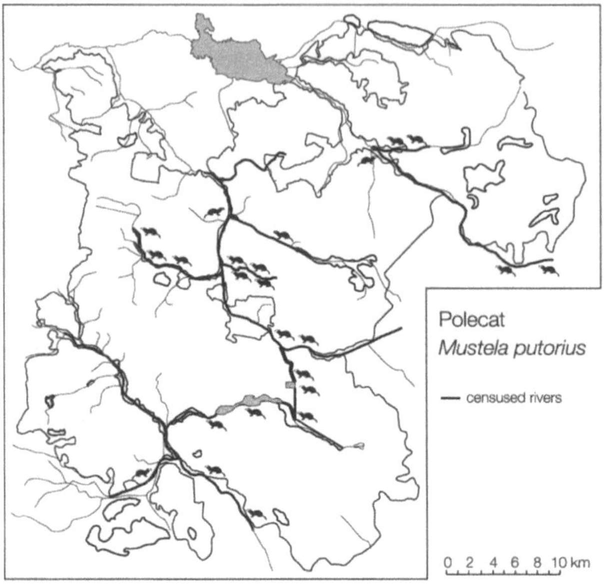

Fig. 3.30. Distribution of polecats on the censused rivers (170 km in total) in Białowieża Primeval Forest in winter 1993/1994. Each *silhouette* denotes one individual. Larger water reservoirs are shaded. Sidorovich et al. (1996)

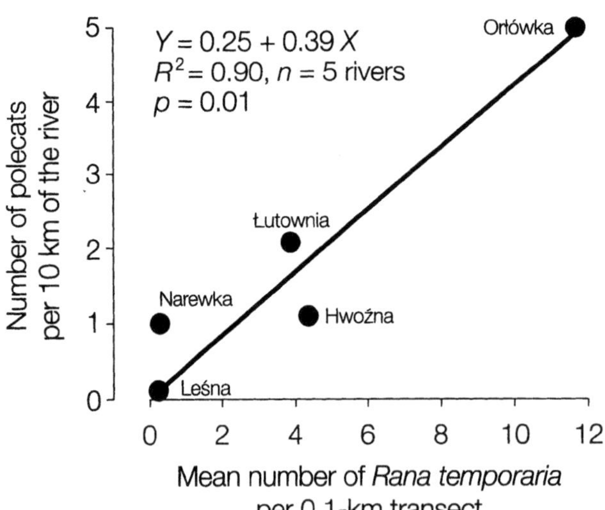

Fig. 3.31. Winter abundance of polecats in relation to numbers of the common frogs (censused in summer) on five rivers in the Polish part of BPF. Data from: Sidorovich et al. (1996) and M.M. Pikulik and co-workers (unpubl. data)

Polecats are as much territorial as they are nomadic. In the cold season, when waters are free of ice at least in some places, and polecats have access to hibernating amphibians, they hold territories along 1–3.1 km stretches of brooks in males (mean 2.3 km) and 0.7–1.7 km in females (mean 1.4 km). There was extensive intersexual overlap of polecat territories (Fig. 3.32). The home range of one radio-tracked male polecat covered 2 km^2. When rivers and bog alderwoods become completely ice-covered, polecats move to drier habitats (primarily the oak-lime-hornbeam forests) to hunt for rodents, and some of them (males and females equally often) roam into the villages, when they sooner or later come into conflict with owners of rabbits and hens. Another period of nomadism (this time of males only) is the mating season in March–April. Some polecats may permanently live and breed in the villages but, in Białowieża Forest, a constant exchange of individuals between forest

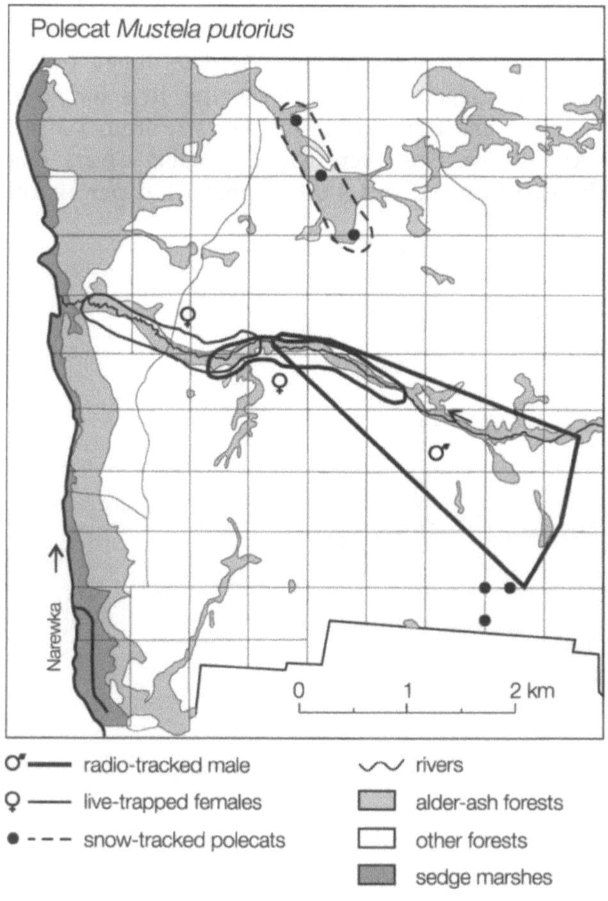

Fig. 3.32. Home ranges of polecats in the south-western part of BNP in winter 1990/1991. Sources: Brzeziński et al. (1992) and authors' unpubl. data

and rural habitats takes place. A male captured and radio-collared on Orłówka River in Białowieża National Park in March 1990 was found in a poultry house in Białowieża village in early May (Brzeziński et al. 1992). Scarce information on polecat litters indicates that the most common litter size is four young per female.

3.12 River Otter (*Lutra lutra*)

River otters were censused on rivers and brooks of Białowieża Forest twice: first, by game wardens in 1795 (Hedemann 1939) and again, by zoologists, in 1994 (Sidorovich et al. 1996). Interestingly, in 1994, about 45 otters were estimated to live in the whole BPF, and they were found in all the places listed as inhabited by otters in 1795. Two centuries ago, a harvest plan based on the cited inventory was 15 otter pelts annually. Today, otter is protected in Poland (but signs of poaching in the Polish part of BPF are not uncommon), and it is a game species in the Belarus Republic.

In Białowieża Forest, spatial distribution of otters is determined by the web of local water sources. The bigger the river, the better it is for otters (Fig. 3.33). On the Narew, Leśna, and other rivers of their size from 1.3 to 5 otters (mean 3) per 10 km of the river were found. Smaller rivers (e.g. Hwoźna, Perevoloka) harboured from 1.3 to 1.9 otters (mean 1.7) per 10 km,

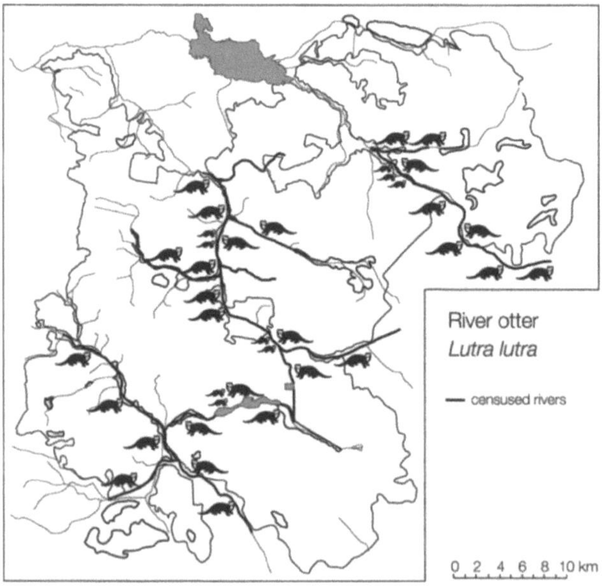

Fig. 3.33. Distribution of otters on the censused rivers in BPF in winter 1993/1994. Each *silhouette* denotes one animal, *small silhouettes* are young shown vis-a-vis females. Sidorovich et al. (1996)

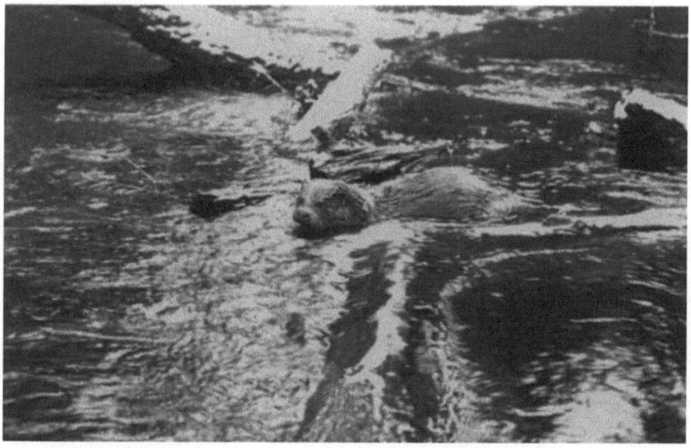

Fig. 3.34. River otter *Lutra lutra*. Photo: V.E. Sidorovich

and very small brooks only 1 otter/10 km on average. In Białowieża Forest as a whole, there were on average 2.2 otters/10 km of river, that is about 3.1 otters/100 km^2 (Sidorovich et al. 1996).

Of all predators living in the temperate forests, the otter is most obligatorily aquatic (Fig. 3.34). It hunts in water and on the banks, rests and breeds in dens and hides close to the river. Only in spring can otters roam somewhat further from the riverbed, to feed on spawning amphibians that concentrate in small ponds and old riverbeds. In winters, when rivers freeze, otters maintain ice-free access to water.

Otters are unusual among predators for their aseasonal breeding; females may give birth to young year round. Fifteen observations of litters showed that Białowieża's otters had from one to three pups (on average two) and no obvious differences in litter size were seen between the warm season and winter.

3.13 American Mink (*Mustela vison*)

Deliberately and widely introduced in the Belarussian SSR since 1953 (Serzhanin 1961), the American mink has rapidly spread and, in many places, occupied the vacant niche of the European mink. In the Polish part of Białowieża Forest, American mink were first observed in 1972 (Ruprecht et al. 1983). Nowadays, it is a dominating semiaquatic predator, inhabiting nearly all flowing waters, banks of ponds and reservoirs. The census conducted in the winter of 1993/1994 (Sidorovich et al. 1996) showed that the biggest rivers (e.g. Narew, Leśna) harboured from 5.7 to 7.5 mink (on average 6.2) per 10 km of the river course, medium-sized rivers (e.g. Hwoźna, Belaya) from 2.5 to 5.6 mink (mean 3.8) per 10 km, and the smallest brooks (e.g. Orłówka, Rudavka) from 0 to 4.3 mink (on average 1.9) per 10 km (Fig. 3.35). In the whole BPF, there are 4.6 mink/10 km of the river course, that is about 6.6 mink/100 km^2 of the forest.

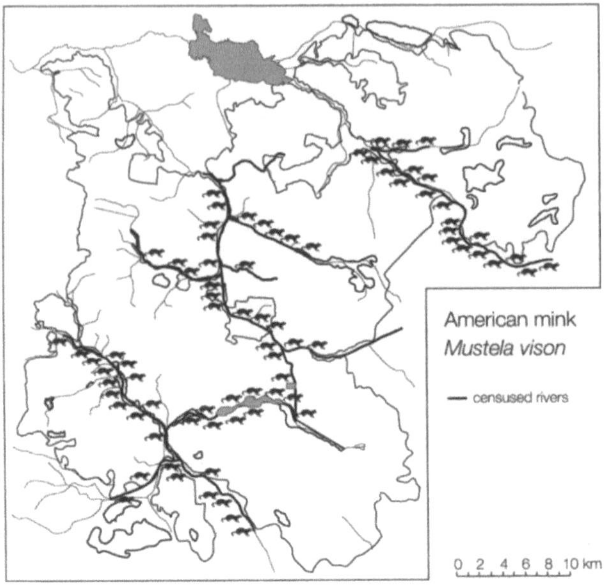

Fig. 3.35. Distribution of American mink on the censused rivers in BPF in winter 1993/1994. Each *silhouette* denotes one animal. Sidorovich et al. (1996)

Fig. 3.36. American mink *Mustela vison*. Photo: V.E. Sidorovich

Mink (Fig. 3.36) often occur near beavers' lodges and dens. By living in the neighbourhood of a beaver settlement, a mink enjoys ice-free access to water in winter. It often digs dens in the thick walls of a beaver's lodge and uses the lodge-hill as a marking place. In Białowieża Forest, there was a strong positive correlation between density of mink and density of beaver settlements (Fig. 3.37). This relationship was manifest also within each class of river size. Therefore, it was actually the beaver numbers (and not only the

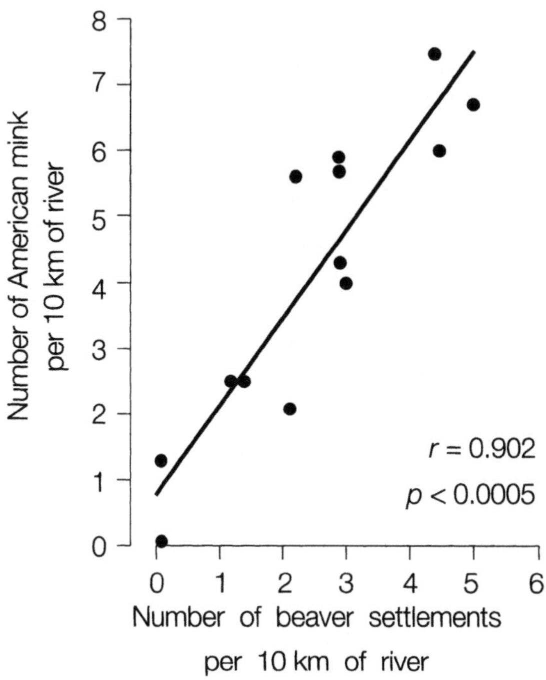

Fig. 3.37. Correlation between abundance of beaver settlements and numbers of American mink on rivers in BPF. Each point denotes one stretch (7–30 km) of a censused river. Sidorovich et al. (1996)

increasing size of the river) that was conducive to higher abundance of mink (Sidorovich et al. 1996).

No data on mink reproduction in Białowieża are available. In the forests of Belarus, Sidorovich (1993) reported that, in summer, on average 3.3 young followed the female.

3.14 Goshawk (*Accipiter gentilis*)

An intruder approaching a goshawk's domain in the Forest is 'welcomed' with its loud, strident territorial voice in summer and winter alike. Goshawks are sedentary and each pair usually has a few nests (50–700 m apart) in a territory. The mean distance between occupied nests of neighbouring pairs was 2.92 km (range 1.65–4.35 km) (Pugacewicz 1996). Thus, the estimated average territory covers 6.7 km^2.

In a nest, built most often in a tall spruce, a female lays eggs in mid-April. By late June/early July, young are fledged (Fig. 3.38), and until mid-August they can be seen in their natal territories. The breeding success of goshawks is highest of all diurnal raptors and varies little between years. In 1985–1994, in the Polish part of the forest, 57 to 71% (on average 65%) of pairs reared at least one young (Pugacewicz 1996). From one to four juveniles (on average 1.1 per breeding pair) were recorded. During most of the breeding season, females tend their broods and obviously protect them from predation that is the major cause of nesting failure in many raptors in Białowieża Forest.

Fig. 3.38. Goshawks *Accipiter gentilis* (an adult bird and chicks) on the nest. Photo: S. Wąsik

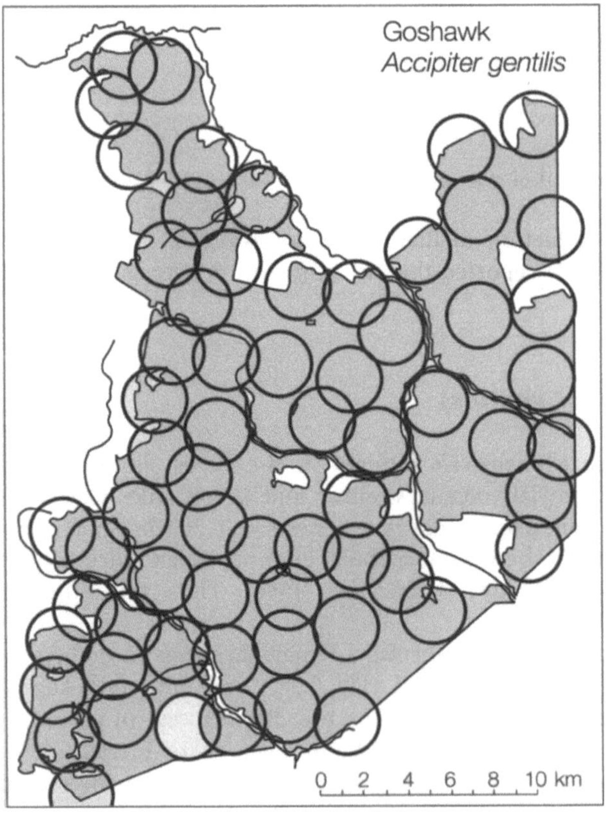

Fig. 3.39. Distribution of schematic 'territories' of goshawks in the Polish part of BPF in 1981-1994. Diameter of 'territory' (2.92 km) based on the mean nearest neighbour distance between occupied nests. Pugacewicz (1996), modified

Goshawks show a clear preference for the taiga-like coniferous and mosaic wet forests with a high predominance of ash-alder-spruce stands. In the Polish part of Białowieża Forest, where goshawk is protected, on average 10.8 pairs/100 km^2 breed (Pugacewicz 1996; Fig. 3.39). In pine-spruce and mixed coniferous stands, its densities are much higher (17.2–33.5 pairs/100 km^2). Interestingly, oak-lime-hornbeam forests, the association most typical of central European deciduous forests, are rather avoided by goshawk (0–6.9 pairs/100 km^2).

In the Belarussian part of BPF, the goshawk population has been suffering persecution since the 1950s, with the most severe period between 1960–1970, when annually, on average, 105 goshawks were shot (Fig. 3.40). Treated as a dangerous enemy of small game (its Russian name *yastreb teterevyatnik*

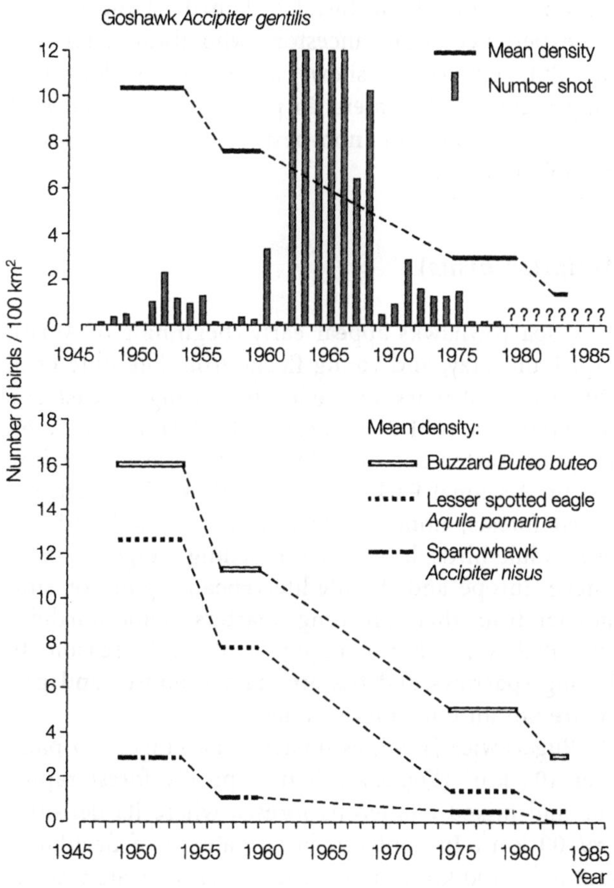

Fig. 3.40. Decline of numbers in four raptor species (due to persecution) in the Belarussian part of BPF between 1948 and 1982. *Horizontal bars* are mean densities censused during a given period. Data from: Gavrin (1953), Golodushko (1961b), and Dackevich (1997). Annual harvest (data available for goshawk, only) from: Korochkina et al. (1980), and Dackevich (1997, and unpubl. data)

means 'black-grouse hawk'), the goshawk has been deliberately eradicated by game wardens. Its density declined over sevenfold, from 5.17 pairs/100 km^2 in 1948–1952 to 0.69 pair/100 km^2 in 1982 (Gavrin 1953, Golodushko 1961b, Dackevich 1997). There are no recent (1990s) data available on goshawk density, but the policy and attitude toward this raptor have not changed. Gavrin (1953), who studied the previously undisturbed goshawk population in 1948–1952, reported its preference for coniferous old growths as breeding habitat. The nearest distance between occupied nests of neighbouring pairs was 2.5 km then. On average 3.5 eggs/nest were laid, and 2.6 juveniles fledged. Golodushko (1961c), having investigated goshawks in 1956–1958, found similar reproductive success: on average 2.7 eggs/nest and 2.2 fledglings/nest.

In justice to ornithologists having worked in the Belarussian part of the forest, it must be said that they recommended at least partial protection of goshawk. However, the administration of the forest ordained or encouraged persecution, endorsing 'the experience of our ancestors, who always shot raptors here to protect game'. But, 'Experience shows' ascertained Golodushko (1961b), 'that overwhelming majority of hunters, in pursuit of so-called pest raptors, destroy masses of rare, good, or non-pest birds, among which goshawks comprise a negligible fraction'.

3.15 Sparrowhawk (*Accipiter nisus*)

In their breeding territories, sparrowhawks appear early (beginning of April). Females lay eggs in late April and May, and young fledge from late June until late July. In 1985–1994, 29 to 77% of pairs succeeded in rearing at least one young (Pugacewicz 1996). From 0.4 to 1.7 juv/nest fledged (range 1–5, on average 1.2). Predation by pine marten has been the main cause of breeding losses of sparrowhawks. In the National Park, where densities of martens are higher than in the exploited forests, none of the five nests found had any success (Pugacewicz 1996). Adult birds leave their breeding ranges in September and head for Western Europe and the Mediterranean region, or simply live a nomadic life not far from their breeding quarters. Some northern sparrowhawks come to the Białowieża Forest region in the cold season. In winter, sparrowhawks chasing sparrows and tits among the houses and gardens of Białowieża village are not an uncommon sight.

In the Polish part of BPF, Pugacewicz (1996) estimated, on average, 16.7 pairs of sparrowhawks breed per 100 km^2 (Fig. 3.41). This typically forest raptor strongly prefers coniferous and mixed coniferous forests, where its densities are from 19 to 33.5 pairs/100 km^2. In contrast, large patches of deciduous stands harbour 10.4 to 11.7 pairs/100 km^2. Sparrowhawk nests, located exclusively in spruce or pine trees, are placed rather low, from 5 to 13 m, on average 10 m. Even in the pristine old growths of Białowieża National Park, where the tallest spruces reach over 50 m, sparrowhawks build nests in low, younger trees. Active nests of neighbouring pairs were 1.1 to 3.3 km (on average 2.17 km) apart (Pugacewicz 1996), the mean breeding territory covered 3.7 km^2.

Fig. 3.41. Distribution of schematic breeding 'territories' of sparrowhawks in the Polish part of BPF in 1981–1994. Diameter of 'territory'=2.17 km (estimated as in Fig. 3.39). Pugacewicz (1996), modified

In the Belarussian part of BPF, sparrowhawks have always been shot as pests. In the late 1960s, the period of most severe persecution, from 7 to 19 sparrowhawks were killed annually (data from Letopis Prirody). Original densities of 1.44 pairs/100 km^2 recorded by Gavrin (1953) in 1948–1952, dropped to 0.23 pairs/100 km^2 in 1974–1979 (Fig. 3.40). During the survey in 1982, no breeding sparrowhawks were found (Dackevich 1997).

3.16 Common Buzzard (*Buteo buteo*)

Soaring flight and territorial calling of buzzards over high canopies are some of the hallmarks of lowland deciduous forests. Indisputable dominant among diurnal raptors of Białowieża Forest, the common buzzard (Fig. 3.42) enjoys its ancestral habitats and pristine living conditions here, and attains very high densities. In the Polish part of BPF, Pugacewicz (1996) recorded, on

Fig. 3.42. Common buzzard *Buteo buteo* in an alder tree. Photo: J. Chyra

average, 58.5 pairs/100 km^2 (Fig. 3.43). Buzzards strongly prefer deciduous (oak-lime-hornbeam and ash-alder) forests, particularly the pristine old growth stands, and mixed deciduous forests (densities 73.6–140.7 pairs/100 km^2). Coniferous and mixed coniferous forests harbour from 36 to 60 pairs of buzzards/100 km^2. Buzzards usually build nests in oak, lime, spruce, or alder trees (Pugacewicz 1996). The mean distance between active nests of neighbouring pairs is 1 km (range 0.32–2.0 km). Thus, the breeding territory typically covers 0.8 km^2.

There seems to be a great variation in the migratory habits of Białowieża's buzzards. Most birds disappear from their breeding territories with the first snowfalls. However, some buzzards are seen in their breeding ranges year-round, particularly in mild winters (such as 1989/1990). Others overwinter in the agricultural landscape west of the forest or undertake migrations to south-western Europe. Buzzards of the eastern race *B. b. vulpinus* may also breed in Białowieża Forest (Gavrin 1953; Pugacewicz 1996), and birds of this race are known to migrate to south-eastern Africa (Alerstam 1990).

However adventurous their migration pattern is, all Białowieża's buzzards are back in their breeding territories in March. Females lay from one to four eggs (on average 2.4) in April (Jędrzejewski et al. 1994b). The breeding success of buzzards depends on the location of their territory, as shown by the study conducted in 1986–1989 in Białowieża National Park that adjoins open

Fig. 3.43. Distribution of schematic breeding 'territories' of buzzards in the Polish part of BPF in 1981–1994. Diameter of 'territory'=1.0 km (estimated as in Fig. 3.39). Pugacewicz (1996), modified

marshes and farmland. Birds that nested near (<1 km) the ecotone of forest and open land feed on highly fluctuating *Microtus* voles. Those which inhabit the deep forest interior rely on more stable bird and forest rodent resources. Reproductive success of 'ecotone' buzzards followed the 3–4 year cycles of voles (Fig. 3.44). In vole crash years, these raptors lay, on average, 1.5 eggs, of which 0.75 juveniles/pair fledge. During the peak of voles, females lay, on average 2.6 eggs, and two to four juveniles (mean 2.4) per pair fledge. Predation on 'ecotone' nests was small (9% of chicks lost). The pairs of buzzards which nest in the forest interior enjoy stable food supply but suffer from steady high predation on the nest, mainly by pine martens and ravens. Of the mean number of 2.4 eggs laid per pair, only 0.7 young fledged; 56% of hatched chicks were predated (Jędrzejewski et al. 1994b). In the period 1985–1994 in the whole Polish part of BPF, an average breeding pair of buzzards produced 0.9 juveniles/year (Pugacewicz 1996). From late June to late July, all juveniles leave nests and by mid September they are gone from their natal territories.

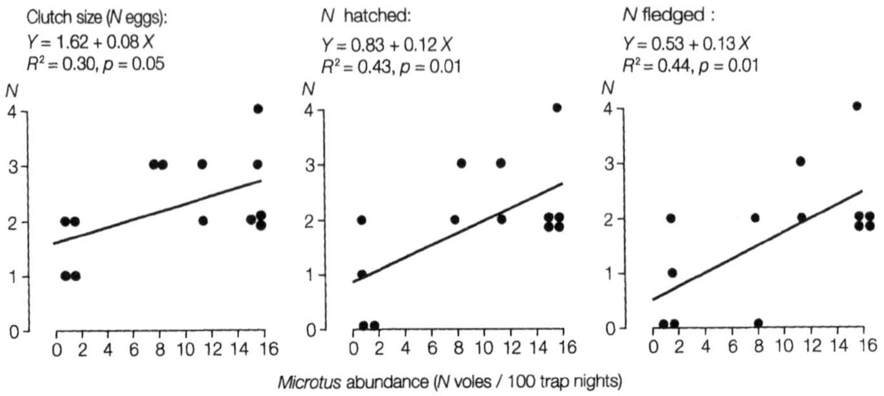

Fig. 3.44. Relationship between *Microtus* spp. abundance in the open river valley and meadows adjoining Białowieża National Park, and the breeding performance of buzzard pairs whose nests were located within 1 km from the forest edge ($n = 13$ 'ecotone' pairs in 1986-1989). (Jędrzejewski et al. 1994b)

In the Belarussian part of BPF, the buzzard shares the dark fate of other raptors. The population declined from 65–75 pairs (8.05 pairs/100 km^2) in the late 1940s to a mere 13 pairs (1.5 pair/100 km^2) estimated to live there in the early 1980s (Fig. 3.40). Gavrin (1953) reported an average breeding success of 2.5 juveniles/nest in 1948–1949, and 2 juv/nest in 1950–1952, when rodent numbers declined. In 1956–1958, Golodushko (1961c) found, on average, 2.4 eggs and 1.8 fledglings per breeding pair.

3.17 Rough-Legged Buzzard (*Buteo lagopus*)

A breeding raptor of tundra, the rough-legged buzzard winters in the boreal and temperate zone. It avoids forests, and stays in open farmlands and sedge marshes in large river valleys. In the whole Podlasie region (eastern Poland), the numbers of rough-legged buzzards counted in January 1984–1993 in river valleys varied from 1 to 22 birds seen per 100 km of the river course. The milder the winter, the higher were numbers recorded. Rough-legged buzzards (28% of all raptors) were the second most numerous diurnal raptor recorded in river valleys in winter (M. Jędrzejewski 1993). Furthermore, surveys have shown that the common buzzard remains the most numerous diurnal raptor in the cold season (50% of birds seen), followed by rough-legged buzzard (28%), goshawk (10%), and sparrowhawk (7%) (Table 3.2). In the large forest tracts, where rough-legged buzzards do not venture, the winter community of diurnal birds of prey is probably dominated by common buzzards and goshawks.

In Białowieża Forest, migrating rough-legged buzzards are seen from the end of September until March, and exceptionally in early April (Gavrin 1953). Only the largest glades with farmlands and meadows in the forest

Table 3.2. Species composition and relative numbers of diurnal raptors wintering in the river valleys of eastern Poland (Podlasie region), as recorded during surveys in January. Data averaged for 1984–1993, recalculated from M. Jędrzejewski (1993)

Species	Mean number of birds/100 km of river course	Percentage
Common buzzard (*Buteo buteo*)	12.7	50.2
Rough-legged buzzard (*Buteo lagopus*)	7.0	27.7
Undetermined buzzard (*Buteo* sp.)	1.1	4.4
Goshawk (*Accipiter gentilis*)	2.5	9.9
Sparrowhawk (*Accipiter nisus*)	1.7	6.7
Marsh harrier (*Circus aeruginosus*)	0.1	0.5
Golden eagle (*Aquila chrysaetos*)	0.1	0.5
Merlin (*Falco columbarius*)	+	+
Kestrel (*Falco tinnunculus*)	+	+

(Białowieża, Skupowo, Masiewo, and Dobrovola glades) offer suitable wintering habitats, each for one to three (rarely six) birds. In some years (e.g. 1982/1983 and 1986/1987), 'invasions' of up to 20 rough-legged buzzards were observed in settlement glades in the Polish part of BPF (E. Pugacewicz, pers. comm.).

3.18 Honey Buzzard (*Pernis apivorus*)

In early May, honey buzzards are back from their winter quarters south of the Sahara and begin courtship flights over their breeding territories in Białowieża Forest. In 1981–1987, on average, 1.36 pairs/10 km^2 bred in the Polish part of BPF (Fig. 3.45). Honey buzzards strongly prefer deciduous forests. Their densities in oak-lime-hornbeam treestands and mosaic deciduous-mixed forests were as high as 2.3 to 3.4 pairs/10 km^2. Coniferous stands, both natural and of secondary origin, are clearly avoided by honey buzzards (densities 0–0.8 pairs/10 km^2) (Pugacewicz 1996). Distances between occupied nests of neighbouring pairs were from 1 to 3.5 km (on average 1.95 km), so the estimated breeding territory covered 3 km^2 (Fig. 3.45). In the Belarussian part of BPF, dominated by pine-spruce forests, honey buzzards are fairly rare. At its best, the density of 0.34 pairs/10 km^2 was recorded in the central part (Khvoinik forest district) in 1948 (Gavrin 1953).

The breeding season for honey buzzards (Fig. 3.46) begins with egg laying in early June and ends in late August, when fully fledged young can be seen in their natal territories. Gavrin (1953) recorded one to two eggs in a clutch (mean 1.83). The breeding success of honey buzzards is rather low. In 1985–1994, Pugacewicz (1996) found that only 35% of pairs succeeded in rearing at least one juvenile. Based on his data, we roughly estimated the average of

Fig. 3.45. Distribution of schematic breeding 'territories' of honey buzzards in the Polish part of BPF in 1981–1987. Diameter of 'territory' = 1.95 km (estimated as in Fig. 3.39). Pugacewicz (1996), modified

Fig. 3.46. Honey buzzards *Pernis apivorus* (an adult bird and chicks) on the nest. Note the wasp nest combs that have been brought by a parent bird to feed young with wasp larvae and which have remained on the nest as bedding. Photo: S. Wąsik

0.5 juveniles fledged per breeding pair. There was a great between-year variation in honey buzzards' breeding success (from 0 to 75%), influenced primarily by the summer weather. Hot and dry summers (e.g. 1991) abounding in wasps and bumblebees, the main food of honey buzzards, are favourable for growth and survival of chicks. However, after cold and rainy summers (e.g. 1994), which cause nearly complete reproductive failure, the number of honey buzzards leaving the forest in autumn is nearly the same as that upon their spring arrival.

3.19 Lesser Spotted Eagle (*Aquila pomarina*)

Old forests, nearly exclusively wet ash-alder and oak-lime-hornbeam stands by open river valleys, and wet pastures are domains of the lesser spotted eagle. Sixty-eight pairs breed in the Polish part of Białowieża Forest, at mean density of 11 pairs/100 km^2 (Pugacewicz 1996). Their distribution is rather uneven; the densest clumps of territories are along water arteries and forest edges (Fig. 3.47). Coniferous forests attracted no breeding pairs. As many as 81% of nests were located closer than 0.5 km from the open area (Pugacewicz 1996). Active nests of neighbouring pairs of lesser spotted eagles were 0.7–3.9 km apart (on average 1.88 km), so the estimated mean size of a breeding territory is 2.8 km^2.

Female eagles lay eggs (usually two) in late April and early May. Fledged juveniles leave nests from late July till late August. They stay a few weeks longer in their natal territories until they reach full flight capability. In 1985–1994, from 27 to 75% of pairs (on average 52%) ended their breeding season with a success, but fratricide meant that, except for one case, all successful pairs produced only one young (Fig. 3.48). In various years, an average breeding pair reared from 0.3 to 0.8 (mean 0.5) juveniles (Pugacewicz 1996). Golodushko (1961c) studied breeding performance of lesser spotted eagles in the eastern part of BPF in 1956–1958. In 35 nests, he reported, on average, 1.83 eggs laid, 1.43 hatchlings, and 0.6 fledglings/nest. By mid-September, eagles leave their breeding territories and head south to the Bosporus and, eventually, to southern Africa (Alerstam 1990).

In the Belarussian part of BPF, lesser spotted eagles used to be common in 1948–1952, when 50–60 pairs (mean density 6.3 pairs/100 km^2) bred, mainly along valleys of the Narew, Leśna, Narewka, Solomenka, Hwoźna, and Rudavka rivers (Gavrin 1953). However, in the following decades the densities declined ninefold: in the 1970s and 1980s, 0.7 pairs/100 km^2 were recorded (Fig. 3.40). Besides persecution, another factor supposedly contributed to that decline, namely the drainage of marshes and amelioration of rivers, which have caused the deterioration of many former breeding habitats of eagles (Dackevich 1997).

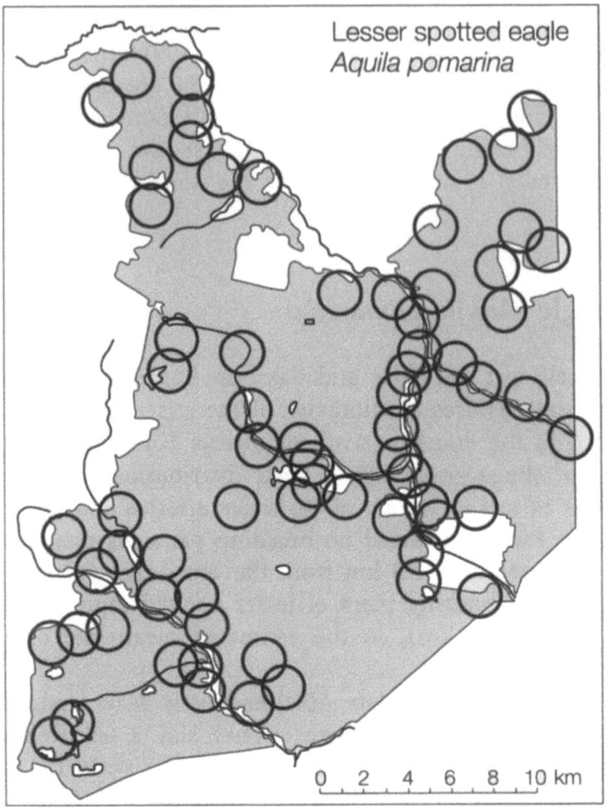

Fig. 3.47. Distribution of schematic breeding 'territories' of lesser spotted eagles in the Polish part of BPF in 1985–1987. Diameter of 'territory' = 1.88 km (estimated as in Fig. 3.39). Pugacewicz (1996), modified

Fig. 3.48. Lesser spotted eagle *Aquila pomarina*, an adult bird and a chick on the nest. Photo: S. Wąsik

3.20 Large and Rare Eagles: Golden Eagle (*Aquila chrysaetos*), Short-Toed Eagle (*Circaetus gallicus*), White-Tailed Eagle (*Haliaeetus albicilla*), and Booted Eagle (*Hieraaetus pennatus*)

Ognev (1907), having visited Białowieża Primeval Forest, wrote: 'Among birds of prey, fairly common are lesser spotted eagles. [...] Much rarer than these, are golden eagles, which most often occur in the Svisloch domain. In earlier years, they were still more frequent but their number declined markedly, when the bounty of 3 roubles was set for each brood'. In fact, in 1889–1902, a total of 48 large raptors, registered under the common name 'eagles' was shot (Karcov 1903). In recent decades, golden eagles have been seen fairly frequently in Białowieża Forest, and in 1951–1961, five birds were shot in the Belarussian part (Dackevich 1997). In 1975, a pair of golden eagles began to nest near the newly constructed Lyackie Reservoir (eastern part of BPF), but after 3 years, the tree with the nest was cut down. Since that time, regular observations in the breeding season indicate that golden eagles still breed somewhere in the Belarussian part of BPF, but so far no nest has been found (Dackevich 1997).

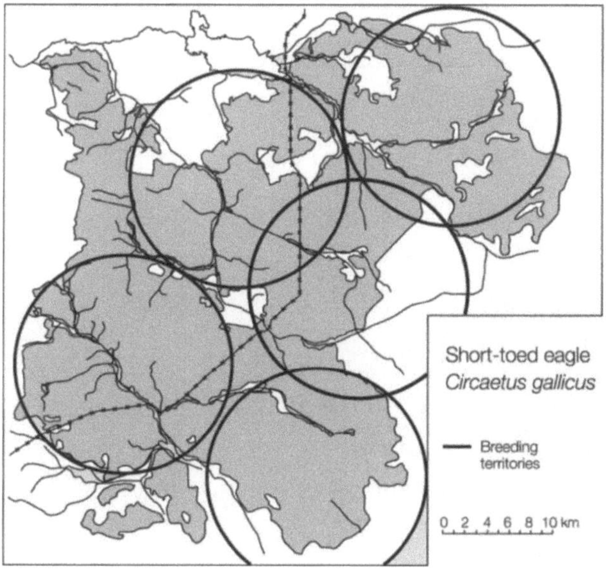

Fig. 3.49. Distribution of schematic breeding 'territories' of short-toed eagles in the whole BPF. Diameter of 'territory' (19.8 km) based on the mean nearest neighbour distance between nests and confimed by observations of birds in the Polish part of BPF. Data on the Polish part from: Pugacewicz (1996), for 1975–1994; Belarussian part: Gavrin (1953) for 1948–1953, but at least two territories (central and northern) existed in 1975–1994, too (Pugacewicz 1996)

Deciduous and mixed forests of temperate Europe are the northernmost breeding quarters of the short-toed eagle. This long-distance migrant, with its winter quarters in the steppes and savannas immediately south of the Sahara (Alerstam 1990), arrives in its breeding territories in April and begins breeding in early May. A female lays only one egg. Observations of four nests (Gavrin 1953; Pugacewicz 1996) indicated a mean breeding success of 0.75 juv/breeding pair. Fully fledged young are seen in late August. By late September, Białowieża's short-toed eagles are again on their way south.

Five pairs of short-toed eagles breed in the forest (Fig. 3.49). Interestingly enough, in BPF breeding sites of this southerly raptor include boreal-type forests: dry pine stands adjoining wet and boggy pine forests, spruce-alder bog forests and spruce forests on peatbog (Gavrin 1953; Pugacewicz 1996). Based on observations of adult birds that carried food to their chick from as far as 14 km from the nest, Pugacewicz (1996) estimated that breeding territories of short-toed eagles covered 120–160 km^2. The average distance between nests of neighbouring pairs (19.8 km) suggests even larger ranges (308 km^2). However, much of each territory's area (e.g. dense and shady deciduous forests) is unsuitable for this reptile hunter.

Construction of large water retention lakes in and near the forest has created a suitable foraging habitat for the white-tailed eagle, which was neither found nesting nor shot here before the 1980s (Gavrin 1953; Dackevich 1997). Nowadays, two pairs live in BPF (both in the Belarussian part), one nest is

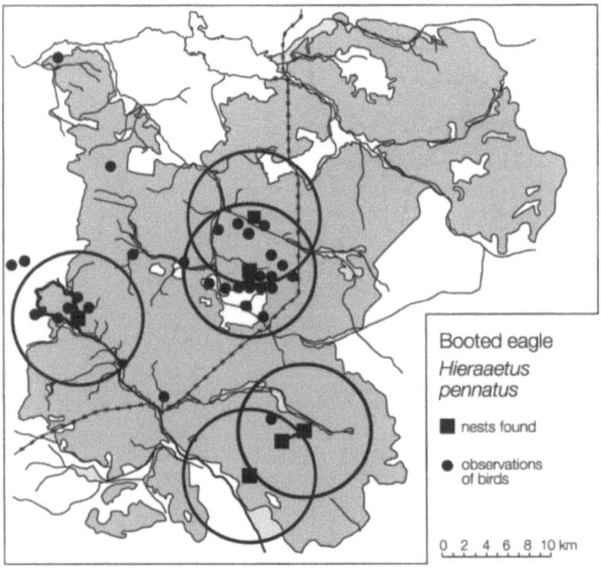

Fig. 3.50. Schematic breeding 'territories' of booted eagles ephemerally nesting in BPF in 1948-1990. Two nests connected with a solid line were occupied in different years. Diameter of 'territory'= 11.95 km (estimated as in Fig. 3.49). Data from: Golodushko (1959) and Pugacewicz (1993)

located near the Lyackie Lake and another pair is frequently seen in the north-eastern part of the forest, not far from Siemianówka Lake. White-tailed eagles are year-round residents. In winters and occasionally in summer, they are seen over valleys of the Hwoźna, Leśna and Narewka rivers in the Polish part of the Forest.

Białowieża Forest is the northern limit of the booted eagle's breeding range. Therefore, nesting of this raptor is ephemeral: since 1948, from zero to four pairs could have bred in any year, always selecting deciduous forests (primarily oak-lime-hornbeam stands) as nesting habitats (Golodushko 1959; Pugacewicz 1993). As judged from the spatial distribution of all observations and nests found, the centres of booted eagles' breeding ranges were, on average, 12 km apart, thus the estimated territories were as large as 112 km^2 and covered both forested and open terrain (Fig. 3.50). Birds arrived in their summer quarters in April and left them in September. Breeding success roughly assessed from the fates of five nests was 1.2 juv/breeding pair (Golodushko 1959; Pugacewicz 1996).

3.21 Black Kite (*Milvus migrans*) and Red Kite (*Milvus milvus*)

In European temperate forests, four species of raptors depend, to various extents, on inland waters: osprey, black kite, white-tailed eagle, and red kite. The physiographic conditions of Białowieża Forest (lack of large bodies of water, the presence of medium- and small-sized rivers only) make it an unsuitable place for ospreys, and a marginal habitat for the other three species. Observations of black kites during the breeding season as well as nests found indicated that tree stands adjoining wide, open marshes in valleys of the Leśna, Narewka, and Narew rivers are nesting habitats of this species in Białowieża Forest. Black kites breed there rather ephemerally (Fig. 3.51). Of three nests checked by Golodushko (1961c), on average 1 juv/pair fledged. Gavrin (1953) presumed that a total of four to five pairs nested in the Belarussian part of BPF, and Pugacewicz (1996) reported a maximum of two pairs from the Polish part. However, the numbers varied from year to year, and in recent decades a declining trend was observed in the whole BPF. Since 1985, despite intense ornithological surveys of raptors, no nests of black kites have been found. Pugacewicz (1996) and Dackevich (1997) believe that the most probable factors of that decline are steady lowering of the water level in the forest rivers (due to large-scale drainage of marshes in the 1950s–1960s) and abandonment of the traditional mowing of grass and pasturing of cattle and horses in the river valleys (in the 1970s). Consequently, rivers are shallow and poor in fish, and the former short-grass meadows regularly and widely flooded in spring have overgrown with tall reeds and willow bushes. Black kites have lost much of their attractive foraging habitat.

In contrast to the black kite, which winters on the African coasts south of the Sahara, the red kite winters in temperate Europe and the Mediterranean

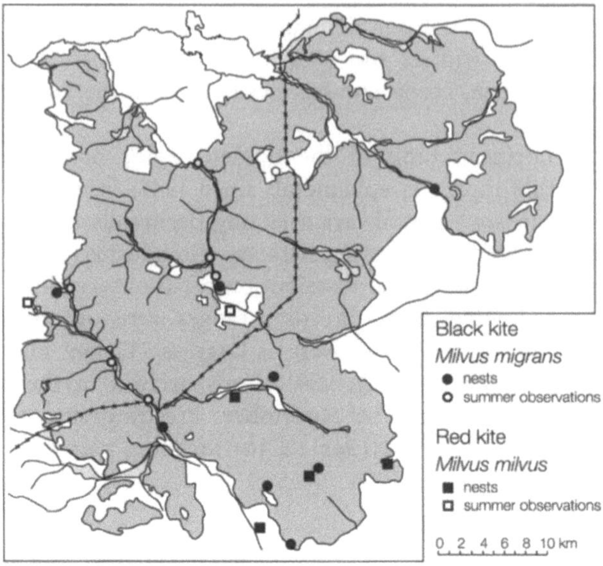

Fig. 3.51. Breeding sites of the black and red kites (both ephemerally breeding in BPF) in 1948–1991. Data from Gavrin (1953), Pugacewicz (1996), and Dackevich (1997)

region (Alerstam 1990). Still rarer than its congeneric, the red kite sporadically breeds in Białowieża Forest. The nesting habitat of this raptor includes forest edges and small woodlots near open river valleys, farmlands, and pastures. In 1948–1952, Gavrin (1953) estimated that four to five pairs nested in the Belarussian part of BPF (Fig. 3.51). In 1956–1958, Golodushko (1961b) found only one breeding pair there. Dackevich (1997) reported that in the 1980s, all former breeding territories of red kites in the south-eastern corner of BPF were unoccupied, but a new pair began to breed near Lyackie Reservoir (Fig. 3.51). In the Polish part, a maximum of two pairs could have nested, but the latest breeding record comes from 1981 (Pugacewicz 1996).

3.22 Marsh Harrier (*Circus aeruginosus*) and Montagu's Harrier (*Circus pygargus*)

Three to four pairs of the marsh harrier are regularly found in the widest (>500 m) fragments of tall reed marshes in the valleys of the Leśna and Narewka rivers in the Polish part of the forest (Pugacewicz 1996). In 1987–1994, the mean density of marsh harriers calculated for the inhabitable area (i.e. marshy river valleys) in the Polish part of BPF was 0.56 pairs/10 km^2 (Pugacewicz 1996). In the Belarussian part, where most of the riverside meadows have been mown or grazed by cattle, the marsh harrier was not found nesting in recent decades (Gavrin 1953; Dackevich 1997). In the spring of 1983, one pair attempted to breed at the Lyackie Lake but was soon shot by a game warden (Dackevich 1997). Thus, the mean density of marsh harriers in the whole forest is 0.24 pairs/100 km^2. After a winter spent in African floodlands and swamplands south of Sahara, marsh harriers arrive in their breeding quarters in April. Up to 4 juv/brood, fully fledged by late August, were observed (Pugacewicz 1996).

In the Polish part of the forest, three to six pairs of Montagu's harrier breed regularly in tall forb meadows on the outer verges of flood plain valleys (the Narewka and the Leśna Rivers) and fields and meadows in the rural glades (Białowieża, Masiewo, Skupowo), i.e. habitats somewhat drier than those preferred by marsh harriers (Pugacewicz 1996). In 1987–1994, the mean density of this species calculated for the inhabitable area was 0.73 pairs/10 km^2 in the Polish part of the forest (Pugacewicz 1996). In the Belarussian part, in the 1950s–1960s, breeding sites (three to seven pairs) were reported in wet meadows with willow shrubs at the south-eastern and eastern peripheries of the forest (Gavrin 1953; Dackevich 1997). In recent decades, breeding by Montagu's harriers became sporadic (Dackevich 1997). Thus, in BPF as a whole, the mean density of Montagu's harrier is about 0.3 pairs/100 km^2. Distant migrant of the same winter range as its congeneric, Montagu's harrier arrive in Białowieża Forest in April and by late August fledged young can be observed over their natal territories. In five observed broods, on average 2.2 young fledged (Gavrin 1953; Pugacewicz 1996).

3.23 Hobby (*Falco subbuteo*)

The rare opportunity to see a hobby may occur in the ecotones of forest and open land, where this falcon vigorously pursues high-flying swallows and swifts until twilight. Hobbies nest in dry coniferous stands on the edges of the Białowieża Forest and at larger clearings in its interior, up to 0.88 km (on average 0.23 km) from the nearest open area (Pugacewicz 1996). They take up old nests of ravens (rarely crows), located in the highest pine trees. In the late 1980s, 16 pairs bred in the Polish part of BPF, i.e. 2.6 pairs/100 km^2 (Pugacewicz 1996). Nests of neighbouring pairs were 4–9 km (on average 6.7 km) apart, thus the mean breeding territory covered 35 km^2 (Fig. 3.52). Consistent with this are observations by Golodushko (1960), who reported that in 80% of cases, hobbies hunted within a 2–3 km radius around the nest. Pugacewicz (1996) observed the farthest hunting distance of 8 km from the nest.

Hobbies arrive at their breeding sites in late April and early May. Females lay two to three eggs in late May and early June. Predation by ravens and crows is the most important factor of losses of eggs and chicks. On average, 0.9 juveniles fledged per breeding pair in the Polish part (Pugacewicz 1996) and 0.75 in the Belarussian part (Golodushko 1961c). Young birds leave their nests in the first half of August. A month later, hobbies migrate south via the Bosporus to their winter quarters with a centre in Zambia (Alerstam 1990).

Fig. 3.52. Schematic breeding 'territories' of hobbies in the Polish part of BPF in 1985-1987. Diameter of 'territory'=6.7 km (estimated as in Fig. 3.39). Pugacewicz (1996), modified

In the Belarussian part, seven to nine pairs of hobbies (on average 0.92 pairs/100 km^2) nested in 1948–1952 (Gavrin 1953), and five to six pairs (0.61/100 km^2) in 1956–1958 (Golodushko 1960, 1961b). During ornithological surveys in 1974–1979 and 1982, hobbies were not recorded. The only recent case of the occurrence of this raptor in the eastern part of BPF is a male shot in May 1983 (Dackevich 1997).

3.24 Tawny Owl (*Strix aluco*)

Territorial calling of tawny owls on chilly nights in autumn and nights of thaws in late winter, as well as the loud screaming of young owlets in late summer are vocal hallmarks of the temperate lowland forests in Europe. This resident territorial bird is an unchalleged dominant in the predator community. It is a woodland animal that lives in any type of forest with trees old enough to find a nesting cavity and perching sites. Also, small parks in the

Fig. 3.53. Tawny owl *Strix aluco* resting during the day in the canopy of hornbeam tree. Photo: S. Wąsik

villages in and around the forest are good enough habitats for the tawny owl. Despite its commoness, the tawny owl (Fig. 3.53), a nocturnal bird, is seldom seen. The rare occasions to observe it in the forest arise in summer, at sunrise. A tumultous mobbing of tits, chaffinches and robins usually points to a tawny owl head popping out of a broken lime trunk at a height of 20 metres.

In the Polish part of Białowieża Forest, on average seven to eight pairs of tawny owls per 10 km^2 were recorded in 1983–1993 (Fig. 3.54). The densities were highest in the pristine old growths of Białowieża National Park, 11–21 pairs/10 km^2, and lowest in the secondary coniferous stands in the exploited forests, 5–6 pairs/10 km^2. Generally, coniferous forests of both natural and secondary origin, harboured from 1.4 to 2.8 times fewer tawny owls per unit area than deciduous (oak-lime-hornbeam) forests (Domaszewicz 1993). Tawny owls are strongly territorial. A pair defends a territory covering from 18 to 64 ha, on average 37 ha (0.37 km^2; Fig. 3.55).

In the Belarussian part of BPF, censuses of tawny owls were done by the transect-count method (Dackevich et al. 1985), which makes the results incomparable to those reported from the Polish part. Nonetheless, it is clear

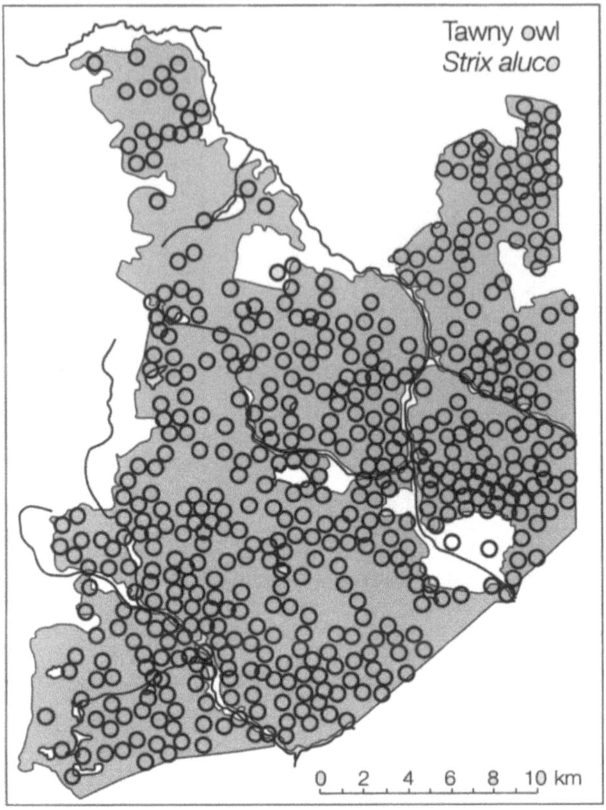

Fig. 3.54. Distribution of schematic 'territories' of tawny owls in the Polish part of BPF in 1983. Diameter of 'territory' (0.68 km) based on the mean size of real territories. Sources: Domaszewicz (1983) and Pugacewicz (1997), modified

that tawny owls were very numerous. In the eastern part of BPF, where the tawny owls' most preferred habitat, the oak-lime-hornbeam forest, covers only 6% of the area, whereas coniferous forests (69%) and bog alderwoods (16%) dominate, owls were found predominantly in the wet alder forests (69% of all tawny owls localised by their territorial calls; Dackevich et al. 1985).

Territories and numbers of tawny owls tend to be rather stable from year to year, though there are two major factors in their variation: numbers of forest rodents, the main prey of tawny owls, and depth of snow cover in winter. The latter acts as a safety cover for terrestrial rodents and, if deep enough, it may virtually cut the owls off from their food resources. In 1986/1987–1991/1992, we surveyed winter densities of tawny owls in Białowieża National Park (Jędrzejewski et al. 1996a). That study, supplemented by Domaszewicz's (1993) census done in the same area in 1992/1993, showed that the tawny owl population was rather stable during moderate and very high densities of

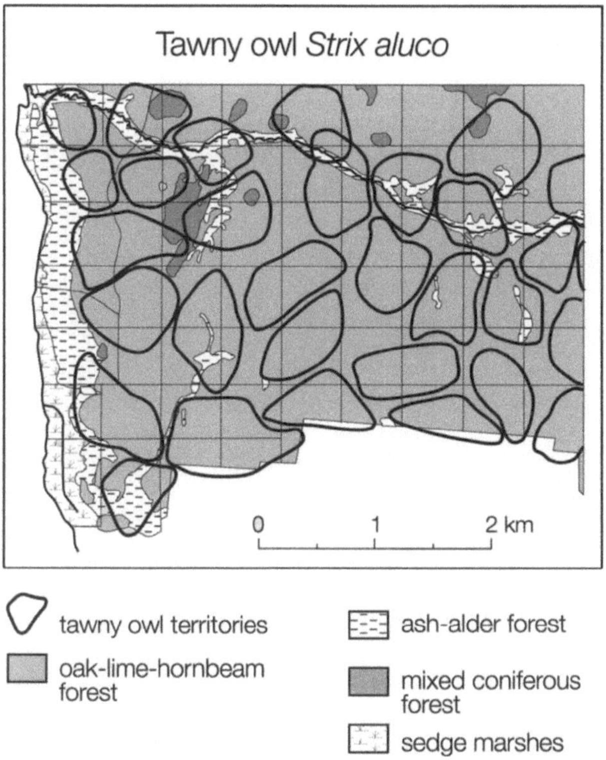

Fig. 3.55. Distribution of territories of tawny owl pairs in the south-western part of Białowieża National Park in 1979. Domaszewicz (1993), modified

forest rodents, and declined by 30% after a deep crash of rodents (Fig. 3.56). Production of young was far more variable than the numbers of residents. Domaszewicz (1993) collected data on the breeding success of owls in BNP in 1983 and 1993. Tawny owls breed in cavities of old trees. Egg laying begins in March and owlets are out of the nest by May. Young, still perching in their natal territories and notoriously crying to get food, were counted in the summer. In 1993 (moderate rodent numbers), fledged young were found for 45% of pairs, and the mean brood size was 2.9 juveniles. If we accept that estimate as typical for all years with moderate numbers of rodents, then in 1988–1989, the numbers of owls in late summer were 1.7 times higher than in winter (Fig. 3.56). In the rodent outbreak year of 1983, food cries of juvenile owls were heard in 80% of tawny owl territories, and on average 3.6 juv/brood were recorded. Therefore, we can estimate that the numbers of tawny owls (adults and young) in late summer 1990 (outbreak of rodents) were as high as 124 owls/10 km^2, i.e. 2.4 times that in winter. In August–September (exceptionally until November) young tawny owls, urged on by the intolerance of resident adults, disperse from their natal territories (Southern 1970).

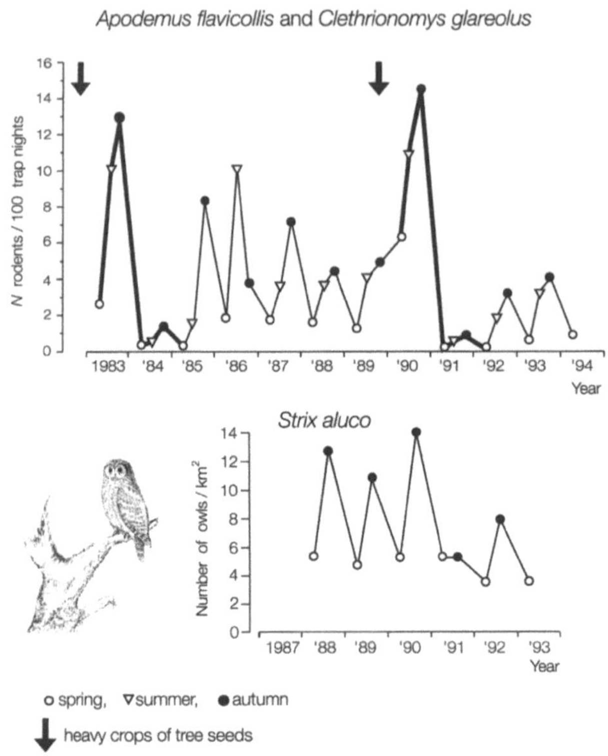

Fig. 3.56. Autumn-winter densities of tawny owls (censused) and their summer numbers (calculated based on data on breeding success from Domaszewicz 1993) in relation to forest rodent abundance in BNP. Data from: Jędrzejewski et al. (1996 a) and Domaszewicz (1993)

Most probably no breeding occurred in the rodent crash year 1991, as observed in Białowieża in 1995, another year of extreme rodent scarcity (W. Jędrzejewski and J. Kupis, unpubl. data) and in other populations of tawny owls in Great Britain (Southern 1970; Petty 1987). In consequence, Białowieża's population of residents declined by 30% in 1992. In both winter and summer, numerical responses of tawny owls to rodent numbers were of the logarithmic type, and thus, the number of owls per 1000 rodents rapidly declined with growing numbers of rodents (Fig. 3.57).

The impact of snow cover on tawny owl numbers was manifest in two exceptionally snowy and severe winters: 1946/1947 and 1995/1996. Gavrin (1953) reported that in 1946/1947 many tawny owls abandoned their territories in the forest and sought chances of survival in the villages. In February 1947, several tawny owls in miserably poor condition (body weights 327–355 g, instead of the normal 460–500 g) were captured in barns. Obviously many other owls died in the forest, as feathers and carcasses were seen on foxes' and martens' trails. In consequence, Gavrin (1953) estimated (based on the mapping of owl calls and searches for nests) that in 1948–1952, only

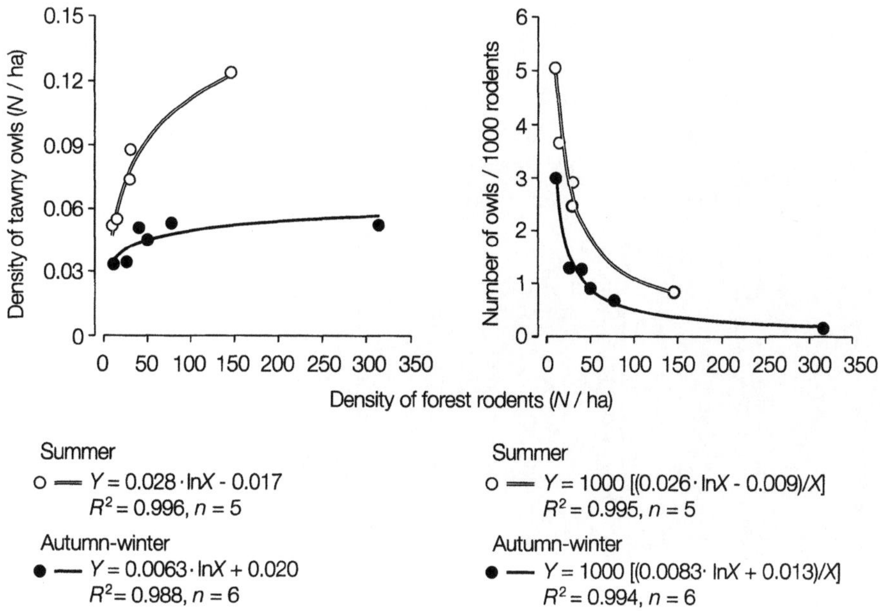

Fig. 3.57. Numerical response of tawny owls to variation in numbers of forest rodents *left panel* and predator:prey ratio (*N* owls per 1000 rodents) *right panel* in BNP in 1987/1988–1993. Data from: Jędrzejewski et al. (1996a) and Domaszewicz (1993)

50–60 pairs of tawny owls (0.7 pair/10 km^2) lived in the Belarussian part of BPF. Virtual decimation of the population was observed again in Białowieża National Park in 1995/1996, when deep snow cover, maximally reaching 70 cm, lasted for over 2 months. Of 15–22 pairs/10 km^2 recorded in 1987–1995, a mere 5 pairs/10 km^2 were found in the spring of 1996 (W. Jędrzejewski and J. Kupis, unpubl. data). During the same winter of 1995/1996, when tawny owls were dying from starvation, forest rodents were enjoying a superabundant crop of hornbeam seeds and were proliferating under the snow cover.

3.25 Pygmy Owl (*Glaucidium passerinum*)

Białowieża Forest is on the southern border of the lowland range of the pygmy owl (Fig. 3.58). Both the geographic range and the local occurrence of this bird is a consequence of its preference for old (>100 years) spruce forests. In the Polish part of Białowieża Forest, 35–40 pairs of pygmy owls are estimated to breed (density 0.65 pairs/10 km^2), but in the National Park, the best-preserved old growth, as many as 1.5–2.9 pairs/10 km^2 were found (Fig. 3.59). Pygmy owls occur in all forest districts of the Belarussian part of BPF, but their density is unknown (Dackevich 1997).

A loud whistling voice emitted at dawn and dusk and diurnal hunting habits makes the pygmy owl an easier subject to observe than other owls and

Fig. 3.58. Pygmy owl *Glaucidium passerinum*. Photo: G. Kłosowski

allowed the mapping of its seasonal shift of habitats (Domaszewicz 1997). In Białowieża National Park, 69% of all observations of owls during the breeding season (March–August) were in coniferous (pine-spruce) forests. In autumn and winter (September–February), 82% of all records of pygmy owls were in deciduous forests (54% in oak-lime-hornbeam forests and 26% in ash-alderwoods). Differences in abundance of passerine birds in these habitats in summer and winter are the most likely reason for the shift of habitats of pygmy owls.

Golodushko and Samusenko (1961) estimated that, during the breeding season, each pair of owls occupied 2.5–4 km^2 and in winter, the area utilised by pygmy owls became larger. Centres of owls' breeding territories mapped by Domaszewicz (1997) in a large contiguous patch of optimal habitat (Białowieża National Park) were 1.07 to 2.25 (on average 1.67) km apart. Thus, the mean estimated territory covers 2.2 km^2. Pygmy owls breed in old nesting cavities of woodpeckers, especially the black woodpecker. Females lay eggs in May. Fledged young are seen in trees (outside their natal cavities) from July through August. A few of our own observations of parents feeding their fledged owlets indicated that, on average, three young per successful brood were reared.

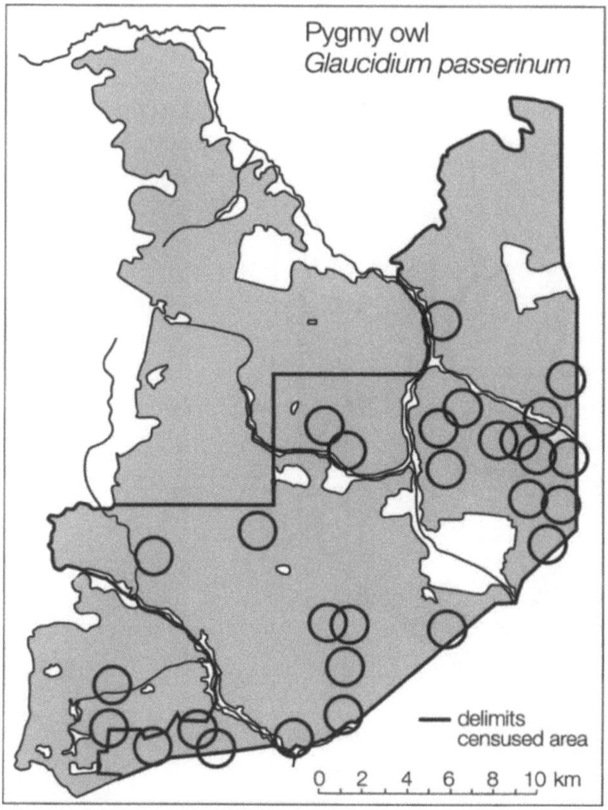

Fig. 3.59. Distribution of schematic breeding 'territories' of pygmy owls in the Polish part of BPF in 1983–1993. Diameter of 'territory' (1.67 km) based on mapping of real breeding territories in BNP (Domaszewicz 1997). Sources: Domaszewicz (1993, 1997) and Pugacewicz (1997), modified

3.26 Tengmalm's Owl (*Aegolius funereus*)

Tengmalm's owl is yet another boreal relic in the temperate forests of Białowieża. This year-round resident occurs in large patches of old coniferous forests and it especially prefers drier forms of old pine forests. Tengmalm's owls breed in cavities. In June, fledglings can be seen out of their natal cavities. In seven observed broods, from three to five owlets (on average 4.3) were recorded (Domaszewicz 1993). About 30 pairs are estimated to inhabit the Polish part of BPF (mean density 5 pairs/100 km^2), but in large contiguous patches of suitable habitats as many as 2.2 pairs/10 km^2 were recorded (Domaszewicz 1993). In such areas, the distance between neighbouring pairs was from 0.1 to 2.6 km, on average 1.16 km (Fig. 3.60). Mean territory size estimated from that distance is 1.1 km^2. Tengmalm's owl occurs in the eastern part of BPF as well, but its densities and spatial distribution are not known (Dackevich et al. 1985).

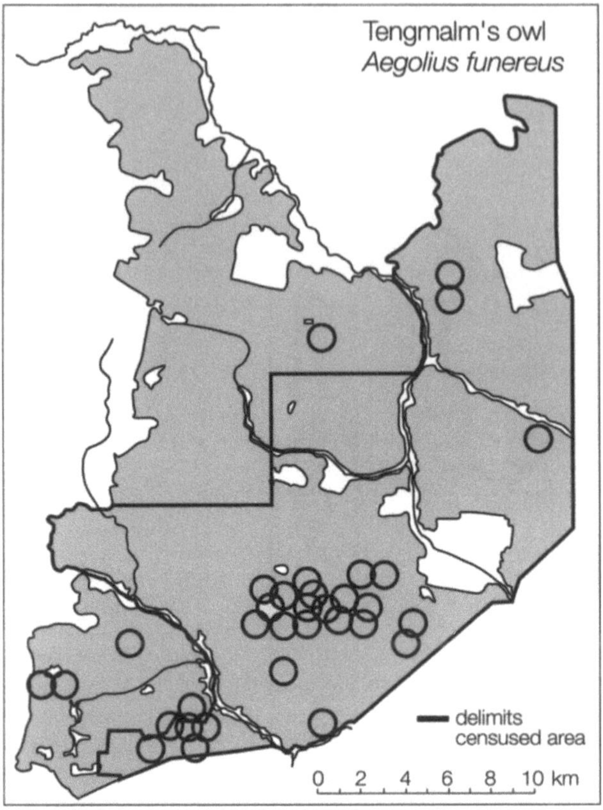

Fig. 3.60. Distribution of schematic 'territories' of Tengmalm's owls in the Polish part of BPF in 1983. Diameter of 'territory' (1.16 km) estimated as the mean nearest neighbour distance in a densely populated southern patch. Sources: Domaszewicz (1993) and Pugacewicz (1997), modified

3.27 Long-Eared Owl (*Asio otus*)

In the primeval forests of Białowieża, the long-eared owl appeared to be a predator of both woodland and open terrain (Dackevich et al. 1985, Domaszewicz 1993). It breeds in the forest (with a strong preference for coniferous stands) but in the neighbourhood of glades, clear-cuts, meadows or sedge marshes (Fig. 3.61). It also bred in the open marshes if there were large enough willow bushes with nests at 2–4 m above ground. In the open area, cyclic abundance of *Microtus* voles is the prime factor determining the numbers of breeding long-eared owls. Moreover, this predator takes advantage of the regular (at 6–9 years) outbreaks of forest rodents. In 1993, a year of moderate numbers of forest rodents and peak abundance of root voles (see Sect. 2.4), 15 pairs of long-eared owls were found in a 323-km^2 study area (4.6 pairs/100 km^2) in the Polish part of BPF. Most of the territories were located

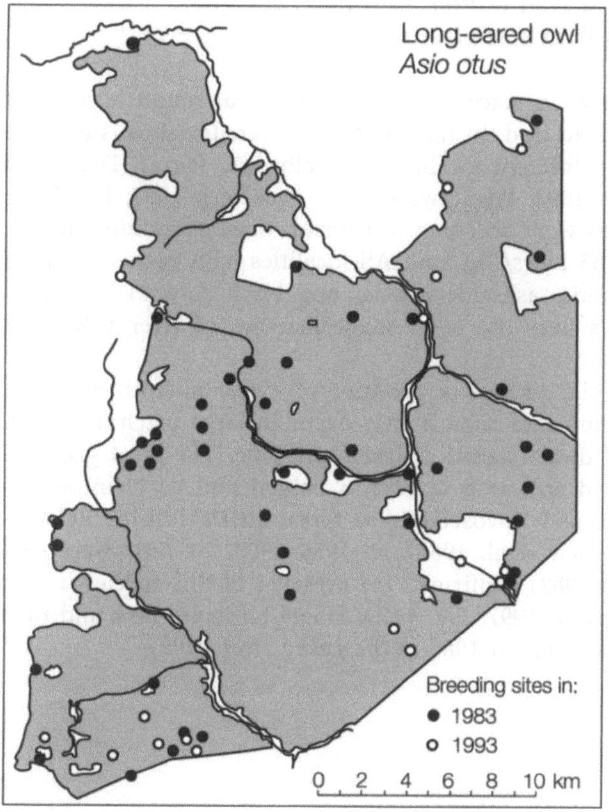

Fig. 3.61. Distribution of breeding sites of long-eared owls in the Polish part of BPF in 1983 (outbreak of forest rodents) and in 1993 (moderate/low numbers of forest rodents). Domaszewicz (1993), modified

on the forest edge. In 1983, a year of forest rodent outbreak, 45 pairs were found in the whole of the Polish part (7.5 pairs/100 km^2), many of them deep in the forest interior, the old growths of Białowieża National Park included (Fig. 3.61).

Long-eared owls occupy old nests of raptors, ravens, crows or magpies. In seven broods, three to four juveniles (on average 3.3) were seen (Domaszewicz 1993). In autumn, owls leave their breeding sites and move to larger open areas; birds wintering on extant meadows and farmlands by the forest can be seen from October through February (Dackevich et al. 1985).

3.28 Large Owls: Eagle Owl (*Bubo bubo*) and Great Grey Owl (*Strix nebulosa*)

Despite its large size and loud voice, the eagle owl, a year-round resident of the forest, is very difficult to find. In the 1970–1990s, territorial calls of eagle owls were heard in eight different localities in Białowieża Forest (Dackevich et al. 1985; Domaszewicz 1993; Pugacewicz 1995; Dackevich 1997; Fig. 3.62). At least five of these sites were breeding territories, thus the minimum reliable density would be 0.35 pairs/100 km^2. All localities with eagle owls were in bog forests (alderwoods, ash-alderwoods, bog birch forests) or mixed coniferous forests, always near the open sedge marshes of river valleys or water reservoirs.

The great grey owl (Fig. 3.63) is a predator of cyclic microtines of the northern taiga. In the temperate zone it may occur in large patches of high peatbogs with loose and dwarf stands of birch and pine. The great grey owl has been repeatedly heard and seen in such a habitat and its vicinities located in the Oshchepskoe and Novoselkovskoe forest districts in the Belarussian part of BPF (Dackevich et al. 1985). In 1956–1960, six owls were shot there, but later censuses (1982) confirmed the presence of this species in that part of the forest (Dackevich 1997; Fig. 3.62). Single birds are now and then observed in the Polish part, e.g. in 1987 in the Leśna river valley.

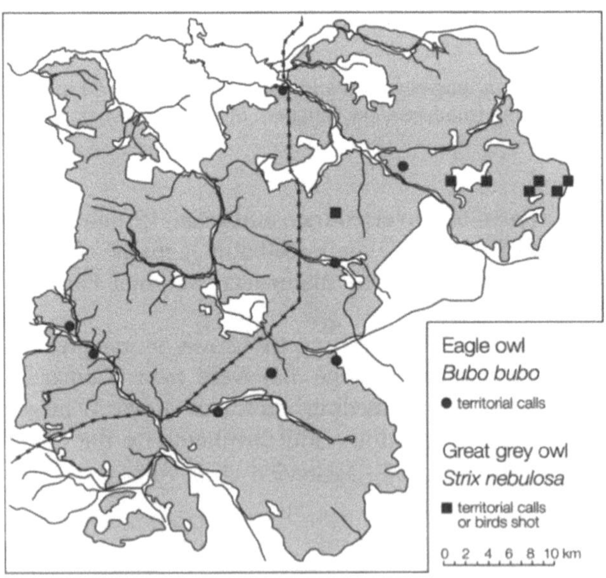

Fig. 3.62. Locations of territorial calls of eagle owls recorded repeatedly in 1982–1991 (data from Dackevich et al. 1985, Domaszewicz 1993), and territorial calls or shot specimens of great grey owls in 1954–1982 (data from Dackevich et al. 1985)

Fig. 3.63. Great grey owl *Strix nebulosa*, a female on a nest located atop a birch stump. Photo: B. Yaminskii

3.29 Predators No Longer Breeding in Białowieża Primeval Forest and Records of Non-Breeding Raptors

Brown Bear (Ursus arctos). Much desired trophy of hunts, the brown bear was still common in Białowieża Forest in the first half of the 19th century. At that time, hunters distinguished 'large black ones, which, for the beauty of their pelts, are called silver bears, and brown ones [...] called ruddies' (Jarocki 1830). Mighty predators, occasionally killing European bison, bears became a target of extermination after 1860. In 1878, the last bear was killed (Karcov 1903). A reintroduction programme that began in 1938 was soon discontinued by the war (Karpiński 1949). Most of several released bears were poached or had to be shot because of conflicts with humans. However, at least two litters were born in the wild after reintroduction. In 1945–1947, a maximum of four bears at a time lived in BPF and a young lone individual was observed there until autumn 1950 (Buchalczyk 1980, Jędrzejewska et al. 1995).

European Mink (Mustela lutreola). Jarocki (1830) reported that minks were numerous on Białowieża's rivers in the early 19th century. In 1891, during a predator extermination programme, two minks were killed (Karcov 1903). Most probably, in the early 20th century, this species was irreversibly lost to the Forest. Worth emphasising is the fact that European minks had become extinct over half century before spontaneous colonisation of the area by American mink (Ruprecht et al. 1983).

Spotted Eagle (Aquila clanga). 'Here and there, the forest is interspersed with fairly large glades, where stacks of hay stay, left for bison. To these glades, especially on early morning and evening glow, eagles fly and rest on haystacks or the largest trees' wrote Ognev (1907) after having visited Białowieża Forest. Of the three *Aquila* species that Ognev could have observed in the early 20th century, the spotted eagle no longer breeds in the forest. The most recent breeding record of this species comes from 1929 (collection of the Nature-Forest Museum in Białowieża). Among 47 lesser spotted eagles, *Aquila pomarina*, shot in the Belarussian part of the forest since 1951, some individuals were characterised by external features and biometrics intermediate between *A. clanga* and *A. pomarina* (Dackevich 1997). Both Gavrin (1953) and Dackevich (1997) concluded that in recent decades, a few surviving spotted eagles hybridised with their smaller congenerics.

Kestrel (Falco tinnunculus). In 1948–1958, a maximum of eight pairs of kestrels nested on the outskirts of the Belarussian part of BPF (near extensive meadows along the upper Hwoźna River) and edges of large settlement glades in its interior (Dobrovola, Tushemla, Rudnya glades) (Gavrin 1953; Dackevich 1997). By the 1970–1980s, not a single breeding pair was found in these formerly occupied habitats (Dackevich 1997). In the Polish part of BPF, the kestrel has not bred for a few decades, either (Pugacewicz 1996). Migrating birds on their way to and from winter quarters in south Africa can occasionally be seen hovering over meadows and stubbles.

Peregrine (Falco peregrinus). The peregrine belongs to Białowieża Forest's past but, hopefully to its future, too. Since the 1960s, no nests of peregrines have been found in Białowieża Forest. In recent years, however, several summer observations of this falcon may forecast its spontaneous return (Pugacewicz 1996; Dackevich 1997). In 1948–1958, four to six pairs of peregrines bred in the Belarussian part of the forest (Fig. 3.64), all in old, clear pine forests bordering with open marshes and farmlands (Gavrin 1953; Golodushko 1961b). Peregrine eyries were old nests of ravens and crows. The mean distance between neighbouring occupied nests was 22 km. This suggests that breeding territories of peregrines were very large (on average 380 km^2). Courtship flights of peregrines, performed over nests at dusk and dawn, were seen in early April. A mean clutch size of 2.2 eggs and mean breeding success of 0.75 fledglings/pair were reported by Golodushko (1961c). In August and September, peregrines left their breeding quarters, but migrants

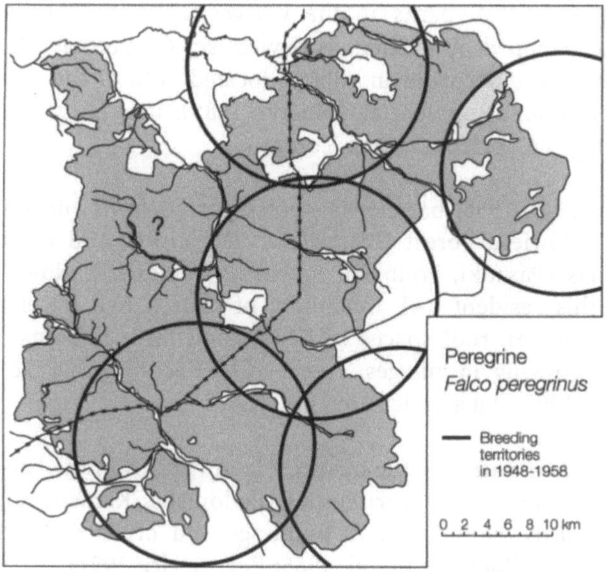

Fig. 3.64. Distribution of schematic breeding 'territories' of peregrine falcons in BPF in 1948–1958. Diameter of 'territory' (22 km) estimated from the mean nearest neighbour distance, confirmed by observations of peregrines' hunting routes. Data from: Gavrin (1953) and Golodushko (1961b)

from northerly areas were occasionally seen over Białowieża Forest until late autumn (Gavrin 1953).

Short-Eared Owl (Asio flammeus). Most likely a former rare breeder in the large marshes of Dziki Nikor and the valley of the Leśna River, the short-eared owl has not been found nesting in Białowieża Forest since the 1950s (Gavrin 1953; Dackevich et al. 1985). In the 1950–1980s, it was occasionally seen during late autumn migrations (Dackevich et al. 1985).

Migrants, nomads or otherwise stray individuals of eight species of raptors were recorded in BPF in recent decades: merlin (*Falco columbarius*), red-footed falcon (*Falco vespertinus*), hen harrier (*Circus cyaneus*), pallid harrier (*Circus macrourus*), osprey (*Pandion haliaetus*), griffon vulture (*Gyps fulvus*), snowy owl (*Nyctea scandiaca*), and hawk owl (*Surnia ulula*) (Gavrin 1953; Borowski and Okołów 1988; Tomiałojć 1995; Dackevich 1997).

3.30 Predators Inhabiting Human Settlements

Stone Marten (Martes foina). From Białowieża village, which is located in the heart of the forest, stone martens were reported in the early 19th century (Jarocki 1830). Nocturnal and very shy, living entirely in farm buildings, stacks of hay, orchards, and small parks in the villages, this predator is rarely seen and sometimes trapped by farmers in devices set for polecats. In recent years, stone martens have occurred in villages located both at the forest's outskirts and in its interior (Ruprecht and Buszko 1985). It is fairly numerous in Białowieża village (A. Zalewski, pers. comm.).

Barn Owl (Tyto alba). In 1949–1986, barn owls were reported from villages located in the heart of Białowieża Forest (Białowieża, Teremiski), and from settlements on its outskirts (Paszuki, Podbiała, Rowbick, Hajnówka). Nowadays, the population of this resident owl, inhabitant of church bell towers, chimneys of abandoned cottages, roof spaces of houses and farm buildings, and cavities in old limes growing in villages, seems to be declining (Dackevich et al. 1985; Tomiałojć 1995; Pugacewicz 1997).

Little Owl (Athene noctua). Few pairs of little owls reside in villages in the Forest (1–2 pairs in Białowieża) and on its peripheries (Podomsha, Kamenyuki, Rozhkovka, Wygon). The owls were found in buildings, old orchards, and small parks in the settlements (Dackevich et al. 1985; Pugacewicz 1997).

Domestic Dogs (Canis familiaris) and Cats (*Felis catus*). The list of Białowieża's predators would not be complete if dogs and cats were set aside. These pets are kept in nearly every farm in the villages located both in the heart of the Forest and on its peripheries. Polish and Belarussian game wardens are obliged to shoot every stray dog and cat they see in the forests. In 1961–1978, in the Belarussian part, the mean annual register was 147 dogs and 75 cats shot (Korochkina et al. 1980). In 1989–1991, the number of cats killed dropped to five per year, but those of dogs remained high – 165 per year (Proekt organizacii i razvitiya lesnogo khozaistva GNPBP 1993). Although some of the shot animals were young mongrels and cats from the villages, killed and brought to the Game Department to get a bounty, these figures convey an idea of how abundant dogs and cats are in the villages of BPF. Fed household scraps, rural dogs and cats supplement their diets with whatever they can hunt in and around villages and nearby woodland. Cats usually penetrate forest ecotones and are rarely seen more than 1 km away from the nearest human settlement. Dogs, especially large ones, are sometimes trained to poach wild boar, and then they can go far into the forest interior. Cases of three to five mongrels of various sizes forming a temporary pack and hunting wild ungulates have been reported but were fairly quickly targeted by game wardens.

3.31 Community of Vertebrate Predators – Species Structure, Densities, and Habitat Distribution

Forty species of raptors and carnivores are reported to have dwelled in Białowieża Primeval Forest in the 20th century: 32 species inhabiting forests, 5 living in marshes, and 3 in human settlements. In the last decade (1985–

1995), there were 29 species of breeding predators found in natural habitats (forests and marshes). They include many raptors and carnivores listed in both Polish and Belarussian Red Data Books (Polish Red Data Book of Animals 1992, Chyrvonaya Kniga Respubliki Belarus 1993). Apart from the profound consequences of this fact which should be considered by administrators of Białowieża Forest, this puts the subject of this book in a proper perspective. Although the local community of predators is not virgin (it has lost a few species and acquired a few others), in comparison to the temporal and spatial scales of human disturbance in other regions of Europe, Białowieża Primeval Forest harbours a nearly pristine community of predators typical of the temperate lowland forests of Europe.

Of these 29 species, we will analyse densities, biomass per unit area, dominance structure and pattern of sharing habitats. Inevitably, the synthesis is inflicted with some simplifications. Various species of predators were censused in different parts of BPF and different years. Thus, we calculated an average density estimate for each species, representative of a period 1985–1995 (in some cases 1980–1995) and of as large a part of BPF as possible. So calculated average densities are presented in Table 3.3.

A notable feature of the predator community inhabiting temperate forests is its seasonal variation. In winter, 17 species (9 carnivores and 8 birds of prey) are present and active all the time. Moreover, two carnivores (badger and raccoon dog) and at least one raptor (buzzard) are present or active for a part of the cold season (Table 3.3). The combined winter density of all species was 60.6 ind/10 km^2, 56% of that number being carnivores and 44% resident raptors (five owls and three diurnal raptors) (Fig. 3.65). In winter, of a total biomass of 56 kg/10 km^2, carnivores dominated (72%).

The predator community changes in spring. Raccoon dogs and badgers wake from their winter sleep in March–April, 12 species of diurnal raptors and 1 owl species return from their winter quarters from late March until early May (Table 3.3). Just at the onset of the breeding season in spring, an average of 78.85 predators/10 km^2 occur, to which carnivores contribute 44%, resident raptors 34%, and migratory raptors 22%. With the arrival of migratory birds, the total biomass of predators increases to 76.56 kg/10 km^2 (Fig. 3.65).

In the late spring, all predators breed and, by May/June, the number of adults together with newly born pups and kittens, newly hatched owlets, eaglets and eyases is seasonally highest. However, this rapid increase in numbers does not translate into equally fast growth of biomass. On average, from one third to two thirds of young perish at an early age due to starvation, intraguild predation, weather adversity and so on. In late August/early September, when young of most species are setting out for independent life and migratory birds will soon leave their breeding territories, nearly 163 predators/10 km^2 are present and constitute 128.65 kg of biomass/10 km^2 (Table 3.3). Again, carnivores dominate by both numbers (57%) and biomass (67%), followed by resident raptors (27 and 20%, respectively), and migratory birds of prey (16 and 13%) (Fig. 3.65). Throughout late summer and autumn, young of most species disperse (exceptions are large predators, the wolf and the lynx), and migratory

Table 3.3. Average densities of predators in Białowieża Primeval Forest in 1985–1995 (in a few cases in 1980–1995)

Predator	Adult body mass (kg)	Present in winter (W)	Mean density (*N*/10 km²)	Biomass in non-breeding season (kg/10 km²)	Mean post-breeding density (*N*/10 km²)	Biomass in post-breeding season (kg/10 km²)
Lynx	18.4	W	0.42	7.73	0.50	8.46
Wolf	32	W	0.23	7.36	0.27	8.00
Red fox	6.1	W	0.97	5.92	2.14	9.49
Raccoon dog	5.9	0.5 W	1.71	10.09	6.81	25.13
Badger	13.2	0.3 W	0.36	4.75	0.50	5.68
Pine marten	1.05	W	5.68	5.96	12.33	12.95
Stoat	0.23	W	2.00	0.46	3.80	0.87
Weasel	0.07	W	18.40	1.29	51.58	3.61
Polecat	0.725	W	4.38	3.18	13.13	9.52
River otter	6	W	0.31	1.47	0.31	1.47
American mink	0.825	W	0.66	0.54	1.75	1.44
Goshawk	1.03	W	2.40	2.47	3.70	3.81
Sparrowhawk	0.2		3.69	0.74	5.90	1.18
Common buzzard	0.81	0.4 W	11.70	9.48	17.11	13.86
Honey buzzard	0.78		2.80	2.18	3.50	2.73
Lesser spotted eagle	1.35		2.20	2.97	2.75	3.71
Golden eagle	4	W	0.02	0.08	0.03	0.12
Short-toed eagle	1.68		0.07	0.12	0.10	0.17
White-tailed eagle	4.5	W	0.03	0.14	0.05	0.23
Booted eagle	0.85		0.08	0.07	0.13	0.11
Marsh harrier	0.61		0.12	0.07	0.36	0.22
Montagu's harrier	0.31		0.14	0.04	0.29	0.09
Hobby	0.235		0.52	0.12	0.75	0.18
Tawny owl	0.49	W	18.10	8.87	29.90	14.65
Pygmy owl	0.065	W	0.72	0.05	1.81	0.12
Tengmalm's owl	0.11	W	0.55	0.06	1.74	0.19
Long-eared owl	0.29		0.50	0.15	1.30	0.38
Eagle owl	2.5	W	0.07	0.18	0.10	0.25
Great grey owl	0.925	W	0.02	0.02	0.03	0.03
Sum: in winter			60.60	56.04		
in spring			78.85	76.56		
in summer/autumn					162.67	128.65

Post-breeding densities are representative of late summer-early autumn (August/September). Adult body masses (averaged male and female weights) based predominantly on data from Białowieża Forest. For calculation of biomasses of the medium-sized and large carnivores (wolf, lynx, otter, fox, raccoon dog) in the post-breeding season, body mass of juveniles was taken as half the adult body mass. Species are listed in the same order as that of their presentation in Chapter 3.
W, species present (and active) in winter; fractions denote an average part of the cold season when animals of a given species are present or active.

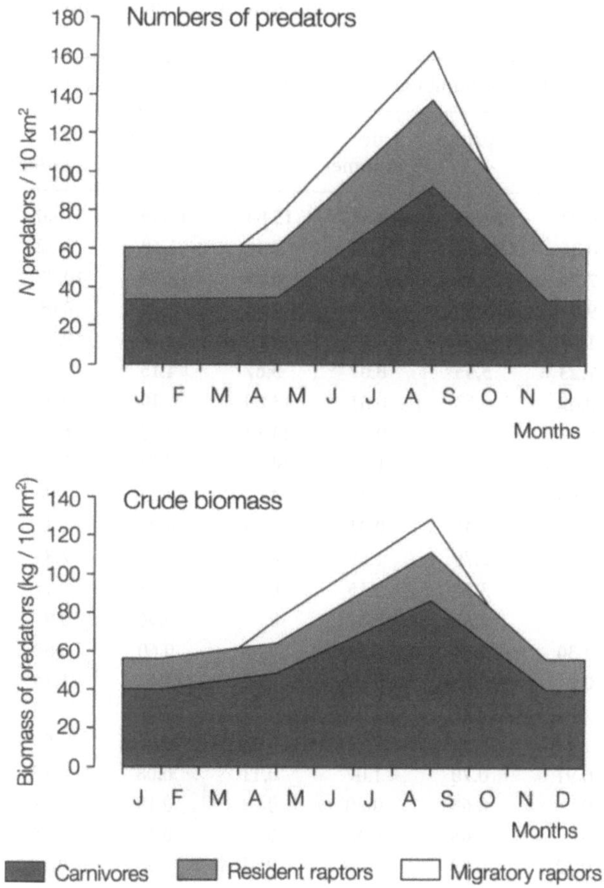

Fig. 3.65. Seasonal variation in the densities and biomass of predators breeding in BPF; data on 29 species in 1985–1995, from Table 3.3

birds of prey leave for their winter resorts. By December, the community of predators shrinks again to 17 species, all at their seasonally lowest numbers.

The tawny owl, comprising 18.4–29.9% of the total number and 11.4–15.8% of all predator biomass, is the unchallenged dominant (Table 3.4). Its top position in the community is still more emphatic, if the densities recorded in the pristine old growth forests (2.5 times higher than the average for the whole BPF) are considered (Table 3.5). Weasel (very numerous but contributing little to the total biomass), common buzzard and pine marten (both with substantial shares to numbers and biomass) also have high ranks in the community. These four dominants together comprise two thirds of all predator numbers and one third of their total biomass (Table 3.4). Generally, resident predators dominate the community; an exception is the buzzard, a very common though partly migratory raptor.

Table 3.4. Percentage shares of each species of carnivore and raptor in the total number and biomass of predators in Białowieża Primeval Forest

Rank	Species	Percentage of numbers			Percentage of biomass		
		Winter	Spring	Late summer	Winter	Spring	Late summer
1	Tawny owl	**29.87**	**22.95**	**18.37**	**15.84**	**11.59**	**11.40**
2	Weasel	**30.36**	**23.34**	**31.70**	2.30	1.68	2.82
3	Common buzzard	**7.72**	**14.84**	**10.52**	**6.77**	**12.38**	**10.77**
4	Pine marten	**9.37**	**7.20**	**7.58**	**10.65**	**7.79**	**10.06**
5	Raccoon dog	1.41	2.17	4.19	**9.00**	**13.18**	**19.54**
6	Polecat	**7.23**	**5.55**	**8.07**	**5.67**	4.15	**7.40**
7	Lynx	0.69	0.53	0.31	**13.80**	**10.10**	**6.58**
8	Wolf	0.38	0.29	0.17	**13.14**	**9.62**	**6.22**
9	Red fox	1.60	1.23	1.32	**10.56**	**7.73**	**7.37**
10	Goshawk	3.96	3.04	2.27	4.41	3.23	2.96
11	Badger	0.18	0.46	0.31	2.54	**6.21**	4.41
12	Lesser spotted eagle	0	2.79	1.69	0	3.88	2.89
13	Honey buzzard	0	3.55	2.15	0	2.85	2.12
14	Sparrowhawk	0	4.68	3.63	0	0.96	0.92
15	Stoat	3.30	2.54	2.34	0.82	0.60	0.68
16	River otter	0.51	0.39	0.19	2.62	1.92	1.14
17	American mink	1.09	0.84	1.08	0.97	0.71	1.12
18	Pygmy owl	1.19	0.91	1.11	0.08	0.06	0.09
19	Tengmalm's owl	0.91	0.70	1.07	0.11	0.08	0.15
20	Long-eared owl	0	0.63	0.80	0	0.19	0.29
21	Hobby	0	0.66	0.46	0	0.16	0.14
22	Eagle owl	0.12	0.09	0.06	0.31	0.23	0.19
23	White-tailed eagle	0.05	0.04	0.03	0.24	0.18	0.17
24	Marsh harrier	0	0.15	0.22	0	0.10	0.17
25	Montagu's harrier	0	0.18	0.18	0	0.06	0.07
26	Short-toed eagle	0	0.09	0.06	0	0.15	0.13
27	Booted eagle	0	0.10	0.08	0	0.09	0.09
28	Golden eagle	0.03	0.03	0.02	0.14	0.10	0.09
29	Great grey owl	0.03	0.03	0.02	0.03	0.02	0.02
Shannon-Wiener diversity index H		1.95	2.33	2.26	2.43	2.62	2.58
H_{max}		3.00	3.37	3.37	3.00	3.37	3.37
Equitability (H/H_{max})		0.65	0.69	0.67	0.81	0.78	0.77

Winter, only resident predators considered.
Spring, pre-breeding densities/biomass of resident and migratory predators.
Late summer, post-breeding densities/biomass of all predators (data from Table 3.3).
The rank of each predator is inversely proportional to its weighed percentage share in the community (averaged percentages of numbers and biomass in the four seasons: winter, late summer and doubled value for spring, presumed to reflect autumn as well). Dominants are in boldface type (>5%).

Table 3.5. Variation in densities of the most common species of predators in the pristine forests of Białowieża National Park. The ratio of max/min recorded numbers of predators feeding on rodents (in winter or summer) was positively related to the predator intrinsic rate of increase ($Y=-0.125+2.557X$, $n=4$, $R^2=0.90$, $p=0.05$). The intrinsic rate of increase was calculated as body mass$^{-0.26}$, according to Fenchel (1974)

Year/season	Density (N ind/10 km^2):				
	Weasel	Tawny owl	Pine marten	Polecat	Red fox
1985/1986, Winter	17.3		5.11	3.47	2.5
1986, Summer	47.6				
1986/1987, Winter	22.3		6.49	5.04	3.5
1987, Summer	47.6				
1987/1988, Winter	27.3	53.0	6.44	3.51	3.7
1988, Summer		87.6			
1988/1989, Winter	20.1	45.2	6.08	6.29	3.3
1989, Summer	41.9	74.7			
1989/1990, Winter		51.4			
1990, Summer	101.7	125.4			
1990/1991, Winter	18.0	52.0	4.44	3.57	5.0
1991, Summer	19.1	52.0			
1991/1992, Winter	5.2	33.0	7.57		
1992, Summer		54.5			
1992/1993, Winter		30.9	3.62		
Main prey	Rodents	Rodents	Rodents	Frogs	Rodents
Ratio of max/min densities	5.3	2.4	2.1	1.8	2.0
Ratio of predator to prey body mass	2.8	19.6	42.0	80.6	122.0
Intrinsic rate of increase	2.00	1.20	0.99	1.09	0.62

Blank spaces, no data for that year/season.

In terms of feeding habits, predators which hunt predominantly for rodents, an abundant and highly productive prey, strongly dominate by both numbers and biomass: tawny owl, weasel, pine marten, buzzard, and red fox fall in this group. In the ranked list, these predators are followed by ungulate-hunters (wolf and lynx), and amphibian-hunters (polecat, otter, and mink). Raptors feeding on birds are generally low ranked (goshawk, sparrowhawk, pygmy owl, hobby, booted eagle). A single reptile eater, the short-toed eagle, is among the rarest predators (Table 3.4). Thus, the relative abundance and position of predators in the community is shaped by the abundance and productivity of their prey (see Sect. 2.11).

The frequency distribution of body masses of predators is log-normal for number of species, number of specimens per unit area, and biomass per unit area (Fig. 3.66). Among the 29 breeding predators of BPF, body mass classes of 0.1–1 and 1.01–10 kg were represented by most species (48 and 35%, re-

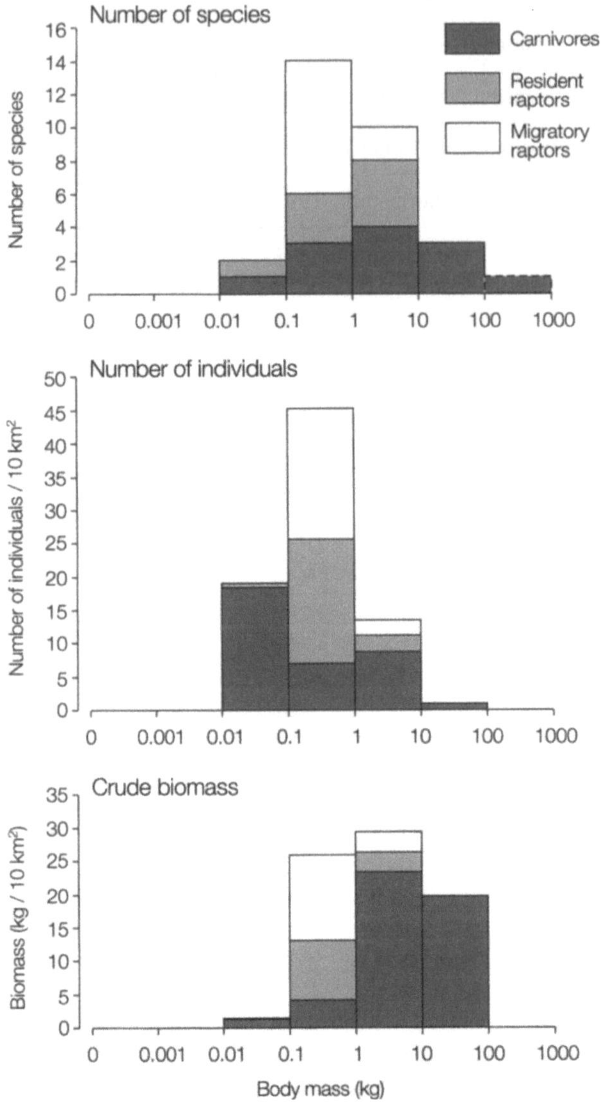

Fig. 3.66. Frequency distributions of Białowieża's predators in classes of body mass. From *top* to *bottom*: number of species ($N=29$), number of all predatory individuals in spring ($N=78.85$ ind/10 km^2), and total biomass in spring (76.56 kg/10 km^2). *Broken line* in the *upper graph* denotes a locally extinct brown bear. Data from Table 3.3

spectively), the greatest number of individuals (57 and 17%), and most biomass (34 and 38%, respectively). An abstract average predator weighed about 1 kg. There were, however, conspicuous differences in the frequences of predatory mammals and birds. Carnivores ranged in size over 450-fold: from a tiny weasel (0.07 kg) to wolf (32 kg), and an abstract average carnivore specimen was 1.4 kg. Raptors were generally smaller. The largest bird of prey was 4.5 kg (white-tailed eagle), i.e. 70 times the smallest one (pygmy owl). Mean body mass of all raptor individuals was 0.6 kg.

To analyse habitat niches of over 20 species, we first reduced the entire habitat variability to six broad types: spruce-pine forests, oak-lime-hornbeam forests, ash-alderwood with the smallest forest brooks, willow-sedge-reed marshes with medium-sized rivers, agricultural land (pastures and fields), and big rivers as such (Fig. 3.67). Secondly, we standardised the available data on 23 species, collected in various parts of the forest. Habitat selection shown for each species in Fig. 3.67 is calculated as if all habitats covered the same area. Thus, with six habitats, a relative percentage occurrence of 17% means selection proportional to availability of a given habitat, values <17% denote avoidance, and from >17 to 100%, a positive selection of a given habitat.

Oak-lime-hornbeam forests and alder-ash forests, the associations indicative of temperate forests, are optimal and the strongly preferred habitat of five species, the geographic ranges of which stretch over the temperate forests: tawny owl, common buzzard, honey buzzard, lesser-spotted eagle, and polecat (Fig. 3.67). Oak-lime-hornbeam forests and coniferous forests (the drier and mixed forms of the latter) are the preferred habitats of badgers, pine martens, and sparrowhawks. A combination of coniferous forests and bog alderwoods or ash-alder forests, which in many places of Białowieża Forest occur side by side (with a fair contribution of spruce in both of them), is a favourite habitat of lynx, goshawk, weasel, and pygmy owl. There are few species that strongly prefer coniferous forests: Tengmalm's owl, hobby, and long-eared owl, the last two cases in close vicinity to open land (Fig. 3.67). Narrow sedge marshes by forest rivers and the rivers as such are the most important non-wooded habitats in Białowieża Forest, crucial for the occurrence of stoat, river otter, mink, and harriers, and widely utilised by a number of forest predators.

A niche breadth index (*B*) was calculated for each predator (according to Levins 1968). With six habitats distinguished, *B* varies from 1 if a predator is found exclusively in one type of habitat, to 6 if it uses all habitats indifferently. The niche breadths of 23 species were from 1 to 4.45 (Fig. 3.67). Nine species appeared to be specialists, restricted to a narrow range of habitats (*B* from 1 to 2.5), 11 species had moderately broad niches, and three predators were typical generalists ($B>4$). In the whole community, the average breadth of habitat niche was 2.84 (SD 0.98). The breadth of a predator's habitat niche was correlated with its rank in the community (Fig. 3.68). Predators with the widest niches are dominants, species inhabiting a narrow range of habitats are those characterised by low densities and biomass.

The most important 'requisites' of a given habitat that make predators prefer or avoid it are prey resources. Examples are the polecat's preference of ash-alder forests, the habitat richest in common frogs; fox's and stoat's selection of open fields and marshes, which abound in *Microtus* voles; or the ample occurrence of tawny owls in oak-lime-hornbeam forests, the optimal hab-

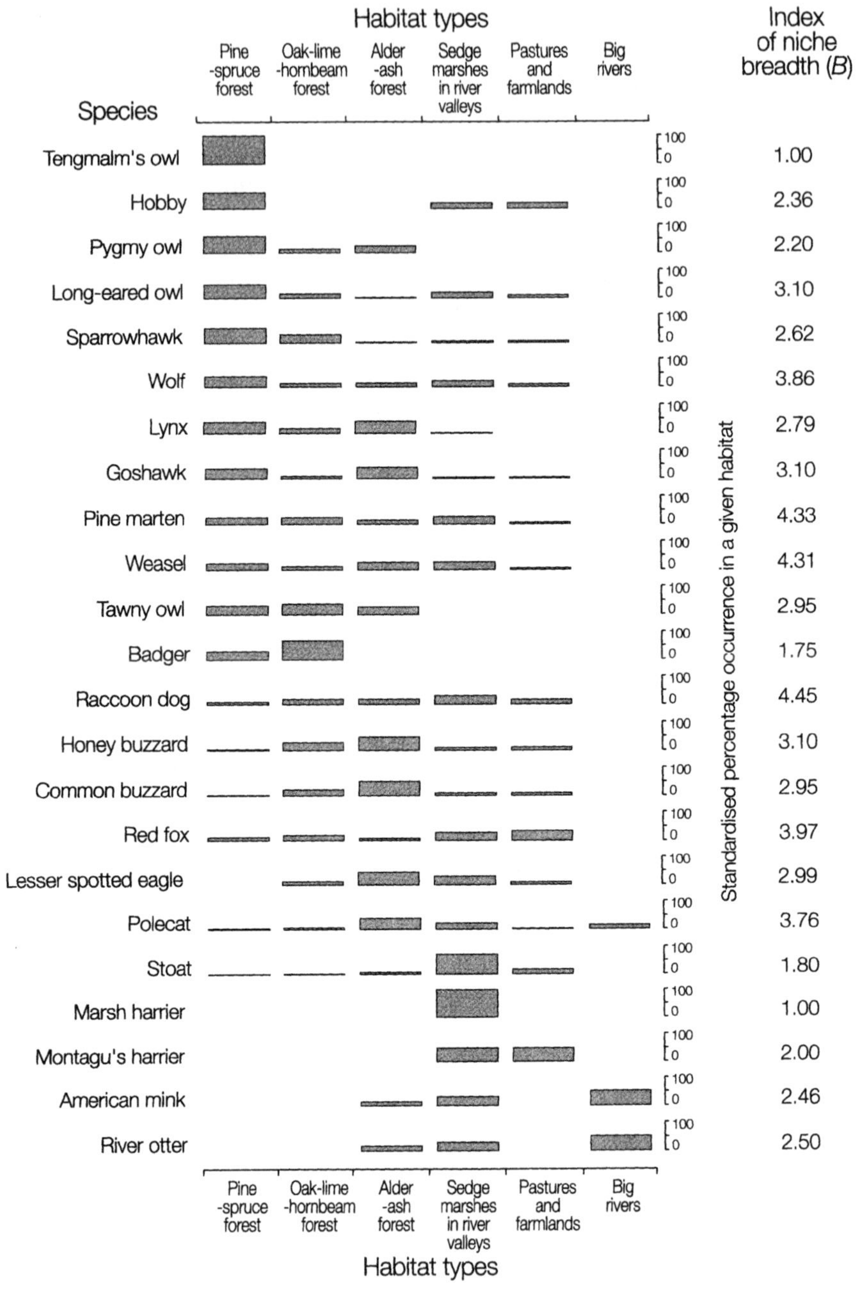

Fig. 3.67. Patterns of habitat selection by 23 species of predators inhabiting BPF. Standardised percentage occurrence in habitats calculated as if all six habitats covered equal areas, thus percentage occurrence <17% means avoidance of a given habitat, and >17% its positive selection by a predator. Data for calculation from: Bunevich (1984), Domaszewicz (1993), Pugacewicz (1996), Sidorovich et al. (1996), and authors' data on snow tracking of carnivores in BNP. Index of niche breadth ($B = 1/\sum p_i^2$, where p_i is a fraction of a given habitat type; after Levins 1968) varies from 1 (the narrowest niche) to 6 (the widest niche possible)

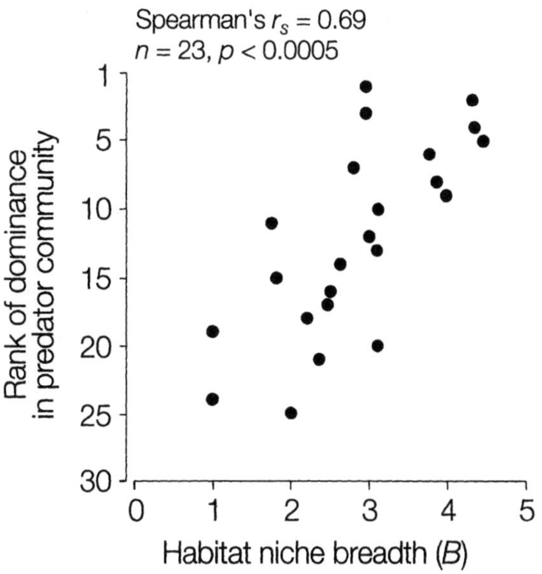

Fig. 3.68. Correlation between the breadths of habitat niche and ranks in the community of 23 species of Białowieża's predators; data from Table 3.4 and Fig. 3.67

itat of yellow-necked mice. This subject will receive detailed treatment in Chapter 4; here, we note in passing that the pattern of habitat selection is shaped predominantly by prey abundance in these habitats, although other features (suitable nesting and denning sites, cover etc.) also play a role.

The data on habitat selection by various predators can be looked at from yet another perspective, in order to explain whether any type of forest harbours more predators than other types. An indirect way to answer this question would be to multiply the average densities and biomasses of predators by their proportional occurrences in various habitats and, then, to sum up the numbers and biomasses of all predators in a given habitat. Calculated estimates of predator densities are similar in coniferous and oak-lime-hornbeam forests, and 30% larger in ash-alder forests. The total biomass of predators per unit area is lowest in coniferous forests, more abundant (by 10%) in oak-lime-hornbeam forests, and largest in ash-alder forests (30% larger than in coniferous forests). The observed trend of growing predator numbers and biomass with growing humidity and richness of forests results from the same trend documented for major prey groups (see Chap. 2). Densities of birds, amphibians, and rodents increased in the same way from coniferous to deciduous, to wet deciduous forests. Also, ungulate numbers were significantly higher in deciduous stands than in coniferous ones.

Overlaps of habitat niches were calculated for 23 species of predators (Fig. 3.69). The frequency distributions of α values for all 253 pairs of predators in summer as well as 120 pairs in winter appeared to be very flat (platykurtic). Neither summer nor winter distributions statistically differed from an even distribution of 20% in each of the five classes of α values (0–0.2, 0.21–0.4, ..., 0.8–1) (G=2.59, df=4, p>0.5 in summer, and G=4.67,

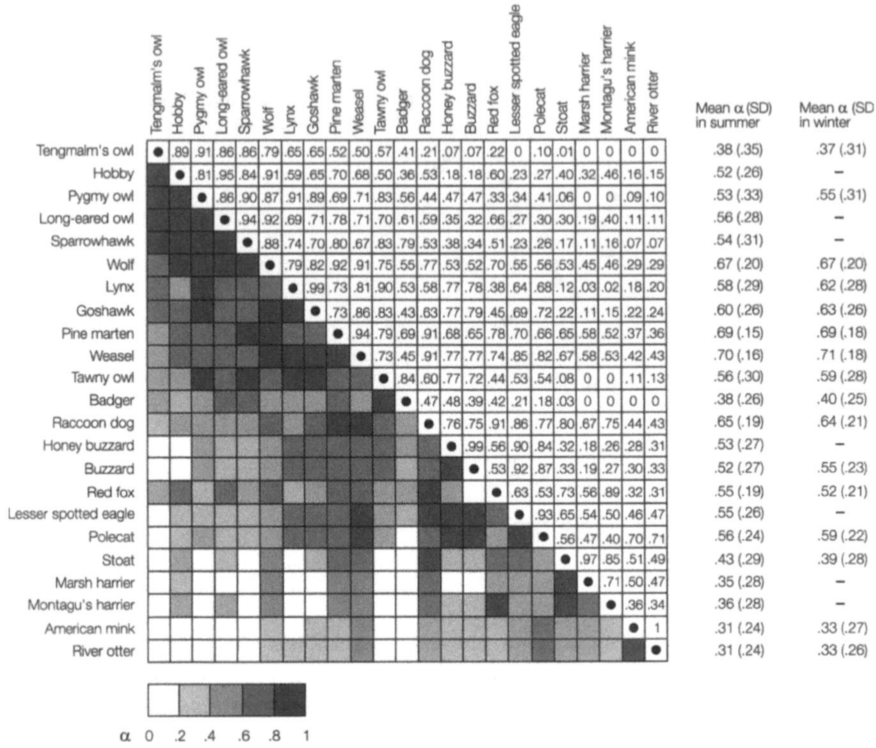

Fig. 3.69. Pattern of habitat niche overlaps among 23 species of predators expressed by Pianka's (1973) index α, which varies from 0 (exclusive niches) to 1 (complete overlap). Niche overlaps were calculated according to the following formula: $\alpha_{ij} = (\sum p_{ia} \cdot p_{ja}) \cdot [(\sum p_{ia}^2) \cdot (\sum p_{ja}^2)]^{-1/2}$, where α_{ij} is the overlap between predators i and j, p_{ia} is the fraction of ath habitat type in the total utilisation of habitats by ith predator, p_{ja} is the fraction of ath habitat type in the total utilisation of habitats by jth predator (values acc. to Fig. 3.67). Mean values of α are calculated for all predators in summer and winter

df=4, $p>0.1$ in winter). The mean α values for the whole community do not vary between summer (0.51, SD=0.12) and winter (0.54, SD=0.13).

Habitat choice is merely one of several dimensions of an animal's niche. Predators sharing the same habitat may hunt for the same prey (e.g. wolves and lynxes for ungulates) or they may live on entirely different resources (e.g. lynxes on ungulates and goshawks on birds). Food niches of predators will be discussed in Chapter 4, and then the data on predators' habitat distribution and diet composition will be put together.

3.32 Factors Shaping Densities and Population Dynamics of Predators

In an attempt to answer the question 'Why do so many or so few predators of a given species live in Białowieża Forest?', one has to consider two elements. First and foremost, one needs to know the sizes of predators' home ranges and, secondly, how densely these home ranges are packed in space. Intuitively, one would say that home range size depends on body mass, since larger animals need a more extensive area to get enough food for survival and reproduction than small ones. In fact, in the six species of carnivores with available data, home ranges clearly increased with increasing body masses, as shown on a double-logarithmic plot in Fig. 3.70. Except for a somewhat lower exponent, the regression equation describing the local conditions of Białowieża Forest did not markedly differ from the general relationship in order Carnivora found by Harestad and Bunnell (1979) or the more detailed scaling for families Mustelidae, Canidae, and Felidae done by Goszczyński (1986 a). In Białowieża, five of the six analysed carnivores feed on mammalian prey. Polecat, the sixth species, relies on amphibians. In any case, they form an ecologically uniform group of predators feeding on abundant and/or very productive prey (see Sect. 2.11).

In contrast to carnivores, home ranges of raptors, when scaled to their body masses, seem to disobey the rule. The scatter plot of points for 11 species (Fig. 3.70) appeared uncomfortably divergent from the relationship for 22 species of diurnal raptors shown by Newton (1979); home ranges of many Białowieża's raptors were extraordinarily large when compared with other species of similar body weight presented in Newton's review.

Why do birds of prey weighing 1–1.5 kg need territories as large as wolves or lynxes? And why do two raptor species of similar sizes, say 0.8 kg, have ranges as different as 0.8 km^2 (common buzzard) and 380 km^2 (peregrine)? The answer came from the analysis of raptors' diets. All the birds of prey (except for the short-toed eagle) fed on two main resources, small mammals and birds, though in various proprotions. The more they relied on birds, the larger the home ranges they needed (Fig. 3.70). The dependence of home range size on the share of bird biomass in total food consumed was very strong; adding raptor body mass as another variable to the regression only increased the total explained variation (R^2) from 0.73 to 0.79. The fairly weak correlation of home range size of bird-eating raptors with their body masses has, nonetheless, an extremely high slope ($a = 164.05$), which showed that with the same rate of increase in body mass, the home ranges of raptors feeding on birds grow 34 times faster than those of mammal-eating predators!

We have shown (Sect. 2.11) that as potential prey of predators in the temperate forests, birds are characterised by a smaller standing crop of biomass and lower productivity than small mammals and amphibians. This is, we think, the clue to explaining the rapid expansion of home ranges with the growing share of birds in raptor diets. Yet another component of that rela-

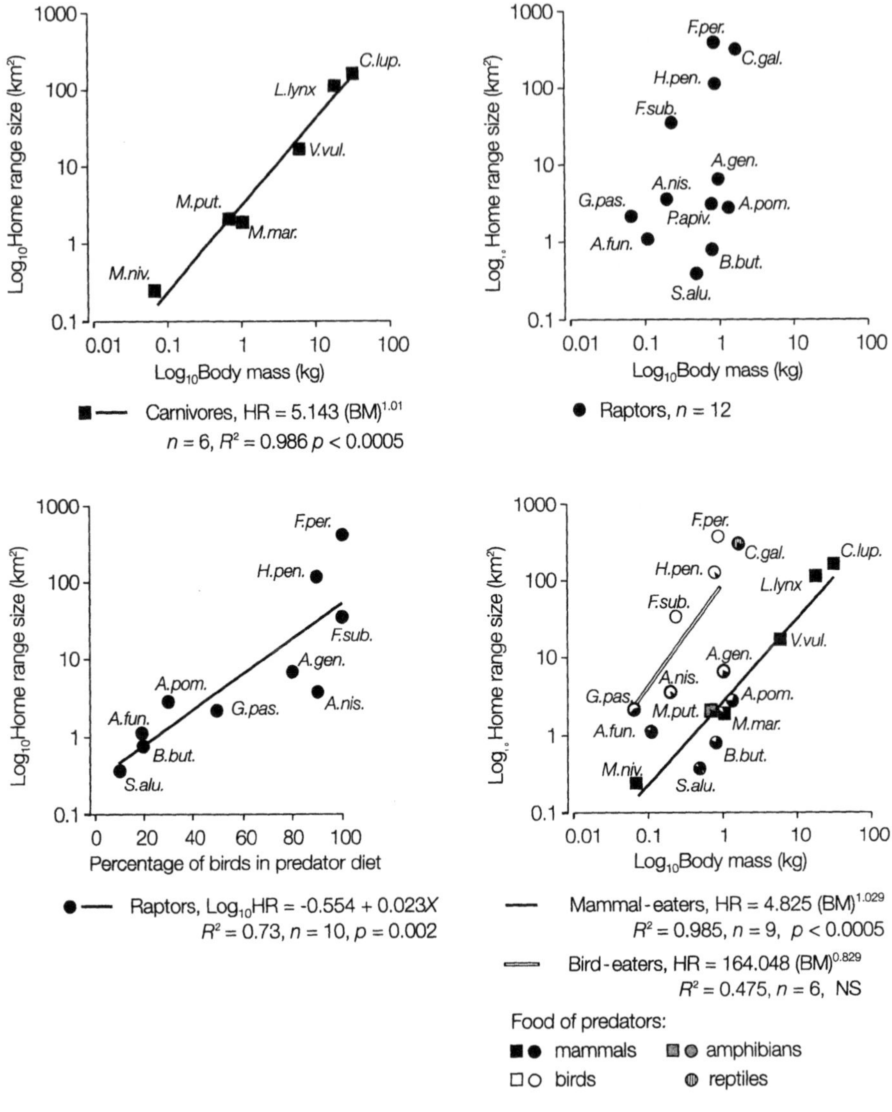

Fig. 3.70. Home range size of predators (abbreviations made of Latin names) inhabiting BPF in relation to their body masses and food. *Upper graphs* body mass – home range size relationships in carnivores *left* and raptors *right*. *Lower graphs, left* home range size of raptors feeding predominantly on small mammals or birds in relation to the share of birds in their diets (honey buzzard *Pernis apivorus*, a wasp-eater, and short-toed eagle *Circaetus gallicus*, a reptile-eater, are not shown). *Lower graphs, right* home range size of all predators in relation to their body masses and food. Circles for each species show relative shares of major food types in predators' diets (in percentages of biomass consumed; data from Chap. 4). Regressions include predators for which a given prey constituted >50% of food taken. Polecat *Mustela putorius* and short-toed eagle are not included in either regression

tionship may be the fact that birds are more difficult for predators to capture than mammals. This factor, however, can only slightly worsen the situation, which is determined by the lower abundance and productivity of birds compared with small mammals. Temeles (1985) reviewed data on 15 species of falcons and hawks and reported that their success was on average 30% when hunting mammals, and 22% when hunting birds (calculated from Table 1 in Temeles 1985).

Interestingly, the home ranges of raptors feeding predominantly on small mammals plotted against their body masses fitted the relationship for Carnivora very well (Fig. 3.70), thus proving that diet shows more significant influence on home range size than taxonomic affinity. Additional support is provided by the cases of polecat and short-toed eagle. The polecat, which relies on amphibians, an abundant and productive prey, fits exactly the home range-body mass relationship for carnivores feeding on mammals. Short-toed eagles which rely on reptiles, a scarce and not very productive resource, hold outstandingly large territories (308 km^2).

In conclusion, the home range sizes of the 17 predators studied in Białowieża Forest are related to their body masses and the share of not very productive and scarce prey in their diets (birds or reptiles): the two variables explained 79% of the total variation in home range sizes (multiple regression analysis; $p<0.0005$), and the contributions of each of the two variables were equally important (semipartial correlations: $sr^2=0.58$ for body mass, and $sr^2=0.44$ for diet composition, computed according to Tabachnick and Fidell 1983). Thus, the larger the predator or the more it feeds on scarce prey, the larger home range it needs and the rarer it is.

Another class of complications are those produced by several species which are rare, despite each individual or pair holding a rather small territory. With no exceptions, these are predators of very narrow habitat niches compared with the array of habitats offered by a pristine temperate forest (see Fig. 3.67): the Tengmalm's owl and the pygmy owl, strongly connected with coniferous taiga-like associations; the river otter and the mink, with semiaquatic habits; the lesser spotted eagle and the harriers, inhabitants of river-side forests or marshy, open river valleys. The densities of these habitat specialists are determined by the amount of inhabitable patches or strips in the whole woodland.

What determines how densities of predators change over time? Again, built into this question are problems of predators' body masses, diets, and variation in densities of their main prey. Before we proceed to analysing these factors, we shall make some definitions. By numerical response we understand the change in predator densities following the change in their food (prey) supply. The time delay of predator numerical response is the time by which the predator reaction lags behind the changes in prey abundance. Predators respond to varying prey abundance by in situ reproduction and mortality as well as by aggregation and dispersal. Whilst probably all species of predators react by varying breeding and survival success, some have additionally developed the adaptation of tracing the changes in prey abundance and gathering or disappearing locally. Notable examples are nomadic raptors feeding on cyclic microtines in the northern Palearctic (e.g. great grey owl, Tengmalm's owl, hawk owl, snowy owl; Hilden and Helo 1981; Löfgren et al. 1986; Litvin and Ovsyanikov 1990; Virkkala 1992; Korpimäki 1994). Importantly, in the temperate zone of Europe, the only species showing manifold year-to-year changes in densities based on nomadic shifts between regions with peaks of *Microtus* voles are the short-eared owl, harriers, and, to some extent, the long-eared owl (Goszczyński 1977; Korpimäki and Norrdahl 1991; Butet and Leroux 1993). All these species are connected with extensive open marshes or farmlands, and in woodlands such as Białowieża Primeval Forest, they occur marginally.

To understand the multiannual variation of predator numbers in Białowieża Forest, we have to remember that four major resources of prey: ungulates, small rodents, amphibians and birds, differ in the magnitude of fluctuations of numbers from year to year. Birds were very stable and predictable prey (maximally twofold changes in spring numbers, see Table 2.15). Ungulates showed tenfold changes on a long-term scale, but much of this due to human exploitation. Small rodents and amphibians were the most variable resources, with 30-fold and 10- to 100-fold changes, respectively, in their densities over several years.

One might expect that predators feeding on birds and rodents would be more stable the more they feed on birds. Except for the tawny owls, no multiannual data on raptor population dynamics are available. However, much information can be drawn from 4–6-year records of the breeding performance of Białowieża's birds of prey (Table 3.6). In five species feeding on vertebrates, year-to-year variation in breeding success increased with the growing share of rodents in their diets: the coefficient of variation was only 7% in goshawks, while for tawny owls it was 80%. Interestingly, the honey buzzard, relying on still more variable prey (wasps, that critically depend on abiotic conditions) also undergoes very wide fluctuations in its breeding success (Table 3.6).

Of all predators that live predominantly on rodents, we have information on the numerical response of four species: weasel, tawny owl, pine marten, and red fox. In reaction to about a 30-fold change in rodent abundance, the numbers of these predators varied from twofold in the fox to 5.3-fold in the weasel. The larger the predator and the lower its intrinsic rate of reproduction, the more stable population dynamics it exhibited (Table 3.5). The polecat, a hunter of greatly fluctuating amphibians, does not diverge from the pattern observed in rodent predators. Its size compared with that of its main prey (the common frog) is of the same order of magnitude as those of fox and marten compared with their prey, and so is the magnitude of the polecat's variation in numbers from year to year (Table 3.5).

An obligatory point of discussion in the analysis of predators' numerical responses is time delay. A mathematically proved, 9-month lag in predator response may be sufficient to produce predator-prey cycles (May 1981). Small rodent ecologists have recently been debating the empirical role of this phenomenon in weasel-microtine cycles in northern Fennoscandia (Henttonen et al. 1987; Hanski et al. 1993; Hanski and Korpimäki 1995). In the predators

Table 3.6. Variability in breeding success of raptors in relation to the share of rodents in their spring-summer diets. Regressions of coefficient of variation (CV%) of breeding success with the percentage biomass of rodents in diet (honey buzzard excluded): $Y=9.772+1.122X$, $n=5$ species, $R^2=0.874$, $p=0.02$ for percentage of successful pairs, and $Y=16.144+1.095X$, $n=5$, $R^2=0.798$, $p=0.04$ for number of juveniles per pair. Data from Domaszewicz (1993), Pugacewicz (1996), and W. Jędrzejewski and J. Kupis (unpubl. data)

Species	*N* years	Percentage of pairs with at least one young fledged			Mean number of fledglings per breeding pair			Percentage biomass of rodents in diet
		Mean	(SD)	CV%	Mean	(SD)	CV%	
Goshawk	6	65	(5)	7	1.1	(0.2)	14	1
Sparrowhawk	6	53	(17)	32	1.2	(0.5)	41	8
Buzzard	6	65	(18)	28	0.9	(0.3)	33	22
Lesser spotted eagle	6	58	(17)	30	0.6	(0.2)	31	24
Tawny owl	4	50	(36)	80	1.5	(1.2)	80	53
Honey buzzard	5	35	(28)	80	?	?	?	–

feeding on rodents in the temperate zone of Europe, two factors are decisive for whether or not they lag behind the fluctuations of their prey. First is the strictly seasonal breeding of predators with no such stringent seasonality in rodents, which begin most of their outbreaks with winter reproduction. For instance, the 1990 outbreak of forest rodents began with their breeding in winter 1989/1990, whereas weasels and tawny owls could only start to reproduce in spring 1990. Thus, predator numerical response to multiannual cycles of rodents is always intervened with autumn–winter periods of no breeding and actual decline in numbers. In other words, the numerical response of predators due to breeding takes place in spring–summer seasons only. In weasels and tawny owls, that response was immediate with no time lag. Also, the decline in weasel and tawny owl numbers was not delayed and followed the decline in rodents promptly. The second factor is the biology of reproduction of a given predator. In the pine marten, a carnivore with delayed implantation, a 1-year lag was found (see Fig. 3.23). Similarly, in the stoat, which exhibits delayed implantation, a delayed numerical response (by 1 year) to changes in prey numbers was found by Kaikusalo (1982) in northern Finland and by Debrot (1983) in Switzerland.

Synthesis of data on Białowieża's predators ends with yet another close look at the frequency distribution of prey and predator biomasses in the classes of body weight (Fig. 3.71). The bimodal distribution of prey biomass (with peaks in both very small and very large animals) strikingly 'gets filled' by a log-normal distribution of predator biomass. This implies that the ratio of predator to prey biomass varies enormously for animals of various sizes (Fig. 3.71). At the two extremes (very small and very large prey), a unit of predator biomass coincides with large amount of prey biomass. In the intermediate size classes (animals of 100 g to 10 kg) predators are nearly as abun-

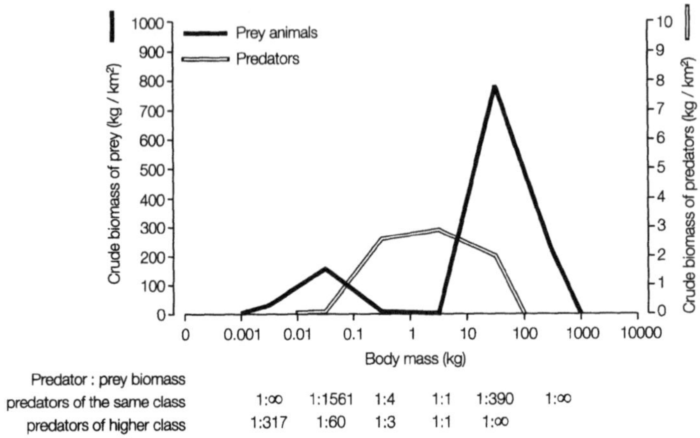

Fig. 3.71. Frequency distribution of the total biomass of prey and predators in BPF (in classes of body masses). Note that the scale of prey abundance is 100 times that of predator abundance. Data schematized from Figs. 2.52 and 3.66

Table 3.7. Ratios of vertebrate predator and prey biomass in the temperate forests of Białowieża; data averaged for 1985–1995

Group of animals	Total biomass (kg/km^2)	
	Winter-spring (non- and pre-breeding season)	Late summer-autumn (post-breeding season)
Ungulates	994	1157
Large predators (wolf and lynx)	1.5	1.6
Ratio of predator: prey biomass	1:663	1:723
Small and medium-sized prey	200	484
Small and medium-sized predators	4.5	11.2
Ratio of predator: prey biomass	1:44	1:43
All groups of prey	1194	1641
All predators	6	12.8
Ratio of predator: prey biomass	1:199	1:128

dant as prey. The pattern is similar when predators are compared with prey of a lower class of body mass ('meal-sized' class of prey). The ratios of predator to prey biomass do not change much between the seasons (Table 3.7). The major difference exists between the large predator-ungulate relationship and the small prey-small predator systems, the latter one characterised by rather high productivity of some prey animals, which thus allow for high predator/prey ratios.

Several intriguing questions emerge from Fig. 3.71 and Table 3.7. Are these predator:prey ratios sufficient explanation for the scarcity of medium-sized herbivores in European temperate forests? Does this say anything about the capabilities of local predators to regulate small, medium-sized and large herbivores? To propose answers to these and other questions about predation we need to dwell on the subject of the composition of predators' diets and their killing rates.

4 Diets and Hunting Habits of Predators

We awoke from a sleep of I don't know how many thousand years.
An eagle flew in the sun again but it didn't mean the same.
Czesław Miłosz, *Notes. Morning*, 1978

4.1 Introduction

For predators, in an overwhelming number of cases, foraging means detecting and seizing another animal. Since all animals, particularly vertebrates, are elusive and shy, hunting is inherently inflicted with high risk of failure and sometimes of injury, too. Yet foraging is the most important of activities for it largely determines where and when a predator abides. In this chapter, the feeding ecology of carnivores and raptors dwelling in Białowieża Primeval Forest receives meticulous attention.

First, we present the composition of each species' diets in the two seasons, spring–summer and (when applicable) autumn–winter. Frequencies of various prey species taken by predators compared with the shares of these animals in the communities of ungulates, birds, rodents and amphibians revealed predators' selectivity. In a few species of predators, the choice of sex and age classes of prey animals is also presented. Furthermore, we aimed at describing predator behaviour when searching for, hunting, utilising, and securing prey. These two aspects, diet composition and foraging behaviours, were then compared among all species in the community by applying the conceptional tools of niche breadth (Levins 1968) and niche overlap (Pianka 1973).

Secondly, the temporal variation in diet composition and/or hunting rates is dealt with. Two time scales are applied to quantify that variability: an annual scale (changes of feeding habits in monthly, bimonthly, or quarterly periods) and a multiannual scale (year-to-year changes). Both short- and long-term variation in predators' diets was related to the observed variation in prey abundance during that time and to abiotic conditions likely to affect predators' feeding behaviour.

Such detailed analysis was expected to create a solid basis for defining and expressing mathematically the functional (dietary) responses of some common species of predators inhabiting Europe's temperate forests. In this book, the measures of functional responses are based on changes in the proportion of prey in predators' diets (see Boutin 1995, and Jędrzejewski et al. 1996a, for discussion of this approach). Since the time when Holling (1959, 1965) had proposed four types of predator functional response, much work has been done to predict consequences of different responses to the total predation impact on prey (Murdoch and Oaten 1975; Andersson and Erlinge 1977; Messier 1995a). The S-shaped response (type III), especially receives much attention. Believed to be typical for generalist predators with relatively stable populations, such a functional response presumably has a strong potential for regulating prey because it would produce predation rates growing with increasing prey densities. The type II response, logarithmic or otherwise rising at a continually decreasing rate, if exhibited by predators with stable populations, yields predation rates that decline with growing densities of prey, i.e. inversely density dependent predation (Murdoch and Oaten 1975; Andersson and Erlinge 1977; Hanski et al. 1991).

Careful analysis of the temporal and, in some species, spatial variation in a predator's feeding habits in response to varying abundance of prey have shaken the fundamental assumptions of the classification of vertebrate predators as generalists or specialists. This classification, elaborated by Andersson and Erlinge (1977), useful in practice due to its simplicity, appeared too rough when it came to the matter of quantifying predator impact on populations at various density levels of prey. We discuss this more broadly at the end of this chapter.

Worth emphasising is the fact that, when analysing predators' fuctional responses, we consider their most important as well as their secondary prey. As proposed by Angelstam et al. (1984), the impact of predators on their secondary (alternative) prey may be shaped by fluctuations of their primary prey. This phenomenon is probably widespread in predators of rodents, and may have profound consequences on the dynamics and numbers of shrews, birds, and amphibians.

An answer to the question 'Who eats whom and in what quantities?' is indispensable if we want to evaluate the role of predators in regulating prey populations. Admittedly, this problem is intriguing in itself as, again and again, it arrests the attention of many researchers, despite the fact that methods of studying predators' diets are neither simple nor easy.

4.2 Methods of Studying Predator Feeding Habits

An early method of assessing what predators eat was to look into the stomachs and alimentary tracts of animals shot during predator control or, less often, for the purpose of zoological studies. This method yielded data (usually auxiliary only) on wolves, foxes, and several species of raptors from the Belarussian part of BPF (Gavrin 1953; Gavrin and Donaurov 1954; Gavrin and Krapivnyi 1968; Bunevich 1988). The fact that the very subject of study was killed made this method inefficient, apart from the obvious ethical matters as we see them today (which, however, were not so obvious a few decades ago).

In the late 1940s, Folitarek (1948) began studying raptor diet by taking the prey brought by adult birds to their chicks before the chicks could consume it. Chicks were treated in a manner resembling the falconry traditions: their heads and beaks were harnessed in a net hood, and their legs were fettered, through a nest floor, to a branch. An investigator climbed the nest daily (in late aftrenoon), collected the prey brought by adult birds, and fed the chicks ad libitum with other food. This method was applied in the Belarussian part of BPF for over 10 years on 12 species of diurnal birds of prey (Folitarek 1948; Gavrin 1953; Golodushko 1958, 1959, 1961 a, b, c). Material collected with this method yielded enormous amounts of detail on species, weight, condition, and health of prey, on hunting efficiency by adult birds, on growth rates and food requirements of (artificially fed) chicks. Though definitely a progression from the analysis of shot birds, Folitarek's method still involved great disturbance of studied birds.

Indirect methods developing in parallel were based on collecting scats of carnivores and pellets of raptors. Learning what a predator had eaten was based on recognition of undigested prey remains, mainly bones, hair, and feathers. In our studies conducted in 1985–1996, analysis followed the standard procedure as elaborated by Lockie (1959) and Goszczyński (1974). Scats and pellets were oven dried at 40–50 °C and then soaked in water for several hours. Carnivore scats were washed through a sieve (1 mm mesh). All recognisable prey parts were separated and identified. We used the key by Pucek (1984) and our own comparative material for identification of mammals' bony remains (skulls, jaws, teeth, bone fragments). Bones of birds were identified according to Moreno (1985, 1986, 1987) and Dr. A. L. Ruprecht's collection of avian skulls, beaks and talons. März (1987) provided a key to the recognition of reptiles (based on bones and scales), amphibians (ilium, scapula, and frontoparietal bones), and fish (skull fragments, jaw apparatus, scales). In addition, amphibian remains were identified based on keys by Böhme (1977) and Engelmann et al. (1985), and fish remains on information given by Brylińska (1991).

In the case of mammalian prey, microscope analysis of hair was done according to the keys by Dziurdzik (1973) and Debrot et al. (1982) and our own comparative microscope slides. Bird feathers were identified to species or genus by comparing them with a reference collection of material and by using März's (1987) key. Parts of beetle exoskeletons were also recognisable. In the cases of badger and raccoon dog scats, microscope searches for earthworm chaetae and their count were performed according to Kruuk (1989).

There are three ways to express the composition of a predator's diet, based on the analysis of its scats or pellets. First, the percentage of occurrence of a given prey in the diet shows that it was found in, say, 70% of all analysed scats or pellets irrespective of whether one or several specimens of it were recovered from a single pellet or scat. This measure approximates the frequency at which a predator feeds on a given prey. Secondly, percentage of prey captured can be calculated. This requires prey specimens recovered from scats or pellets to be counted with reasonable accuracy, which was only possible in the case of raptors. Their regurgitated pellets contain teeth, jaws, skulls and other parts of vertebrate skeletons (Fig. 4.1). If the prey was identified by microscopic analysis of hair, we assumed that it represented one specimen. The third measure is percentage of food biomass consumed. It expresses the part that each prey plays in the total fresh weight of food con-

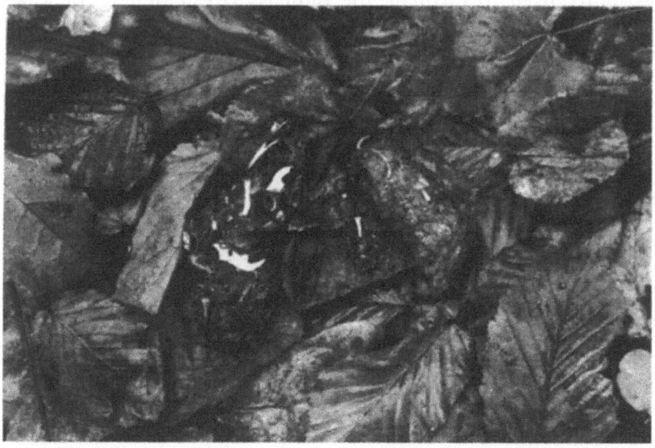

Fig. 4.1. Tawny owl pellets on the forest floor in Białowieża National Park. Note the undigested bone, hair, and feather remains of prey. Photo: K. Zub

sumed by a predator. Translating the content of scats or pellets into biomass of food was done differently in carnivores and raptors. For carnivores, the method was based on coefficients of digestibility, i.e. the ratio of fresh weight of a given prey to the dry weight of its remains in scats, empirically found for the main prey types. In our analysis, dry prey remains were weighed and then multiplied by adequate coefficients of digestibility to obtain an estimate of the fresh weight of food consumed. The applied coefficients (determined in a series of enclosure experiments) were taken from the works of Lockie (1961), Goszczyński (1974), Fairley et al. (1987), Lode (1990), and Roger et al. (1991) and are listed in Table 4.1.

A slight modification of this method was necessary in analysing scats of badgers and raccoon dogs that ate many earthworms. From each scat, after drying, but prior to soaking in water, a small sample (0.1 g) was taken for a microscope count of earthworm chaetae. The biomass of consumed earthworms was calculated assuming 1000 chaetae per earthworm (after Kruuk 1989) and a mean body mass of 2.5 g per one earthworm (our own data from BPF).

In raptors, calculation of the biomass of consumed food was done by multiplying the number of prey specimens recovered from pellets by their body masses (after Sokołowski 1979; Pucek 1984; Pikulik 1985; Pikulik et al. 1988; Yalden 1985; Juszczyk 1987; März 1987; and our own data). This was not done, however, in the cases of large prey (i.e. prey of a size significantly exceeding one meal) in order not to overestimate their share in a raptor's diet. We calculated the mean fresh weight of food represented by pellets containing remains of small prey only (SP weight). This value approximated stomach capacity or one large meal in a given raptor species. We assumed that for one meal, a raptor consumed a portion equal to SP weight from each large prey. The respective mean SP weights for various raptors are given in Table 4.2.

Table 4.1. Coefficients of digestibility used to estimate the biomass (fresh weight) of food eaten by carnivores, based on dry weight of prey remains recovered from their scats

Species or group of prey	Coefficients of digestibility and sources[a]				
	Wolf, lynx	Fox, marten	Badger, raccoon dog	Polecat	Otter, mink
Shrew, mole, bat	23_G	23_G	23_G	$15.2_{R,LD}$	5_F
Small rodent, squirrel, weasel	23_G	23_G	23_G	17.8_R	9_F
Medium-sized mammal	50_G	50_G	50_G		25[b]
Wild ungulate	118_G				
Carcass of wild boar[c]		118_G	118_G	118_G	
Carcass of Cervidae[d]		15_J	15_J	15_J	
Domestic ungulate[e]	118_G	118_G	118_G		
Bird	35_G	35_G	35_G	12.4_R	12_F
Amphibian, reptile	18_F	18_F	41.3_R	41.3_R	18_F
Fish			25_F	25_F	25_F
Gastropoda, Crustacea	5_L	5_L	7_F	7_F	7_F
Insects	5_L	5_L	5_L	5_L	5_L
Fruit, seed, mushroom		14_L	14_L		
Other plant material	4_J	14_L	4_J	14_L	4_J

[a] Sources of data are given as subscripts: G, Goszczyński (1974); L, Lockie (1961); J, Jędrzejewski and Jędrzejewska (1992, and unpubl. data); F, Fairley et al. (1987); R, Roger et al. (1991); LD, Lode (1990).
[b] Empirical data are lacking; this coefficient was calculated based on relative degrees of digestibility of this and other food types in the red fox.
[c] All wild boar scavenged upon by medium-sized carnivores were assumed to have been whole carcasses of boar dead from starvation/disease.
[d] All cervids scavenged upon by medium-sized carnivores were assumed to have been leftovers of wolf and lynx kills.
[e] Domestic ungulates consumed by medium-sized carnivores were taken as carcasses of dead animals.

Since all methods of analysing the composition of predators' diets are inaccurate and inflicted with errors, whenever possible, we took two or three measures (percentages of occurrence, prey specimens and biomass). In the studies on raptors where percentage of prey in the diet was given (e.g. Golodushko 1958; Ruprecht and Szwagrzak 1987), we calculated the percentage of food biomass. Belarussian studies on mammalian predators gave the percentage of occurrence, only. Furthermore, the researchers usually pooled the samples of scats, stomachs, and prey remains found on trails and at dens.

Analysis of scats, pellets or prey brought to nests was the chief method in studying the diets of Białowieża's predators. Yet three other methods added extra information on the subject. The first and foremost was the search for wolf and lynx kills: the remains of ungulates. Conducted for several decades in both Polish and the Belarussian part of BPF, this method yielded invaluable data on the species, age, and sex composition of ungulates seized by

Table 4.2. Mean biomass of all prey represented in a pellet containing the remains of small prey only (i.e. prey of size not exceeding one meal). The values for given raptors were used to deal with findings of large prey remains in pellets (see Sect. 4.2)

Raptor species	*N* pellets containing exclusively the remains of small prey	Mean fresh weight of all small prey recovered from a pellet (g)
Tawny owl	686	64
Tengmalm's owl	81	29
Pygmy owl	117	31
Common buzzard	228	66
Lesser spotted eagle	85	67
Goshawk	36	98
Sparrowhawk	60	47
Hobby	72	28

large predators. Assisted in our hard field work by many volunteers, who intensively searched for wolf kills, we were able to estimate killing rates and consumption time of prey by wolves.

Second was the snow tracking of mammalian predators. In 1985–1993 we snow tracked nine species of mammalian predators in the pristine forests of Białowieża National Park. A person followed the fresh trail of a predator, measured its length by pacing and recorded all activities that could be read from the tracks and signs in the snow. All the habitat requisites utilised by a predator were also recorded. These data threw light on searching modes, hunting activities, and foraging behaviour by carnivores.

Thirdly, radio-tracking of some carnivores allowed an insight into their hunting and foraging behaviour. For this purpose, besides daily localisations, the collared carnivores were radio-tracked continually for one to several days, taking precise locations at 15- to 30-min intervals. In the case of weasels, direct observations of their hunting were possible (Jędrzejewski et al. 1992b). In lynxes, radio-tracking made it possible for us to find consecutive prey, to measure the time the lynx spent feeding and searching and, in consequence, to estimate their killing rates. Also, radio-tracking of wolves yielded indirect information on their rate of ungulate kills.

The amount of information available on foraging ecology differs among various predators. Nonetheless, the basic data on diet composition (list of prey species with their frequencies) is available for as many as 29 out of the 40 species recorded as breeders in Białowieża Primeval Forest.

4.3 Lynxes – Stalking the Roe Deer

During the last 50 years, feeding conditions for lynxes in Białowieża Forest have changed in many aspects, e.g. a manifold increase in ungulate numbers and the decline of hare and tetraonid birds. In 1947–1951, when Gavrin (unpubl., cited after Geptner and Sludskii 1972) analysed stomachs and scats of lynxes, this predator predominantly hunted brown hare, then roe deer and other ungulates, and tetraonids (Table 4.3). Proportions of the three groups of prey were as follows: per one killed ungulate, lynxes successfully hunted 1.8 hare and 0.2 grouse (hazel hen and capercaillie). In 1986–1995, when we analysed scats and prey remains of lynxes in the western part of BPF, ungulates were the most common prey of lynxes (Table 4.3). Per one ungulate killed, lynxes hunted only 0.1 hare and 0.02 grouse. Besides ungulates and hare, Białowieża's lynxes fed on a variety of other animals (squirrels, birds, raccoon dogs, pine martens, small rodents), which, however, always formed a minor addition in terms of food biomass. No big differences in the lynx diet between the warm and the cold season were found (Table 4.3).

A similar picture emerged from the analysis of 172 prey remains found on trails of radio-tracked and snow tracked lynxes in 1987–1996 (Table 4.4). In both spring–summer and autumn–winter seasons, 85% of kills were ungulates, predominantly roe deer (62%) and red deer (22%). While roe deer was the most common prey of all sex/age classes of lynx, the shares of very large and small-sized prey varied. Adult male lynxes hunted significantly more red deer and fewer small prey than subadults and adult female lynxes with kittens. Brown hare was captured more often by subadults and single adult females (19% of their prey) than by other lynxes. Finally, family groups (females with kittens) killed many other small prey (birds and mammals). Tracks in snow indicated that some of those small prey were captured by the kittens.

The lynx's strong reliance on ungulates as main prey is typical of the whole zone of temperate forests. By reviewing studies on lynx diet from 15 localities in the Palaearctic region, we have shown that the relative contributions of ungulates and hare in lynx diet are determined by latitude (Jędrzejewski et al. 1993b). The share of hare is highest in the north and declines towards the south. The opposite trend is observed in ungulate occurrence in lynx diet; south of 52°–54° N ungulates dominate lynx diets.

Coexisting with five species of wild ungulates and a few domestic ones that have occurred in Białowieża Forest, the lynx hunted for three species only: roe deer, red deer, and wild boar. Unlike wolves, lynxes were never reported to raid domestic ungulates, although sporadically they would capture dogs, cats, and poultry around settlements located in the forest. In the early 19th century, when red deer were temporarily eradicated, lynxes relied mainly on roe deer, hare, and moose calves (Jarocki 1830). All studies done in Białowieża showed that roe deer, the smallest ungulate, was strongly preferred by lynxes and still more so by females and subadults than by adult male lynxes (Figs. 4.2 and 4.3). Wherever roe deer are abundant enough, red deer are hunted by lynxes less often than would have been expected from its share in the community. It was

Table 4.3. Diet composition of lynx in Białowieża Primeval Forest in spring-summer (16 April-30 September) and autumn-winter (1 October-15 April)

Item	Eastern part (1947-1951)		Western part (1986-1995)			
	Spring-summer	Autumn-winter	Spring-summer		Autumn-winter	
	% Occ	% Occ	% Occ	% Bio	% Occ	% Bio
Undetermined shrew (Soricidae)	-	1.3	-	-	0.8	+
Brown hare (*Lepus europaeus*)	60.0	38.9	16.7	1.0	11.0	6.0
Red squirrel (*Sciurus vulgaris*)	1.2	3.2	-	-	0.8	+
Bank vole (*Clethrionomys glareolus*)	5.0	0.8	-	-	0.8	+
Pine vole (*Pitymys subterraneus*)	-	-	-	-	0.8	0.1
Undetermined vole (*Microtus* sp.)	12.5	22.2	8.3	0.1	0.8	+
Undet. mouse (*Apodemus* sp.)	1.3	1.3	-	-	-	-
Undetermined small rodent	10.0	8.7	-	-	-	-
Domestic dog (*Canis familiaris*)	-	-	8.3	3.4	0.8	+
Raccoon dog (*Nyctereutes procyonoides*)	-	-	-	-	1.6	0.1
Pine marten (*Martes martes*)	-	-	-	-	0.8	2.3
Wild boar (*Sus scrofa*)	3.7	4.0	-	-	3.9	1.0
Red deer (*Cervus elaphus*)	-	1.6	-	-	5.5	10.9
Roe deer (*Capreolus capreolus*)	12.5	28.6	8.3	16.4	8.7	9.6
Undetermined deer (Cervidae)	-	-	75.0	77.4	74.0	69.4
Ungulates total	**16.2**	**34.2**	**83.3**	**93.8**	**91.3**	**90.9**
Hazel hen (*Bonasa bonasia*)	6.3	3.2	-	-	1.6	0.1
Capercaillie (*Tetrao urogallus*)	1.3	-	-	-	-	-
Domestic hen (*Gallus gallus*)	-	-	-	-	0.8	0.3
Undetermined owl (Strigidae)	-	-	-	-	0.8	+
Undetermined thrush (*Turdus* sp.)	-	-	8.3	1.3	-	-
Jay (*Garrulus glandarius*)	-	1.6	-	-	-	-
Undetermined bird	1.3	2.4	-	-	0.8	+
Bird egg	1.3	-	-	-	-	-
Birds total	**10.0**	**7.1**	**8.3**	**1.3**	**3.9**	**0.5**
Amphibian	-	-	-	-	0.8	+
Insect	-	-	-	-	0.8	+
Plant material	2.5	-	16.7	0.4	12.6	0.1
Total numbers of scats analysed	80	126	12		127	
Biomass consumed per 1 scat (g)	?	?	-	1591	-	529

Eastern part, Belarussian part (after Gavrin, unpubl., cited after Geptner and Sludskii 1972). Western part, Polish part, data combined from Jędrzejewski et al. (1993b) and Okarma et al. (1997). %Occ, percentage occurrence in scats. %Bio, percentage of biomass consumed.

only in the National Park, where roe deer were very scarce that lynxes relied predominantly on red deer. Wild boars were avoided by lynxes, and found as prey of adult male lynxes only (Table 4.4).

The functional (dietary) response of lynxes was calculated based on spatial variation in deer abundance in the Polish part of BPF (Fig. 4.4). The share of roe deer among lynx prey increased very fast with increasing density of this ungulate, and its functional response was of type II (asymptotic curve, according to Holling 1959). In contrast, the proportion of red deer among lynx prey grew in a linear manner (type I response). The relative proportions of the two ungulates in the total density of deer were more important for lynx selection of prey than their densities (Fig. 4.4). The share of roe deer among lynx prey grew very fast with an increasing proportion of this species among all deer. In contrast, the share of red deer in lynx prey increased only when this species heavily outnumbered roe deer.

Lynxes do not particularly select any age or sex class from the roe deer population (Table 4.5). Of the much larger red deer, lynxes clearly chose fawns and almost entirely avoided adult stags. Also, the physical condition

Table 4.4. Species composition of lynx prey (in numbers and percentages) found during radio-tracking and snow tracking in 1987/1988–1996 (Okarma et al. 1997)

Prey species	Prey of lynx of a given sex/age class				Total prey found		
	Adult M	Adult F with kittens	Subadults and single adult F	Unknown single lynx	Sp–Su	Au–Wi	Sum *N* (%)
Wild boar	2 (3%)	–	–	–	–	2	2 (1%)
Red deer	22 (36%)	10 (18%)	4 (8%)	2	6	32	38 (22%)
Roe deer	36 (59%)	34 (61%)	34 (69%)	2	34	72	106 (62%)
Ungulates total	**60 (98%)**	**44 (79%)**	**38 (77%)**	**4**	**40**	**106**	**146 (85%)**
Brown hare	1 (2%)	5 (9%)	9 (19%)	–	5	10	15 (9%)
Domestic dog	–	–	1 (2%)	–	–	1	1 (0.5%)
Pine marten	–	2 (3%)	–	–	1	1	2 (1%)
Red squirrel	–	–	–	1	–	1	1 (0.5%)
Hazel hen	–	1 (2%)	1 (2%)	1	1	2	3 (2%)
Long-eared owl	–	1 (2%)	–	–	–	1	1 (0.5%)
Thrush (*Turdus* sp.)	–	2 (3%)	–	–	–	2	2 (1%)
Great spotted woodpecker	–	1 (2%)	–	–	–	1	1 (0.5%)
Small prey total	**1 (2%)**	**12 (21%)**	**11 (23%)**	**2**	**7**	**19**	**26 (15%)**
Total *N* prey	61	56	49	6	47	125	172
Percentage	100	100	100				100

Sp–Su, spring–summer. Au–Wi, autumn–winter.

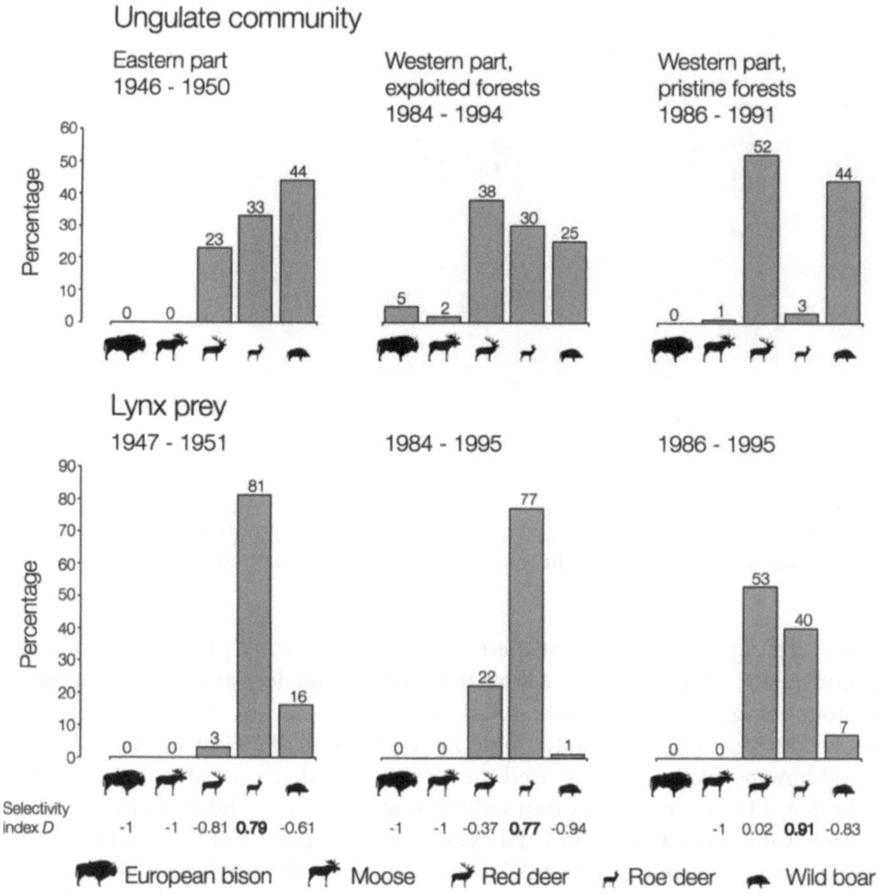

Fig. 4.2. Choice of ungulate prey by lynx in relation to varying species structure of the ungulate community in Białowieża Primeval Forest. Selectivity index *D* after Jacobs (1974): $D=(r-p)/(r+p-2rp)$, where r is a fraction of a given species in a total number of ungulates killed by lynx, and p is the fraction of a species in an ungulate community. *D* ranges from −1 (complete avoidance) to 0 (killing proportional to occurrence) to 1 (maximum positive selection). Data from: Gavrin (unpubl. data, cited in Geptner and Sludskii 1972), Jędrzejewski et al. (1993 b), Okarma et al. (1995) and authors' unpubl. data

(approximated by percentage content of marrow fat) of roe deer and red deer killed by lynxes in autumn–winter differred markedly. Femur bones of 54 roe deer had, on average, 50% of marrow fat content. In 21 red deer killed by lynxes, this index averaged 34%. Although no random samples from deer populations were available, it is probable that lynxes selected weak individuals from the population of large prey, the red deer (Okarma et al. 1997). Mean body mass of an ungulate killed by lynxes was 28.6 kg.

Except for females rearing kittens and male-female short contacts, lynxes live solitary lives. Solitary hunts are the rule even in adult females unless the

Fig. 4.3. Adult male roe deer killed by the lynx 'Trofim' (a radio-collared adult male); Hajnówka State Forestry District, BPF, late spring 1995. Typically, lynxes drag prey to concealment and cover it with grass, litter, or snow. Photo: W. Jędrzejewski

young are big enough to assist their mother in capturing prey. The success of hunting critically depends on a lynx approaching the deer as closely as possible without being seen by its target, and then, a short chase. In 1986–1989, we snow tracked lynxes in Białowieża National Park for a total of 26 km and have experienced how perfectly the lynx used the structure of dense, pristine woodland to stalk deer (Table 4.6). They often walked along and under fallen trees and close to their root plates to remain partly concealed. Frequent walking on fallen logs and climbing the root plates of fallen trees widened the lynx's field of vision for spotting prey. When not hunting, lynxes moved on forest paths and frozen streams that made walking in snow less strenuous. Daily hunting efforts (estimated from snow tracking) included one failed attack on deer, two unsuccessful attacks on medium-sized prey, seven attacks on small prey (one successful), and two feeds on a previously killed deer (Table 4.6).

Lynxes' foraging behaviour (as studied by radio-tracking in 1991–1996; Okarma et al. 1997) consists of the repeated series of killing a prey, feeding on a carcass and staying near it for some days, leaving it to search and hunt for another one, killing, and so on. Time spent both searching and consuming were noticeably different between age/sex groups of lynx. Females leading kittens consumed deer during a rather short time and in 19% of all successful hunts on ungulates, family groups utilised two prey at a time, both killed by a female lynx in very short temporal and spatial distance (Figs. 4.5 and 4.6). Obviously, killing two or more prey at a time (earlier reported by Kossak 1989) is not uncommon in female lynxes leading kittens. Moreover, females (and occasionally kittens, too) killed small animals (hare, hazel hens) while still feeding on an ungulate carcass (Fig. 4.6). The more kittens in a litter, the shorter was the consumption time (i.e. time spent at a carcass)

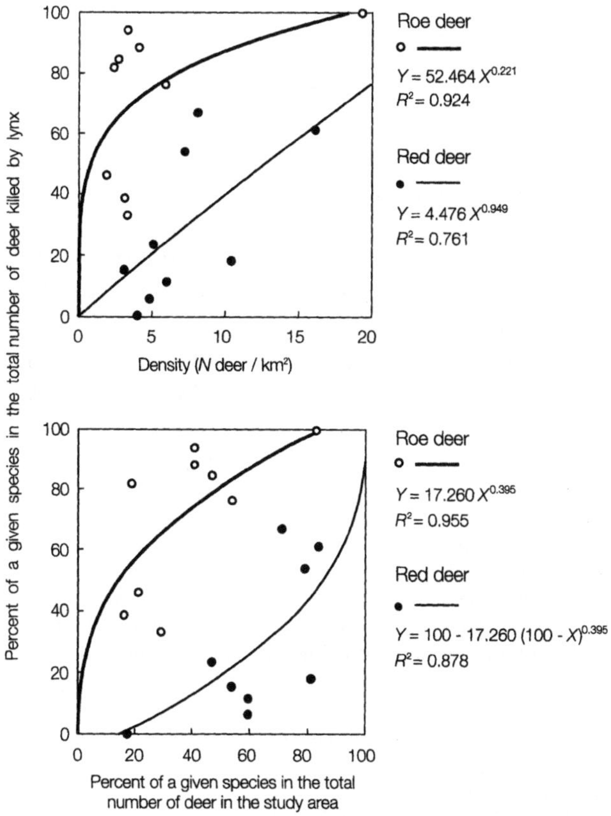

Fig. 4.4. Functional (dietary) responses by lynx to spatially varying densities of roe deer and red deer *upper panel* and to varying percentages of each species in the total number of deer *lower panel* in the Polish part of BPF. Each point denotes one area (game hunting district or part of the National Park) covering from 20 to 75 km^2. Okarma et al. (1997)

and the searching time of family groups (Fig. 4.7, Table 4.7). A female with three young utilised an ungulate during 38 hours and killed a new one after 10 hours, on average.

Consecutive kills by females with kittens, for instance by female Sonia shown in Fig. 4.5, were usually more concentrated in space than those of a solitary lynx. In spring and summer, when kittens are small and a female has to bring them along to each new kill, consecutively killed ungulates were on average 1.4 km apart (straight-line distance; Okarma et al. 1997). In autumn-winter, when kittens are more mobile and often assist their mothers in hunting, that distance increased to 3.8 km, on average. Female lynxes appeared to be impressively effective killers, an ability no doubt achieved by the food demands of their growing young.

Adult male lynxes, an example being the large male Trofim shown in Fig. 4.5, moved over extensive areas when searching for prey. Quite commonly, a male would leave one prey before finishing consuming it, and roam far away to kill another prey in a distant place. He would feed on it for some time, and then

Table 4.5. Choice of sex and age classes of roe and red deer killed by lynxes

Living population or lynx prey	Age/sex class of ungulates			Source
	Adult M (%)	Adult F (%)	Juv (%)	
		Roe deer (*Capreolus capreolus*)		
Population	25	46	29	Okarma et al. (1995)
Lynx prey (N=129)	17	58	25	Okarma et al. (1995)
Lynx prey (N=92)	23	41	36	Okarma et al. (1997)
		Red deer (*Cervus elaphus*)		
Population	27	46	27	Okarma et al. (1995)
Lynx prey (N=43)	–	33	67	Okarma et al. (1995)
Lynx prey (N=36)	–	39	61	Okarma et al. (1997)
Lynx prey (N=29)	10	21	69	Shostak and Bunevich (1986)

Sources: Okarma et al. (1995) – western part of BPF, exploited forests, prey remains found by game wardens in 1984–1994. Okarma et al. (1997) – western part (exploited and pristine forests), data collected during snow tracking and radio-tracking of lynxes in 1986–1991. Shostak and Bunevich (1986) – eastern part of BPF, red deer killed by lynxes found by game wardens before 1985 (no data on the structure of living population to compare).

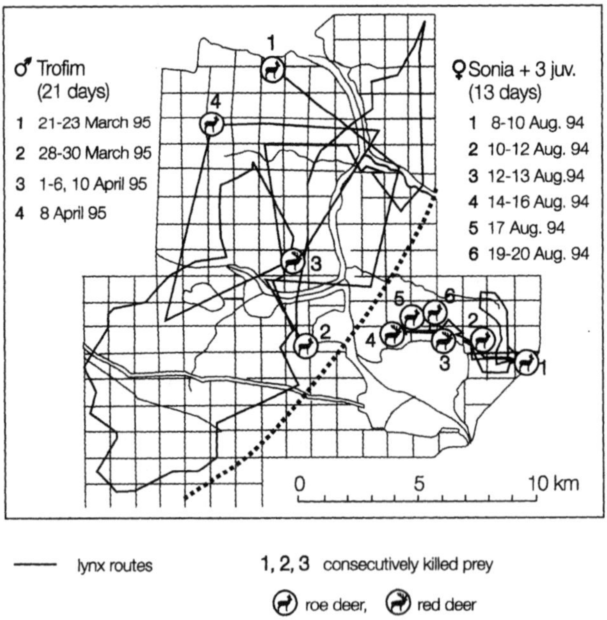

Fig. 4.5. Schematic routes of movements and consecutive ungulate prey of two lynxes, an adult male and an adult female leading three kittens, as determined by radio-tracking and search for prey remains after the lynx had left it. Source: W. Jędrzejewski, K. Schmidt, and H. Okarma, unpubl. data

return to the previous kill to finish it (Fig. 4.6). Ungulates killed in sequence were, on average, only 1.8 km apart in spring-summer, but as far as 7.5 km in winter, when male lynxes spend much time walking in search of females and markedly expand their ranges. The mean consumption time of deer by adult male lynxes was 75 h and searching time 38 h (Fig. 4.7, Table 4.7).

Table 4.6. Lynx hunting behaviour and activities associated with modes of searching for prey as revealed by 25.9 km of snow tracking of individual lynxes in BNP in 1986–1989. Figures in parentheses indicate unsuccessful attacks. All activities were recalculated per 10 km of trail, which approximates the daily movement distance of lynx (Goszczyński 1986a). Source: Jędrzejewski et al. (1993b)

Activity	Number (*n*/10 km of trail)	Total length (m/10 km)	Total *n* of prey
Modes of movement and area searching			
Walking along a fallen tree	40	126	
Crouching under a fallen tree	68	88	
Closely approaching the root plate of uprooted tree	75		
Walking on a fallen log	185	274	
Climbing the root plate of an uprooted tree	3		
Following: forest path		1681	
frozen stream		58	
Searching for prey			
Following: ungulate trail	42.0	710	
hare trail	1.0	44	
Inspecting: wild boar lair	1.2		
red fox den	0.4		
raccoon dog den	1.6		
Attacks and feeding			
Attacks on: red deer	0.4		1 (1)[a]
roe deer	0.4		1 (1)[b]
raccoon dog	0.8		2 (2)[c]
red squirrel	0.8		2 (2)[d]
woodpecker	0.4		1
small rodent	6.0		15 (?)[e]
Feeding on red deer killed earlier	2.0		2[f]
Drinking water in bog alderwood	1.2		

[a] Lynx stalked a group of deer by staying hidden behind fallen trees and attacked by jumping over fallen logs.
[b] Lynx attacked a roe der lying (sleeping?) on the ground and chased it 30 m before giving up.
[c] Both raccoon dogs escaped into a cavity at the bottom of tree.
[d] Both attacks at trees, squirrels escaped up the tree.
[e] Attacks on rodents by scratching through snow.
[f] In one case, two lynxes fed on a female red deer on six consecutive days; in another case two lynxes returned to the female red deer.

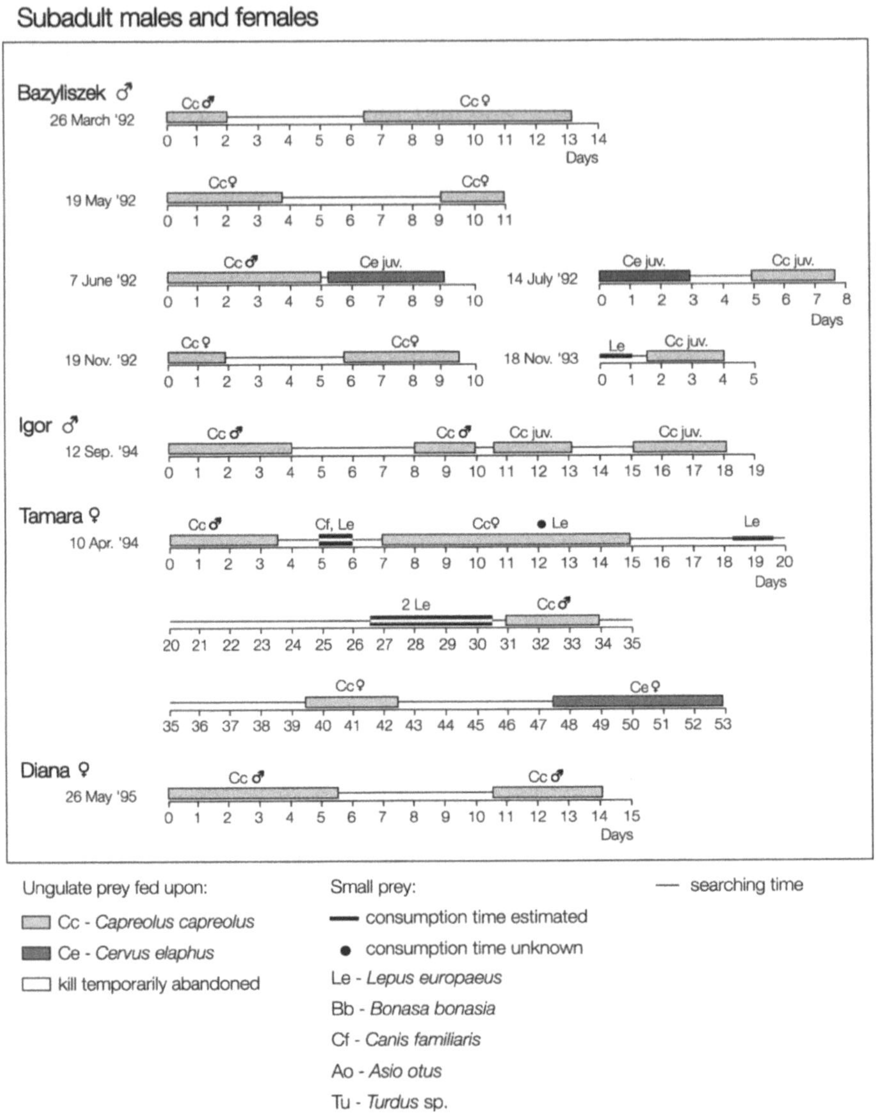

Fig. 4.6. Sequences of killing and utilisation of prey by three sex/age groups of radio-collared lynxes. Lengths of bars denote the time of lynx utilisation of a given prey. *Parallel bars* are two kills located close to each other and utilised by a lynx concurrently. Lengths of lines mark the time of searching and hunting for ungulates. Okarma et al. (1997)

Subadult lynxes seemed the least effective hunters. They stayed a long time at each kill (about 100 h, on average, Figs. 4.6 and 4.7) and consumed the kill as completely as possible. They never managed to have two kills at a time. Their long search times for ungulates (mean 105 h, Table 4.7) were often interspersed with catches of hares (Figs. 4.6 and 4.7). Mean distance be-

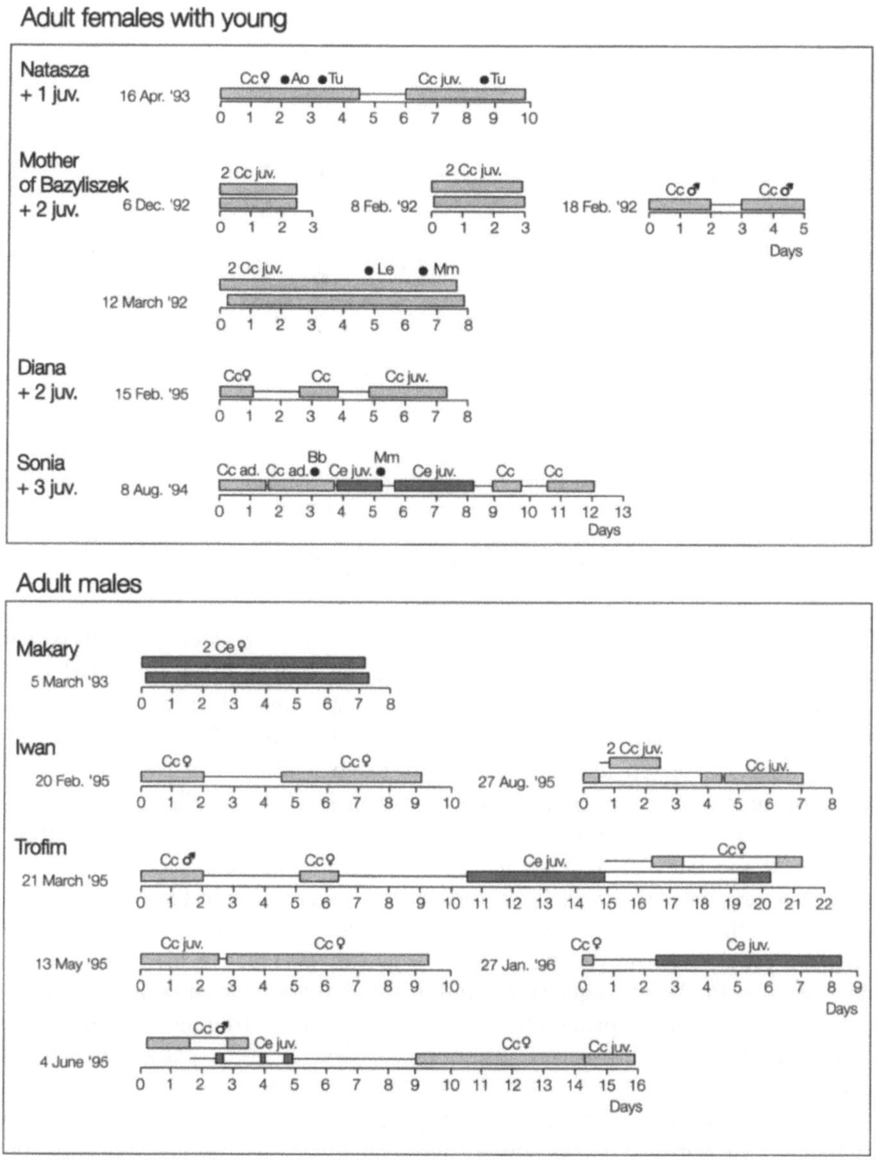

tween consecutive ungulate prey killed by subadult lynxes was fairly long (4.9 km) in spring–summer, the first season of their independent lives, and it declined (to 3.8 km) in autumn–winter, presumably with increasing hunting experience (Okarma et al. 1997).

Generally, in all lynxes, consumption time of prey grew logarithmically with prey size, from an average of 1.3 days in the case of brown hare to 3.8 days for a female red deer (Fig. 4.8). In the majority of seized roe deer

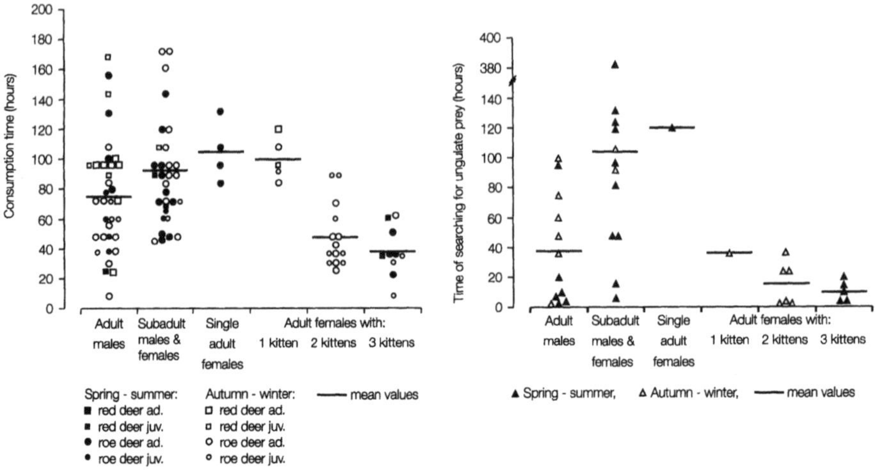

Fig. 4.7. *Left panel* consumption times of ungulate prey by lynx of various age, sex, and reproductive status. Each point denotes the time a lynx stayed at and fed upon one prey. *Right panel* lynx searching times for ungulates. Each point = the time from leaving the remains of one deer to killing another deer. Okarma et al. (1997)

(80%), the lynx ate nearly all edible tissue, 19% of carcasses were classified as half-eaten, and only 1% was abandoned by lynxes after they had consumed a small portion of meat only. In red deer, a much larger prey, the proportion of completely consumed kills was notably lower (36%); 59% of carcasses were half eaten and 5% were only eaten a little. The two major reasons for incomplete use of prey were scavengers (see details in Sect. 5.10) and decay of meat in summer. How do lynxes cope with these adversities? We traced the fates of 31 kills in detail (Jędrzejewski et al. 1993b). Having seized an ungulate, the lynx finds the nearest place suitable for concealing it, and drags the carcass there; 80% of roe deer but only 30% of red deer (heavier prey) were moved from 2 to 75 m (on average 14 m) from the kill site, usually to dense vegetation, under a fallen log, under a low, ground-touching branches of spruce etc. In addition, the lynx frequently covered the carcass with grass, dry leaves, litter, moss, deer hair, or snow (in 68% of cases).

All these efforts were, however, only partly effective. In the pristine forests of Białowieża National Park, 80% of lynx kills were fed upon by one to five species of scavengers (two on average), the most common being the wild boar recorded at 40% of prey. Wild boar was the only species that could appropriate a fresh kill from a lynx and consume it completely. In summer, after 3 days, killed deer carcass is repulsively fetid, entirely covered with eggs and larvae of flies, and a lynx may abandon it.

Based on proportions of roe and red deer among lynx prey, and the average consumption and search times, the killing rates of three groups of lynxes (adult males, subadults, and females with litters of one to three kittens) could have been calculated (Table 4.7). Adult male lynxes killed, on average, 6.3 deer/

Table 4.7. Mean killing rates by lynxes of various sex/age classes of roe deer and red deer in BPF in 1991–1996. Five variants of family size (female with kittens) cover the whole range of observed individual variation (females with one, two, or three juveniles) and year-to-year variation in the mean breeding success of the lynx population during the study period (used for calculation of lynx predation impact on deer; Chap. 5). (Okarma et al. 1997)

Parameter	Adult M lynx	Subad lynx, single ad F	Adult females with the following number of kittens:					Mean for all
			1.0	1.4	1.8	2.0	3.0	
Roe: red deer ratio	1.64	8.50	3.40	3.40	3.40	3.40	3.40	2.80
Searching time (h)	37.7	105.4	43.1	30.8	22.0	18.6	8.1	54.0
Consumption time of roe deer (h)	66.5	93.3	77.9	66.5	56.8	52.5	35.4	72.2
N roe deer killed/month	3.9	3.2	4.4	5.5	6.8	7.5	12.2	4.0
Consumption time of red deer (h)	91.9	100.0	97.8	83.5	71.3	65.9	44.5	90.7
N red deer killed/month	2.4	0.4	1.3	1.6	2.0	2.2	3.6	1.5
Total *N* deer killed/month	6.3	3.6	5.7	7.1	8.8	9.7	15.8	5.5
Mean kill rate (*N* days/deer)	4.7	8.3	5.2	4.2	3.4	3.1	1.9	5.4

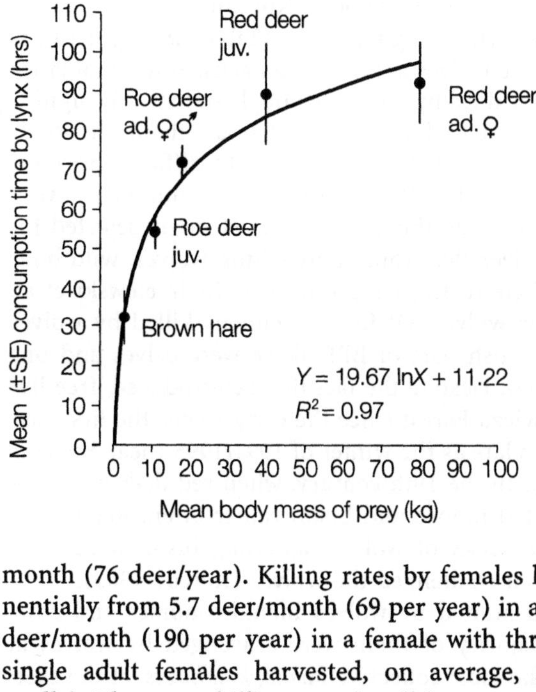

Fig. 4.8. Mean (±SE) consumption times by lynx in relation to the mean body mass of prey. Okarma et al. (1997)

month (76 deer/year). Killing rates by females leading kittens increased exponentially from 5.7 deer/month (69 per year) in a female with one young to 15.8 deer/month (190 per year) in a female with three young. Subadult lynxes and single adult females harvested, on average, 3.6 deer/month (43 deer annually). The mean killing rate for all lynxes was one deer every 5.4 days.

In 1985, when we began our study on predators in Białowieża, we were told a story that happened during a severe winter a few years earlier. In the dark early morning, several people waiting for a train in Białowieża Towarowa station saw a panicked red deer running from the forest straight to the station as if seeking rescue by being close to people, as witnesses reported. When a few men walked back along the deer tracks in snow, they learnt that several dozen metres before the station, a lynx had jumped off the deer's back and retreated to the forest! It must have attacked the deer in the forest and 'ridden' it for nearly 100 m before giving up. Having heard the story, we were sceptical about its authenticity. After 10 years of studying lynxes' amazing adaptations to hunt ungulates, we give it a full credit.

4.4 Wolves - Running After Red Deer

All species of wild and domestic ungulates of Białowieża Forest are potential and actual prey to wolves (Table 4.8), but no species is preferred as strongly as the red deer (Fig. 4.9). Depending on red deer abundance relative to other ungulates, it made up 11 to 83% of all ungulates killed by wolves (Fig. 4.10). Calves are a special target of wolves' hunts, taken more often than expected from their share in the deer population (Table 4.9). Does are killed roughly in proportion to their abundance but stags are notably avoided by wolves. Gavrin and Donaurov (1954) and Bunevich (1988) noted that stags were successfully hunted by wolves after the rut, in October–November.

Roe deer, the second most important ungulate prey (Table 4.8, Fig. 4.10), is small enough to fall as prey to lone wolves as well as to packs. Some avoidance of bucks and positive selection of does by wolves (though statistically significant in neither case) was reported (Table 4.9). Wild boar, thanks to their tusks, active defence, and aggregative behaviour are truly difficult prey for wolves, hunted in fair numbers only when their share in the community is very high or when red deer are scarce. Even then, piglets are strongly selected for (Table 4.9). Whilst red and roe deer flee from approaching wolves, wild boar may stand and form a circle, their tusks facing outwards (Jędrzejewski et al. 1992 a). Moose are rare prey for wolves. Of four specimens killed by wolves in the exploited forests of the Polish part of BPF, three were calves and one was a female (Okarma et al. 1995). Despite the fact that reintroduced, free living bison have roamed in Białowieża Forest since the early 1950s, the first case of bison killed by wolves came as late as the winter of 1994/1995 (near Hwoźna River in the Polish part of BPF). In the 19th century, when red deer was temporarily eradicated, wolves hunted moose and bison rather often. Reports by game wardens emphasised that packs of wolves harassing bison were large and the bison often stood or even challenged the wolves (Karcov 1903).

The list of domestic animals hunted by wolves includes horses and cows (although the overwhelming majority of them were scavenged by wolves at the carrion discarding sites of the local collective farms, *kolkhozes*, at the eastern outskirts of the forest), goats, sheep, swine, geese and dogs (Table 4.8).

Table 4.8. Diet composition of wolves in Białowieża Primeval Forest in spring-summer (Sp-Su is April-August in the Belarussian studies, and May-September in the Polish study) and autumn-winter (Au-Wi)

Item	Eastern part				Western part			
	1946-1950		1979-1985		1986-1996			
	Sp-Su	Au-Wi	Sp-Su	Au-Wi	Sp-Su		Au-Wi	
	%Occ	%Occ	%Occ	%Occ	%Occ	%Bio	%Occ	%Bio
Eastern hedgehog (*Erinaceus concolor*)	-	-	-	-	-	-	0.3	0.2
Mole (*Talpa europaea*)	-	0.1	-	-	-	-	-	-
Common shrew (*Sorex araneus*)	-	-	-	-	-	-	0.9	+
Undet. shrew (Soricidae)	-	0.1	-	-	-	-	-	-
Insectivores total	**-**	**0.2**	**-**	**-**	**-**	**-**	**1.2**	**0.2**
Brown hare (*Lepus europaeus*)	18.5	16.3	2.7	-	3.0	0.9	1.2	0.1
Beaver (*Castor fiber*)	-	-	-	-	3.0	2.3	-	-
Red squirrel (*Sciurus vulgaris*)	1.0	0.2	-	-	-	-	0.3	+
Bank vole (*Clethrionomys glareolus*)	-	-	-	-	1.5	+	1.5	+
Undet. vole (*Microtus* sp.)	-	-	-	-	-	-	1.7	0.1
Undet. mouse (*Apodemus* sp.)	-	-	-	-	1.5	+	-	-
Undetermined small rodent	3.5	4.4	-	-	1.5	+	0.6	+
Small rodents total	**3.5**	**4.4**	**-**	**-**	**3.0**	**+**	**3.2**	**0.1**
Lynx (*Lynx lynx*)	-	0.2	-	-	-	-	-	-
Red fox (*Vulpes vulpes*)	-	0.1	2.7	-	-	-	-	-
Raccoon dog (*Nyctereutes procyonoides*)	-	-	2.7	-	-	-	1.2	0.2
Badger (*Meles meles*)	-	0.1	2.7	-	-	-	-	-
Pine marten (*Martes martes*)	-	0.1	-	-	-	-	-	-
Wild boar (*Sus scrofa*)	24.4	19.6	5.4	11.6	43.3	17.4	35.2	12.7
Moose (*Alces alces*)	-	0.1	-	2.3	-	-	-	-
Red deer (*Cervus elaphus*)	4.9	6.7	43.2	58.1	20.9	24.5	9.3	10.7
Roe deer (*Capreolus capreolus*)	12.8	17.1	18.9	22.5	4.5	1.9	3.2	3.9
Undetermined deer (Cervidae)	-	-	-	-	68.7	52.8	80.2	69.0

Table 4.8 (continued)

Item	Eastern part				Western part			
	1946–1950		1979–1985		1986–1996			
	Sp–Su	Au–Wi	Sp–Su	Au–Wi	Sp–Su		Au–Wi	
	%Occ	%Occ	%Occ	%Occ	%Occ	%Bio	%Occ	%Bio
European bison (*Bison bonasus*)	–	–	–	–	–	–	0.6	1.8
Mouflon (*Ovis ammon*)[a]	–	–	2.7	–	–	–	–	–
Wild ungulates total	**42.2**	**43.6**	**70.3**	**94.6**	**100.0**	**96.6**	**98.6**	**98.1**
Cattle carcass	5.9	6.4	10.8	1.6	–	–	0.6	0.7
Domestic horse carcass	9.6	12.2	–	–	–	–	0.6	0.5
Domestic swine	2.2	3.2	2.7	2.3	–	–	–	–
Domestic sheep	12.1	6.3	–	–	–	–	–	–
Domestic goat	0.2	0.3	–	–	–	–	–	–
Domestic dog	3.2	4.9	5.4	1.6	–	–	–	–
Domestic goose	0.7	0.7	–	–	–	–	–	–
Domestic animals total	**34.1**	**33.5**	**18.9**	**5.4**	**–**	**–**	**1.2**	**1.2**
Great spotted woodpecker (*Dendrocopos major*)	–	–	–	–	–	–	0.3	+
Nuthatch (*Sitta europaea*)	–	–	–	–	–	–	0.3	+
Hazel hen (*Bonasa bonasia*)	–	0.1	–	–	–	–	–	–
Tawny owl (*Strix aluco*)	–	–	–	–	–	–	0.6	+
Undetermined large bird	–	–	–	–	1.5	0.1	–	–
Undet. medium-sized bird	–	–	–	–	–	–	0.9	+
Undetermined small bird	0.7	0.7	–	–	–	–	–	–
Bird egg	–	–	–	–	1.5	+	–	–
Birds total	**0.7**	**0.7**	**–**	**–**	**3.0**	**0.1**	**1.7**	**+**
Undetermined reptile	–	–	–	–	1.5	+	0.3	+
Undetermined anuran	–	–	–	–	4.5	+	0.3	+
Plant material	2.0	0.8	5.4	–	28.4	+	32.6	0.1
Total *N* scats or biomass (kg)	1063 Scats and 48 stomachs		32 Scats, 50 stomachs, 82 prey remains		67	98.2	344	518.2

Data for the Belarussian (eastern) part presented as percentage of occurrence in scats (%Occ), recalculated from Gavrin and Donaurov (1954) and Bunevich (1988). Data for the Polish (western) part presented as percentage occurrence in scats and percentage of consumed biomass (%Bio), from Jędrzejewski et al. (in press).

[a] Wolves killed some mouflons that were kept in a small enclosure.

Fig. 4.9. Remnants of an old male red deer killed and consumed by a pack of wolves; reed marsh in the juncture of the Orłówka and the Narewka rivers, Białowieża National Park, early spring 1993. Photo: M. Brzeziński

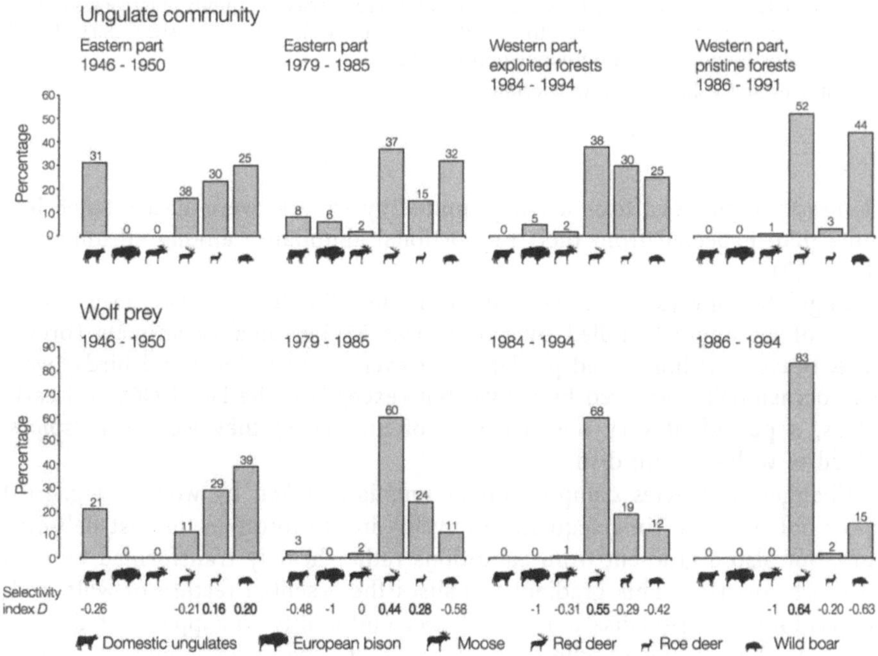

Fig. 4.10. Choice of ungulate prey by wolves in relation to varying species structure of ungulate community in Białowieża Primeval Forest. Selectivity index *D* as in Fig. 4.2. Data from: Gavrin and Donaurov (1954), Bunevich (1988), Jędrzejewska et al. (1994, 1997), and Okarma et al. (1995)

Table 4.9. Choice of sex and age classes of ungulates by wolves

Living population or wolf prey	Age/sex class of ungulates			Source
	Adult M (%)	Adult F (%)	Juv (%)	
	Red deer (*Cervus elaphus*)			
Population	27	46	27	Okarma et al. (1995)
Wolf prey (N=335)	14	49	37	Okarma et al. (1995)
Wolf prey (N=33)	39[a]		61	Jędrzejewski et al. (1992a)
Wolf prey (N=50)	14	31	55	Bunevich (1988)
	Roe deer (*Capreolus capreolus*)			
Population	25	46	29	Okarma et al. (1995)
Wolf prey (N=106)	15	64	21	Okarma et al. (1995)
	Wild boar (*Sus scrofa*)			
Wolf prey (N=34)	32[a]		68	Jędrzejewski et al. (in press)

Sources: Okarma et al. (1995) – western part of Białowieża Forest (exploited forests); population structure censused in late winter, wolf prey are carcasses found; Jędrzejewski et al. (1992a) – prey remains recovered from wolf scats; Bunevich (1988) – eastern part of BPF, prey remains found, no data on population structure.

[a] Adult deer and wild boar of undetermined sex.

Domestic ungulates (those actively hunted by wolves) were always taken less often than expected from their proportional abundance among all ungulates (Fig. 4.10).

Ungulates always dominated the wolf diet (Table 4.8). The mean body mass of an ungulate killed by wolves was 55 kg. Smaller animals (brown hare, beaver, medium-sized predators, or even small rodents and birds) were also occasionally captured by wolves but (except for the late 1940s and early 1950s, a period of very low densities of ungulates) they were just 'spices' added to wolves' main dish.

Changes in species composition of ungulates taken by wolves (Fig. 4.10) show not only the time sequence of events in the forest in the last half-century, but also a gradient from conditions rather heavily transformed by man to nearly pristine. This gradient visualises the essential feature of wolf feeding ecology: its specialisation on red deer and ability to supplement the diet with all other species of wild (and domestic, if necessary) ungulates living in the temperate forests. The main factor moulding wolf selection of ungulate species was the density of red deer. The results of three studies covering a span of 45 years and an over tenfold increase in red deer densities (Gavrin and Donaurov 1954; Bunevich 1988; Jędrzejewski et al., in press) allowed us

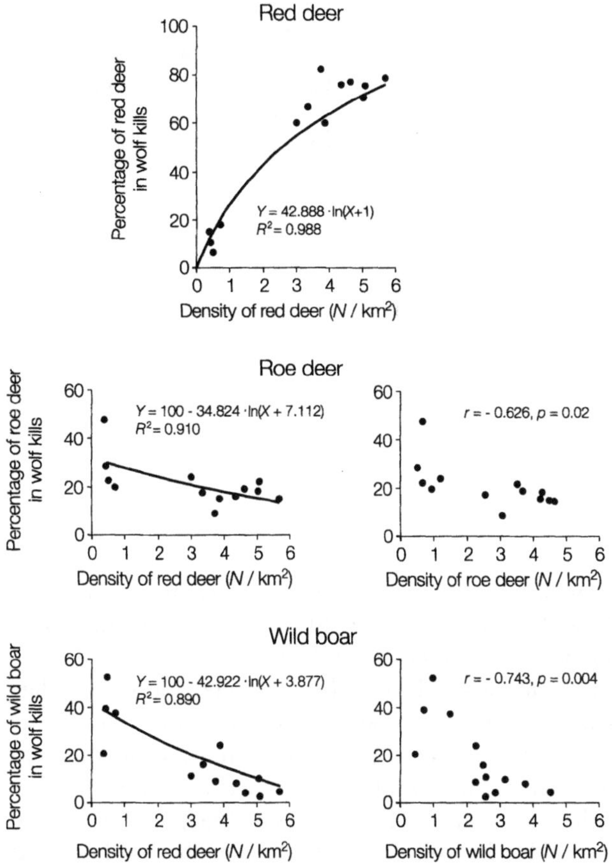

Fig. 4.11. Functional (dietary) responses of wolves to changes in densities of red deer, roe deer, and wild boar. Each point denotes 1 year (between 1946 and 1993). Jędrzejewski et al. (in press)

to analyse wolf prey selection in relation to ungulate densities. Wolves showed a strong functional response (of the logarithmic type) to an increase in red deer densities from an average 0.5 ind/km^2 in 1947–1950 to over 5 ind/km^2 in the early 1990s (Fig. 4.11). Furthermore, with a growing density of red deer, wolves hunted fewer roe deer and wild boar, despite the fact that densities of these two ungulate species increased as well. Thus, it was the density of red deer that determined the proportions of other (alternative) prey among wolf kills. The decrease in the wolf's hunting rate for alternative prey with a growing density of red deer was more rapid in the case of wild boar (sixfold decline) than roe deer (twofold decline) (Fig. 4.11). The roe deer is a more stable component of wolf diet than the wild boar.

The regressions shown in Fig. 4.11 gave an empirical basis for a generalised model of changes in wolf prey composition in response to varying densities of red deer (Fig. 4.12). When having little access to their favourite prey, wolves hunted for wild boar, then roe deer and other (mainly domesticated)

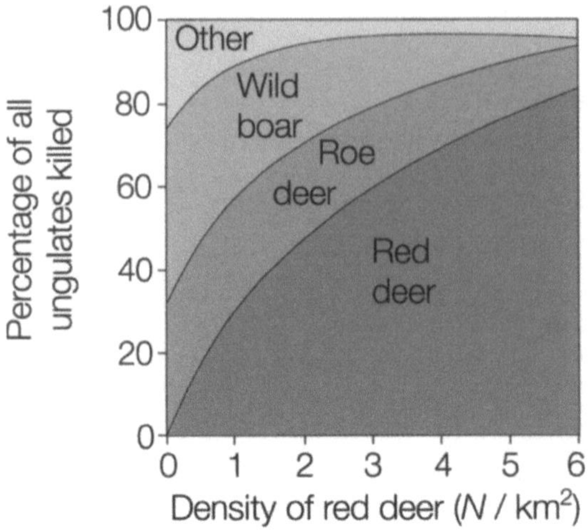

Fig. 4.12. General model of functional responses of wolves to variation in red deer densities in Białowieża Primeval Forest based on regressions from Fig. 4.11. Jędrzejewski et al. (in press)

ungulates. With an increasing density, red deer immediately became the wolves' main prey, supplemented by roe deer, wild boar and other ungulates (in 1985–1991, these other prey being moose and bison).

There are also some extrinsic factors involved in the selection of ungulate species by wolves. In spring–summer, the share of wild boar in the wolf diet increased compared with autumn–winter (Table 4.8); conducive to this was the presence of a large number of susceptible piglets in a wild boar population. In severe winters with deep snow (1946/1947, 1949/1950, 1985/1986 and 1986/1987), the share of Cervidae in wolf prey compared with wild boar was notably higher than in mild winters (1947/1948, 1948/1949, 1987/1988 and 1988/1989; Gavrin and Donaurov 1954; Jędrzejewski et al. 1992a), a plausible explanation being that red deer, wolves' favourite prey, are more difficult to capture in snowless winters.

How do wolves get their prey and utilise it? Wolves seem to be primarily persistent runners. Our radio-tracking of two packs in the Polish part of BPF showed that these predators are very mobile. They covered from a few to over 20 km each night, stopping for some hours only if having killed an ungulate or for diurnal rests (Fig. 4.13). Ungulates consecutively killed by wolves were usually far apart. Snow tracking and detailed recording of wolves' behaviour from their tracks and signs in snow showed that they often went the easiest routes possible: forest roads and paths, beds of frozen brooks and rivers (Table 4.10). They clearly avoided dense, wind-fallen fragments of tree stands. In searching for prey, wolves inspected ungulates' lairs and foraging places, and followed their trails. Finally, when they spotted potential prey, the typical hunting mode was a chase.

Wolves do not kill ungulates easily. Out of a total 25 winter hunts fully documented by snow tracking, only 9 ended with success (36%). In all other attempts, ungulates escaped, although in few cases with superficial wounds

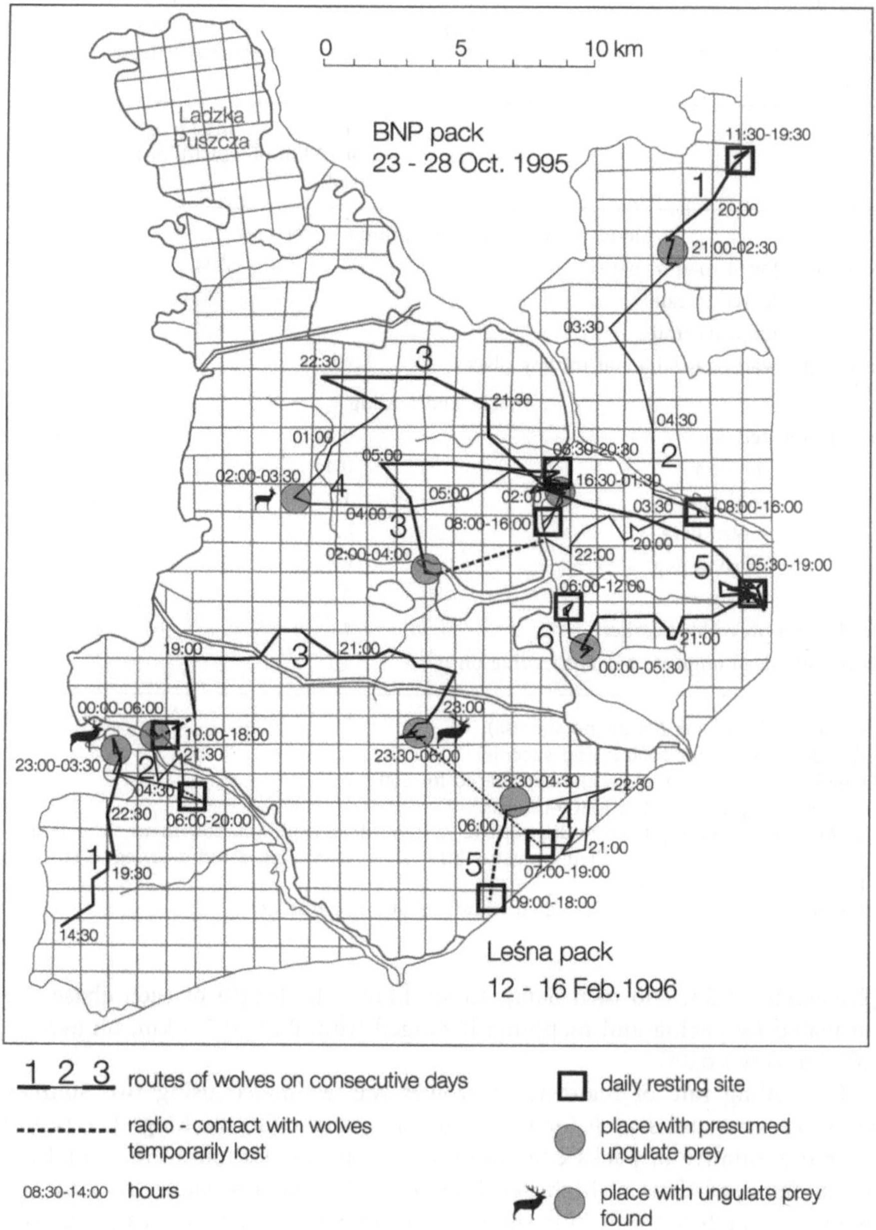

Fig. 4.13. Examples of the routes of two wolf packs during four to six days of continuous radio-tracking sessions, with presumed and actual prey locations. Source: W. Jędrzejewski, H. Okarma, and K. Schmidt (unpubl.)

Table 4.10. Wolf hunting behaviour and activities associated with modes of searching for prey as revealed by 24.8 km of snow tracking of wolf packs in BNP in 1986–1993. Figures in parentheses indicate unsuccesful attacks. All activities were recalculated per 10 km of trail

Activity	Number (*n*/10 km of trail)	Total length (m/10 km)	Total *n* of prey
Modes of movement and searching for prey			
Following: forest road or path		3734	
frozen stream		552	
ungulate trail		577	
Inspecting wild boar lairs and rooting places	13.6		
Attacks and feeding			
Attacks on: red deer	0.4		1 (1)[a]
roe deer	0.4		1 (1)[b]
wild boar	1.2		3 (2)[c]
brown hare	0.4		1[d]
raccoon dog	0.4		2 (2)[e]
rodent	0.4		1 (?)
Feeding on prey killed earlier	1.6		4[f]
Inspecting dead ungulate but not feeding on it	1.2		3[g]

[a] Red deer chased for 206 m (no success).
[b] Roe deer chased for 1500 m (no success).
[c] One boar attacked in a lair, another chased for 400 m.
[d] Brown hare captured after a 500-m chase.
[e] Wolves unsuccessfully tried to dig out two raccoon dogs from a hollow in an alder tree.
[f] Wolves visited remains of their previous kills: a red deer skin, a red deer leg, a red deer head, and roe deer bones.
[g] Carcasses of wild boar that had died from starvation or disease.

(Bunevich 1986a, and own data). In six hunts, the length of each chase was estimated by pacing and mapping; it ranged from 0.21 to 1.5 km, on average 0.65 km (own data).

The killing rate of Białowieża's wolves was estimated using two sources: snow tracking and search for prey remnants, and radio-tracking. The results are representative of packs containing five to six wolves, on average, and living under conditions of high densities of wild ungulates. Snow tracking and intensive search for wolf kills showed that on 56% of days wolves seized one ungulate, on 10% of days they killed two prey, and on 30% of days they captured no large prey (Fig. 4.14). The killing rate was, on average, 0.78 ungulates per pack per day, or one ungulate killed every 1.32 days. During 5 days of continuous radio-tracking, we found two kills and two other presumed kill sites for the Leśna pack of wolves (Fig. 4.13). Within 6 days of our continuous radio-tracking of a BNP wolf pack, the wolves most probably killed five ungulates (one of which was found). The killing rate estimated, based on

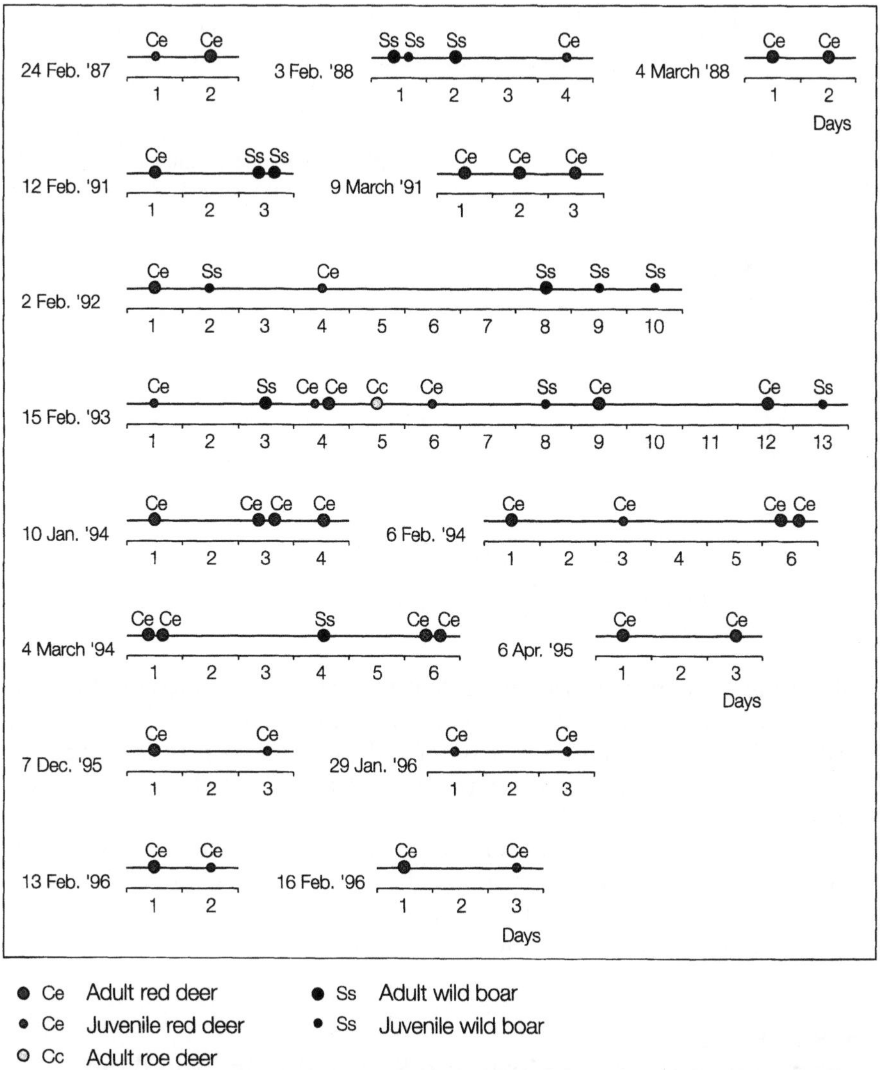

Fig. 4.14. Minimum reliable number of ungulate prey killed by a pack of wolves during the time spans marked by horizontal lines (data from intensive snow tracking of wolves). Dates denote the first day of each sequence. Jędrzejewski et al. (in press)

11 such series of continuous radio-tracking was, on average, 0.78 ungulates per pack daily, and one ungulate killed every 1.4 days (Jędrzejewski et al., in press). The mean daily killing rate per wolf was 0.14 ungulates.

A pack of wolves consumes a killed ungulate very fast, usually within 1 day, often taking small pieces aside to hide for later use. Of 39 1-day-old kills, 72% were consumed at least by 75% (Fig. 4.15). Wolves frequently return to prey killed earlier and finish them off (Table 4.10). Mean utilisation of ungulate prey by wolves (only consumable parts) was 91% (Jędrzejewski

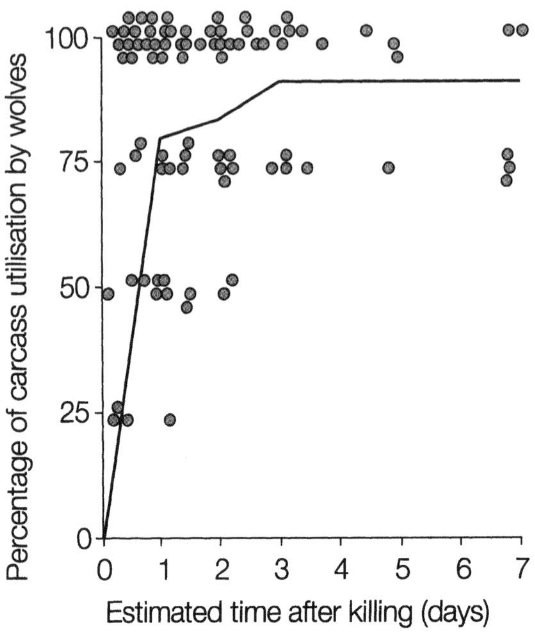

Fig. 4.15. Rate of consumption of ungulate prey by wolves (each point denotes one prey, $N=85$ fresh carcasses of ungulates found in the winter seasons of 1986/1987–1995/1996). The line joins mean values of carcass utilisation after 1, 2, and ≥3 days after killing

et al., in press). With the assumption that the consumable fraction of an ungulate carcass is 75% (Messier and Crete 1985), of which 91% is utilised by wolves, and that the mean body mass of wolves' ungulate prey is 55 kg, one can calculate that one wolf consumed 5.3 kg of food biomass daily (Jędrzejewski et al., in press).

Eight species of scavengers were recorded at leftovers of wolf kills. The most typical (and most common by numbers) were ravens, followed by foxes and wild boars (see Sect. 5.10). All that was left for scavengers were bones, intestines, skin, the head and a little soft tissue.

4.5 Red Foxes – Hunting for Voles in Meadows and Scavenging in Woodland

The red fox, one of the best-studied predators in Europe, has also been thoroughly investigated in Białowieża Primeval Forest. Food habits of foxes were studied here for a total of 16 years (Gavrin and Krapivnyi 1968; Bunevich 1986b; Jędrzejewski and Jędrzejewska 1992). In Białowieża, foxes consume a wide spectrum of foods. However, there is always one predominant item in their diets, *Microtus* voles, which made up, on average, 40% of the biomass consumed (Table 4.11). Moreover, three other resources are important to foxes: ungulate carcasses (26% of food biomass), forest rodents (14%), and brown hare (13%).

Microtus voles are hunted by foxes year-round with little seasonal variation (Fig. 4.16). The share of forest rodents (bank vole and yellow-necked

Table 4.11. Diet composition of the red fox in spring-summer (Sp–Su, 16 April–30 Sept.) and autumn-winter (Au–Wi, 1 Oct.–15 April) seasons in Białowieża Primeval Forest

Item	Eastern part				Western part			
	1946–1952		1979–1983		1985–1989			
	Sp–Su	Au–Wi	Sp–Su	Au–Wi	Sp–Su		Au–Wi	
	%Occ	%Occ	%Occ	%Occ	%Occ	%Bio	%Occ	%Bio
Mole (*Talpa europaea*)	–	–	–	–	5.0	0.4	3.4	0.9
Common shrew (*Sorex araneus*)	–	–	–	–	–	–	1.9	0.2
Lesser shrew (*Sorex minutus*)	–	–	–	–	–	–	0.4	+
Shrew (*Sorex* sp.)	–	–	–	–	1.6	0.4	0.7	0.4
Water shrew (*Neomys fodiens*)	–	–	–	–	–	–	0.4	+
Undetermined (Soricidae)	2.7	1.3	4.3	1.9	0.8	0.1	4.5	0.8
Insectivores total	**2.7**	**1.3**	**4.3**	**1.9**	**6.6**	**0.9**	**10.8**	**2.4**
Brown hare (*Lepus europaeus*)	13.7	9.7	6.1	8.2	19.8	14.7	14.6	11.8
Red squirrel (*Sciurus vulgaris*)	1.4	1.0	0.7	1.7	0.8	0.2	0.4	0.3
Muskrat (*Ondatra zibethicus*)	–	–	0.3	0.4	–	–	–	–
Bank vole (*Clethrionomys glareolus*)	5.3	5.3	5.6	14.3	17.4	6.7	19.8	8.9
Pine vole (*Pitymys subterraneus*)	–	–	–	–	0.8	0.4	0.4	0.1
Water vole (*Arvicola terrestris*)	2.5	2.5	–	–	0.8	0.8	–	–
Root vole (*Microtus oeconomus*)	–	–	–	–	14.1	10.7	21.6	16.2
Short-tailed vole (*Microtus agrestis*)	–	–	–	–	0.8	0.2	1.9	1.3
Common vole (*Microtus arvalis*)	–	–	–	–	8.3	13.2	9.0	8.1
Undet. vole (*Microtus* sp.)	53.9	67.3	29.0	32.6	37.2	15.2	28.4	14.8
Yellow-necked mouse (*Apodemus flavicollis*)	–	–	–	–	4.1	4.3	5.6	3.1
Striped field mouse (*Apodemus agrarius*)	–	–	–	–	–	–	0.8	0.3
Undet. mouse (*Apodemus* sp.)	4.0	4.2	2.2	–	5.8	3.3	5.2	1.4
Birch mouse (*Sicista betulina*)	–	–	–	–	0.8	+	–	–

Table 4.11 (continued)

Item	Eastern part				Western part			
	1946–1952		1979–1983		1985–1989			
	Sp-Su	Au-Wi	Sp-Su	Au-Wi	Sp-Su		Au-Wi	
	%Occ	%Occ	%Occ	%Occ	% Occ	% Bio	% Occ	% Bio
Harvest mouse (*Micromys minutus*)	–	–	–	–	0.8	+	0.8	0.1
Common rat (*Rattus norvegicus*)	–	–	–	–	–	–	0.8	0.5
Fat dormouse (*Glis glis*)	–	–	–	–	–	–	0.4	+
Undetermined rodent	3.0	–	–	–	–	–	1.1	+
Rodents total	**68.7**	**77.6**	**36.8**	**42.3**	**76.0**	**55.1**	**78.0**	**54.8**
Weasel (*Mustela nivalis*)	–	–	–	–	–	–	0.8	+
Wild boar (*Sus scrofa*) carcass	–	–	–	–	26.5	19.9	23.9	20.5
Undetermined Cervidae	–	–	–	–	18.2	1.5	30.6	5.5
Undet. wild ungulate carcass	–	–	14.8	32.8	–	–	–	–
Domestic sheep carcass	–	–	–	–	1.7	3.2	0.8	2.4
Domestic cow carcass	–	–	–	–	–	–	0.4	0.1
Undet. domestic ungulate carcass	–	–	12.0	7.3	–	–	–	–
Undet. ungulate carcass	8.1	10.3	–	–	–	–	–	–
Ungulate carcasses total	**8.1**	**10.3**	**26.8**	**40.1**	**38.0**	**24.5**	**47.8**	**28.5**
Capercaillie (*Tetrao urogallus*)	0.7	0.3	–	–	–	–	–	–
Black grouse (*Lyrurus tetrix*)	0.1	–	–	–	–	–	–	–
Partridge (*Perdix perdix*)	0.4	0.1	–	–	–	–	–	–
Hazel hen (*Bonasa bonasia*)	1.1	1.5	–	–	0.8	0.1	0.4	+
Undet. tetraonid bird	–	–	1.7	0.2	–	–	–	–
Domestic hen (*Gallus gallus*)	1.1	0.7	10.8	0.4	4.1	1.2	1.9	0.9
Wild duck (*Anas* sp.)	0.4	0.3	0.3	1.1	–	–	–	–
Undetermined wader	–	–	1.7	–	–	–	–	–
Woodpecker (*Dendrocopos* sp.)	–	–	0.3	–	–	–	1.1	0.3
Hawfinch (*Coccothraustes coccothraustes*)	–	–	–	–	0.8	0.1	–	–
Tawny owl (*Strix aluco*)	–	–	–	–	1.7	0.2	–	–
Wren (*Troglodytes troglodytes*)	–	–	–	–	–	–	0.4	+
Robin (*Erithacus rubecula*)	–	–	–	–	0.8	+	–	–
Song thrush (*Turdus philomelos*)	–	–	–	–	1.7	0.1	–	–
Thrush (*Turdus* sp.)	–	–	–	–	3.3	0.4	0.4	+

Table 4.11 (continued)

Item	Eastern part				Western part			
	1946–1952		1979–1983		1985–1989			
	Sp–Su	Au–Wi	Sp–Su	Au–Wi	Sp–Su		Au–Wi	
	%Occ	%Occ	%Occ	%Occ	%Occ	%Bio	%Occ	%Bio
Jay (*Garrulus glandarius*)	–	–	–	–	–	–	1.5	0.3
Tit (*Parus* sp.)	–	–	–	–	–	–	0.4	+
Chaffinch (*Fringilla coelebs*)	–	–	–	–	1.7	0.1	–	–
Bullfinch (*Pyrrhula pyrrhula*)	–	–	–	–	0.8	0.1	–	–
Yellowhammer (*Emberiza citrinella*)	–	–	–	–	–	–	0.4	0.1
Undet. small passerine	7.1	4.7	14.6	3.5	13.2	0.4	6.0	0.1
Undet. large bird	–	–	–	–	3.3	0.1	0.4	+
Undetermined bird	1.0	0.7	7.2	3.0	–	–	–	–
Bird egg	–	–	–	–	0.8	0.3	–	–
Birds total	**11.9**	**8.3**	**36.6**	**8.2**	**32.2**	**3.1**	**12.7**	**1.8**
Lizard (*Lacerta* sp.)	–	–	–	–	0.8	+	–	–
Undetermined reptile	6.8	1.1	0.3	0.2	–	–	–	–
Undetermined anuran	0.5	0.2	–	–	7.4	0.2	2.6	+
Fish	–	–	0.7	0.4	–	–	0.4	+
Snail	–	–	–	–	1.7	+	0.4	+
Earthworm	–	–	–	–	–	–	1.9	+
Beetle	6.9	5.0	7.3	1.7	–	–	–	–
Gryllotalpa gryllotalpa	4.1	1.9	0.3	0.4	–	–	–	–
Undetermined insect	2.6	1.6	3.5	1.4	10.7	0.1	2.2	+
Invertebrates total	**13.6**	**8.5**	**11.1**	**3.5**	**12.4**	**0.1**	**4.5**	**+**
Soft fruit	3.5	2.9	0.7	–	0.8	+	0.4	+
Tree seed (hazel, hornbeam)	–	–	–	–	5.8	0.1	1.9	+
Green plant	4.0	5.9	2.3	4.1	20.7	1.2	11.2	0.3
Other plant material	0.4	–	2.0	7.0	1.7	+	4.5	+
Plant material total	**7.9**	**8.5**	**5.0**	**11.1**	**27.3**	**1.4**	**15.7**	**0.4**
Total *N* scats or prey biomass (g)	1508 Scats and 42 stomachs		576[a]	463[a]	121	4920	268	16769

Eastern (Belarussian part) – data for 1946–1952 recalculated from Gavrin and Krapivnyi (1968), and data for 1979–1983 taken from Bunevich (1986b). Western (Polish) part – data for 1985–1989 from Jędrzejewski and Jędrzejewska (1992).

%Occ, percentage of occurrence in scats (or scats and stomachs); %Bio, percentage of biomass consumed.

[a] The sample also includes stomachs and prey remains collected at the entrances of breeding dens.

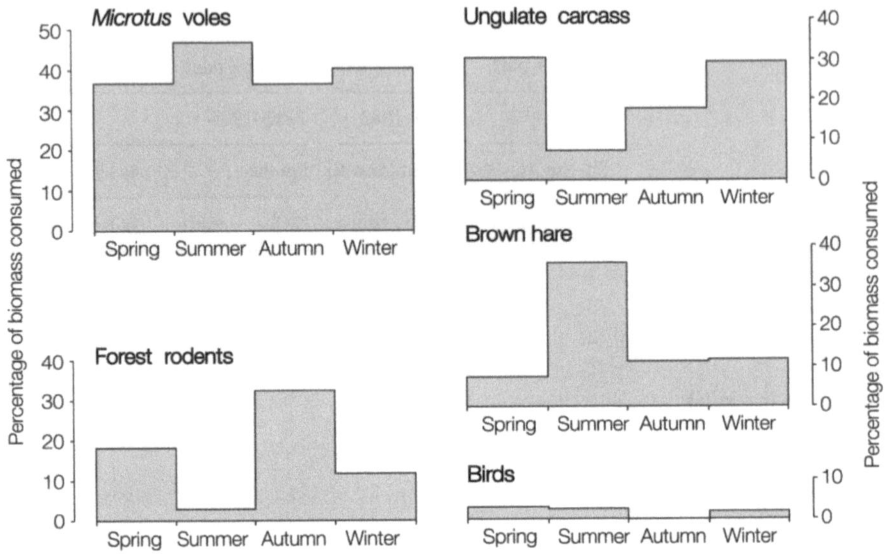

Fig. 4.16. Seasonal variation in the shares of five food resources in red fox diet in Białowieża National Park. Data (pooled for 1985/1986–1989) from Jędrzejewski and Jędrzejewska (1992)

mouse) in the fox's diet is highest in autumn, and that of brown hare in summer, when young hare are abundant. Ungulate carcasses are consumed by foxes in substantial amounts during any season except summer. Birds are frequently captured in spring and summer, when they are most numerous (Fig. 4.16). Yet, compared with fox diets described from agricultural landscapes and the forest-farmland mosaic (where poultry is common and readily taken by foxes), birds constituted a negligible share in the total food biomass taken by foxes in Białowieża woodland (cf. Goszczyński 1974, 1986b). Moreover, Gavrin and Krapivnyi (1968) observed that many small birds were remnants of goshawk's and sparrowhawk's prey, scavenged by foxes near the raptors' nests. In severe winters, foxes may feed on dead birds. For instance, in January 1950, when, during some days, the minimum temperature dropped to −40 °C, dead tits, thrushes, woodpeckers, and yellowhammers were seen on the forest floor and in meadows (Gavrin and Krapivnyi 1968). Foxes, with their good senses of smell and sight, could find frozen birds with no difficulty.

Year-to-year variation in fox feeding on *Microtus* voles was influenced by fluctuations in vole numbers (Fig. 4.17), and depth of snow cover. Voles, the inhabitants of sedge marshes, meadows, fallows, and fields in the forest, exhibit 3–4-year cycles of abundance with peak numbers up to 20 times higher than the crash numbers (see Fig. 2.26, Sect. 2.4). The consequent range of fluctuations of the *Microtus* share in the total biomass consumed by foxes was only twofold (Fig. 4.17). Indeed, foxes specialise in voles and are devoted to hunting them even when they are scarce.

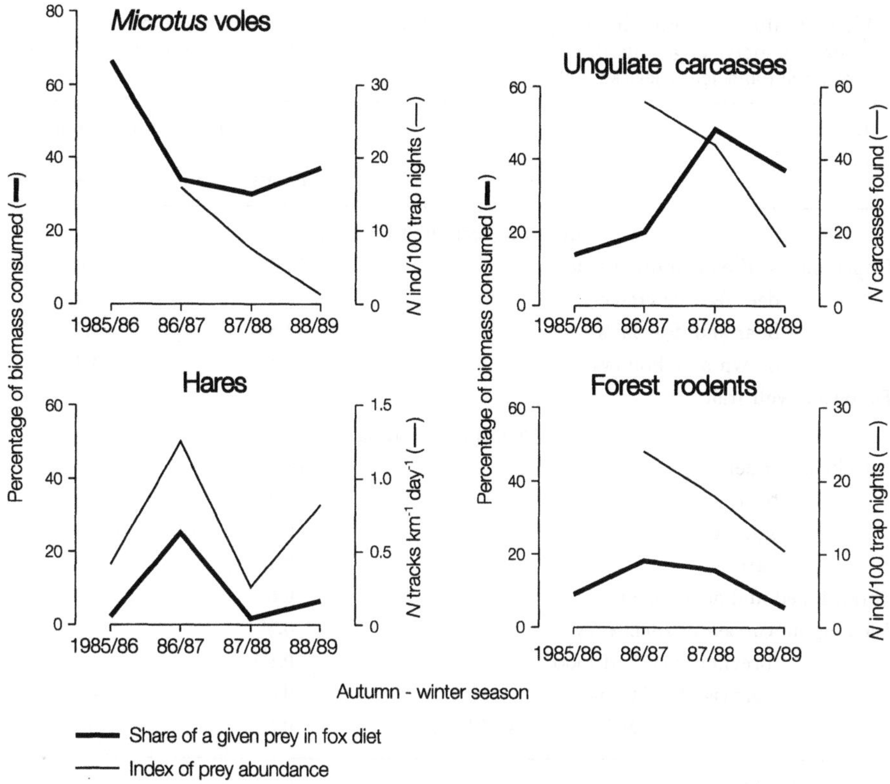

Fig. 4.17. Year to year variation in foxes' diet (four main prey) in relation to the abundance of food resources in autumn-winter periods. Jędrzejewski and Jędrzejewska (1992)

Deep snow cover seems to be a more serious obstacle than low numbers of voles for successful vole hunting by foxes. These predators capture rodents by scratching and digging through snow (69% of attacks as shown by snow tracking) and pounces (31% of attacks) (Jędrzejewski and Jędrzejewska 1992). The deeper the snow, the more difficult it is for a fox to hunt successfully. In the snowy winters of 1985/1986 and 1986/1987, the mean weight of rodent remains per one scat declined fivefold, from 3 g at no snow cover to 0.6 g when snow was 50 cm deep (Jędrzejewski and Jędrzejewska 1992).

Foxes compensated for lower numbers and availability of voles by an increased consumption of three other food resources: hares, forest rodents, and ungulate carcasses. Which buffer food was exploited most intensely was determined by relative abundances and availabilities of those resources. Hare and forest rodents were hunted proportionally to their numbers (Fig. 4.17), whereas ungulate carcasses were the last food choice. The abundance of carrion in the forest was positively related to the severity of winter, but foxes utilised it in substantial quantities only when other foods failed (e.g. in 1987/1988 and 1988/1989). Foxes feed on intact carcasses of ungulates that die

Table 4.12. Modes of searching for prey and hunting by the red fox in two types of habitats (forests and open areas) in BPF as revealed by 72 km of snow tracking in 1986–1990. (Jędrzejewski and Jędrzejewska 1992, and unpubl.)

Activity	Number per 10 km of trail	
	Forest	Open area
Modes of searching for prey		
Inspecting: wild boar rooting places	10.0	38.0
deer feeding craters	0.5	2.2
boar and deer lairs	2.6	3.2
brown hare hollows	0.3	3.2
Following wolf trail	0.7	-
Attacks and feeding		
Attacks on: rodent	18.9	49.6
bird	0.3	-
squirrel	0.1	-
hare	0.4	-
Shrew killed and abandoned	0.6	1.3
Feeding on: carcass of small prey	0.3[a]	-
previously cached food	0.8[b]	-
carcass of wild boar	0.8	-
remnant of deer killed by wolf/lynx	0.6	-

[a] One bird, one squirrel.
[b] Two hares, four pieces of deer remains.

from starvation and disease, mainly wild boar, and the leftovers of wolf and lynx kills, predominantly red and roe deer (see Sect. 5.10).

Snow tracking of foxes showed that in the open marshes, meadows and fields, their foraging was restricted to hunting mainly small rodents and hares (Table 4.12). In woodland, attacks on rodents were also the most frequent foraging activity of foxes. However, forests offered more diverse resources, and foxes searched for ungulate carcasses (by visiting ungulates' lairs and feeding places as well as by following wolf trails), attempted attacks on birds and squirrels, and fed on prey cached earlier (Table 4.12). From snow tracking data, we estimated that in winter foxes were successful on at least 14% of their attacks on rodents.

As its diet composition and hunting modes show, the fox feels at home in the forest-open area ecotones. Forested terrain is sought for location of dens (see Chap. 3). In autumn and winter, foxes frequently forage in woodland where they hunt for forest rodents and feed on ungulate carcasses. In spring and summer, they venture more into open areas to hunt for voles and hares. The ratio of food biomass of forest origin to that found by foxes in open meadows and marshes was 1:1 in autumn, winter and spring, but 1:8 in favour of open area food in summer (Jędrzejewski and Jędrzejewska 1992). In conclusion, the fox foraging strategy can be defined as specialisation on *Mi-*

crotus voles with adaptation to utilise other resources (hare, forest rodents, ungulate carcasses) buffering the low numbers and availability of voles.

4.6 Raccoon Dogs – Scavenging on Ungulate Carcasses

When snow tracking raccoon dogs in Białowieża National Park, we could hardly believe that we were recording the trail of a predator. Raccoon dogs often moved in pairs, a male and a female, gnawing leaf buds off low shrubs and tree saplings, scratching through snow and litter to collect invertebrates, and digging in mud to find frogs (Table 4.13). In such a 'gathering' mode of foraging, raccoon dogs acquired small quantities of poor quality food. Only occasionally did they truly gorge, having found carrion of dead wild boar. In winter, these predators are primarily scavengers, actively searching for ungulates that die from severe winter conditions, disease or starvation, and remnants of wolf and lynx kills. Data from snow tracking suggested that raccoon dogs fed on an ungulate carcasses, on average, 1.5 times daily.

Primeval temperate forests, which abound in ungulates, inevitably abound in their carrion. This subject will be discussed in detail in Section 5.10. Here, suffice it to say that the supply of carrion depends on many factors. Wolves and lynxes constantly leave remnants of large prey: red deer, roe deer, and less often, wild boar. This fairly predictable source of carrion is of poor quality though. Usually, only bones, hide and a little soft tissue remain for scavengers, which, moreover, have to risk fatal contact with a host predator. Severe winters are another providers of ungulate carrion, in an unpredictable but generous manner. Wild boar in particular are susceptible to death from starvation and disease during winters with hard frosts and deep snow as well as those after poor acorn crops. Fewer ungulate carcasses are available in summer than in winter, but – to a good nose – they are easier to find then.

Analysis of scats, stomachs, and prey remains showed the prime role of ungulate carcasses in the raccoon dog's diet in both autumn–winter (56% of food biomass consumed) and spring–summer (29%; Table 4.14). The second most important prey were amphibians (represented predominantly by the common frog), which constituted 22% of food biomass in spring–summer and 16% in autumn–winter. Frequent feeding on frogs as well as capturing other vertebrates of wet habitats (e.g. water vole) confirms the habitat distribution of raccoon dogs described earlier: their preference of open sedge marshes by forest rivers (see Fig. 3.67). In summer, raccoon dogs also capture frogs in oak-lime-hornbeam and alder-ash forests.

Moreover, three other peculiarities of raccoon dog foraging are worth mentioning. First, it is their feeding on invertebrates (13% of food biomass in spring–summer), particularly earthworms and ground beetles (Table 4.14). Raccoon dogs gather invertebrates at night from the forest floor, and swallow them with quite a lot of plant debris and litter. Of all native carnivores, only the badger feeds on earthworms and ground insects in that way, but it relies on them as nearly exclusive food. Secondly, among small mammals eaten by raccoon dogs, shrews are nearly as common as rodents (Table 4.14). This is rather unusual for carnivores, most of which avoid catching shrews or leave carcasses of killed shrews uneaten (see fox and pine marten). Finally, a peculiar feature of the raccoon dog is its notorious consumption of any plant material to be found on the ground: green plants, tree seeds, grain put in the forest as fodder for ungulates, stomach contents of dead ungulates, and – if anything else fails – faeces of wild boar (Tables 4.13 and 4.14, and Bunevich and Dackevich 1985).

Between year variation in raccoon dog diet was shaped predominantly by the severity of winters (Fig. 4.18). During harsh winters, raccoon dogs relied nearly exclusively on ungulate carcasses (76% of food biomass, of which wild boar carrion made up 67%). In the moderate and mild winters, carrion constituted 49% of food biomass, and the raccoon dog diet was supplemented with increasing shares of amphibians, rodents, and shrews, but also plant material and invertebrates. The pattern of substituting one food source for

Table 4.13. Foraging activities of raccoon dogs in winter, as revealed by snow tracking in the Białowieża National Park. Total length of snow tracking 17.05 km (in winters 1986/1987–1988/1989). Activities recalculated per day, assuming daily movement distance in winter of 3 km (according to Bunevich 1983 a)

Activity	Number per day when active	Total number recorded (notes)
Scavenging: on dead wild boar	0.7 (5.2%)	4 Lairs (with 1, 1, 5, and 7 dead boar)
on remains of wolves' prey	0.5 (3.9%)	2 Red deer females (fresh) and 1 male (old remnant bones)
on remains of lynx's prey	0.2 (1.3%)	1 Roe deer (a head)
Attacks on: rodent	2.3 (16.9%)	13 Attacks, 4 bank voles killed
hazel hen	0.2 (1.3%)	1 Failed attack on resting bird
Digging in mud (river banks) for frogs	0.5 (3.9%)	3 Attempts, results unclear
Feeding on invertebrates:		
by scratching litter	0.5 (3.9%)	3
by licking tree trunks	0.2 (1.3%)	1
Feeding on vegetation:		
mushrooms	0.2 (1.3%)	1
leaf buds of low shrubs	1.4 (10.4%)	8
grass	0.2 (1.3%)	1
other plants on forest floor	1.8 (13.0%)	10
Consumption of wild boar scats	0.4 (2.6%)	2
Eating an old bird egg found under the litter	0.2 (1.3%)	1
Scratching through snow and litter	4.4 (32.5%)	25 Attempts (search for rodents? invertebrates? vegetation?), results unclear
Total number of foraging attempts	13.7 (100%)	77 Attempts, at least 39 successful

Table 4.14. Diet composition of the raccoon dog in Białowieża Primeval Forest in spring-summer (Sp–Su, 16 April–30 Sept.) and autumn-winter (Au–Wi, 1 Oct.–15 April)

Item	Eastern part (1977–1982)		Western part (1986–1995)			
	Sp–Su	Au–Wi	Sp–Su		Au–Wi	
	%Occ	%Occ	%Occ	% Bio	%Occ	%Bio
Mole (*Talpa europaea*)	–	–	3.5	0.2	2.2	0.8
Common shrew (*Sorex araneus*)	–	–	3.5	0.4	2.2	0.4
Undetermined shrew (*Sorex* sp.)	–	1.9	–	–	5.0	1.6
Water shrew (*Neomys fodiens*)	1.3	–	–	–	2.8	0.1
Undetermined Soricidae	2.6	2.5	20.7	6.2	17.9	1.7
Insectivores total	**3.9**	**4.4**	**27.7**	**6.8**	**29.6**	**4.6**
Brown hare (*Lepus europaeus*)	–	1.9	–	–	1.1	0.2
Red squirrel (*Sciurus vulgaris*)	–	0.6	–	–	–	–
Bank vole (*Clethrionomys glareolus*)	–	1.3	13.8	5.8	16.2	2.7
Water vole (*Arvicola terrestris*)	–	1.3	6.9	0.6	–	–
Root vole (*Microtus oeconomus*)	–	2.5	–	–	3.9	3.1
Undet. vole (*Microtus* sp.)	–	–	3.5	0.2	17.3	3.6
Undetermined Microtidae	18.4	8.9	–	–	–	–
Yellow-necked mouse (*Apodemus flavicollis*)	–	–	–	–	2.2	0.4
Undet. mouse (*Apodemus* sp.)	–	–	–	–	5.0	1.0
Undetermined rodent	21.1	12.7	6.9	0.2	3.4	0.2
Rodents total	**39.5**	**26.7**	**31.0**	**6.8**	**45.8**	**11.0**
Badger (*Meles meles*) cub	1.3	–	3.5	4.7	–	–
Weasel (*Mustela nivalis*)	–	–	–	–	0.6	0.1
Wild boar (*Sus scrofa*) carcass	–	5.7	34.5	25.8	37.4	47.3
Roe deer (*Capreolus capreolus*) carcass	1.3	8.3	–	–	–	–
Red deer (*Cervus elaphus*) carcass	1.3	5.1	–	–	–	–
Undetermined Cervidae carcass	–	–	41.4	2.8	45.3	6.6
Domestic sheep carcass	–	–	–	–	1.1	2.2
Undetermined ungulate carcass	14.5	34.4	–	–	–	–
Ungulate carcasses total	**17.1**	**53.5**	**69.0**	**28.6**	**69.3**	**56.1**
Swan (*Cygnus* sp.)	–	–	–	–	0.6	0.6
Undet. wild duck (*Anas* sp.)	–	–	3.5	1.8	–	–
Hazel hen (*Bonasa bonasia*)	2.6	–	–	–	–	–
Woodcock (*Scolopax rusticola*)	1.3	–	–	–	–	–
Thrush (*Turdus* sp.)	–	–	–	–	0.6	+
Jay (*Garrulus glandarius*)	–	–	3.5	8.6	0.6	0.5
Undetermined large bird	–	–	10.3	1.7	–	–

Table 4.14 (continued)

Item	Eastern part (1977–1982)		Western part (1986–1995)			
	Sp–Su	Au–Wi	Sp–Su		Au–Wi	
	%Occ	%Occ	%Occ	%Bio	%Occ	%Bio
Undet. medium-sized bird	–	–	–	–	1.1	0.1
Undetermined small passerine	10.5	3.2	27.6	0.9	6.7	0.3
Undetermined bird	9.2	1.3	–	–	–	–
Bird egg	1.3	–	–	–	–	–
Birds total	**23.7**	**4.5**	**44.8**	**13.0**	**9.5**	**1.5**
Lizard (*Lacerta* sp.)	?	?	–	–	1.1	0.4
Undetermined reptile	?	?	–	–	2.2	0.2
Reptiles total	**2.6**[a]	**0.6**[a]	**–**	**–**	**3.3**	**0.6**
Common toad (*Bufo bufo*)	–	–	–	–	0.6	0.4
Undet. toad (*Bufo* sp.)	–	–	3.5	1.3	1.7	1.5
Common frog (*Rana temporaria*)	–	–	17.2	16.2	8.4	11.2
Undetermined frog (*Rana* sp.)	18.4	4.5	3.5	0.2	3.9	0.6
Undetermined anuran	–	–	34.5	4.2	20.1	2.5
Amphibians total	**18.4**	**4.5**	**55.2**	**21.9**	**32.4**	**16.2**
Fish	–	0.6	–	–	–	–
Snail	5.3	–	27.6	0.6	22.4	0.2
Earthworm	Not analysed		48.3	7.9	39.1	5.1
Carabus sp.	–	–	10.3	0.4	3.4	0.1
Calosoma sp.	–	–	3.5	+	–	–
Undetermined Carabidae beetles	–	–	17.2	0.6	5.6	0.1
Geotrupes sp.	–	–	10.3	3.0	0.6	+
Undetermined Scarabaeidae	–	–	3.5	+	–	–
Sylphidae beetle	–	–	–	–	0.6	+
Curculionidae beetle	–	–	–	–	0.6	+
Other and undet. Coleoptera	38.2[b]	8.9[b]	–	–	1.1	+
Coleoptera larvae	–	–	13.8	0.1	7.8	0.1
Heteroptera	–	–	–	–	1.1	+
Bumblebee nests (*Bombus* sp.)	–	–	3.5	+	–	–
Wasp nests (Vespidae)	–	–	6.9	0.1	3.9	+
Hymenoptera[c]	10.5	5.7	–	–	–	–
Trichoptera larvae	–	–	–	–	1.1	+
Other and undetermined insects	17.1	–	48.3	0.2	23.5	0.1
Invertebrates total	**65.8**	**12.7**	**89.7**	**12.9**	**58.1**	**5.7**
Blueberry (*Vaccinium myrtillus*)	21.1	–	–	–	–	–
Raspberry (*Rubus* sp.) fruit	–	–	6.9	0.5	1.7	+

Table 4.14 (continued)

Item	Eastern part (1977–1982)		Western part (1986–1995)			
	Sp–Su	Au–Wi	Sp–Su		Au–Wi	
	%Occ	%Occ	%Occ	%Bio	%Occ	%Bio
Undetermined soft fruit	–	–	–	–	0.6	+
Oak (*Quercus robur*) seeds	–	1.9	10.3	0.3	10.6	1.4
Seeds of other trees and shrubs	–	–	10.3	+	7.8	0.2
Bulb roots of plants	–	–	3.5	0.6	2.2	0.3
Leaf buds of shrubs and trees	–	–	–	–	6.7	+
Grasses and sedges	21.1	23.6	86.2	3.0	39.7	0.6
Vegetal matter from boar faeces and stomachs of dead ungulates	9.2	52.9	17.2	0.7	27.4	1.3
Other plant material	3.9	0.6	6.9	0.1	1.1	+
Plant material total	**55.3**	**79.0**	**89.7**	**5.2**	**69.3**	**3.9**
Raccoon dog hair	2.6	5.1	13.8	0.1	17.3	0.1
Total *N* scats or prey biomass (g)	76[d]	157[d]	29	2374 g	179	12880 g

Eastern part: 1977–1982 (compiled from Bunevich 1983 a, Bunevich and Dackevich 1985). Western part: data for 1986–1995 (W. Jędrzejewski, unpubl.). Denotations as in Table 4.11.

[a] Two grass snakes (*Natrix natrix*) and one viviparous lizard (*Lacerta vivipara*).

[b] Mainly Carabidae beetles, *Geotrupes* sp. and Elateridae.

[c] Mainly wasp nests.

[d] The entire sample (N=233) included 40 stomachs, 175 scats, and 18 prey remains found during snow tracking.

another in the gradient of severe to mild winters was the same as that between autumn–winter and spring–summer seasons.

4.7 Badgers and Earthworms

Every night from late March till late October, badgers go out of their setts to slowly, systematically dodge the litter on the forest floor in their search for earthworms. It is a rather unusual evolutionary adaptation for such a large-bodied predator to feed on so tiny a prey. However, this adaptation is not as paradoxical as it may seem. In the vegetation season of 1962, Utenkova and Nazarova (1968) surveyed earthworms in various deciduous tree stands and sedge marshes in Białowieża Forest. They found, on average, 47 earthworms/m^2 in late spring, 56 in mid-summer, and 39 in early autumn. Given 2.5 g as a mean body mass of an earthworm, 1 km^2 of the forest soil is the store of 97–140 tons of these invertebrates! This is by two orders of magnitude more than all vertebrate prey groups, from tiny shrews to bison, taken together.

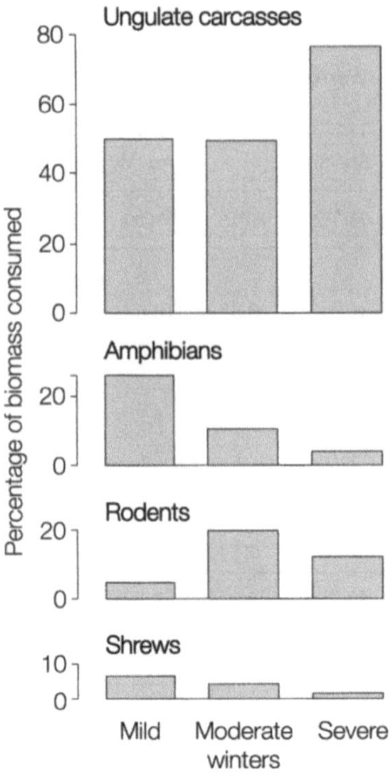

Fig. 4.18. Variation in autumn–winter diet of raccoon dog (four main food resources) in Białowieża National Park in relation to the severity of winter. Mild winters: 1988/1989–1990/1991; moderate winters: 1987/1988, 1992/1993, and 1993/1994; severe winters: 1985/1986 and 1986/1987

However, as food to homeothermic animals, earthworms have a major drawback: they are available from spring through autumn, only. Badgers cope with the seasonal food shortage by a winter sleep.

In the diet of Białowieża's badgers, earthworms made up, on average 70% of the consumed biomass (Table 4.15). Their share was higher in spring (83% of biomass between 1 March–15 May), than in summer-autumn (56% between 16 May–30 November), which reflects the seasonal decline in earthworm abundance in the soil (by 30% from mid-summer to early autumn, according to Utenkova and Nazarova 1968).

There are three groups of prey taken by badgers increasingly as summer proceeds and earthworms become less available: amphibians (increase in badger's diet from 8% of biomass in spring to 23% in summer-autumn), fruit (0.2 to 6%), and insects (1 to 3%). Among amphibians eaten by badgers (15% of biomass, i.e. the second most important prey), toads, mainly the common toad *Bufo bufo*, dominated and formed 41.3% of all identified remains, whereas in the forest community of terrestrial amphibians it comprised 22% only (see Sect. 2.10). Thus, unlike other carnivores, badgers did not avoid toads, indeed they took them more than expected at random. Insects do not weigh much in the total biomass eaten by badgers, nevertheless, they are one of the commonest prey actively searched for. Badgers eat ground insects, predominantly dung beetles in the genus *Geotrupes* (51% of all imagines of insects) and flightless

Table 4.15. Diet composition of badgers in Białowieża Primeval Forest in spring, summer and autumn (March–November)

Item	Eastern part (1978–1984)	Western part (1986–1995)	
	% Occ	% Occ	% Bio
Eastern hedgehog (*Erinaceus concolor*)	–	8.5	2.4
Mole (*Talpa europaea*)	–	2.8	1.1
Common shrew (*Sorex araneus*)	–	0.7	0.3
Undetermined Soricidae	–	3.6	0.2
Insectivores total	**–**	**14.9**	**4.0**
Bank vole (*Clethrionomys glareolus*)	.	2.8	0.2
Pine vole (*Pitymys subterraneus*)	.	3.5	0.3
Common vole (*Microtus arvalis*)	.	1.4	0.4
Undetermined vole (*Microtus* sp.)	.	2.8	0.3
House mouse (*Mus musculus*)	.	0.7	0.1
Undetermined rodent	.	0.7	+
Rodents total	**38.1**	**9.9**	**1.3**
Wild boar (*Sus scrofa*) carcass	.	0.7	+
Deer (Cervidae) carcass	.	13.5	0.3
Ungulate carcasses total	**7.9**	**13.5**	**0.3**
Song thrush (*Turdus philomelos*)	.	0.7	+
Undetermined medium-sized bird	.	2.1	0.1
Undetermined small bird	.	2.8	0.3
Bird egg	.	0.7	+
Birds total	**4.8**	**5.7**	**0.4**
Lizard (*Lacerta* sp.)	–	0.7	+
Slowworm (*Anguis fragilis*)	–	0.7	+
Reptiles total	**–**	**1.4**	**0.1**
Undetermined toad (*Bufo* sp.)	.	13.5	6.4
Common frog (*Rana temporaria*)	.	10.6	5.5
Moor frog (*Rana arvalis*)	.	1.4	0.4
Undetermined frog (*Rana* sp.)	.	7.1	1.6
Undetermined amphibian	.	19.2	1.3
Amphibians total	**34.9**	**41.8**	**15.2**
Fish	–	0.7	+
Snail	Present	12.8	0.1
Carabidae beetle	.	61.0	0.6
Dung beetles (*Geotrupes* sp.)	.	49.7	0.8
Sylphidae	.	5.0	+
Other Coleoptera	.	14.2	0.1

Table 4.15 (continued)

Item	Eastern part (1978–1984) % Occ	Western part (1986–1995) % Occ	% Bio
Nest of wasps (Vespidae)	.	8.5	0.2
Dytiscus sp.	.	0.7	+
Undetermined insect	.	4.3	+
Larvae of insects	.	11.4	0.2
Earthworm (Lumbricidae)	Not analysed	90.8	70.0
Invertebrates total	**85.8**	**98.6**	**72.1**
Oak (*Quercus robur*) seeds	9.5	5.0	0.5
Wild raspberry (*Rubus* sp.)	.	7.8	2.8
Other seeds and fruits	.	3.6	+
Mushrooms	–	2.1	0.1
Plant litter (leaves, needles, grass)	.	88.7	3.2
Plant material total	**12.7**	**90.8**	**6.6**
N scats and biomass consumed (g)	?	141	19881 g

Eastern part: Bunevich (1986 c; data for 1978–1984).
Western part: own data (1986–1995). Bunevich (1986 c) did not conduct the microscope analysis of prey remains to find chaetae of earthworms.
Dot, lack of numerical information on the contribution of that prey.

Fig. 4.19. Underground nest of wasps dug out by a foraging badger; coniferous forest in Białowieża National Park, compartment 318, late summer 1996. Photo: W. Jędrzejewski

Carabidae beetles (37%) (Table 4.16). Badgers capture markedly more insects in summer-autumn than in spring, the increase in percentage of occurrence in badgers' scats was from 45 to 58% in *Geotrupes* sp., from 52 to 77% in car-

Table 4.16. Insects recovered from scats of badgers in Białowieża National Park (1986–1995)

Taxonomic group of insects	Numbers (and percentages)		
	Imagines	Larvae	Nests
Hymenoptera (total)	9		4 (100%)
Vespidae (total)	9		4
Vespa crabro L.	2		1
Other Vespidae	7		3
Coleoptera (total)	544	3	
Undetermined Coleoptera	5	3	
Carabidae (total)	203 (36.6%)	4	
Calosoma sp.	20		
Carabus sp.	114		
Undetermined Carabidae	69	4	
Silphidae (total)	16 (2.9%)		
Silpha sp.	14		
Nicrophorus vespillo (L.)	2		
Scarabaeidae (total)	292 (52.6%)		
Geotrupes sp.	281		
Other Scarabaeidae	11		
Melolonthidae (total)	3	35 (47.9%)	
Melolontha melolontha L.	2	35	
Cetonia sp.	1		
Elateridae (total)	1		
Dytiscidae (total)	1		
Dytiscus marginalis L.	1		
Coccinelidae (total)	1		
Coccinella sp.	1		
Cerambycidae (total)	15 (2.7%)		
Curculionidae (total)	7		
Undetermined insects	2	31	
Total	555 (100%)	73 (100%)	4 (100%)

abid beetles, and from 1% to as much as 21% in Vespidae. The underground nests of wasps are frequently dug up by badgers in late summer and autumn (Fig. 4.19). The increase in badgers' consumption of dung beetles and carabids towards autumn is caused by the decline of earthworms. Nonetheless, in the case of *Geotrupes* beetles, conducive to this is the second, autumnal peak of their abundance which regularly occurs in early September (Borowski 1960).

The whole array of foods taken by badgers have one thing in common: they are animals or plants to be found on the ground, just below or just above the ground, and most of the prey are slow movers. Even among small mammals captured by badgers, quite uncommon but entirely ground-living species predominate: the pine vole and the hedgehog. Unlike many other predators, badger's feeding does not involve strain and risk, but it does take time.

4.8 Pine Martens – Generalistic Feeding Habits of Bank Vole Specialists

All carnivores of temperate forests in Europe are terrestrial with two exceptions, the pine marten and the weasel. Although climbing a few metres up trees as it does, the weasel always rests and usually hunts on the ground. The pine marten feels equally at home on the ground and in tree canopies a few dozen meters high. Most of martens' resting sites are arboreal (Zalewski 1997). In searching for food, martens exploit the ground, where they search for rodents, shrews, frogs, and ungulate carcasses. They also climb shrubs and trees for birds, fruit, nests of wild bees and wasps, and squirrels.

Despite such agility and a wide spectrum of prey taken, pine martens feed primarily on rodents, which make up, on average, 74% of their food biomass in spring-summer and 83% in autumn-winter (Table 4.17). Consumption of forest rodents by martens peaks in autumn and winter months and is lowest in June–September (Fig. 4.20). *Microtus* voles were most numerous in the marten diet in April–May and again in July–September. Among non-rodent prey of marten, birds are most important and preyed upon mainly in May–

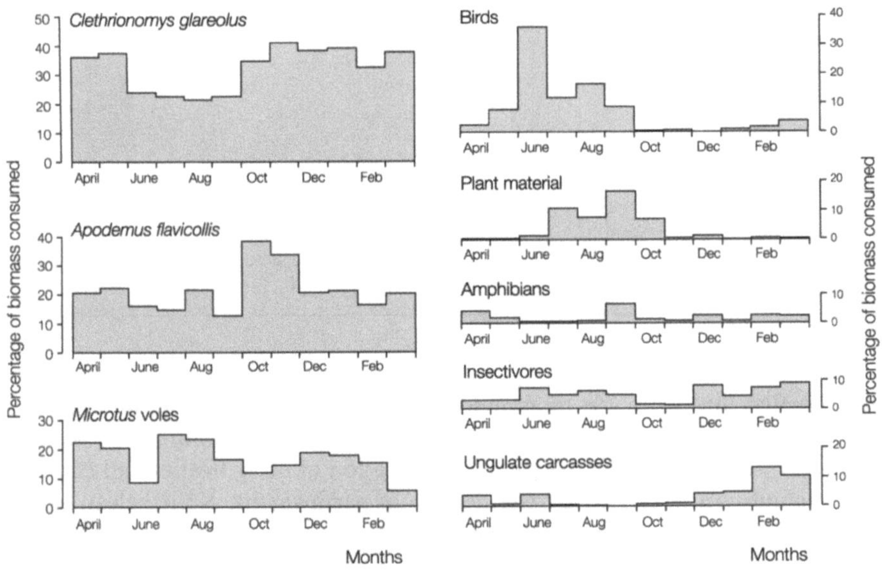

Fig. 4.20. Monthly changes in percent biomass of eight main groups of food resources in pine marten diet in Białowieża National Park; data pooled for 1985/1986–1991/1992. Jędrzejewski et al. (1993 c)

Table 4.17. Food of the pine marten in spring–summer (Sp–Su, 16 April–30 Sept.) and autumn–winter (Au–Wi, 1 Oct.–15 April) in Białowieża Primeval Forest

Item	Eastern part		Western part			
	Sp–Su	Au–Wi	Sp–Su		Au–Wi	
	%Occ	%Occ	%Occ	%Bio	%Occ	%Bio
Eastern hedgehog (*Erinaceus concolor*)	–	–	–	–	0.1	0.3
Mole (*Talpa europaea*)	0.5	0.2	2.5	1.5	3.7	2.0
Common shrew (*Sorex araneus*)	–	–	1.7	1.5	2.5	1.0
Lesser shrew (*Sorex minutus*)	–	–	0.3	0.3	0.5	0.2
Undetermined shrew (*Sorex* sp.)	–	–	1.5	0.4	1.5	0.4
Water shrew (*Neomys fodiens*)	–	–	0.5	0.2	–	–
Undet. water shrew (*Neomys* sp.)	–	–	0.2	+	0.3	0.1
Undetermined Soricidae	4.2	3.8	1.7	0.4	1.4	0.7
Insectivores total	**4.7**	**4.0**	**8.1**	**4.3**	**9.8**	**4.7**
Undetermined bat (Chiroptera)	0.5	–	0.2	+	–	–
Brown hare (*Lepus europaeus*)	2.1	1.6	–	–	0.4	0.5
Red squirrel (*Sciurus vulgaris*)	5.3	3.5	0.3	0.1	0.6	0.8
Bank vole (*Clethrionomys glareolus*)	27.9	28.3	36.6	31.5	43.9	37.5
Water vole (*Arvicola terrestris*)	–	–	0.3	0.3	0.2	0.1
Pine vole (*Pitymys subterraneus*)	–	–	4.5	3.0	5.1	4.8
Root vole (*Microtus oeconomus*)	–	–	4.5	4.9	2.6	2.7
Short-tailed vole (*Microtus agrestis*)	–	–	1.2	0.9	1.0	1.1
Common vole (*Microtus arvalis*)	–	–	0.7	0.6	1.2	1.5
Undetermined vole (*Microtus* sp.)	51.6	50.0	17.2	12.2	13.2	9.7
Common rat (*Rattus norvegicus*)	–	–	–	–	0.1	+
Harvest mouse (*Micromys minutus*)	0.5	–	–	–	0.6	0.3
Striped field mouse (*Apodemus agrarius*)	–	–	–	–	0.5	0.3
Yellow-necked mouse (*Apodemus flavicollis*)	5.3	5.3	10.8	10.8	13.6	12.7
Undet. mouse (*Apodemus* sp.)	–	–	11.3	8.5	18.5	12.2
Birch mouse (*Sicista betulina*)	–	0.2	0.8	0.4	0.3	0.1
Fat dormouse (*Glis glis*)	–	0.2	–	–	0.1	+
Undet. dormouse (Gliridae)	–	–	–	–	0.1	+
Undetermined small rodent	–	–	2.5	1.1	0.6	0.4
Small rodents total	**85.3**	**84.0**	**77.0**	**74.2**	**84.4**	**83.4**
Polecat (*Mustela putorius*)	–	–	0.2	0.5	–	–
Weasel (*Mustela nivalis*)	–	–	0.5	0.4	0.1	+
Domestic cat (*Felis catus*)	–	–	–	–	0.4	0.6
Wild boar (*Sus scrofa*) carcass	.	.	1.3	0.4	2.4	2.9

Table 4.17 (continued)

Item	Eastern part		Western part			
	Sp–Su	Au–Wi	Sp–Su		Au–Wi	
	%Occ	%Occ	%Occ	%Bio	%Occ	%Bio
Roe deer (*Capreolus capreolus*) carcass	.	.	–	–	–	–
Undetermined Cervidae carcass	–	–	4.8	0.9	8.5	1.5
Domestic livestock carcass	–	–	–	–	0.1	0.1
Ungulate carcasses total	**2.1**	**1.4**	**6.0**	**1.3**	**10.6**	**4.5**
Hazel hen (*Bonasa bonasia*)	1.6	1.0	–	–	–	–
Undetermined *Columba* sp.	–	–	0.2	0.1	–	–
Woodpecker (*Dendrocopos* sp.)	–	0.6	0.3	0.7	0.3	0.2
Wren (*Troglodytes troglodytes*)	–	–	0.3	0.2	–	–
Robin (*Erithacus rubecula*)	–	–	0.2	0.1	0.1	+
Blackbird (*Turdus merula*)	–	–	0.2	0.1	–	–
Songthrush (*Turdus philomelos*)	–	–	0.8	1.0	0.1	0.1
Undetermined thrush (*Turdus* sp.)	–	–	0.3	0.1	–	–
Spotted flycatcher (*Muscicapa striata*)	–	–	–	–	0.1	+
Flycatcher (*Ficedula* sp.)	–	–	0.5	0.2	0.1	0.2
Blue tit (*Parus caeruleus*)	–	–	0.3	0.5	–	–
Nuthatch (*Sitta europaea*)	–	–	0.3	0.2	0.2	+
Jay (*Garrulus glandarius*)	–	–	0.3	0.2	–	–
Starling (*Sturnus vulgaris*)	–	–	0.2	0.1	–	–
Hawfinch (*Coccothraustes coccothraustes*)	–	–	0.2	0.1	–	–
Undetermined medium-sized bird	–	–	1.3	1.0	0.2	0.1
Undetermined small passerine	17.4	4.3	17.2	8.2	3.5	1.0
Bird egg	0.5	0.2	1.3	0.2	0.1	+
Birds total	**19.0**	**6.1**	**23.3**	**13.0**	**4.5**	**1.6**
Lizard (*Lacerta* sp.)	0.5	0.2	0.2	+	0.3	+
Undetermined reptile	4.2	0.8	–	–	–	–
Reptiles total	**4.7**	**1.0**	**0.2**	**+**	**0.3**	**+**
Undetermined frog (*Rana* sp.)	–	–	0.7	0.3	2.1	0.6
Undetermined anuran	–	0.4	9.9	1.3	12.9	1.4
Amphibians total	**–**	**0.4**	**10.6**	**1.6**	**15.0**	**2.0**
Fish	–	–	–	–	0.3	+
Snail	–	–	2.0	+	2.1	+
Carabidae beetle	–	–	4.1	0.1	3.7	0.1
Undetermined beetle	7.9	2.4	2.8	0.1	1.4	+

Table 4.17 (continued)

Item	Eastern part		Western part			
	Sp–Su	Au–Wi	Sp–Su		Au–Wi	
	%Occ	%Occ	%Occ	%Bio	%Occ	%Bio
Larvae of beetles	–	–	0.7	+	2.5	+
Nest of wasps	–	–	10.1	0.4	5.5	0.2
Nest of bees	–	–	1.3	0.1	1.2	0.1
Nest of bumblebees	–	–	2.3	0.1	0.5	0.2
Other and undetermined insects	6.8	1.8	1.7	+	1.2	+
Earthworm	Not studied		1.0	+	0.5	+
Invertebrates total	**14.7**	**4.2**	**23.5**	**0.8**	**16.5**	**0.6**
Raspberry (*Rubus* sp.)	–	–	4.8	1.2	2.9	0.1
Vaccinium myrtillus, V. vitis-idaea, Oxycoccus quardipetalus berries	9.5	2.4	–	–	–	–
Rowan (*Sorbus aucuparia*) fruit	–	–	4.3	1.5	1.9	0.3
Malus sp. and *Crataegus* sp. fruit	0.5	0.4	–	–	–	–
Other soft-tissue fruit	–	–	1.7	0.1	1.1	+
Oak (*Quercus robur*) seeds	–	–	0.3	+	0.9	0.1
Hazel (*Corylus avellana*) nuts	–	–	0.5	+	0.7	0.1
Other tree seeds	–	–	0.7	+	1.2	0.1
Other plant material	2.6	–	10.4	0.8	4.0	0.3
Mushrooms	–	–	2.3	0.2	3.4	0.3
Plant material total	**12.6**	**2.8**	**22.2**	**3.8**	**14.1**	**1.3**
Total *N* scats or biomass (g)	190	492	604	8810 g	1131	18392 g

Eastern part – Dackevich (1979; period of data collecting not stated). Western part – Jędrzejewski et al. (1993 c); data for 1985–1992.
Denotations as in Table 4.11.

Dot, lack of numerical information on the contribution of that prey.

September (up to 37% of biomass consumed in June), when migratory songbirds are abundant and all birds reproduce. Interestingly, 50% of all birds identified from marten scats were hole-nesters, an obviously vulnerable prey to martens. Dackevich (1979), who studied the diet of pine martens in the Belarussian part of the forest, identified hazel hens and woodpeckers among birds eaten by martens (Table 4.17).

In late summer and early autumn, martens consumed much vegetal matter (up to 17% of food biomass in September), particularly raspberries, mushrooms, and rowan fruit (Fig. 4.20). In the eastern part of the forest, martens most frequently consumed blueberries, cowberries, and cranberries (Table 4.17). Other important food resources to martens were insectivores (moles

Table 4.18. Modes of marten foraging in the cold season as revealed by snow tracking of individual pine martens in 1985/1986–1991/1992 in BNP. The total length of snow tracking was 64.1 km (Jędrzejewski et al. 1993 c)

Activity	Number/km of trail
Modes of area searching	
Walking: along fallen logs and root plates of fallen trees	6.5 (21) [a]
under fallen logs or branches	15.7 (25) [a]
on log and root plate of a fallen tree	21.9 (40) [a]
round the base of a standing tree	6.1
Penetrating the cavity of a hollow tree	0.8
Climbing a growing tree	1.0 [b]
Number of loops on own trail	2.7
Attacks and foraging	
Attacks on rodents (total)	4.1
by scratching snow	3.6
by jumping or chasing prey	0.3
by digging into rodent burrow	0.2
Attacks on: squirrel	0.05
bird	0.03
weasel	0.02
shrew	0.05 [c]
Digging in mud for frogs	0.05
Scavenging on ungulate carcasses	0.06 [d]
Feeding on previously cached food	0.05 [e]
Foraging on plant material and mushrooms	0.02

[a] In parentheses: mean distance of a given area searching mode (in m/km of trail).
[b] The mean straight line distance of off-ground walking was 4.2 m.
[c] Includes 0.03 shrews/km killed and left uneaten.
[d] One red deer killed by wolves, one roe deer killed by lynx, one dead red deer.
[e] One bank vole and one shrew.

and shrews), frogs, and the carcasses of ungulates. Martens scavenged on carcasses most intensely during late winter, the harshest season (Fig. 4.20). They preferred to feed on the remains of wolf kills rather than on ungulates that had died from disease and/or starvation (see Sect. 5.10). Also worth mentioning are insects, which made up a fairly small portion of the biomass of marten food (maximally 3% in September), but were very frequently eaten; insect remains were found in 51% of scats collected in September; Jędrzejewski et al. 1993 c). Nearly half of all invertebrates taken by martens were Hymenoptera insects: wasps *Vespula vulgaris*, *V. rufa*, *V. germanica*, *Dolichovespula media*, and *D. saxonica*, wild honey bees *Apis mellifera*, bumblebees *Bombus* sp., and hornets *Vespa crabro*. This list includes crown-nesting, hole-nesting, and ground-nesting species, and shows that martens were eager to get both insects and honey, whenever available.

Month-to-month variation in marten diet (Fig. 4.20) shows that in spring and summer, martens become markedly more arboreal than during the cold season. In June and September, up to 40% of food biomass is found at higher 'floors' of the forest. In contrast, in autumn–winter, they obtained most of their food on the ground. Snow tracking of pine martens conducted in Białowieża National Park revealed that on each kilometre of its trail, a marten walked on or over the logs and root plates of wind-fallen trees 22 times on average (Table 4.18). It climbed a standing tree once per kilometre and moved in the canopies for an average of 4.2 m straight-line distance before going down again. In winter, most hunting efforts were attacks on rodents (92% of all attacks), mostly by scratching through snow. At least 12% of attacks were successful. With a mean daily movement distance of 7.2 km (as revealed by snow tracking and radio-tracking), pine marten attempted 30 daily attacks on rodents, of which at least four were successful. Moreover, on average, every second day a marten scavenged on an ungulate carcass.

Deep snow cover makes it difficult for martens to hunt bank voles, the folivores staying and foraging in subnivean spaces. During the seven winters of our study (1985/1986–1991/1992), the average dry weight of bank voles in marten scats dropped from 0.31 g/scat at no snow cover to 0.04 g/scat at snow >41 cm deep, that is nearly eightfold (Jędrzejewski et al. 1993 c). Interestingly, the snow-caused decrease in marten's consumption of yellow-necked mice (which often gather tree seeds from the snow surface) and *Microtus* voles was rather weak.

Multiannual variation in martens' diet composition was still greater than seasonal changes and it was driven by fluctuations in numbers of forest ro-

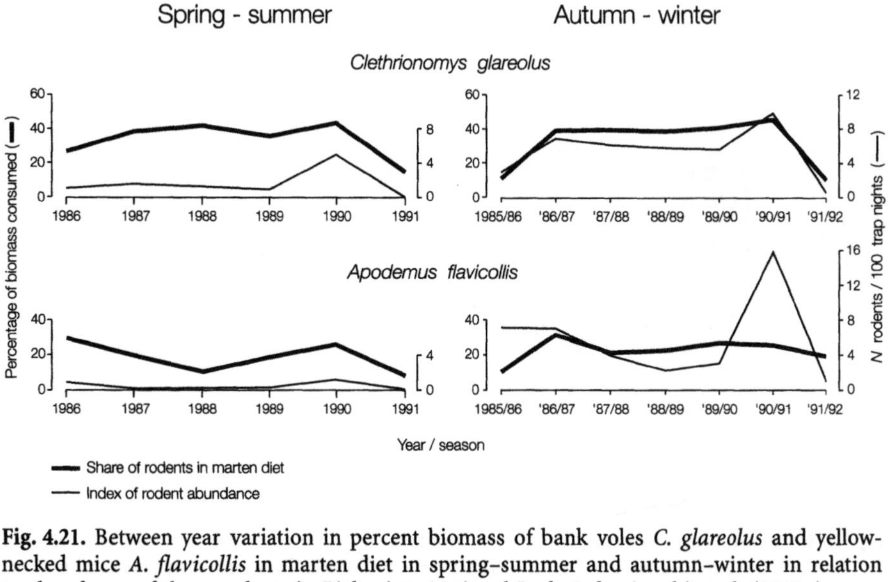

Fig. 4.21. Between year variation in percent biomass of bank voles *C. glareolus* and yellow-necked mice *A. flavicollis* in marten diet in spring–summer and autumn–winter in relation to abundance of these rodents in Białowieża National Park. Jędrzejewski et al. (1993 c)

dents, primarily the bank vole – martens' favourite and most important prey species. Percent biomass of bank voles in marten diet in autumn–winter correlated with the summer-autumn numbers of these rodents (Fig. 4.21). The share of yellow-necked mice in the cold season diet of martens was not related to the abundance of mice. In spring, when the two rodent species were scarce, martens showed a dietary response to the numerical changes in both the bank vole and the mouse (Fig. 4.21).

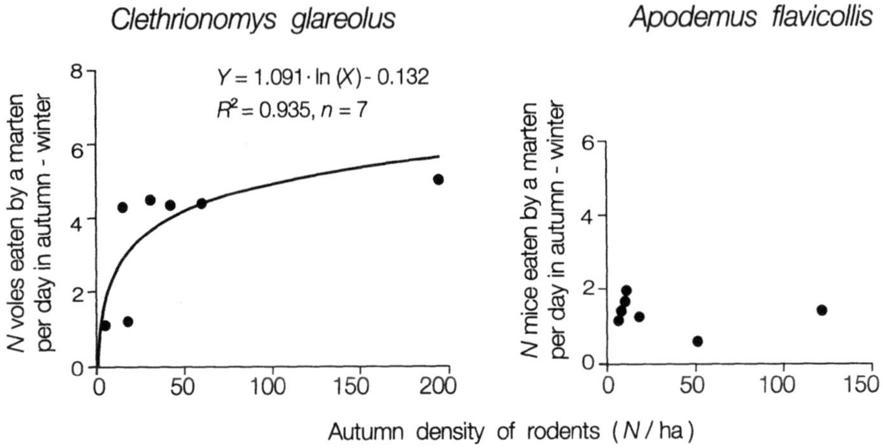

Fig. 4.22. Functional (dietary) response of pine martens in autumn–winter (1 Oct.–15 April) to variation in autumn (September/October) densities of bank voles and yellow-necked mice. Data for 1985/1986–1991/1992, each point denotes 1 year. Zalewski et al. (1995)

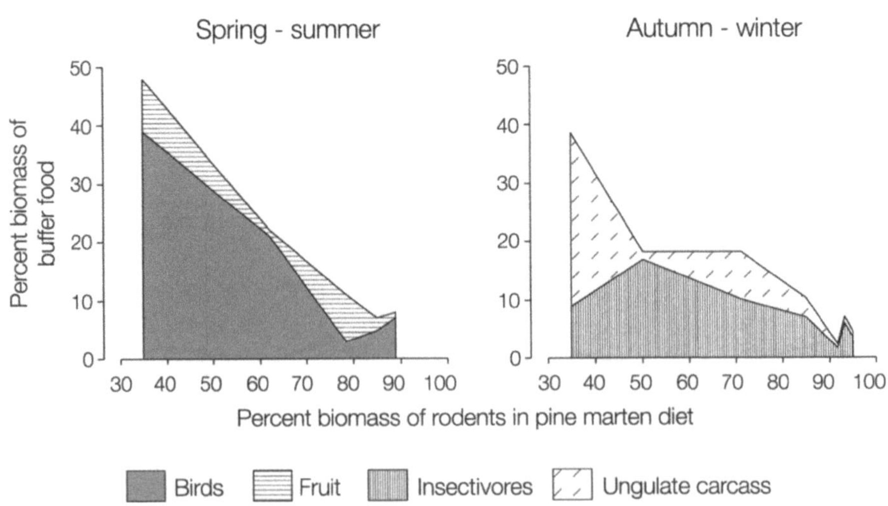

Fig. 4.23. Relationship between the percent biomass of rodents and percent biomass of buffer food resources in marten diet in 1985/1986–1991/1992. Data from: Jędrzejewski et al. (1993 c)

In 1985–1991, autumn densities of forest rodents (voles and mice combined) varied from 11 to 315 ind/ha, i.e. 30 times. However, their share in martens' diet ranged from 29% of biomass consumed to 71%; 2.5-times only (Jędrzejewski et al. 1993c, Zalewski et al. 1995). Even during crash numbers of rodents, martens continued to hunt them rather intensively, and supplemented their diet with secondary resources. Martens showed a distinct functional (dietary) response to changes in bank vole numbers; it was clearly an asymptotic curve (type II, according to Holling 1959; Fig. 4.22). Similar, manifold fluctuations of yellow-necked mouse numbers evoked no functional response from martens. This fact, together with high predominance of bank voles among forest rodents captured by martens, shows a specialisation of these predators in *Clethrionomys glareolus*, a species occurring widely in all types of temperate and boreal forests, throughout the geographic range of the pine marten.

Martens compensated for the scarcity of rodents by increased consumption of buffer resources. Interestingly, they utilised different alternative foods in spring–summer (birds and fruit) and in autumn–winter (insectivorous mammals and ungulate carcasses). The shares of these resources in marten diet grew with its declining consumption of rodents (Fig. 4.23). Birds in particular seem to suffer much heavier predation when rodents are too scarce to provide martens with enough food.

4.9 Stoats - Digging for Rodents

Information we have about the stoat (Fig. 4.24) represents its secondary habitats, the forests of Białowieża National Park, and not its prime living space, open marshes in the river valleys. In the forests, rodents constituted 83% of prey biomass and amphibians made up the rest (Table 4.19). Strong positive selection of *Microtus* voles (14% of all rodents captured versus 7% in forest

Fig. 4.24. Stoat *Mustela erminea* with a killed vole. Photo: S. Wąsik

Table 4.19. Diet composition of stoats in Białowieża Forest. Data represent autumn-winter seasons of 1987–1988 and forests of Białowieża National Park. $N=18$ prey specimens recovered from 15 scats

Item	Percentage of prey	Percentage of biomass consumed
Bank vole (*Clethrionomys glareolus*)	38.9	31.0
Undetermined vole (*Microtus* sp.)	11.1	12.0
Yellow-necked mouse (*Apodemus flavicollis*)	27.8	40.4
Rodents total	**77.8**	**83.4**
Undetermined amphibian (*Anura* sp.)	22.2	16.6

Table 4.20. Modes of area searching and hunting by stoat in the forests of Białowieża National Park as revealed by snow tracking in 1986/1987 and 1987/1988. Total length of snow tracking 8.8 km. All activities recalculated per 1 km of trail

Activity	Number (*N*/km of trail)	Length (m/km of trail)
Modes of movement and searching		
Walking: along and under fallen logs and branches	18.0	41.6
on fallen logs, branches and root plates of fallen trees	12.5	17.8
around bottoms of standing trees	29.7	
under snow cover	0.1	0.3
Hunting and foraging modes		
Penetrating potential rodent holes	9.3	
including: holes in trunks and root plates	7.6	
Attacks on rodents	8.6	
including: by digging into holes at tree bottoms	5.4	
by digging in snow	1.2	
by chasing a rodent	0.2	1.7[a]
Digging in mud for frogs	0.2	

[a] One successful chase (7 m long) and an unsuccessful one (8 m).

rodent community, see Sect. 2.4) also highlights the feeding specialisation of stoats living in the marshes. Snow tracking of stoats in the forests revealed that they hunted by a systematic search of fallen logs and standing trees, especially by penetrating all kinds of holes and cavities at the bottom of trees (Table 4.20). Stoats often dug at an entrance to enlarge it. Only rarely did they chase a rodent.

4.10 Weasels - Rodent Specialists

No other carnivore in Białowieża Forest is as strongly specialised on rodents as weasel; 88% of food biomass of this predator is made up of rodents (Table 4.21). In years of very high numbers of rodents (>300 ind/ha), weasels exclusively captured rodents. In years of rodent crashes, weasels were not able to survive on any buffer prey (Jędrzejewski et al. 1995); occasional captures of shrews, moles and birds could not substitute for the lack of their main prey. In hunting rodents though, weasels are true masters. Snow tracking we have done in Białowieża National Park revealed that they systematically searched the forest floor: fallen logs, branches, bottoms of standing trees and root plates of fallen trees, as well as cavities and crevices up the tree trunks (Table 4.22). They penetrated as many as 26 potential rodent holes per 1 km of their movement and in four of them they attacked rodents.

In autumn–early winter 1990, when rodents were abundant after a heavy tree seed crop, we made many visual observations of radio-collared male weasels (Jędrzejewski et al. 1992b). Diurnal, active from sunrise to sunset, weasels appeared fairly easy and very enjoyable targets of observations. After a radio-collared weasel had entered a burrow or tree cavity inhabited by rodents, they immediately escaped through 'emergency exits'. If a hunt was suc-

Table 4.21. Diet composition of weasels in Białowieża Forest. Data include 12 stomachs (1961–1991), 10 scats (1989–1991) and 36 successful attacks by radio-collared weasels (1990–1991). Seasons and years pooled. $N=69$ prey specimens

Prey item	Percentage of prey	Percentage of biomass consumed
Mole (*Talpa europaea*)	1.4	3.9
Undetermined shrew (*Sorex* sp.)	8.7	3.1
Undetermined insectivore	2.9	1.1
Insectivores total	**13.0**	**8.1**
Bank vole (*Clethrionomys glareolus*)	29.0	21.9
Undetermined vole (*Microtus* sp.)	5.8	5.9
Yellow-necked mouse (*Apodemus flavicollis*)	42.0	57.9
Undetermined rodent	2.9	2.5
Rodents total	**79.7**	**88.2**
Nuthatch (*Sitta europaea*)	1.5	1.5
Bird egg	1.5	0.1
Birds total	**3.0**	**1.6**
Undetermined amphibian (*Anura* sp.)	2.9	2.1
Carabidae beetle	1.4	+

Table 4.22. Modes of area searching and hunting by weasels, as surveyed by snow tracking in Białowieża National Park in 1986/1987–1988/1989. Total length of snow tracking 3.9 km. All activities recalculated per 1 km of trail

Activity	Number (*N*/km of trail)	Length (m/km of trail)
Modes of movement and area searching		
Walking: along fallen logs and branches	36.9	124
under fallen logs and branches	104.6	292
around bottoms of standing trees	36.2	27
Moving under snow cover		39
Climbing standing trees	0.5	
Hunting and foraging modes		
Penetrating potential rodent holes	26.3	
including: holes at tree bottoms	11.6	
cavities and hollow trees	2.3	
Attacks on rodents	3.9	
including attacks:		
in holes at bottoms of standing trees	2.3	
at the root plates of fallen trees	0.25	
under fallen logs and branches	0.25	
by digging in snow	0.5	

cessful, a weasel usually retreated to the rodent nest just explored and often stayed there overnight (Fig. 4.25).

Autumnal observations of 60 attacks on rodents by radio-collared weasels yielded information complementary to that gained by snow tracking in winter. During their systematic search, weasels attacked rodents in underground dens or tunnels (33% of attacks, mostly on bank voles), on the ground at fallen logs or among vegetation (45%, equally often on voles and yellow-necked mice), and in cavities located 1–4 m up the old trees (22% of attacks, all on mice, Fig. 4.26). In fact, the proportion of rodents encountered by weasels in underground dens was still higher, because most of the rodents captured on the ground were those that had been chased by a weasel out of their dens. No wonder that in spring and summer, when rodents reproduce, 70% of bank voles killed by weasels were nestlings (Jędrzejewska and Jędrzejewski 1990). Usually, a successful hunt underground or in cavity was rapid and a weasel was briefly seen carrying seized prey back to its burrow. Only on a few occasions, did we observe more adventurous pursuits. In six attacks, all on yellow-necked mice, a weasel chased a mouse for 5 to 20 meters. Exceptionally, the chase might have lasted as long as 2 minutes. In one case, the weasel chased a female mouse which had a few offspring attached to her nipples. She climbed a lime tree and managed to escape but lost some of the babies, which fell off as she entered a crevice in the bark 0.5 m up the trunk.

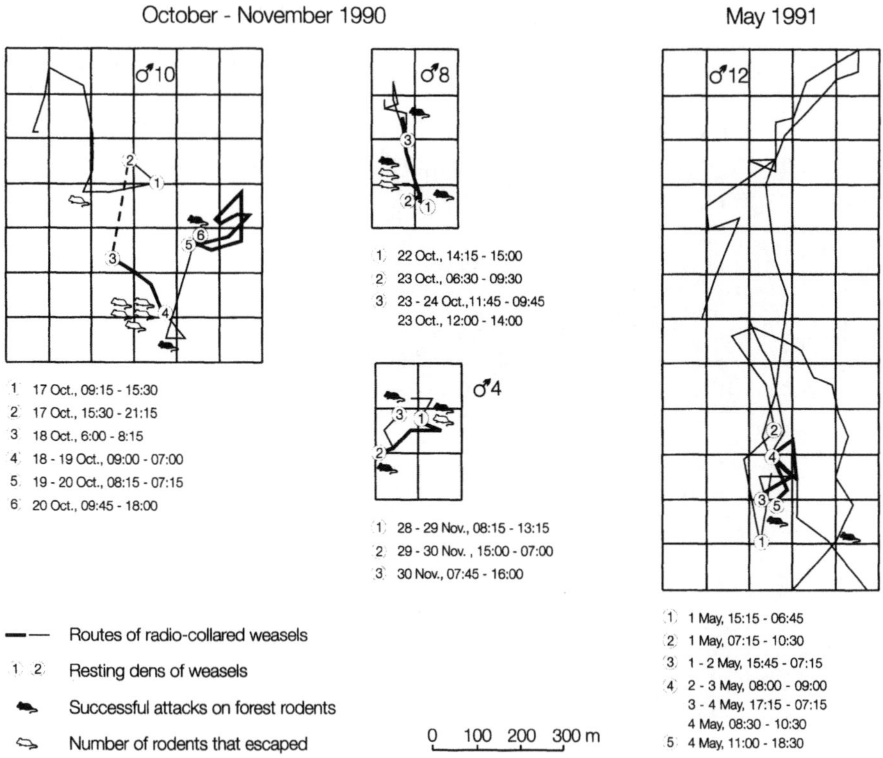

Fig. 4.25. Routes, resting sites, and attacks on rodents by four radio-collared weasels in 1990 (high densities of rodents) and 1991 (very low densities of rodents). (W. Jędrzejewski and B. Jędrzejewska, unpubl. data)

In 1990, despite very high densities of rodents, only 42% of weasel attacks were successful (Jędrzejewski et al. 1992b). In 41 cases, a solitary rodent was a weasel's target, and in 19 cases, huddles of two to six rodents were attacked. Except for one hunt that ended with two prey killed, weasels killed one rodent of the huddle or none. Thus, rodents staying in pairs or groups had, on average, 20% higher chance of surviving weasel attacks than solitary rodents.

To fulfill their daily food requirements (30–36 g), a weasel has to capture one to two forest rodents per day (Fig. 4.27). However, with their small elongated bodies and short fur, weasels are very sensitive to cold and wet weather. Our observations of weasels hunting rodents under seminatural conditions in the outdoor enclosures (Jędrzejewska and Jędrzejewski 1989) showed that, in summer, weasels went out hunting on 90% of days and killed, on average, two rodents per day. In winter (mean ambient temperature -7 °C), weasels hunted on 60% of days only, and on other days they stayed in well-insulated nests (Fig. 4.27). This necessary reduction in hunting time was balanced by extensive surplus killing and storing of buffer caches of food. On average, 3.1 rodents/day with hunt were killed at an ambient temperature of

Fig. 4.26. A radio-collared male weasel climbing a tree trunk in search of yellow-necked mice which rest in cavities and hollows up to 4 m high; Białowieża National Park, autumn 1990. Photo: K. Zub

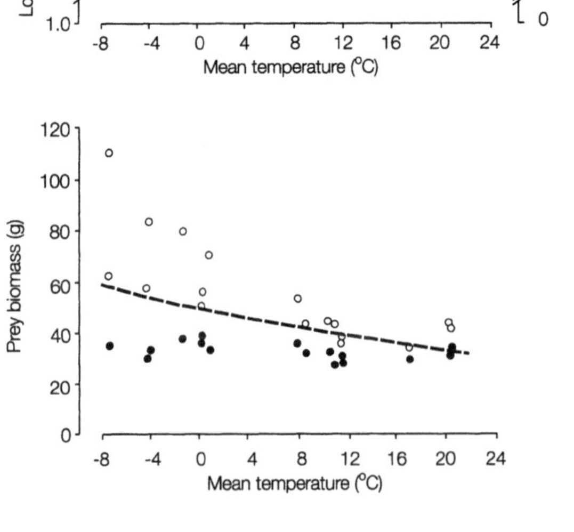

Fig. 4.27. *Upper panel* relationship between percentage of days with hunting and the mean temperature in 17 enclosure experiments with weasels and small rodents. Every point is a mean from one experiment. *Lower panel* effect of seasonal changes in temperature on weasels' hunting rate. *Solid points* mean daily consumption by weasels. *Open points* mean amount of prey caught per day with hunt. *Broken line* designated killing rate predicted from daily consumption and percentage of days with hunt. Actual killing rate was usually higher than predicted, particularly so in the cold season. Jędrzejewska and Jędrzejewski (1989)

–5 to 5 °C. As a rule, vast stores (6 to 14 times the daily food requirement) accumulated in weasels' nests in early winter. On even colder days (<–5 °C), weasels predominantly lived on previously stored food.

The pattern of storage accumulation and diminution was alike in all weasels in these experiments. Preferably, they ate the most recently caught rodents first. If the daily kill exceeded the immediate needs of a predator, it usually gnawed off and ate the heads and fore parts of several prey and left the rest. During a prolonged series of very frosty days, the weasel survived many days relying exclusively on the cache. When the sudden onset of hard frost (an overnight drop of 12 °C) found two weasels with almost empty larders, they barely managed to supply themselves with enough food, and at the price of being exposed to extreme cold temperatures. They caught one or two voles per day, lost weight and finally died, despite the abundance of highly vulnerable prey in the enclosures.

Severe winters and crashes of rodent numbers are not the sole obstacles to weasels' otherwise dolce vita. Perhaps equally important is the rodents' evolutionary gain of remote recognition of the weasel and adequate reactions to its hunting technique. We studied bank vole's antipredatory behaviour in series of enclosure and laboratory experiments and observed them in the forest during radio-tracking of weasels (Jędrzejewska and Jędrzejewski 1990; Jędrzejewski and Jędrzejewska 1990; Jędrzejewski et al. 1992b, 1993d). Bank voles recognised the scent of weasels and avoided places and areas tainted with it. Antipredatory behaviours were diverse and included ground escape by running, arboreal escape (climbing vegetation), escape out of burrows, staying immobile, and plasticity in diel activity rhythm. That these life-saving tactics are successful, was proved by weasels' relatively low hunting success (31% in seminatural enclosures and 42% under natural conditions). To us, one of the most memorable symbols of predator-prey relationships in Białowieża Forest has been the sight of a little bank vole sitting calm and immobile atop a stinging nettle while watching a weasel, which excitedly searched the ground vegetation in a nearby place, still smelling a vole.

4.11 Polecats' Cold Meals of Frogs

In their 'linear' territories along forest brooks and ash-alder forests, polecats slowly walk up and down the watercourse, foraging on frogs and covering daily 1.1 km, on average (Brzeziński et al. 1992). Anurans, mainly the common frog, comprised 94% of polecat food biomass in the cold season, and 60% in spring–summer (Table 4.23). Alternative prey for polecats were small mammals: forest rodents that were hunted year-round, and shrews, which were captured in substantial numbers in spring–summer. Of the two dominant species of forest rodents, polecats captured more yellow-necked mice than bank voles. Among Soricidae captured, water shrews were the most numerous (Table 4.23). Weber (1989) found, on the basis of enclosure experiments, that polecats captured frogs more easily than rodents. They could find and excavate frogs from 30 cm of soil, and locate them even if snow cover was up to 1 m thick.

Table 4.23. Diet of polecats in spring-summer (1987–1989) and autumn-winter (1986–1991) in Białowieża National Park (Polish part of BPF). Autumn-winter data after Jędrzejewski et al. (1993a)

Item	Spring-summer		Autumn-winter	
	%Occ	%Bio	%Occ	%Bio
Mole (*Talpa europaea*)	–	–	0.9	0.1
Common shrew (*Sorex araneus*)	2.5	1.0	0.5	0.1
Lesser shrew (*Sorex minutus*)	1.3	0.6	–	–
Undetermined shrew (*Sorex* sp.)	2.5	2.4	–	–
Water shrew (*Neomys fodiens*)	3.8	2.5	0.9	0.1
Undetermined water shrew (*Neomys* sp.)	2.5	1.3	–	–
Undetermined shrew (*Sorex* or *Neomys*)	11.4	3.5	3.2	0.2
Insectivores total	**24.0**	**11.3**	**5.4**	**0.5**
Bank vole (*Clethrionomys glareolus*)	10.1	4.3	5.0	1.0
Pine vole (*Pitymys subterraneus*)	6.3	4.2	–	–
Root vole (*Microtus oeconomus*)	1.3	1.8	–	–
Short-tailed vole (*Microtus agrestis*)	1.3	1.4	–	–
Undetermined vole (*Microtus* sp.)	7.6	3.9	–	–
Harvest mouse (*Micromys minutus*)	–	–	0.5	0.2
Yellow-necked mouse (*Apodemus flavicollis*)	5.1	2.7	10.4	3.0
Undetermined rodent	–	–	0.9	+
Rodents total	**25.3**	**18.3**	**16.7**	**4.2**
Cervidae carcass	–	–	1.4	0.8
Woodpecker (*Dendrocopos* sp.)	1.3	0.1	–	–
Blackbird (*Turdus merula*)	1.3	0.1	–	–
Undetermined passerines	11.4	1.4	1.4	+
Birds total	**12.7**	**1.6**	**1.4**	**+**
Common frog (*Rana temporaria*)	15.2	18.3	18.5	27.4
Moor frog (*Rana arvalis*)	–	–	1.4	1.2
Undetermined frog (*Rana* sp.)	36.7	41.6	67.6	65.7
Amphibians total	**51.9**	**59.9**	**86.9**	**94.3**
Stickleback (*Gasterosteus aculeatus*)	5.1	4.9	–	–
Undetermined small fish	5.1	0.1	1.8	0.1
Fish total	**10.1**	**5.0**	**1.8**	**0.1**
Snail	3.8	2.6	–	–
Dytiscus sp.	12.7	0.5	4.1	0.1
Geotrupes sp.	2.5	0.1	–	–
Other and undetermined insects	22.8	0.2	5.9	+

Table 4.23 (continued)

Item	Spring–summer		Autumn–winter	
	%Occ	%Bio	%Occ	%Bio
Earthworm	8.9	+	0.5	+
Invertebrates total	**45.6**	**3.4**	**10.4**	**0.1**
Plant material	3.8	0.5	0.5	+
Total *N* scats or biomass (g)	79	1165 g	222	5543 g

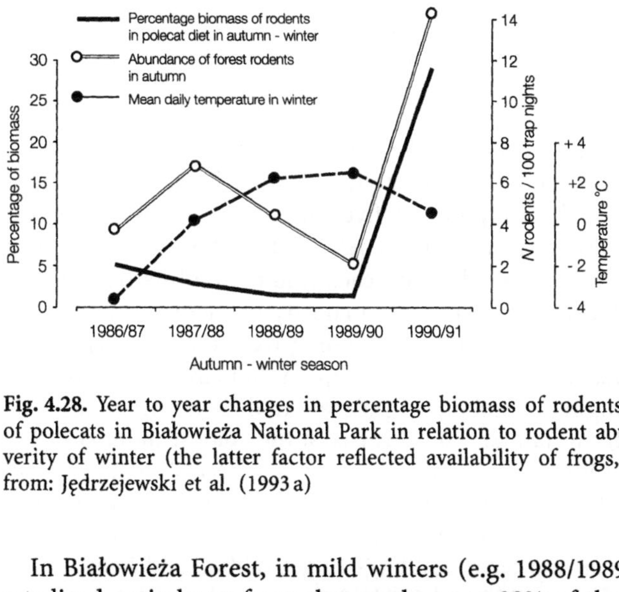

Fig. 4.28. Year to year changes in percentage biomass of rodents in the autumn–winter diet of polecats in Białowieża National Park in relation to rodent abundance in autumn and severity of winter (the latter factor reflected availability of frogs, polecat's main prey). Data from: Jędrzejewski et al. (1993 a)

In Białowieża Forest, in mild winters (e.g. 1988/1989 and 1989/1990), polecats lived entirely on frogs that made up to 98% of their food biomass. Moreover, polecats may store excess frogs captured at favourable conditions and exploit the stores when conditions become too harsh for successful hunting. Nonetheless, some conditions force polecats to prey on alternative resources. First, harsh winters and dry summers, each in a different way, make frogs unavailable to polecats. Secondly, forest rodents when very numerous can be easy and rewarding prey for polecats. In consequence, the proportions of frogs and rodents in the winter diet of polecats depended on the combination of winter temperature and rodent abundance (Fig. 4.28). In winter 1990/1991, after an outbreak of forest rodents, 29% of food biomass taken by polecats was rodents. In the severe winter of 1986/1987, when rodent numbers were rather low, frogs constituted 88% of polecat food, rodents 5%, and Cervidae carcass 6% (Fig. 4.28).

Snow tracking done in Białowieża National Park showed that, in riverside habitats, the polecat mostly searched for amphibians (Table 4.24). It waded

Table 4.24. Different foraging activities of polecats in wet habitats (bog alderwoods and riverside ash-alder forests) and drier habitats (oak-lime-hornbeam and spruce-pine forests), as revealed by snow tracking in 1986/1987–1990/1991. Total length of snow tracking: 6.8 km in wet habitats and 4.2 km in dry habitats. (Jędrzejewski et al. 1993 a)

Activity	Number or length (m) per 1 km of trail	
	Wet habitats	Drier habitats
Foraging on stream:		
wading in water or walking on ice	11.1	2.6
length of movement on ice	325.2 m	21.0 m
looking under banks	10.0	1.7
Hunting rodents: all attacks on rodents	4.9	14.8
including attacks by digging	2.9	12.2
Moving under snow cover	12.4 m	7.1 m

into water or walked on ice 11 times per kilometre of trail, and looked under the banks of the brook ten times. In this habitat, hunting for small mammals was also undertaken, though rarely (about five times per day). When forest brooks got ice-covered, polecats began to penetrate drier forests, where they predominantly hunted for rodents. Then, on each kilometre of its trail, a predator attempted 14.8 attacks on rodents, the majority of them (82%) by digging into rodent burrows at the bottoms of standing trees (Table 4.24).

During harsh winters, some polecats end up in human settlements. In Białowieża village, the highest numbers of polecats raiding domestic animals were captured by farmers in January and February, the coldest months of the year (Brzeziński et al. 1992). Conducive to this was polecats' adaptation to get its food by digging (frogs and rodents, and especially yellow-necked mice from their dens at tree bottoms). With scarcity or little access to their natural resources, polecats dug into farm buildings to get rabbits and hens.

In the cold seasons of 1993/1994 and 1994/1995, we made a pilot survey of the diet of polecats living on other rivers in Białowieża Forest, particularly those with open sedge marshy valleys (W. Jędrzejewski and V.E. Sidorovich, unpubl. data). Although material was rather small (26 scats), it was not different from that collected earlier in the pristine forests of Białowieża National Park. As much as 77% of food biomass consisted of amphibians, 20% rodents (*Microtus* voles and bank voles), and 3% insectivores (water shrews). Thus, in all types of its pristine habitats, polecats are specialised for hunting frogs.

4.12 River Otters – Between Fishing and 'Frogging'

In shallow, slowly flowing rivers and brooks of Białowieża Forest, fish was the staple food of otters year-round, constituting 51% of prey biomass consumed in spring-summer, and 40% in autumn-winter (Table 4.25). However,

Table 4.25. Diet composition of otter in spring-summer (1 April–30 Sept.) and autumn-winter (1 Oct.–30 March) in Białowieża Primeval Forest. Data for 1988–1996 after Brzeziński et al. (1993) and V.E. Sidorovich and W. Jędrzejewski (unpubl. data)

Item	Spring-summer		Autumn-winter	
	%Occ	%Bio	%Occ	%Bio
Water shrew (*Neomys* sp.)	–	–	1.3	0.1
Bank vole (*Clethrionomys glareolus*)	–	–	0.3	+
Mouse (*Apodemus* sp.)	–	–	0.3	0.1
Mammals total	**–**	**–**	**2.0**	**0.2**
Mallard (*Anas platyrhynchos*)	1.1	1.7	0.7	0.5
Wild duck (*Anas* sp.)	3.3	0.8	0.6	0.4
Undetermined bird	5.6	1.1	–	–
Birds total	**10.0**	**3.6**	**1.3**	**0.9**
Lizard (*Lacerta* sp.)	–	–	0.3	+
Common toad (*Bufo bufo*)	2.2	3.2	7.2	5.1
Common frog (*Rana temporaria*)	3.3	4.2	48.7	37.7
Edible frog (*Rana esculenta*)	–	–	1.6	0.6
Undetermined frog (*Rana* sp.)	15.6	20.5	15.7	6.1
Undetermined anuran	23.3	6.2	15.7	8.0
Anuran spawn	–	–	0.3	0.1
Amphibians total	**43.3**	**34.1**	**82.4**	**57.6**
Pike (*Esox lucius*)	8.9	4.3	14.1	5.6
Common bream (*Abramis brama*)	–	–	0.7	0.4
Roach (*Rutilus rutilus*)	8.9	10.1	8.2	5.1
Ide (*Leuciscus idus*)	–	–	2.3	0.5
Gudgeon (*Gobio gobio*)	1.1	2.0	0.7	0.1
Ide or gudgeon	–	–	2.6	1.4
Red eye (*Scardinius erythrophthalmus*)	–	–	0.3	0.1
Tench (*Tinca tinca*)	–	–	0.3	0.1
White bream (*Blicca bjoerkna*)	1.1	0.1	1.0	0.3
Crucian carp (*Carassius carassius*)	1.1	4.6	3.6	5.6
Undetermined Cyprinidae	7.8	5.0	2.0	1.9
Mud loach (*Misgurnus fossilis*)	–	–	1.6	0.8
Undetermined Cobitidae	1.1	0.1	0.7	0.3
Burbot (*Lota lota*)	–	–	1.6	1.4
Stickleback (*Gasterosteus aculeatus*)	–	–	3.9	2.5
Perch (*Perca fluviatilis*)	8.9	8.7	17.0	10.5
Ruff (*Gymnocephalus cernuus*)	–	–	2.3	0.7
Undetermined fish	38.9	15.7	12.4	2.5
Fish total	**65.6**	**50.7**	**55.6**	**39.8**

Table 4.25 (continued)

Item	Spring–summer		Autumn–winter	
	%Occ	%Bio	%Occ	%Bio
Crayfish (*Astacus* sp.)	1.1	0.1	7.5	0.5
Gastropoda	6.7	0.7	0.3	+
Dytiscus sp.	66.7	9.6	19.3	0.9
Dragonflies (Odonata)	3.3	+	–	–
Silpha sp.	–	–	0.3	+
Undetermined beetle	1.1	+	–	–
Larvae of insects	24.4	1.2	–	–
Undetermined insect	1.1	+	1.0	+
Arachnida	1.1	+	–	–
Invertebrates total	**80.0**	**11.6**	**25.5**	**1.5**
Plant material	–	–	0.7	+
Total *N* scats or biomass (g)	90	1670 g	306	9639 g

amphibians were as important as fish, and contributed 34% to otters' food in spring–summer and 58% in autumn–winter. Large water beetles, *Dytiscus* sp., were eaten in masses in the warm season (10% of biomass). The clear seasonal difference in the role of amphibians in otter diet is a consequence of frogs' biology. Throughout spring, summer and autumn, common frogs occur in large numbers in deciduous forests, but from October to March, they aggregate in small rivers and brooks to hibernate and spawn, and then become more available to otters.

The general picture of otter diet composition shown in Table 4.25 includes data from all types of Białowieża Forest rivers, from the very small (Orłówka) where otters ventured only temporarily, to small (Łutownia, Hwoźna) and medium-sized ones (Narewka, Leśna), where otters live permanently. The share of fish in the autumn–winter diet of otters was smallest from small rivers (10–13% of biomass) and grew with increasing river size (to 76% on the Leśna River). Otters compensated for the scarcity of fish in small brooks by preying more on amphibians (Fig. 4.29). This shift in otter's diet was made possible by higher abundance of frogs on small rivers than on bigger ones (Fig. 4.30).

Between year variation in otter diets was rather small. We have compared the material collected from Narewka River in the winters of 1990/1991, 1993/1994, and 1994/1995. The share of fish ranged from 49 to 65% of food biomass, and that of amphibians from 33 to 49%. However, we have to emphasise that the three winters studied were mild, and we have no information on otters feeding during very harsh winters.

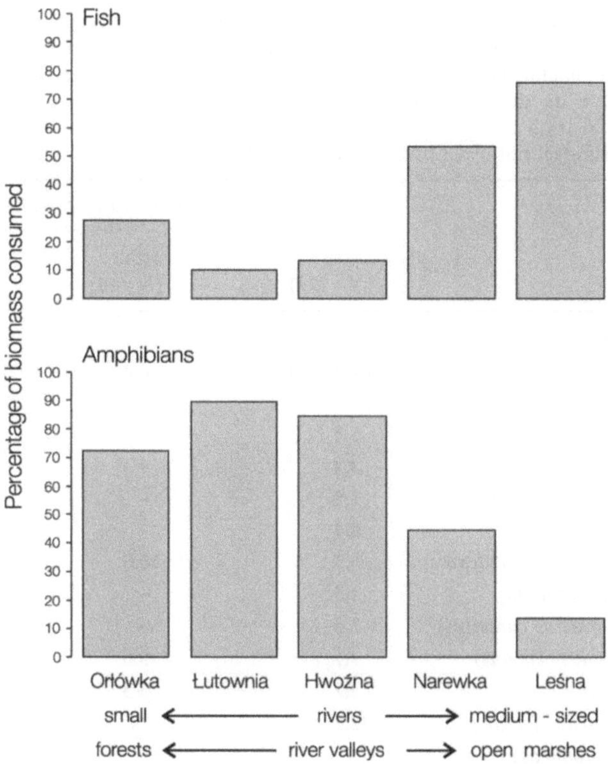

Fig. 4.29. Contributions of fish and amphibians to autumn-winter diet of river otters inhabiting rivers that differ in size and type of valley (forested versus open marsh). Data for the Polish part of BPF, from Brzeziński et al. (1993) and W. Jędrzejewski, V.E. Sidorovich, and B. Jędrzejewska (unpubl.)

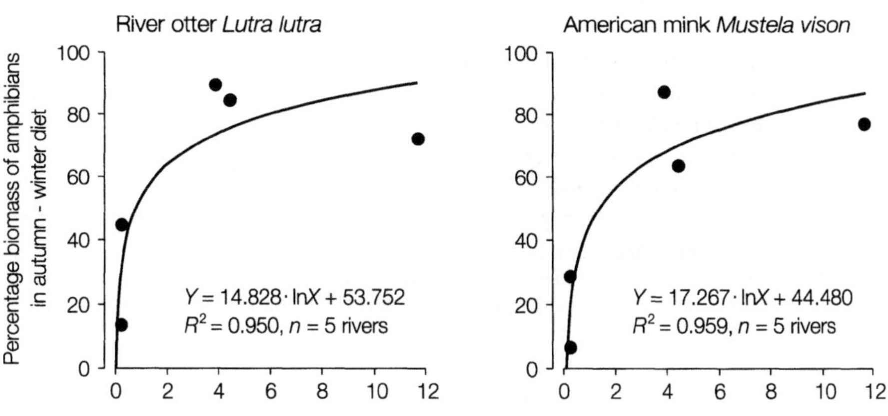

Fig. 4.30. Dietary response of river otters and American mink on various rivers in Białowieża Forest to spatial variation in abundance of the common frog *Rana temporaria*. Each point = one river (see Fig. 4.29)

Table 4.26. Percentage species structure of fish taken by otters and American mink (expressed as percentage of all fish remains identified from scats, with an assumption that an occurrence of a given species in one scat represented one individual), compared with the list of fish species reported from water bodies of Białowieża Forest. Abundance of fish was ranked based on Dackevich's (1983) information on rarity/commonness of all species in Narew, Narewka, and Leśna rivers, and artificial lakes in the Belarussian part of BPF. Among fish captured by otters, undetermined Cyprinidae, not listed in the table, formed 5.7%

Rank in fish community	Species of fish	Percentage of fish identified from scats	
		Otter (N=227)	Mink (N=18)
1	Perch (*Perca fluviatilis*)	26.5	–
2	Pike (*Esox lucius*)	22.5	22.2
3	Roach (*Rutilus rutilus*)	14.5	16.7
4	Ide (*Leuciscus idus*)	5.7	–
5	Gudgeon (*Gobio gobio*)	2.2	–
6	White bream (*Blicca bjoerkna*)	1.8	–
7	Tench (*Tinca tinca*)	0.4	–
8	Stickleback (*Gasterosteus aculeatus*)	5.3	50.0
9	Spiny loach (*Cobitis taenia*)	1.3[a]	–
10	Crucian carp (*Carassius carassius*)	5.3	–
11	Ruff (*Gymnocephalus cernuus*)	3.1	5.6
12	Burbot (*Lota lota*)	2.2	5.5
13	Mud loach (*Misgurnus fossilis*)	2.2	–
14	Bleak (*Alburnus alburnus*)	–	–
15	Dace (*Leuciscus leuciscus*)	–	–
16	Red eye (*Scardinius erythrophthalmus*)	0.4	–
17	Sunbleak (*Leucaspius delineatus*)	–	–
18	Chub (*Leuciscus cephalus*)	–	–
19	Barbel (*Barbus barbus*)	–	–
20	Common bream (*Abramis brama*)	0.9	–
21	Goldfish (*Carassius auratus gibelio*)	–	–
22	Wels (*Silurus glaris*)	–	–
23	Eel (*Anguilla anguilla*)[b]	–	–
24	Bitterling (*Rhodeus sericeus amarus*)[b]	–	–
25	Carp (*Cyprinus carpio*)[c]	–	–

[a] This figure includes all undetermined Cobitidae.
[b] Extinct in the 1960s, after the amelioration of river beds.
[c] It was released into man-made water reservoirs a few times but always perished.

What species of fish do otters capture? A thorough faunistic review of the fish community in the water bodies of Białowieża Forest written by Dackevich (1983) enabled us not only to present the list of species, but also to rank them according to their occurrence in the three main river systems (Leśna, Narew, Narewka) and some artificial lakes in BPF. The rivers of

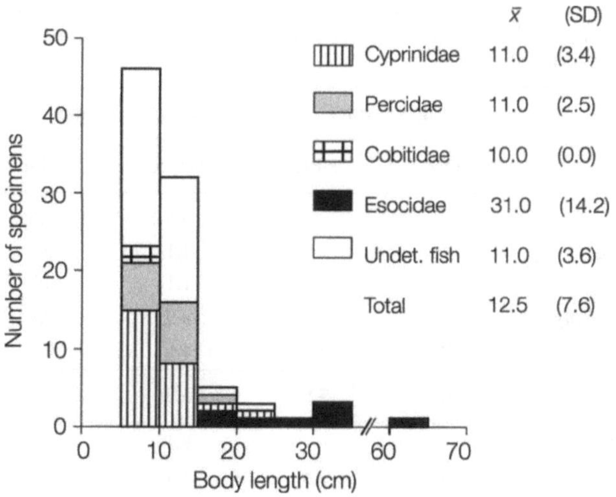

Fig. 4.31. Size of fish eaten by otters on Narewka and Hwoźna rivers. Total number of fish with estimated body length = 91. Brzeziński et al. (1993)

Białowieża Forest freeze partly or completely in winter. In summer, they harbour a rich community of water plants that fill the water body, making the water non-transparent and poor in oxygen. The rivers are thus inhabited by a community of fish resistant to oxygen deficiency and typical of nearly stagnant waters. The most widespread fish are perch, pike, roach, ide and gudgeon (Table 4.26). Except for pike, most of them are small species. Otters preyed on fish proportionally to their abundance, taking perch, pike, roach, and ide most frequently. By measuring the size of fish vertebrae recovered from otter scats, we estimated the size (body lengths) of fish captured by otters (Fig. 4.31). Cyprinids, perch, and cobitids were, on average, 10–11 cm long. Only pikes were considerably larger (mean length 31 cm).

The thriving otter population in Białowieża Forest shows that unpolluted small forest rivers supporting amphibians, small fish and water beetles are a significant otter habitat. In the past, such rivers with shallow, slowly flowing waters and flood plain valleys were the most numerous water courses in European lowlands. The example of Białowieża Forest shows that in this type of habitat, otters are doing fine by fishing and 'frogging'.

4.13 American Minks – Hunting Fish and Frogs in Summer, Frogs and Rodents in Winter

In the warm season, whenever we saw a mink in Białowieża Forest, it was at water: swimming in a river or dodging in reeds at the water's edge. In spring and summer, mink is truly a semiaquatic predator, relying on fish (40% of food biomass) and frogs (32%, mainly the common frog). It also supple-

Table 4.27. Diet composition of American mink in spring-summer (16 April–30 Sep.) and autumn-winter (1 Oct.–15 April) in Białowieża Primeval Forest. Data for 1987–1995 (V.E. Sidorovich and W. Jędrzejewski, unpubl. data)

Item	Spring-summer		Autumn-winter	
	%Occ	%Bio	%Occ	%Bio
Mole (*Talpa europaea*)	–	–	6.3	2.6
Common shrew (*Sorex araneus*)	–	–	1.1	0.3
Water shrew (*Neomys fodiens*)	–	–	1.1	0.3
Undetermined water shrew (*Neomys* sp.)	–	–	2.9	0.9
Undetermined Soricidae	28.0	5.5	4.6	0.6
Insectivores total	**28.0**	**5.5**	**16.0**	**4.7**
Bank vole (*Clethrionomys glareolus*)	–	–	9.1	4.4
Water vole (*Arvicola terrestris*)	4.0	3.5	3.4	4.7
Root vole (*Microtus oeconomus*)	–	–	9.1	7.6
Undetermined vole (*Microtus* sp.)	12.0	5.7	27.4	14.3
Pine vole (*Pitymys subterraneus*)	–	–	0.6	0.4
Undetermined mouse (*Apodemus* sp.)	–	–	0.6	0.1
Birch mouse (*Sicista betulina*)	–	–	0.6	+
Rodents total	**16.0**	**9.2**	**48.6**	**31.5**
American mink (*Mustela vison*)	12.0	6.1	–	–
Spotted crake (*Porzana porzana*)	–	–	2.3	1.9
Undetermined bird	8.0	0.8	2.3	0.5
Birds total	**8.0**	**0.8**	**4.6**	**2.4**
Common frog (*Rana temporaria*)	12.0	13.3	24.0	28.5
Undetermined frog (*Rana* sp.)	24.0	18.8	21.7	16.7
Undetermined anuran	–	–	13.1	5.9
Amphibians total	**36.0**	**32.1**	**56.6**	**51.1**
Pike (*Esox lucius*)	–	–	2.3	1.4
Roach (*Rutilus rutilus*)	4.0	12.5	1.1	1.8
Burbot (*Lota lota*)	4.0	2.3	–	–
Stickleback (*Gasterosteus aculeatus*)	4.0	2.5	4.6	3.1
Ruff (*Gymnocephalus cernuus*)	4.0	2.3	–	–
Undetermined fish	24.0	20.7	5.7	3.4
Fish total	**36.0**	**40.3**	**11.4**	**9.7**
Crayfish (*Astacus* sp.)	–	–	0.6	0.3
Gastropoda	4.0	0.2	–	–
Dytiscus sp.	24.0	3.8	6.3	0.3
Larvae of insects	20.0	1.9	–	–
Invertebrates total	**40.0**	**5.9**	**6.9**	**0.6**

Table 4.27 (continued)

Item	Spring-summer		Autumn-winter	
	%Occ	%Bio	%Occ	%Bio
Plant material	4.0	0.1	0.6	+
Total *N* scats or biomass (g)	25	130 g	175	1889 g

ments its diet with small quantities of rodents, shrews and water beetles (Table 4.27). In autumn–winter, mink becomes markedly more terrestrial. Its fur is less water resistant than that of otter, so mink tries to reduce swimming and wading in water. The share of fish in its autumn–winter diet drops to a mere 10% of the food biomass. In that season mink prey on frogs which have aggregated in rivers for hibernation (51% of food biomass), and rodents (31%). While polecats hunt more rodents in summer when frogs, their main food, are dispersed throughout forests and are more difficult to capture, minks, in the contrast, hunt more rodents in winter when fish become less available.

Most of our data covered 2 years (1994–1995), so between-year variation in mink diet (e.g. mink response to varying winter conditions, water level in summer etc.) remains unknown. However, the spatial variation in autumn-winter feeding by mink is well represented in that material. The rivers studied were arranged from very small (Orłówka) to medium-sized ones (Leśna). Along the same gradient, the valleys of rivers changed from forested ones to wide open sedge-reed marshes. In these habitats, the feeding habits of mink in autumn–winter shifted distinctly (Fig. 4.32). On the small Orłówka and Łutownia Rivers, minks lived on frogs (77 and 88% of food biomass, respectively), the share of which declined to 6% on the Leśna River. Such a change in mink diet was caused by the varying abundance of frogs: lowest on big rivers and highest on small ones (Fig. 4.30). Living on rivers with a scarce supply of frogs, minks hunted for rodents. The latter prey made only 2.5% of mink food in the forested valley of Orłówka and grew to 59% in the open valley of the Leśna River (Fig. 4.32). Minks hunted mainly *Microtus* voles (especially the root vole) and water voles. In contrast to polecats, which managed to dig out forest rodents when brooks froze, minks hunted rodents of open marshes almost exclusively.

Fish played a minor role in the autumn–winter diet of minks and, interestingly, they were captured mainly in the biggest and the smallest rivers (Fig. 4.32). Although the number of fish identified to species is small (Table 4.26), it is clear that minks captured different species in the Narewka and Leśna Rivers (roach and pike) compared with the small Orłówka brook (mainly tiny sticklebacks). Fishing for sticklebacks in the Orłówka River in autumn–winter seemed to be a necessity forced on minks by a lack of *Microtus* voles in that forested habitat. Snow tracking of individual minks on the

Orłówka River showed that most of their foraging activities were focused at finding frogs and fish, with few attempts to capture rodents (Table 4.28). Undoubtedly, a flexible feeding behaviour, based on adaptations to utilise aquatic (fish), semiaquatic (frogs), and terrestrial prey (rodents) was one of the American mink's keys to successful colonisation of Europe once introduced few decades ago.

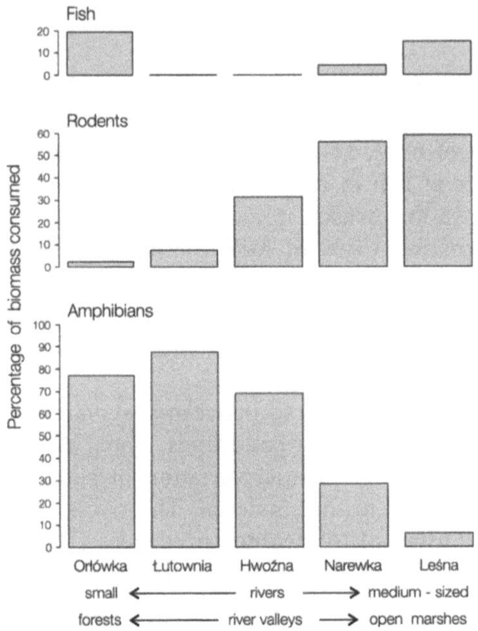

Fig. 4.32. Contributions of fish, rodents, and amphibians to autumn–winter diet of American mink inhabiting rivers that differ in size and type of valley. Data for the Polish part of BPF, from W. Jędrzejewski, V.E. Sidorovich and B. Jędrzejewska (unpubl.)

Table 4.28. Foraging activities of American mink on the Orłówka river in Białowieża National Park, as revealed by snow tracking in 1987/1988 and 1988/1989. Total length of snow tracking, 2.1 km

Activity	Number (*N*/km of trail)	Length (m/km of trail)
	Hunting for amphibians and fish	
Walking on ice-covered river	16.5	126.6
Searching under the river banks	7.3	11.8
Going into water	10.2	
Swimming	0.5 [a]	75.8
	Hunting for rodents	
Number of attacks on rodents	1.5	

[a] One long swim of 157 m.

4.14 Goshawks - Year-Round Pursuit of Birds

Although to inhabitants of villages in and around Białowieża Forest it symbolises an enemy of their chickens, the goshawk is primarily a woodland raptor that hunts under the forest canopy. Goshawks are elusive predators, very difficult to observe. Their hunting actions are usually seen post factum; a handful of jay feathers scattered on litter or snow. In the 1940–1960s, Białowieża's goshawks were the subject of interest as predators of game birds, and the stomachs of shot birds as well as prey collected at their nests provided the first information on goshawk diet in the eastern part of BPF (Table 4.29). In the breeding seasons of 1993 and 1994, we gathered pellets and prey remains from 20 nests of goshawks in the western part of BPF. All the material showed that goshawks are primarily bird eaters (Table 4.30). In winter and summer alike, deep in the forest interior as well as near the forest edge, birds made up about 86% of food biomass consumed by goshawks. In winter, no other predator of the temperate forests relies on birds so heavily. In the remaining 14% of goshawk's food, the red squirrel, an arboreal large rodent, strongly dominated (11% of consumed biomass). Small terrestrial rodents and shrews constituted usually less than 3% of animals captured (Tables 4.29 and 4.30).

Goshawks captured birds of at least 45 species, but just few groups of birds formed the bulk of their diet. Jays constituted from 6 to 26% of goshawk's food, and were one of the essential resources to this raptor in all seasons and the whole forest (Tables 4.29 and 4.30). In spring and summer, thrushes (mainly blackbirds and song thrushes, but also redwings and fieldfares) made up from 13 to 31% of goshawk's food biomass. In the predominantly coniferous eastern part of the forest, the hazel hen formed 22% of biomass in spring–summer, and 28% in autumn–winter (Table 4.29). In the more deciduous western part, goshawks captured hazel hen occasionally (0.2% of biomass), but they took fair numbers of woodpeckers (7–14% of biomass; Fig. 4.33).

All birds important to goshawks are large or medium-sized. Of them, the jay is particularly selected for. In our study, jays made up at least 26% of all birds captured by goshawks in spring–summer, whereas their share in the breeding community of forest birds did not exceed 1% (Tomiałojć et al. 1984). In the cold season (October–March), jays made up 7% of all birds or bird flocks observed in mature coniferous forests (goshawk's preferred habitat), woodpeckers 27.6%, tits 29.4%, and goldcrests 16% (calculated from Wołk 1985). Therefore, jays and woodpeckers were practically the only medium-sized birds which were fairly abundant in winter. These two are also vocally very active. Typical sounds of the forest on winter days are woodpeckers drumming and jays screeching, making these birds - perhaps to goshawks, too - very conspicuous.

Year-round, goshawks prefer to hunt the prey that is their essential and relatively common food resource in winter, a period critical for survival. Furthermore, they seem well adapted to hunt a particular size of year-round

Table 4.29. Diet composition of goshawk in spring-summer (April–Sept.) and autumn-winter (Oct.–March) seasons in the eastern part of Białowieża Primeval Forest in 1946–1960 (data compiled from Gavrin 1953; V.A. Dackevich, unpubl. data)

Item	Spring-summer		Autumn-winter	
	%Prey	%Bio	%Prey	%Bio
Brown hare (*Lepus europaeus*)	2.9	3.7	6.6	7.2
Red squirrel (*Sciurus vulgaris*)	8.8	11.2	9.2	10.2
Undetermined small rodent	4.4	1.3	–	–
Ungulate carrion	1.5	1.9	–	–
Mammals total	**17.6**	**18.1**	**15.8**	**17.4**
Mallard (*Anas platyrhynchos*)	1.5	1.8	1.3	1.4
Hazel hen (*Bonasa bonasia*)	17.6	22.4	25.0	27.6
Black grouse (*Lyrurus tetrix*)	1.5	1.9	2.6	2.9
Partridge (*Perdix perdix*)	–	–	9.2	10.2
Domestic hen (*Gallus gallus*)	5.9	7.5	14.5	16.0
Tawny owl (*Strix aluco*)	–	–	1.3	1.5
Undetermined small owl	–	–	1.3	1.4
Black woodpecker (*Dryocopus martius*)	–	–	2.7	2.9
Undet. woodpecker (*Dendrocopos* sp.)	1.5	1.2	–	–
Blackbird (*Turdus merula*)	13.2	17.1	–	–
Song thrush (*Turdus philomelos*)	13.2	12.0	–	–
Undetermined thrush (*Turdus* sp.)	1.5	1.6	–	–
Jay (*Garrulus glandarius*)	4.4	5.6	13.2	14.5
Jackdaw (*Corvus monedula*)	1.5	1.8	–	–
Starling (*Sturnus vulgaris*)	2.9	3.0	–	–
Bullfinch (*Pyrrhula pyrrhula*)	–	–	3.9	1.1
Yellowhammer (*Emberiza citrinella*)	–	–	1.3	0.4
Undetermined small passerine	13.2	5.2	–	–
Other or undetermined birds	–	–	7.9	2.7
Birds total	**77.9**	**81.1**	**84.2**	**82.6**
Sand lizard (*Lacerta agilis*)	1.5	0.2	–	–
Viviparous lizard (*Lacerta vivipara*)	1.5	0.2	–	–
Reptiles total	**3.0**	**0.4**	**–**	**–**
Undetermined frog (*Rana* sp.)	1.5	0.4	–	–
Amphibians total	**1.5**	**0.4**	**–**	**–**
Total *N* prey or biomass (g)	68	5250 g	76	6753 g

%Prey, percentage of all vertebrates found in stomachs of shot birds and in the nests of breeding pairs.
%Bio, percentage of total biomass of prey.

Table 4.30. Spring-summer food of goshawks in the Polish part of BPF in 1993-1994 (B. Jędrzejewska, W. Jędrzejewski, E. Pugacewicz and K. Zub, unpubl.)

Item	Forest interior			Forest-farmland ecotone		
	%Occ	%Prey	%Bio	%Occ	%Prey	%Bio
Eastern hedgehog (*Erinaceus concolor*)	-	-	-	1.0	0.3	0.4
Mole (*Talpa europaea*)	1.9	0.6	0.6	-	-	-
Insectivores total	**1.9**	**0.6**	**0.6**	**1.0**	**0.3**	**0.4**
Brown hare (*Lepus europaeus*)	2.3	0.8	1.1	2.5	0.8	1.1
Red squirrel (*Sciurus vulgaris*)	23.2	7.9	11.2	24.6	8.2	10.7
Bank vole (*Clethrionomys glareolus*)	6.1	2.6	0.6	2.5	0.8	0.2
Water vole (*Arvicola terrestris*)	0.4	0.1	0.2	-	-	-
Root vole (*Microtus oeconomus*)	-	-	-	0.5	0.2	0.1
Common vole (*Microtus arvalis*)	-	-	-	0.5	0.2	+
Undetermined vole (*Microtus* sp.)	0.8	0.3	0.1	3.4	1.1	0.3
Harvest mouse (*Micromys minutus*)	0.4	0.1	+	-	-	-
Undetermined mouse (*Apodemus* sp.)	1.9	0.6	0.3	-	-	-
Forest dormouse (*Dryomys nitedula*)	0.8	0.3	0.1	0.5	0.2	0.1
Undetermined small rodent	0.4	0.1	+	-	-	-
Small rodents total	**9.5**	**4.2**	**1.3**	**6.9**	**2.4**	**0.7**
Marten (*Martes* sp.)	-	-	-	0.5	0.2	0.2
Polecat (*Mustela putorius*)	-	-	-	0.5	0.2	0.2
Ungulate carcass	-	-	-	1.0	0.3	0.4
Mammals total	**31.2**	**13.5**	**14.2**	**33.0**	**12.4**	**13.3**
Mallard (*Anas platyrhynchos*)	-	-	-	0.5	0.2	0.2
Goshawk (*Accipiter gentilis*)	1.1	0.4	0.6	0.5	0.2	0.2
Sparrowhawk (*Accipiter nisus*)	4.2	1.4	2.0	5.4	1.8	2.4
Undetermined hawk (*Accipiter* sp.)	1.9	0.6	0.9	1.0	0.3	0.4
Hazel hen (*Bonasa bonasia*)	0.8	0.3	0.4	-	-	-
Domestic hen (*Gallus gallus*)	-	-	-	1.5	0.5	0.6
Partridge (*Perdix perdix*)	-	-	-	5.9	2.0	2.6
Lapwing (*Vanellus vanellus*)	0.4	0.1	0.2	0.5	0.2	0.2
Green sandpiper (*Tringa ochropus*)	-	-	-	0.5	0.2	0.2
Black-headed gull (*Larus ridibundus*)	0.8	0.3	0.4	1.0	0.3	0.4
Wood pigeon (*Columba palumbus*)	1.1	0.4	0.6	0.5	0.2	0.2
Rock dove (*Columba livia f. domestica*)	-	-	-	5.4	2.1	2.8
Undetermined *Columba* sp.	6.8	2.3	3.3	28.1	9.3	12.2
C#uckoo (*Cuculus canorus*)	0.4	0.3	0.4	0.5	0.2	0.2
Tawny owl (*Strix aluco*)	1.9	0.6	0.9	3.4	1.1	1.5
Long-eared owl (*Asio otus*)	0.4	0.1	0.2	0.5	0.2	0.2

Table 4.30 (continued)

Item	Forest interior			Forest-farmland ecotone		
	%Occ	%Prey	%Bio	%Occ	%Prey	%Bio
Owl (*Strix* or *Asio*)	–	–	–	0.5	0.2	0.2
Wryneck (*Jynx torquilla*)	0.8	0.3	0.1	–	–	–
Green woodpecker (*Picus viridis*)	0.4	0.1	0.2	0.5	0.2	0.2
Black woodpecker (*Dryocopus martius*)	1.5	0.5	0.7	–	–	–
Great spotted woodpecker (*Dendrocopos major*)	2.3	0.8	0.9	2.0	0.7	0.7
Middle spotted woodpecker (*Dendrocopos medius*)	1.5	0.5	0.4	0.5	0.2	0.1
White-backed woodpecker (*Dendrocopos leucotos*)	0.4	0.1	0.1	–	–	–
Woodpecker (*Dendrocopos* sp.)	29.7	10.6	11.4	18.7	6.2	6.2
Sky lark (*Alauda arvensis*)	–	–	–	3.0	1.0	0.5
Pipit (*Anthus* sp.)	0.8	0.3	0.1	–	–	–
Robin (*Erithacus rubecula*)	0.4	0.1	+	–	–	–
Blackbird (*Turdus merula*)	16.0	5.7	8.2	12.8	4.4	5.9
Song thrush (*Turdus philomelos*)	12.2	4.2	4.2	3.9	1.3	1.2
Redwing (*Turdus iliacus*)	0.4	0.1	0.1	0.5	0.2	0.2
Fieldfare (*Turdus pilaris*)	–	–	–	4.9	1.6	2.2
Undetermined thrush (*Turdus* sp.)	12.9	4.4	5.7	8.9	2.9	3.5
Warbler (*Phylloscopus* sp.)	0.4	0.1	+	–	–	–
Pied flycatcher (*Ficedula hypoleuca*)	0.4	0.1	+	–	–	–
Great tit (*Parus major*)	–	–	–	0.5	0.2	+
Undetermined tit (*Parus* sp.)	1.5	0.5	0.1	–	–	–
Nuthatch (*Sitta europaea*)	0.4	0.1	+	–	–	–
Golden oriole (*Oriolus oriolus*)	3.0	1.0	1.1	3.4	1.1	1.1
Jay (*Garrulus glandarius*)	48.3	18.7	26.4	32.0	12.7	16.7
Magpie (*Pica pica*)	–	–	–	1.5	0.5	0.6
Nutcracker (*Nucifraga caryocatactes*)	1.9	0.6	0.9	1.0	0.5	0.6
Crow (*Corvus corone*)	0.8	0.3	0.4	1.0	0.5	0.6
Raven (*Corvus corax*)	–	–	–	0.5	0.2	0.2
Rook (*Corvus frugilegus*)	0.4	0.1	0.2	–	–	–
Undetermined corvid (*Corvus* sp.)	0.8	0.3	0.4	3.4	1.1	1.5
Starling (*Sturnus vulgaris*)	0.8	0.3	0.3	8.9	3.4	3.7
Chaffinch (*Fringilla coelebs*)	3.4	1.2	0.4	3.4	1.1	0.4
Greenfinch (*Carduelis chloris*)	0.4	0.1	0.1	–	–	–
Hawfinch (*Coccothraustes coccothraustes*)	3.4	1.2	0.9	1.0	0.3	0.2
Undetermined large bird	2.3	0.9	1.3	3.9	1.3	1.7
Undetermined small bird	31.9	10.9	10.9	43.3	14.4	13.3
Birds total	**96.6**	**71.2**	**85.5**	**98.0**	**74.7**	**86.2**

Table 4.30 (continued)

Item	Forest interior			Forest-farmland ecotone		
	%Occ	%Prey	%Bio	%Occ	%Prey	%Bio
Grass snake (*Natrix natrix*)	0.4	0.1	0.2	–	–	–
Undetermined frog (*Rana* sp.)	1.1	0.4	0.1	–	–	–
Carabidae beetle	11.4	4.5	+	5.4	1.8	+
Geotrupes sp.	1.1	0.4	+	3.0	1.0	+
Chrysomelidae beetle	1.1	0.6	+	2.0	0.7	+
Dytiscus sp.	0.8	0.3	+	–	–	–
Other beetles	7.2	2.7	+	11.8	5.4	+
Hornet (*Vespa crabro*)	0.4	0.1	+	0.5	0.2	+
Other and undetermined insects	10.6	6.1	+	6.9	3.9	+
Invertebrates total	**27.8**	**14.8**	**+**	**20.7**	**12.9**	**+**
Total *N* pellets and prey remains, *N* prey specimens or biomass (g)	263	770	53381 g	203	613	45693 g

%Occ, percentage occurrence in pellets; %Prey, percentage of all prey specimens recovered from pellets; %Bio, percentage of biomass consumed. Forest interior, food composition of pairs breeding in the forest >1 km from the nearest open area. Forest-farmland ecotone, pairs breeding <1 km from the forest edge.

Fig. 4.33. Middle spotted woodpecker *Dendrocopos medius*. All woodpeckers are strongly selected out of the bird community by sparrowhawks, goshawks, and buzzards. Photo: S. Wąsik

resident avian (or otherwise arboreal) prey. This important, positively selected prey can be hazel hens, red squirrels, or - as shown by diet composition of the 'ecotone' pairs of goshawks - domestic pigeons, chickens, and partridges (Tables 4.29 and 4.30). Since bird numbers are rather stable from year to year (see Sect. 2.8), multiannual variation in goshawk diet was negligible, uncomparably smaller than that in predators relying on rodents.

4.15 Sparrowhawks - Raptors Specialised on Small Forest Birds

The forests and other habitats of Białowieża are home to rich communities of birds. The number of species breeding in an average 25 ha of the forest often reaches 50 (Tomiałojć et al. 1984). The overwhelming majority of species and individuals are small and very small passerines (see Sect. 2.8). Despite their vocal and visual conspicuousness, forest birds are not easy prey. From the ground level to canopies 50 m above the ground, the complicated structure of a primeval forest seems to offer endless possibilities for birds' foraging, nesting, and escape tactics. In the temperate forests of central Europe, there are only three species of predators that are truly specialised on forest birds: the goshawk, the sparrowhawk, and the pygmy owl. All other bird eaters either hunt birds in addition to the main prey (e.g. buzzard, pine marten, tawny owl), or capture predominantly non-forest birds (e.g. peregrine, hobby, booted eagle). The sparrowhawk hunts on forest birds proper.

Gavrin (1953), who collected 208 prey items brought to nests by sparrowhawks in the Belarussian part of the forest in 1947–1953, found that 98% of them were birds belonging to 48 species. They ranged in size from a goldcrest to a turtle dove and a hazel hen chick. Most numerous, however, were chaffinch (8% of all prey), oriole (6%), thrushes (all species combined 12%), tree pipits, tits, yellowhammers, and bullfinches. Small rodents made up only 2% of sparrowhawks' prey.

In the spring–summer seasons of 1993 and 1994, we studied the diet composition of sparrowhawks in the western part of the forest. Fifty-two species of birds were found in pellets and among prey remains of these raptors. Birds made up 77–93% of prey specimens and 86–97% of food biomass taken by the sparrowhawk (Table 4.31). Other prey, all of minute importance, were rodents, insectivores, bats, frogs, and insects.

Sparrowhawks nested both in the forest interior and near the ecotone of forests and farmlands. Pairs breeding deep in the forests relied nearly exclusively on birds. Sparrowhawks nesting near the forest edge went out to the open fields markedly more often and hunted small mammals there (predominantly *Microtus* voles), which formed 20% of all captured prey. Interestingly, both 'forest' and 'ecotone' pairs hunted predominantly forest birds and the species composition of their avian prey did not differ (Table 4.31). Among 887 birds captured by sparrowhawks and identified to species or genus, the most numerous were: great tit (13%), chaffinch (10%), blackbird (10%), robin (9%), songthrush (7%), great spotted woodpecker (6%), blue tit

Table 4.31. Spring-summer diet of sparrowhawk in the Polish part of BPF in 1993–1994 (B. Jędrzejewska, W. Jędrzejewski, E. Pugacewicz and K. Zub, unpubl.). Explanations as in Table 4.30

Item	Forest interior			Forest-farmland ecotone		
	%Occ	%Prey	%Bio	%Occ	%Prey	%Bio
Mole (*Talpa europaea*)	–	–	–	0.5	0.2	0.3
Common shrew (*Sorex araneus*)	–	–	–	3.3	1.3	0.4
Lesser shrew (*Sorex minutus*)	–	–	–	2.9	1.0	0.1
Insectivores total	**–**	**–**	**–**	**5.3**	**2.5**	**0.8**
Undetermined bat (Chiroptera)	0.4	0.1	+	–	–	–
Bank vole (*Clethrionomys glareolus*)	7.8	2.7	1.7	11.5	4.1	2.4
Root vole (*Microtus oeconomus*)	–	–	–	5.3	2.0	1.9
Short-tailed vole (*Microtus agrestis*)	–	–	–	2.4	1.0	0.8
Common vole (*Microtus arvalis*)	–	–	–	17.2	8.7	5.7
Undetermined vole (*Microtus* sp.)	0.8	0.3	0.2	2.4	0.8	0.6
Undetermined Microtidae	0.8	0.3	0.2	–	–	–
House mouse (*Mus musculus*)	–	–	–	0.5	0.2	0.1
Common rat (*Rattus norvegicus*)	–	–	–	1.4	0.5	0.8
Yellow-necked mouse (*Apodemus flavicollis*)	–	–	–	0.5	0.2	0.2
Wood mouse (*Apodemus sylvaticus*)	0.4	0.1	0.1	–	–	–
Undetermined mouse (*Apodemus* sp.)	0.4	0.2	0.1	–	–	–
Birch mouse (*Sicista betulina*)	0.4	0.1	+	–	–	–
Undetermined rodent	–	–	–	0.5	0.2	0.1
Rodents total	**9.9**	**3.7**	**2.4**	**36.8**	**17.7**	**12.7**
Ungulate carcass	–	–	–	0.5	0.2	0.3
Common snipe (*Gallinago gallinago*)	–	–	–	0.5	0.2	0.3
Woodcock (*Scolopax rusticola*)	0.4	0.1	0.2	0.5	0.2	0.3
Green sandpiper (*Tringa ochropus*)	0.8	0.3	0.5	1.0	0.3	0.6
Rock dove (*Columba livia f. domestica*)	0.4	0.1	0.2	–	–	–
Wood pigeon (*Columba palumbus*)	–	–	–	0.5	0.2	0.3
Undetermined *Columba* sp.	0.4	0.2	0.3	2.9	1.7	2.8
Nightjar (*Caprimulgus europaeus*)	–	–	–	0.5	0.2	0.3
Swift (*Apus apus*)	–	–	–	0.5	0.2	0.2
Wryneck (*Jynx torquilla*)	–	–	–	0.5	0.2	0.2
Great spotted woodpecker (*Dendrocopos major*)	1.2	1.1	2.0	2.9	1.2	2.0
Middle spotted woodpecker (*Dendrocopos medius*)	1.2	0.4	0.7	1.0	0.3	0.6
Lesser spotted woodpecker (*Dendrocopos minor*)	0.8	0.3	0.2	1.0	0.3	0.2
Woodpecker (*Dendrocopos* sp.)	15.2	5.3	8.3	12.0	4.3	6.4

Table 4.31 (continued)

Item	Forest interior			Forest-farmland ecotone		
	%Occ	%Prey	%Bio	%Occ	%Prey	%Bio
Three-toed woodpecker (*Picoides tridactylus*)	1.6	0.6	1.0	–	–	–
Sky lark (*Alauda arvensis*)	2.1	0.9	1.2	3.8	1.5	2.0
Sand martin (*Riparia riparia*)	0.4	0.1	0.1	–	–	–
Swallow (*Hirundo rustica*)	0.8	0.3	0.2	2.4	1.5	1.0
Tree pipit (*Anthus trivialis*)	0.8	0.3	0.2	–	–	–
Pipit (*Anthus* sp.)	2.5	0.9	0.6	1.0	0.3	0.2
Alauda arvensis or *Anthus* sp.	1.6	0.6	0.6	–	–	–
Yellow wagtail (*Motacilla flava*)	0.4	0.1	0.1	1.4	0.5	0.3
White wagtail (*Motacilla alba*)	–	–	–	1.0	0.3	0.3
Wagtail (*Motacilla* sp.)	1.2	0.4	0.3	–	–	–
Wren (*Troglodytes troglodytes*)	0.8	0.3	0.1	1.4	0.5	0.1
Dunnock (*Prunella modularis*)	2.1	0.7	0.6	0.5	0.2	0.1
Robin (*Erithacus rubecula*)	16.0	6.6	4.1	13.9	4.9	2.9
Thrush nightingale (*Luscinia luscinia*)	–	–	–	0.5	0.2	0.1
Blackbird (*Turdus merula*)	12.3	5.4	9.4	10.0	5.1	8.4
Fieldfare (*Turdus pilaris*)	–	–	–	1.0	0.3	0.6
Song thrush (*Turdus philomelos*)	8.6	4.7	8.1	8.1	3.2	5.3
Undetermined thrush (*Turdus* sp.)	7.0	2.4	4.2	5.3	1.9	3.1
Icterine warbler (*Hippolais icterina*)	0.8	0.3	0.1	0.5	0.3	0.2
Whitethroat (*Sylvia communis*)	0.4	0.1	0.1	0.5	0.2	0.1
Blackcap (*Sylvia atricapilla*)	0.8	0.3	0.2	0.5	0.2	0.1
Undetermined warbler (*Sylvia* sp.)	2.5	0.9	0.5	1.9	1.0	0.5
Wood warbler (*Phylloscopus sibilatrix*)	1.6	0.6	0.2	–	–	–
Chiffchaff (*Phylloscopus collybita*)	0.4	0.1	+	–	–	–
Greenish warbler (*Phylloscopus trochilus*)	–	–	–	0.5	0.2	0.1
Warbler (*Phylloscopus* sp.)	3.7	1.7	0.6	3.3	1.5	0.5
Goldcrest (*Regulus regulus*)	0.4	0.1	+	–	–	–
Collared flycatcher (*Ficedula albicollis*)	0.8	0.3	0.1	–	–	–
Pied flycatcher (*Ficedula hypoleuca*)	2.5	0.9	0.4	1.4	0.5	0.2
Titmouse (*Aegithalos caudatus*)	0.4	0.2	+	–	–	–
Great tit (*Parus major*)	13.6	5.9	3.9	7.7	4.1	2.6
Blue tit (*Parus caeruleus*)	7.0	2.7	1.1	2.9	1.0	0.4
Undetermined tit (*Parus* sp.)	24.3	8.5	4.9	6.7	2.4	1.3
Nuthatch (*Sitta europaea*)	5.3	2.0	1.8	4.8	1.7	1.4
Tree creeper (*Certhia familiaris*)	0.4	0.3	0.1	–	–	–
Golden oriole (*Oriolus oriolus*)	0.4	0.2	0.3	2.4	0.8	1.4
Red-backed shrike (*Lanius collurio*)	2.9	1.2	1.3	2.4	1.0	1.1
Jay (*Garrulus glandarius*)	4.1	2.3	3.9	9.1	4.4	7.2

Table 4.31 (continued)

Item	Forest interior			Forest-farmland ecotone		
	%Occ	%Prey	%Bio	%Occ	%Prey	%Bio
Nutcracker (*Nucifraga caryocatactes*)	1.2	0.4	0.7	–	–	–
Undetermined corvid (*Corvus* sp.)	–	–	–	0.5	0.2	0.3
Starling (*Sturnus vulgaris*)	0.8	0.3	0.5	1.4	0.5	0.8
Chaffinch (*Fringilla coelebs*)	15.2	8.0	7.3	12.4	6.5	5.6
Greenfinch (*Carduelis chloris*)	0.4	0.1	0.2	–	–	–
Goldfinch (*Carduelis carduelis*)	–	–	–	0.5	0.2	0.1
Finch (*Carduelis* sp.)	1.6	0.6	0.4	–	–	–
Bullfinch (*Pyrrhula pyrrhula*)	1.2	0.4	0.4	–	–	–
Hawfinch (*Coccothraustes coccothraustes*)	4.5	2.0	3.5	9.6	3.6	5.9
Yellowhammer (*Emberiza citrinella*)	6.2	2.2	2.4	5.3	1.9	1.9
Reed bunting (*Emberiza schoeniclus*)	0.4	0.1	0.1	0.5	0.2	0.1
Undetermined small bird	52.3	18.2	19.3	42.6	15.4	15.6
Birds total	**99.2**	**93.0**	**97.5**	**78.5**	**77.4**	**86.0**
Undetermined frog (*Rana* sp.)	–	–	–	1.0	0.3	0.2
Carabidae beetle	2.1	1.2	+	0.5	0.2	+
Geotrupes sp.	0.4	0.2	+	–	–	–
Other Coleoptera beetles	2.9	1.4	+	3.3	1.7	+
Undetermined insects	1.2	0.4	+	–	–	–
Invertebrates total	**6.6**	**3.2**	**+**	**3.8**	**1.9**	**+**
Total *N* pellets and prey remains, *N* prey specimens or biomass (g)	243	698	19085 g	209	588	16862 g

(5%), and hawfinch (4%). All these birds are very common or even dominating in the commmunity of breeding forest birds (see Sect. 2.8). Tits (*Parus* spp.) were particularly selected for: they formed 22.5% of birds captured by sparrowhawks compared with a mean of 10% in the forest bird community. Sparrowhawks clearly avoided numerous but very small birds (body mass <13 g): wood warblers, collared flycatchers, goldcrests, and wrens. Birds typical of open areas (sky larks, wagtails, swallows, yellowhammers and the like) contributed 11% to the total number of birds captured by the 'ecotone' pairs, and 6% to the total captured by the 'forest' pairs. This showed that Białowieża's sparrowhawks are true specialists on forest birds. They hunt by a very fast patrolling flight from just above the forest floor vegetation to several metres high. During the flight, sparrowhawks masterly manoeuvre between trees and shrubs (Pugacewicz 1996).

In winter, most of locally breeding sparrowhawks are gone, and migratory birds from the north are seen predominantly near villages and settlements.

Gavrin (1953) reported frequent observations of sparrowhawks hunting on house sparrows, tree sparrows, yellowhammers, tits and, rarely, bullfinches. Stomachs of four sparrowhawks shot in the winters of 1948/1949–1952/1953 in the eastern part of the forest exclusively contained bird remains.

4.16 Buzzards - Shifting from Voles to Birds and Moles

Paradoxically, the large amount of material on buzzards' diets collected in Białowieża Forest between 1946–1994, for a total of 18 years, showed such a diversity of prey that defining patterns and mechanisms of buzzards' foraging seemed a hopeless task. Suffice it to say that various studies reported rodents, birds (Fig. 4.34), moles, frogs, or reptiles, each as dominant prey in some years or some parts of BPF. Only when analysed against the background of habitats and abundance of various prey animals, did the material reveal the feeding strategy of the buzzard.

In the autumn–winter seasons, buzzards, if wintering in or near Białowieża Forest, hunt in the open areas: farmlands and fallow land by and within the forest, and open marshes in river valleys. Their food consists of rodents, nearly exclusively *Microtus* voles which made, on average 86% of prey captured and 85% of food biomass (Table 4.32). The role of *Microtus* in buzzards' diet declines from spring (April) till the end of summer (August), when voles constitute less than 10% of food biomass (Fig. 4.35). Meanwhile, buzzards increasingly prey on birds, moles (each making up over 30% of their food biomass in summer), and forest rodents. Thus, from the pre-breeding to breeding season, buzzards shift their hunting and foraging activity from open areas to their territories located in woodland. Not only pairs nesting deep in the forest interior but also those living close to the forest edge relied largely on birds, moles, and forest rodents (Table 4.32). When hunting under the forest canopy, a buzzard employs two tactics: it sits on a

Fig. 4.34. Common buzzard feeding on a passerine bird. Photo: S. Wąsik

Table 4.32. Diet composition of the common buzzard in spring–summer (16 April–30 Sept.) and autumn–winter (1 Oct.–15 April) seasons in the western (Polish) part of BPF in 1986–1994 (based on analysis of pellets and prey remains; from Jędrzejewski et al. 1994b; W. Jędrzejewski, unpubl.). Forest interior and river valleys – food of pairs breeding either deep in the forest (>1 km from the nearest open farmland) or near the narrow marshy valleys of the forest rivers. Other explanations as in Table 4.30

Item	Autumn–winter			Spring–summer					
				Forest interior and river valleys			Forest-farmland ecotone		
	%Occ	%Prey	%Bio	%Occ	%Prey	%Bio	%Occ	%Prey	%Bio
Eastern hedgehog (*Erinaceus concolor*)	–	–	–	6.3	2.0	4.2	2.1	0.6	1.4
Mole (*Talpa europaea*)	1.1	0.5	1.5	51.3	16.5	32.4	30.8	8.9	19.3
Common shrew (*Sorex araneus*)	–	–	–	4.4	1.5	0.4	1.3	0.5	0.1
Lesser shrew (*Sorex minutus*)	3.4	1.6	0.2	0.7	0.2	+	–	–	–
Undetermined shrew (*Sorex* sp.)	–	–	–	0.9	0.4	0.1	0.3	0.2	+
Water shrew (*Neomys fodiens*)	–	–	–	0.7	0.2	0.1	0.5	0.2	0.1
Undetermined Soricidae	–	–	–	5.6	1.7	0.4	3.2	0.9	0.3
Insectivores total	**4.5**	**2.1**	**1.7**	**60.8**	**22.5**	**37.6**	**34.6**	**11.2**	**21.2**
Brown hare (*Lepus europaeus*)	–	–	–	–	–	–	1.1	0.3	0.7
Red squirrel (*Sciurus vulgaris*)	–	–	–	0.5	0.1	0.3	7.0	2.0	4.7
Bank vole (*Clethrionomys glareolus*)	4.5	2.1	1.7	49.0	15.8	8.6	27.9	8.3	5.0
Water vole (*Arvicola terrestris*)	–	–	–	0.2	0.1	0.2	1.6	0.5	1.1
Pine vole (*Pitymys subterraneus*)	–	–	–	4.4	1.4	0.8	3.5	1.0	0.6
Root vole (*Microtus oeconomus*)	33.0	20.9	25.4	1.6	0.7	0.5	8.0	3.2	2.9
Short-tailed vole (*Microtus agrestis*)	12.5	6.4	6.9	–	–	–	0.8	0.2	0.2
Common vole (*Microtus arvalis*)	58.0	54.5	48.5	3.5	1.2	0.8	8.3	3.6	2.4
Undetermined vole (*Microtus* sp.)	9.1	4.3	4.4	23.1	7.3	5.2	45.0	13.5	10.5
Undetermined small Microtidae	–	–	–	1.6	0.5	0.4	0.5	0.2	0.1
House mouse (*Mus musculus*)	–	–	–	–	–	–	0.3	0.1	+

Table 4.32 (continued)

Item	Autumn–winter			Spring–summer					
				Forest interior and river valleys			Forest-farmland ecotone		
	%Occ	%Prey	%Bio	%Occ	%Prey	%Bio	%Occ	%Prey	%Bio
Harvest mouse (*Micromys minutus*)	1.1	0.5	0.2	0.7	0.2	0.1	0.8	0.2	0.1
Yellow-necked mouse (*Apodemus flavicollis*)	1.1	0.5	0.8	–	–	–	0.5	0.2	0.2
Undet. mouse (*Apodemus* sp.)	4.5	2.1	3.1	5.6	1.7	1.7	8.6	2.5	2.7
Birch mouse (*Sicista betulina*)	–	–	–	–	–	–	0.5	0.2	+
Undetermined small rodent	1.1	0.5	0.2	7.0	2.2	1.3	7.0	2.1	2.0
Small rodents total	**97.7**	**91.4**	**91.1**	**76.0**	**28.8**	**18.1**	**83.6**	**33.4**	**25.8**
Domestic dog (*Canis familiaris*)	–	–	–	–	–	–	0.3	0.1	0.2
Weasel (*Mustela nivalis*)	–	–	–	0.7	0.2	0.4	1.9	0.5	1.1
Domestic cat (*Felis catus*)	1.1	0.5	1.7	–	–	–	–	–	–
Wild boar (*Sus scrofa*) carcass	–	–	–	2.8	0.9	1.9	1.1	0.3	0.7
Deer (Cervidae) carcass	1.1	0.5	1.7	1.6	0.5	1.1	2.4	0.7	1.6
Wild duck (*Anas* sp.)	–	–	–	–	–	–	0.3	0.1	0.2
Hazel hen (*Bonasa bonasia*)	–	–	–	0.2	0.1	0.2	2.1	0.6	1.4
Partridge (*Perdix perdix*)	–	–	–	0.2	0.1	0.2	–	–	–
Undetermined *Columba* sp.	–	–	–	1.2	0.4	0.8	0.8	0.2	0.5
Tawny owl (*Strix aluco*)	–	–	–	1.2	0.4	0.8	0.8	0.2	0.5
Hoopoe (*Upupa epops*)	–	–	–	0.2	0.1	0.2	–	–	–
Woodpecker (*Dendrocopos* sp.)	–	–	–	4.7	1.6	3.4	1.9	0.5	1.3
Pipit (*Anthus* sp.)	–	–	–	0.5	0.1	0.1	–	–	–
Wren (*Troglodytes troglodytes*)	–	–	–	0.7	0.2	0.1	0.5	0.2	+
Dunnock (*Prunella modularis*)	–	–	–	0.2	0.1	+	1.1	0.3	0.2
Robin (*Erithacus rubecula*)	–	–	–	1.2	0.4	0.2	0.5	0.2	0.1

Table 4.32 (continued)

Item	Autumn–winter			Spring–summer					
				Forest interior and river valleys			Forest-farmland ecotone		
	%Occ	%Prey	%Bio	%Occ	%Prey	%Bio	%Occ	%Prey	%Bio
Blackbird (*Turdus merula*)	1.1	0.5	1.7	6.1	2.0	4.2	6.2	1.8	4.1
Songthrush (*Turdus philomelos*)	–	–	–	13.8	4.4	8.2	9.9	2.8	5.8
Redwing (*Turdus iliacus*)	–	–	–	–	–	–	0.3	0.1	0.2
Undet. thrush (*Turdus* sp.)	–	–	–	7.7	2.4	5.0	3.8	1.1	2.4
Warbler (*Phylloscopus* sp.)	–	–	–	0.9	0.3	0.1	0.3	0.1	+
Undetermined tit (*Parus* sp.)	–	–	–	2.1	0.7	0.4	1.3	0.4	0.2
Jay (*Garrulus glandarius*)	–	–	–	3.3	1.0	2.2	1.9	0.5	1.3
Starling (*Sturnus vulgaris*)	–	–	–	–	–	–	2.7	0.9	2.2
Chaffinch (*Fringilla coelebs*)	–	–	–	1.4	0.4	0.4	1.6	0.5	0.4
Hawfinch (*Coccothraustes coccothraustes*)	–	–	–	0.5	0.1	0.3	3.2	0.9	1.8
Yellowhammer (*Emberiza citrinella*)	1.1	0.5	0.8	–	–	–	1.1	0.3	0.3
Undetermined large bird	–	–	–	0.7	0.2	0.5	0.3	0.1	0.2
Undetermined small passerine	3.4	1.6	1.4	33.8	10.7	6.5	54.4	16.4	11.0
Bird egg	–	–	–	0.5	0.1	+	3.2	0.9	0.2
Birds total	**4.5**	**2.7**	**3.8**	**63.2**	**25.8**	**33.5**	**75.9**	**29.1**	**34.4**
Grass snake (*Natrix natrix*)	–	–	–	5.1	1.6	3.4	4.6	1.3	3.1
Undetermined reptile	–	–	–	2.1	0.7	0.9	3.8	1.1	1.7
Reptiles total	**–**	**–**	**–**	**7.2**	**2.3**	**4.3**	**8.3**	**2.4**	**4.7**
Common frog (*Rana temporaria*)	–	–	–	0.2	0.1	+	1.1	0.5	0.3
Moor frog (*Rana arvalis*)	–	–	–	0.5	0.1	0.1	–	–	–
Undetermined frog (*Rana* sp.)	–	–	–	1.6	0.5	0.3	0.3	0.1	0.1

Table 4.32 (continued)

Item	Autumn–winter			Spring–summer					
				Forest interior and river valleys			Forest-farmland ecotone		
	%Occ	%Prey	%Bio	%Occ	%Prey	%Bio	%Occ	%Prey	%Bio
Common toad (*Bufo bufo*)	–	–	–	2.3	0.8	1.2	2.9	0.9	1.5
Undetermined amphibian	–	–	–	2.3	0.9	1.0	7.8	2.2	2.8
Amphibians total	**–**	**–**	**–**	**7.0**	**2.4**	**2.6**	**12.1**	**3.7**	**4.7**
Snail	–	–	–	0.7	0.2	+	1.6	0.5	+
Carabidae beetle	–	–	–	8.2	3.1	+	2.1	1.0	+
Geotrupes sp.	–	–	–	3.0	1.0	+	0.5	0.2	+
Silpha sp.	–	–	–	0.9	0.3	+	0.8	0.2	+
Chrysomelidae beetle	–	–	–	1.2	0.4	+	0.3	0.1	+
Other and undetermined insects	3.4	1.6	+	34.7	11.0	+	45.3	13.0	+
Larvae of insects	–	–	–	0.5	0.1	+	2.4	1.0	+
Invertebrates total	**3.4**	**1.6**	**+**	**45.5**	**16.1**	**0.1**	**49.9**	**16.0**	**0.1**
Plant material	2.3	1.1	0.1	0.7	0.3	+	0.8	0.2	+
Total *N* pellets and prey remains, *N* prey specimen or biomass (kg)	88	187	3.99 kg	429	1373	4.27 kg	373	1301	3.67 kg

high branch or fallen tree close to the ground and watches for bird or mammal movements, or it hunts while flying 2–3 m above the forest floor and trying to flush prey.

Out of two common species of woodland rodents, the bank vole and the yellow-necked mouse, buzzards nearly exclusively hunted bank voles. Strictly nocturnal mice were rather free from the risk of predation by these diurnal raptors. Buzzards captured birds of over 30 species, but they particularly selected for thrushes, *Turdus* spp., migratory forest birds of medium body size. Of all birds captured by buzzards, 20–30% were thrushes, which contributed, on average, 9% to the forest community of breeding birds (see Sect. 2.8). As many as 44% of *Turdus* remains identified from buzzards' pellets were juvenile birds. Buzzards nesting near the forest-open area ecotones added starlings and other ecotone (but still forest breeding) birds to their diet (Table 4.32), but did not make any serious use of the numerous species of birds dwelling in meadows, marshes and farmland.

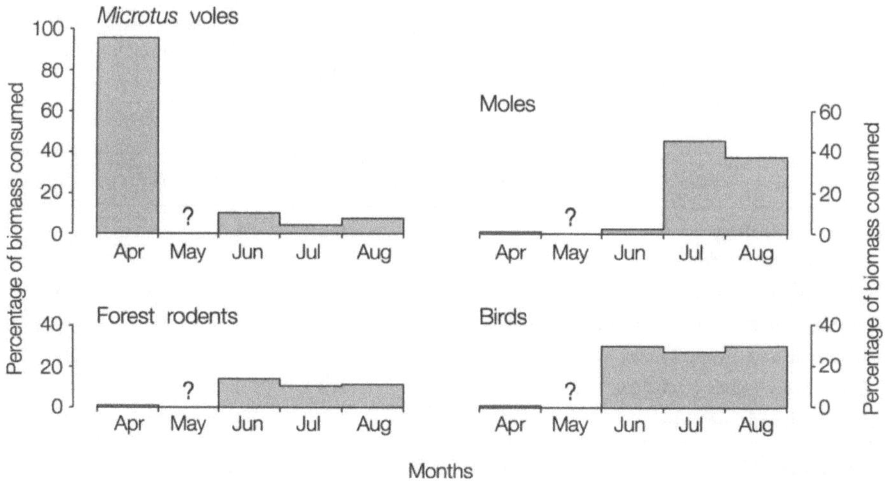

Fig. 4.35. Changes in percentage biomass of four main groups of prey in the common buzzard diet during the breeding season (1993–1994). ? no data. Source: W. Jędrzejewski and co-workers (unpubl. data)

There was a distinct spatial variation in buzzard diet composition. Pairs nesting at the outermost peripheries of the forest, adjoining a deforested landscape, continued to rely on *Microtus* voles, in the breeding season, too (Fig. 4.36, Table 4.33). The role of voles in the spring–summer diet of buzzards declined with increasing distance of a breeding territory from farmlands and meadows. In the diet of 'ecotone' pairs, voles constituted 20% of captured prey, whereas in the forest pairs they constituted 10% only (Table 4.32).

Multiannual variation in buzzard diet was chiefly caused by fluctuations of rodent numbers. This holds true for both field and forest rodents (Fig. 4.36). During the outbreak of forest rodents in 1959, due to a heavy crop of oak seeds, rodents made up 57% of prey specimens captured by buzzards. In the crash year of forest rodents, their share in buzzards' food dropped to a mere 10% of prey captured (Fig. 4.36, Table 4.33).

Since numbers of forest birds are rather stable from year to year, birds provide a predictable food resource for buzzards. In consequence, the proportion of birds in a buzzard's diet in spring–summer was fairly stable. Nonetheless, birds could not fully substitute for the lack of rodents, when the latter crashed to extremely low numbers. Concurrent crashes of *Microtus* (exhibiting 3–4-year cycles) and forest rodents (showing 4–7 years of noncyclic dynamics with moderate densities followed by 2 years of outbreak-crash triggerred by a heavy crop of tree seeds) occurred, for instance, in 1956 and 1991. In 1956, the staple foods of buzzards were birds, reptiles, and amphibians (Fig. 4.36, Table 4.33). A detailed analysis of prey brought to nests by various pairs showed that some of them hunted mainly reptiles (maximally 48% of prey) or frogs (maximally 28% of prey) (Golodushko

Table 4.33. Food of the common buzzard in spring–summer seasons in the Belarussian part of BPF in 1946–1959. Data based on prey brought by adults to the nests (years 1946 and 1956) are from Folitarek (1948) and Golodushko (1958). Data based on stomach content of shot buzzards (year 1959) are from V.A. Dackevich (unpubl.)

Item	1946		1956		1959	
	%Prey	%Bio	%Prey	%Bio	%Prey	%Bio
Eastern hedgehog (*Erinaceus concolor*)	–	–	0.2	3.3	–	–
Mole (*Talpa europaea*)	1.4	2.3	2.4	2.7	2.8	6.8
Common shrew (*Sorex araneus*)	3.3	0.7	–	–	–	–
Undetermined shrew (*Sorex* sp.)	–	–	0.8	0.1	–	–
Water shrew (*Neomys fodiens*)	–	–	0.3	+	–	–
Undetermined Soricidae	–	–	–	–	11.1	2.7
Insectivores total	**4.7**	**3.0**	**3.7**	**6.1**	**13.9**	**9.5**
Brown hare (*Lepus europaeus*)	–	–	1.1	9.8	–	–
Red squirrel (*Sciurus vulgaris*)	0.3	1.8	2.9	12.4	–	–
Bank vole (*Clethrionomys glareolus*)	3.9	1.8	4.5	1.4	7.4	5.1
Water vole (*Arvicola terrestris*)	14.6	33.1	1.1	1.6	5.5	14.8
Short-tailed vole (*Microtus agrestis*)	–	–	0.5	0.2	–	–
Common vole (*Microtus arvalis*)	55.6	28.9	0.5	0.2	–	–
Undetermined vole (*Microtus* sp.)	–	–	–	–	0.9	0.7
Yellow-necked mouse (*Apodemus flavicollis*)	–	–	–	–	2.9	3.5
Undetermined small rodent	–	–	3.0	1.0	40.7	29.7
Small rodents total	**74.1**	**63.8**	**9.6**	**4.4**	**57.4**	**53.8**
Stoat (*Mustela erminea*)	0.3	1.0	–	–	–	–
Weasel (*Mustela nivalis*)	–	–	0.3	0.3	–	–
Domestic duckling	–	–	0.3	1.2	–	–
Buzzard (*Buteo buteo*) or goshawk (*Accipiter gentilis*) chick	–	–	–	–	0.9	2.5
Hazel hen (*Bonasa bonasia*)	0.8	9.5	0.3	2.0	1.9	4.9
Corncrake (*Crex crex*)	–	–	–	–	0.9	2.5
Lapwing (*Vanellus vanellus*)	–	–	0.8	2.9	–	–
Green sandpiper (*Tringa ochropus*)	–	–	0.3	0.5	–	–
Stock dove (*Columba oenas*)	–	–	0.3	1.3	–	–
Little owl (*Athene noctua*)	0.3	1.4	–	–	–	–
Tengmalm's owl (*Aegolius funereus*)	–	–	0.3	0.6	–	–
Nightjar (*Caprimulgus europaeus*)	–	–	0.5	0.6	–	–
Great spotted woodpecker (*Dendrocopos major*)	0.3	0.6	1.0	1.6	–	–
Woodpecker (*Dendrocopos* sp.)	–	–	0.3	0.3	–	–
Sky lark (*Alauda arvensis*)	1.6	1.7	1.0	0.7	1.9	2.8
Wood lark (*Lullula arborea*)	–	–	0.3	0.1	–	–

Table 4.33 (continued)

Item	1946		1956		1959	
	%Prey	%Bio	%Prey	%Bio	%Prey	%Bio
Tree pipit (*Anthus trivialis*)	3.3	1.8	4.2	1.6	–	–
Robin (*Erithacus rubecula*)	–	–	0.3	0.1	–	–
Whinchat (*Saxicola rubetra*)	–	–	0.5	0.2	–	–
Blackbird (*Turdus merula*)	–	–	2.1	3.9	1.9	4.9
Song thrush (*Turdus philomelos*)	1.1	2.1	4.0	5.1	0.9	2.5
Great reed warbler (*Acrocephalus arundinaceus*)	–	–	0.3	0.2	–	–
Jay (*Garrulus glandarius*)	–	–	3.2	10.3	–	–
Chaffinch (*Fringilla coelebs*)	0.5	0.4	0.3	0.1	–	–
Siskin (*Carduelis spinus*)	0.3	0.1	0.3	0.1	–	–
Undetermined small passerine	–	–	1.0	0.7	1.8	2.6
Birds total	**8.2**	**17.6**	**21.6**	**34.1**	**10.2**	**22.7**
Viviparous lizard (*Lacerta vivipara*)	1.4	0.4	0.3	0.1	0.9	0.4
Lizard (*Lacerta* sp.)	–	–	0.3	0.1	–	–
Slowworm (*Anguis fragilis*)	4.1	4.0	28.5	18.3	5.6	7.9
Grass snake (*Natrix natrix*)	–	–	0.3	0.4	0.9	2.5
Viper (*Vipera berus*)	2.2	6.0	1.8	3.4	–	–
Reptiles total	**7.7**	**10.4**	**31.2**	**22.3**	**7.4**	**10.8**
Fire-belly (*Bombina* sp.)	–	–	1.1	0.3	–	–
Common toad (*Bufo bufo*)	0.3	0.4	6.9	5.8	–	–
Edible frog (*Rana esculenta*)	–	–	0.8	0.7	–	–
Common frog (*Rana temporaria*)	4.4	2.0	9.6	3.0	–	–
Moor frog (*Rana arvalis*)	–	–	0.8	0.2	–	–
Undetermined frog (*Rana* sp.)	–	–	1.6	0.5	4.6	3.0
Amphibians total	**4.7**	**2.4**	**20.8**	**10.5**	**4.6**	**3.0**
Geotrupes sp.	–	–	2.4	+	–	–
Gryllotalpa gryllotalpa	–	–	–	–	2.8	0.1
Undetermined insect	–	–	6.4	0.1	3.7	0.1
Invertebrates total	**–**	**–**	**8.8**	**0.1**	**6.5**	**0.2**
Total *N* prey specimens or biomass (kg)	363	13.28 kg	375	20.44 kg	108[a]	2.67 kg

[a] Total number of prey found in 43 stomachs of shot buzzards.

1958). In 1992, when rodents were very scarce, the most important prey of buzzards in the dense mature forests of Białowieża National Park (with reptiles hardly occurring) were birds (38% of vertebrates captured) and moles (28%).

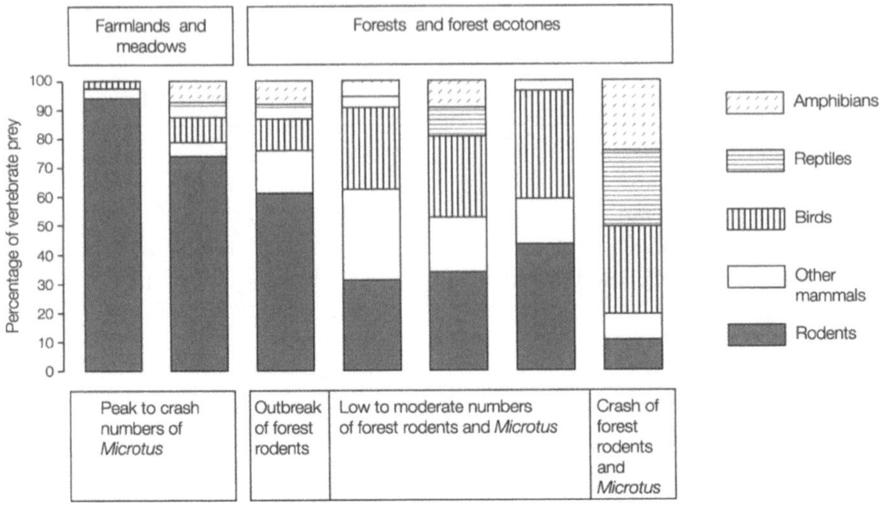

Fig. 4.36. Multiannual and spatial variation in the diet composition (six major groups of vertebrate prey) of common buzzards in BPF in relation to qualitative information on abundance of rodents and type of buzzard's breeding habitat. Data for 1946–1994 from: Folitarek (1948), Gavrin (1953, and unpubl. data cited in Golodushko 1958), Golodushko (1958, 1961a), V.A. Dackevich (unpubl. data), and authors' unpubl. data

The majority of the diverse vertebrate prey of buzzards have one thing in common: they can be captured on the ground, cases in point being rodents, moles, shrews, frogs, and reptiles. Also, among birds falling as prey to buzzards, ground-foraging species (blackbird, song thrush) are the most numerous. In autumn and winter, when birds, moles, frogs, and reptiles become scarce or unavailable, and small mammals are difficult to capture because of snow cover, buzzards leave their forest territories.

4.17 Honey Buzzards – A Misleading Name for Wasp-Eaters

The English name honey buzzard, no doubt poetic, is far from the truth. Much closer to reality is the Latin name of this raptor, *apivorus*, the bee-eater, and the Polish name, *trzmielojad*, the bumblebee-eater. But only its Russian name is perfectly accurate: *osoed*, the wasp-eater. Honey buzzards are specialised for finding the delicate 'paper' nests of wasps, especially *Dolichovespula saxonica* and *Vespula vulgaris* (Table 4.34). Wasps of the Vespinae subfamily, to which all wasps identified from honey buzzards' food remains belong, lead a strongly seasonal life (Puławski 1967). In autumn, young female wasps become fertilised and are the only members of a colony to survive winter. In spring (late March or April), each female becomes a queen-founder of a new colony, for which she begins the construction of a 'paper' nest located among twigs of dense bushes, under branches and boughs of trees, in tree cavities, or underground. All spe-

Table 4.34. Diet composition of the honey buzzard in the western part of BPF in the spring–summer seasons of 1993–1994. Figures in parentheses are numbers of identified combs (in wasp nests) or clusters (in a bumblebee nest)

Item	Prey items (*N*)	Percentage of prey
Mole (*Talpa europaea*)	1	3.6
Undetermined shrew (Soricidae)	1	3.6
Polecat (*Mustela putorius*)	1	3.6
Pigeon (*Columba* sp.)	1	3.6
Vertebrates total	**4**	**14.4**
Nest of bumblebee (*Bombus pascuorum*)	1 (4)	3.6
Nest of wasp (*Vespula vulgaris*)	3 (10)	10.7
Nest of undetermined wasp (*Vespula* sp.)	2 (20)	7.1
Nest of wasp (*Dolichovespula media*)	1 (1)	3.6
Nest of wasp (*Dolichovespula saxonica*)[a]	14 (?)	50.0
Nest of wasp (*Dolichovespula* sp.)[b]	3 (10)	10.6
Invertebrates total	**24**	**85.6**
Total number of prey items	28	100

[a] Including two underground nests, and 11 aboveground nests.
[b] Including one underground nest.

cies prefer dry, warm places. As soon as the first worker wasps emerge (a full cycle from egg to imago lasts about 1 month), they quickly add new combs, widen the walls, and enlarge the nest, so by the end of summer it contains several to over a dozen combs with larvae and pupae. These are the honey buzzard's main food. The buzzard cuts the combs with its beak, takes them to its nest and feeds its young by giving them larvae one at a time. Empty combs remain on the nest as bedding.

In late summer, the eggs laid by a queen wasp develop into a few workers and many females and males. In autumn, soon after the fertilisation of young females, the colony disintegrates and all members, except for queens-to-be, die (Pławski 1967). In autumn and winter, in the temperate forests, there is no food for a raptor specialised on wasps.

Honey buzzards begin breeding late (see Sect. 3.18), because the peak abundance of wasp larvae occurs in mid-summer. Furthermore, the supply of this type of food is tremendously dependent on summer weather. During cool and rainy summers, not only do the larvae of wasps develop slower but also many underground and hanging nests are damaged by rains. In such years, honey buzzards hardly rear any young. However, in hot, dry summers, like that of 1992, these raptors relish in abundant food. Suffice it to say that in the summer of 1992, the attics and walls of our wooden house in Białowieża village were inhabited by five colonies of wasps!

Honey buzzards may occasionally capture vertebrates: moles, shrews, birds or amphibians (Table 4.34). Golodushko (1961c) cited the unpublished data collected by S.S. Folitarek in the eastern part of Białowieża Forest in 1946. In the total weight of 9144 g of food remains found on honey buzzards' nests, 94.8% were nests of *Vespula vulgaris* wasps, 3.9% were amphibians, 1.2% birds, and 0.1% small mammals.

4.18 Lesser Spotted Eagles – Shifting from Amphibians to Birds, to Mammals

Lesser spotted eagles hunt by low cruising flights or by slow walks across meadows, sedge marshes, river valleys, fallow lands by the forest, and – rarely – across the forest floor in clear deciduous forests. They also use the sit-and-wait tactic and watch prey from a tall perch on the forest edge. The food they get is truly diverse (mammals, birds, reptiles, amphibians, and insects), but most of these prey are ground-living animals.

In Białowieża Forest, the diet of lesser spotted eagles was studied for a total of 11 years, between 1946–1994. Only in two studies are the detailed lists of prey available (Table 4.35). These lists are impressively long. Among 27 species of mammals captured by eagles, most numerous were voles in the genus *Microtus*, moles, common shrews, brown hares, hedgehogs, and small mustelids. All are numerous in the open river valleys, meadows, and farmlands by the forest. Of 30 species of birds that fell as prey to lesser spotted eagles, most common were sky larks, song thrushes, partridges, blackbirds, corncrakes, and – in the eastern part of the Forest – hazel hens and woodcocks. All these and many other birds eaten by eagles are ground-nesters or ground-foragers. Eagles not only capture adult birds and fledglings, but may also frequently eat very small chicks, the remains of which can be recovered neither among prey brought to the nest for young eaglets nor in the pellets. V.A. Dackevich, who analysed 19 stomachs of eagles shot between 1947–1959 in the eastern part of the forest, noted that one stomach contained a full clutch of five tiny (about 1-day-old) sky larks, swallowed intact by the eagle. Gavrin (1953) reported that all corncrakes, spotted crakes, hazel hens, and black grouse that fell as prey to eagles were small hatchlings. Amphibians captured by eagles were dominated by the common frog. Among insects, the ground living, flightless carabids were most often eaten (Table 4.35).

Eagles strongly prefer wet forests along open river valleys as nesting habitats but they also breed on the ecotone of forest and relatively drier farmlands, fallow lands and meadows. A few pairs regularly have breeding territories deep in the forest interior, in wet forests with no large clearings (see Sect. 3.19). In 1993–1994, we collected pellets and prey remains under the nests located in these three habitats. Interestingly, the composition of food taken by eagles in all three habitats was not much different (Table 4.35). *Microtus* voles were important food everywhere. Of insectivorous mammals, moles dominated the eagle's diet in the forest-farmland ecotone, whereas the

Table 4.35. Diet composition of the lesser spotted eagle in spring–summer in the Belarussian (eastern) and Polish (western) parts of BPF

Item	Eastern part (1956)		Western part (1993–1994)					
			River valleys and forests			Forest-farmland ecotone		
	%Prey	%Bio	%Occ	%Prey	%Bio	%Occ	%Prey	%Bio
Eastern hedgehog (*Erinaceus concolor*)	3.8	47.9	2.0	0.6	1.5	8.5	2.8	5.6
Mole (*Talpa europaea*)	0.6	0.7	16.3	5.1	11.6	26.1	8.5	15.8
Common shrew (*Sorex araneus*)	–	–	7.8	5.4	1.6	1.2	0.5	0.1
Masked shrew (*Sorex caecutiens*)	–	–	0.2	0.1	+	–	–	–
Lesser shrew (*Sorex minutus*)	–	–	2.2	1.0	0.1	–	–	–
Undet. shrew (*Sorex* sp.)	1.0	0.1	1.2	0.3	0.1	–	–	–
Water shrew (*Neomys fodiens*)	–	–	0.2	0.1	+	–	–	–
Mediterranean water shrew (*Neomys anomalus*)	–	–	0.2	0.1	+	–	–	–
Undet. water shrew (*Neomys* sp.)	–	–	0.2	0.1	+	0.2	0.1	+
Undetermined Soricidae	–	–	0.7	0.2	0.1	0.2	0.1	+
Insectivores total	**5.4**	**48.7**	**29.8**	**13.0**	**15.1**	**32.1**	**12.0**	**21.6**
Brown hare (*Lepus europaeus*)	–	–	5.9	1.8	4.5	9.0	2.9	5.9
Red squirrel (*Sciurus vulgaris*)	–	–	1.0	0.3	0.8	0.5	0.2	0.3
Bank vole (*Clethrionomys glareolus*)	5.4	1.7	8.5	3.0	1.9	5.1	1.6	0.9
Water vole (*Arvicola terrestris*)	0.7	1.0	6.6	2.0	6.3	4.2	1.3	3.4
Pine vole (*Pitymys subterraneus*)	–	–	4.6	3.9	2.5	–	–	–
Root vole (*Microtus oeconomus*)	1.0	0.5	9.0	3.1	3.0	5.3	2.0	1.6
Short-tailed vole (*Microtus agrestis*)	4.8	2.1	2.0	0.8	0.6	0.9	0.3	0.2
Common vole (*Microtus arvalis*)	2.9	1.1	15.6	6.5	4.6	22.4	9.6	5.5
Undet. vole (*Microtus* sp.)	0.4	0.2	57.6	18.2	14.3	53.8	18.7	11.9
Undet. small Microtidae	–	–	1.5	0.5	0.4	2.1	0.7	0.4
House mouse (*Mus musculus*)	0.4	0.1	0.7	0.2	0.1	–	–	–

Table 4.35 (continued)

Item	Eastern part (1956)		Western part (1993–1994)					
			River valleys and forests			Forest-farmland ecotone		
	%Prey	%Bio	%Occ	%Prey	%Bio	%Occ	%Prey	%Bio
Common rat (*Rattus norvegicus*)	–	–	0.5	0.2	0.4	2.3	0.8	1.5
Harvest mouse (*Micromys minutus*)	–	–	0.5	0.1	+	–	–	–
Striped field mouse (*Apodemus agrarius*)	–	–	–	–	–	0.2	0.1	0.1
Yellow-necked mouse (*Apodemus flavicollis*)	–	–	0.2	0.1	0.1	1.4	0.8	0.7
Wood mouse (*Apodemus sylvaticus*)	–	–	0.5	0.2	0.2	0.5	0.1	0.1
Undet. mouse (*Apodemus* sp.)	–	–	0.2	0.1	0.1	3.2	1.1	0.9
Forest dormouse (*Dryomys nitedula*)	–	–	–	–	–	0.5	0.1	0.1
Fat dormouse (*Glis glis*)	–	–	0.5	0.1	0.4	–	–	–
Undetermined small rodent	4.8	1.8	0.7	0.2	0.2	0.5	0.2	0.1
Small rodents total	**20.4**	**8.5**	**91.2**	**39.2**	**35.1**	**90.5**	**37.4**	**27.4**
Polecat (*Mustela putorius*)	–	–	0.5	0.1	0.4	–	–	–
Stoat (*Mustela erminea*)	–	–	0.2	0.1	0.2	3.0	1.0	2.0
Weasel (*Mustela nivalis*)	0.4	0.5	2.7	0.8	1.8	2.6	0.8	1.5
Mustela nivalis or *M. erminea*	–	–	2.0	0.6	1.5	3.2	1.1	2.1
Mustelidae total	**0.4**	**0.5**	**5.4**	**1.6**	**3.9**	**8.8**	**2.9**	**5.6**
Ungulate carcass	–	–	2.2	0.7	1.7	1.6	0.5	1.1
Teal (*Anas crecca*)	–	–	–	–	–	0.2	0.1	0.2
Hazel hen (*Bonasa bonasia*)	0.7	5.0	–	–	–	–	–	–
Domestic hen (*Gallus gallus*)	0.2	3.9	–	–	–	0.2	0.1	0.2
Partridge (*Perdix perdix*)	–	–	2.0	0.6	1.5	6.2	2.0	4.1
Water rail (*Rallus aquaticus*)	–	–	–	–	–	0.5	0.1	0.3
Corncrake (*Crex crex*)	0.2	0.6	–	–	–	2.8	0.9	1.8
Woodcock (*Scolopax rusticola*)	0.4	2.2	–	–	–	–	–	–
Wood pigeon (*Columba palumbus*)	–	–	–	–	–	0.5	0.1	0.3

Table 4.35 (continued)

Item	Eastern part (1956)		Western part (1993–1994)					
			River valleys and forests			Forest-farmland ecotone		
	%Prey	%Bio	%Occ	%Prey	%Bio	%Occ	%Prey	%Bio
Undetermined *Columba* sp.	–	–	1.2	0.4	0.9	0.7	0.2	0.4
Tawny owl (*Strix aluco*)	–	–	–	–	–	0.2	0.1	0.2
Middle spotted woodpecker (*Dendrocopos medius*)	0.2	0.2	–	–	–	–	–	–
Woodpecker (*Dendrocopos* sp.)	–	–	0.2	0.1	0.2	–	–	–
Sky lark (*Alauda arvensis*)	–	–	3.2	1.0	1.2	9.0	2.9	3.3
Wood lark (*Lullula arborea*)	0.2	0.1	–	–	–	–	–	–
Tree pipit (*Anthus trivialis*)	1.3	0.5	–	–	–	–	–	–
Pipit (*Anthus* sp.)	–	–	–	–	–	0.5	0.1	0.1
Wagtail (*Motacilla* sp.)	–	–	–	–	–	0.7	0.2	0.1
Robin (*Erithacus rubecula*)	–	–	–	–	–	1.6	0.5	0.3
Thrush nightingale (*Luscinia luscinia*)	0.2	0.1	–	–	–	–	–	–
Whinchat (*Saxicola rubetra*)	–	–	0.5	0.1	0.1	0.7	0.2	0.1
Blackbird (*Turdus merula*)	0.2	0.4	2.2	0.7	2.5	0.7	0.2	0.7
Song thrush (*Turdus philomelos*)	0.4	0.6	2.4	0.7	2.0	3.7	1.2	2.6
Undet. thrush (*Turdus* sp.)	–	–	2.2	0.7	2.0	3.0	1.0	2.3
Undetermined tit (*Parus* sp.)	–	–	–	–	–	0.5	0.1	0.1
Tree creeper (*Certhia familiaris*)	–	–	0.2	0.1	+	–	–	–
Red-backed shrike (*Lanius collurio*)	–	–	–	–	–	0.5	0.2	0.1
Golden oriole (*Oriolus oriolus*)	–	–	–	–	–	1.4	0.5	1.0
Jay (*Garrulus glandarius*)	–	–	0.2	0.1	0.2	0.2	0.1	0.2
Crow (*Corvus corone*)	–	–	0.2	0.1	0.2	–	–	–
Starling (*Sturnus vulgaris*)	–	–	0.2	0.1	0.2	0.5	0.2	0.4

Table 4.35 (continued)

Item	Eastern part (1956)		Western part (1993–1994)					
			River valleys and forests			Forest-farmland ecotone		
	%Prey	%Bio	%Occ	%Prey	%Bio	%Occ	%Prey	%Bio
Chaffinch (*Fringilla coelebs*)	–	–	–	–	–	1.4	0.5	0.3
Finch (*Carduelis* sp.)	–	–	0.5	0.1	0.1	–	–	–
Yellowhammer (*Emberiza citrinella*)	–	–	1.0	0.3	0.3	1.2	0.5	0.4
Reed bunting (*Emberiza schoeniclus*)	–	–	0.2	0.1	0.1	0.7	0.2	0.1
Undetermined large bird	–	–	0.7	0.2	0.6	0.9	0.5	0.8
Undetermined small bird	0.4	0.3	28.8	8.8	17.2	19.4	6.3	9.9
Birds total	**4.4**	**13.9**	**43.7**	**14.2**	**29.4**	**46.4**	**19.0**	**30.3**
Slowworm (*Anguis fragilis*)	0.4	0.3	–	–	–	–	–	–
Sand lizard (*Lacerta agilis*)	0.2	+	–	–	–	–	–	–
Viviparous lizard (*Lacerta vivipara*)	1.3	0.3	–	–	–	–	–	–
Lizard (*Lacerta* sp.)	0.8	0.2	–	–	–	0.5	0.1	0.1
Grass snake (*Natrix natrix*)	–	–	3.7	1.1	3.6	5.8	1.9	4.8
Viper (*Vipera berus*)	1.7	3.2	–	–	–	–	–	–
Undetermined reptile	–	–	1.0	0.3	1.1	0.2	0.1	0.2
Reptiles total	**4.4**	**4.0**	**4.6**	**1.4**	**4.7**	**6.2**	**2.1**	**5.1**
Fire-belly (*Bombina* sp.)	0.4	0.1	–	–	–	–	–	–
Common toad (*Bufo bufo*)	0.2	0.2	–	–	–	–	–	–
Edible frog (*Rana esculenta*)	8.1	7.1	–	–	–	–	–	–
Common frog (*Rana temporaria*)	46.3	14.9	6.8	3.5	2.7	3.9	1.7	1.1
Moor frog (*Rana arvalis*)	6.7	2.0	2.0	0.9	0.5	0.9	0.3	0.2
Undetermined frog (*Rana* sp.)	0.4	0.1	5.6	1.8	1.4	6.0	2.0	1.2
Amphibians total	**62.1**	**24.4**	**13.7**	**6.2**	**4.6**	**10.6**	**4.0**	**2.5**
Snail	–	–	0.2	0.1	+	0.2	0.1	+
Carabidae beetle	–	–	21.7	9.1	0.1	16.4	6.3	0.1
Silpha sp.	–	–	10.2	3.2	+	9.2	3.2	+

Table 4.35 (continued)

Item	Eastern part (1956)		Western part (1993–1994)					
			River valleys and forests			Forest-farmland ecotone		
	%Prey	%Bio	%Occ	%Prey	%Bio	%Occ	%Prey	%Bio
Coccinella sp.	–	–	2.2	0.7	+	2.3	0.8	+
Dytiscus sp.	–	–	2.4	0.7	+	0.5	0.2	+
Geotrupes sp.	–	–	6.1	1.9	+	2.5	0.8	+
Melolontha sp.	–	–	–	–	–	2.3	0.8	+
Undetermined insect	2.9	+	13.9	5.9	+	11.8	4.4	+
Larvae of insects	–	–	–	–	–	0.7	2.4	+
Invertebrates total	**2.9**	**+**	**38.8**	**21.6**	**0.2**	**34.9**	**19.0**	**0.2**
Total *N* pellets and prey remains, *N* prey specimens or biomass (kg)	480	25.32 kg	410	1332	35.68 kg	433	1333	43.90 kg

Eastern part: Golodushko (1958; data based on prey brought to nests by adult birds); western part: B. Jędrzejewska, W. Jędrzejewski, E. Pugacewicz, and K. Zub (unpubl. data based on analysis of pellets and prey remains). River valleys and forests, pairs nesting along marshy valleys of forest rivers and in the forests; forest-farmland ecotone, pairs nesting at the edge of forest and open farmland. Other explanations as in Table 4.30.

common shrew was most important in the marshy river valleys. Sky larks, partridges, thrushes, and corncrakes were the most frequently captured birds in the forest-farmland ecotone, thrushes, skylarks, and partridges in the river valleys, and thrushes in the forests. These differences in the species of prey obviously reflected their availability in a given habitat. Generally, in all pairs of eagles, there was a notable avoidance of forest rodents and birds, which showed that eagles rarely hunt under canopies.

The roles of the chief food resources to eagles (mammals, birds, and amphibians) varied from spring through late summer (Fig. 4.37). Amphibians were caught intensively by eagles in April (27% of their food biomass), when frogs concentrated in the river valleys and marshes for lekking and spawning. Later (June–August), common frogs dispersed into forests and formed a negligible addition (2–3%) to the eagles' diet. It must be noted that the role of amphibians in the eagle's diet may be underestimated due to a high degree of bone digestion. Consumption of birds peaked in June (47% of food biomass) when juvenile birds were abundant. The share of mammals increased from 46% in April to over 60% in July–August.

A synthesis of all information on eagles' diets between 1946–1994 revealed that small mammals were the most important food of these raptors. In the study by S.S. Folitarek (unpubl. data, cited in Golodushko 1958), mammals made 79% of all vertebrate prey brought to nests by eagles during 1946–1948

(Fig. 4.38). Large amounts of data collected by Gavrin (1953) between 1947–1953 showed that, in a pooled sample of a few years, mammals (mainly voles) constituted 54% of vertebrate prey, and amphibians 42%. In 1956, a population crash of both *Microtus* voles in the open areas, and forest rodents meant that mammals formed only 27% of vertebrates captured by eagles (Golodushko 1958). As much as 64% of eagles' prey was made of amphibi-

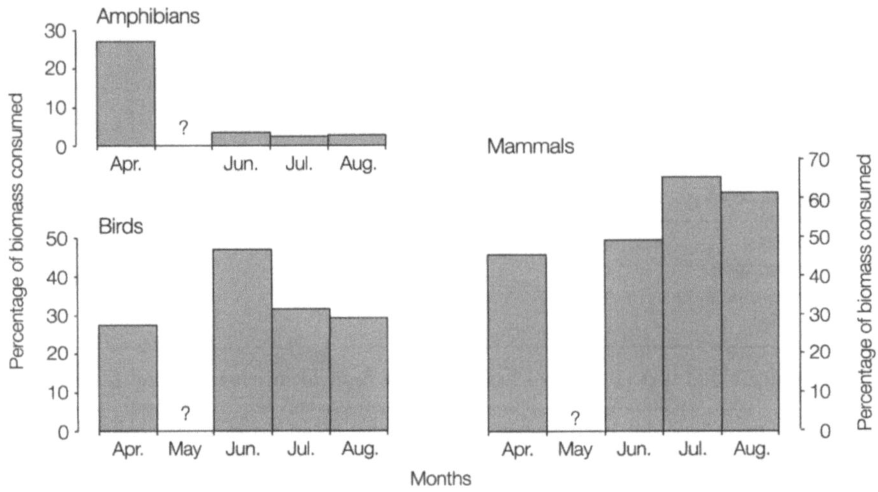

Fig. 4.37. Changes in percentage biomass of three main prey resources in the diet of lesser spotted eagles in BPF during the breeding season of 1994. ? no data. Source: W. Jędrzejewski and co-workers (unpubl. data)

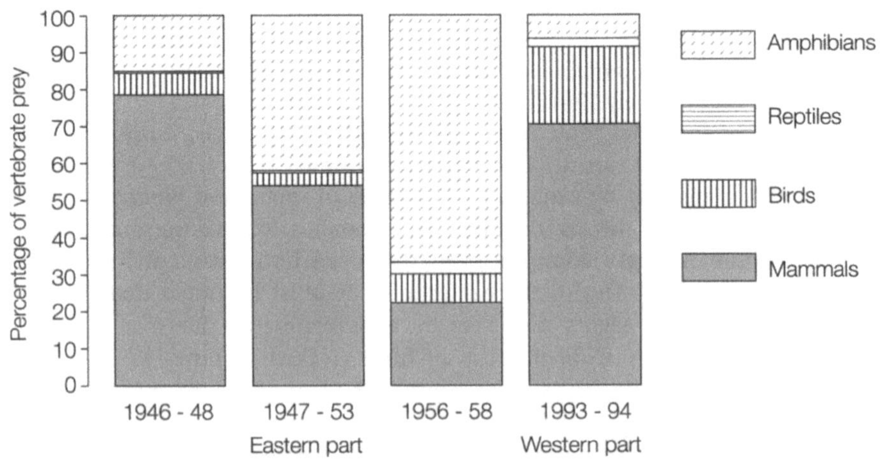

Fig. 4.38. Variation in the diet composition (four prey groups) of lesser spotted eagles in BPF in spring–summer 1946–1994. Data from: S. S. Folitarek (unpubl. data presented in Golodushko 1958), Gavrin (1953), Golodushko (1958, 1961 a), and authors' unpubl. data

ans, their crucial resource then (Fig. 4.38). During the 2 years of our study, *Microtus* voles in river valleys were very abundant (peak year) in 1993, and scarce in 1994. However, the share of mammals among vertebrate prey of eagles was stable: 70 and 71%, respectively. Interestingly, birds constituted the second most important prey group for lesser spotted eagles. Although the data from 1946–1994 could not have been quantitatively related to indices of prey abundance, they show a fair plasticity of eagles' feeding habits.

4.19 Short-Toed Eagles - Specialisation on Snakes and Snake-Like Lizards

The short-toed eagle, another name for which is snake eagle (but *zmeeyad*, a viper-eater in Russian and *gadożer*, a reptile-eater in Polish) lives on reptiles. In 1948–1958, the stomachs of seven birds shot in the eastern part of Białowieża Forest contained predominantly snake-like lizards, the slowworms (35% of food biomass) and vipers (29%) (Table 4.36). Gavrin's collection of prey brought by eagles to nests showed that most of them were vipers, other snakes, and lizards (totalling 89% of food biomass), but also some frogs and small mammals. The reptile community of the eastern part of BPF in the 1950s, i.e. before the drainage of large open marshes, was dominated by viviparous lizards, vipers, and slowworms (Bannikov and Belova 1956). In 1993, we gathered some pellets and prey remains from an eagle nest in the Polish part of the forest, where the viviparous lizard, grass snake, and slowworm are the most common species of reptiles (see Sect. 2.9). The grass snake was the dominant prey of the short-toed eagle (nearly 60% of biomass consumed), but quite a few small mammals and insects were also eaten (Table 4.36).

Short-toed eagles are specialists on snakes and snake-like lizards. These cold-blooded, predatory animals live in Białowieża Forest at very low density. Moreover, they exhibit greatly reduced activity compared with warm-blooded birds and mammals. After winter hibernation, they come out rather late. In Białowieża Forest, the first spring observations of vipers were on 10 April, on average (range 15 March–5 May), and of slowworms on 26 April (range 18 April–3 May) (Dackevich 1977). This is why short-toed eagles begin their breeding season as late as May.

4.20 Booted Eagles - Hunting on Medium-Sized Birds of Open Land and Forests

The only investigation on the diet of booted eagles was conducted in the eastern part of BPF in the 1950s (Gavrin 1953; Golodushko 1959) by collecting pellets and prey remains brought by adults to hatchlings. Most of the prey (90%) were medium-sized birds of 25 species, predominantly lapwings, waders, rollers, and other inhabitants of open meadows, marshes, and fields (Table 4.37). In the forests, booted eagles frequently captured thrushes, orioles and woodpeckers.

Table 4.36. Diet composition of the short-toed eagle in Białowieża Primeval Forest in spring–summer

Item	Eastern part				Western part	
	(1948–1958)		(1950, 1954)		(1993)	
	%Prey	%Bio	%Prey	%Bio	%Prey	%Bio
Eastern hedgehog (*Erinaceus concolor*)	5.3	22.1	–	–	3.8	14.3
Undetermined shrew (Soricidae)	–	–	–	–	5.7	1.3
Insectivores total	**5.3**	**22.1**	**–**	**–**	**9.5**	**15.6**
Root vole (*Microtus oeconomus*)	–	–	–	–	5.7	4.2
Common vole (*Microtus arvalis*)	–	–	–	–	9.4	5.1
Undetermined vole (*Microtus* sp.)	–	–	–	–	9.4	6.2
Rodents total	**–**	**–**	**–**	**–**	**24.5**	**15.5**
Undetermined small mammal	–	–	4.3	2.6	–	–
Song thrush (*Turdus philomelos*)	–	–	–	–	1.9	3.8
Birds total	**–**	**–**	**–**	**–**	**1.9**	**3.8**
Sand lizard (*Lacerta agilis*)	5.3	2.0	–	–	–	–
Viviparous lizard (*Lacerta vivipara*)	5.3	0.8	–	–	–	–
Undetermined lizard (*Lacerta* sp.)	–	–	10.9	4.2	–	–
Slowworm (*Anguis fragilis*)	31.6	34.9	19.6	21.9	–	–
Grass snake (*Natrix natrix*)	–	–	4.3	11.8	24.5	59.5
Viper (*Vipera berus*)	26.3	29.1	45.7	51.2	–	–
Undetermined reptile	–	–	–	–	1.9	4.6
Reptiles total	**68.5**	**66.8**	**80.5**	**89.1**	**26.4**	**64.1**
Undetermined frog (*Rana* sp.)	–	–	15.2	8.3	–	–
Amphibians total	**–**	**–**	**15.2**	**8.3**	**–**	**–**
Beetle	5.2	0.1	–	–	22.6	0.3
Undetermined insects and larvae	10.5	0.2	–	–	13.2	0.2
Wasp nest	5.3	9.1	–	–	–	–
Invertebrates total	**21.0**	**9.4**	**–**	**–**	**35.8**	**0.5**
Plant material	5.2	1.7	–	–	1.9	0.5
Total *N* prey specimens or biomass (g)	19	601 g	46	1437 g	53	1858 g

Data for eastern part from V.A. Dackevich (unpubl. material on stomach content of shot birds, 1948–1958), and V.F. Gavrin (calculated from Gavrin 1953, and Bannikov and Belova 1956; material based on prey remains brought by adult birds to nests, 1950 and 1954). Western part – own data (analysis of pellets and prey remains). Explanations as in Table 4.30.

Table 4.37. Diet composition of booted eagles during breeding season (spring–summer) in the Belarussian part of BPF. Data combined for 1951, 1952 and 1956. (Gavrin 1953; Golodushko 1959)

Item	%Prey	%Bio
Red squirrel (*Sciurus vulgaris*)	2.8	4.7
Hamster (*Cricetus cricetus*)	1.0	2.0
Muskrat (*Ondatra zibethicus*)	1.0	5.0
Water vole (*Arvicola terrestris*)	2.0	1.1
Yellow-necked mouse (*Apodemus flavicollis*)	2.0	0.4
Mammals total	**8.8**	**13.2**
Domestic duckling	2.0	3.4
Hazel hen (*Bonasa bonasia*)	1.0	2.8
Partridge (*Perdix perdix*)	2.0	4.7
Domestic chicken (*Gallus gallus*)	2.0	6.7
Spotted crake (*Porzana porzana*)	2.0	1.3
Corncrake (*Crex crex*)	1.0	1.1
Lapwing (*Vanellus vanellus*)	17.5	24.2
Common snipe (*Gallinago gallinago*)	1.0	0.7
Black-tailed godwit (*Limosa limosa*)	1.0	1.6
Green sandpiper (*Tringa ochropus*)	1.0	0.6
Common sandpiper (*Tringa hypoleucos*)	2.9	0.9
Nightjar (*Caprimulgus europaeus*)	1.0	0.4
Roller (*Coracias garrulus*)	3.9	3.8
Hoopoe (*Upupa epops*)	1.0	0.4
Wryneck (*Jynx torquilla*)	1.0	0.2
Grey-headed woodpecker (*Picus canus*)	1.0	0.8
Great spotted woodpecker (*Dendrocopos major*)	4.9	2.7
Wood lark (*Lullula arborea*)	2.0	0.3
Wheatear (*Oenanthe oenanthe*)	2.9	0.6
Blackbird (*Turdus merula*)	3.9	2.7
Song thrush (*Turdus philomelos*)	2.0	0.9
Undetermined thrush (*Turdus* sp.)	3.9	1.9
Great reed warbler (*Acrocephalus arundinaceus*)	2.0	0.5
Spotted flycatcher (*Muscicapa striata*)	1.0	0.1
Golden oriole (*Oriolus oriolus*)	4.9	2.4
Lesser grey shrike (*Lanius minor*)	2.0	0.9
Jay (*Garrulus glandarius*)	5.9	7.1
Starling (*Sturnus vulgaris*)	2.0	1.1
Undetermined bird	11.5	11.2
Birds total	**90.2**	**86.1**
Viper (*Vipera berus*)	1.0	0.7
Total *N* prey specimens or biomass (g)	102	14 881 g

During the whole period of rearing young, parents brought, on average, 1.7 prey items/day to a nest with two young, the mean weight of one prey item being 163 g. Fully fledged juveniles consumed 300 g prey/day. Golodushko (1959) mentioned that the hunting success of adult eagles was markedly higher on sunny days than on cloudy and rainy ones. Also, he reported that in the nests of booted eagles nest bedding (twigs of oak, spruce, and birch) was changed for fresh material daily, and birds brought to the nest had been plucked and eviscerated. Quite a contrast, commented Golodushko, to the dirty and fetid nests of another bird eater, the peregrine.

4.21 Black Kites – 'River Ospreys' of Temperate Forests, and Red Kites – Unspecialised Hunters and Scavengers

In Białowieża Forest, the black kite nested exclusively along rivers (Leśna, Narew, Narewka) and fish formed the bulk of its prey (58% of prey specimens; Table 4.38). Gavrin (1953), who studied the food brought to the nest by adult kites, reported that pike and roach were most frequently captured. Some fish may have been taken dead from the water surface or river banks, but many must have been captured alive. In the forest rivers, pikes began

Table 4.38. Spring-summer diet of the black kite in the Belarussian part of BPF in 1946–1953. (Gavrin 1953)

Item	Percentage of prey
Voles and mice	2.7
Common rat (*Rattus norvegicus*)	8.8
Other mammals (hegdehog, *Erinaceus concolor*, mole, *Talpa europaea*, young hare, *Lepus europaeus*)	8.3
Mammals total	**19.8**
Thrush (*Turdus* sp.)	1.1
Golden oriole (*Oriolus oriolus*)	1.1
Jay (*Garrulus glandarius*)	1.1
Rook (*Corvus frugilegus*)	1.1
Other birds (young domestic chicken and ducks, pipit, *Anthus* sp., sky larks, *Alauda arvensis*, and others)	15.6
Birds total	**20.0**
Amphibians (anurans)	2.0
Fish (mainly pike, *Esox lucius* and roach, *Rutilus rutilus*)	58.2
Total *N* prey specimens	91

their spawning season on 30 March, on average (range 12 March–16 April), and roach on 28 April (range 16 April–9 May) (Dackevich 1977). For spawning, pike migrate up the rivers and brooks to the very shallow flood waters (20–50 cm deep) of rivers or even penetrate small canals in the meadows and marshes (Brylińska 1991). Spawning sites of roach are located in the same areas that few weeks earlier were left by spent pike, but in still shallower, quiet flood waters, separated from the main river course by reeds and other tall, marshy vegetation. During the spawning season, roaches often jump above the water surface (Brylińska 1991). Pike and roach spawn in the same places from year to year, which make them a predictable food resource for black kites. In some years (for instance in June 1949, June 1955, July–August 1980), mass mortality of fish in Białowieża rivers was observed as a result of midsummer floods caused by unusually heavy rainfalls (Dackevich 1983). In such cases, black kites obviously utilised dead fish. Kites were also successful hunters of small and medium-sized mammals and birds. Each of these two groups made up 20% of prey items (Table 4.38). Thus, in the pristine temperate forests, the black kite appeared a 'river osprey', able to supplement its diet with mammals and birds.

Unlike its congeneric, the red kite did not rely on aquatic food. It occasionally consumed dead fish on river banks, but its main prey were mammals (46% of prey: small and medium-sized rodents, hare, and ungulate carrion), and

Table 4.39. Spring–summer diet of the red kite in the eastern part of BPF in 1946–1953. (Gavrin 1953)

Item	Percentage of prey
Brown hare (*Lepus europaeus*)	12.8
Voles and mice	12.8
Other mammals (hegdehog, *Erinaceus concolor*, muskrat, *Ondatra zibethica*, red squirrel, *Sciurus vulgaris*, ungulate carrion)	20.5
Mammals total	**46.1**
Sparrowhawk (*Accipiter nisus*)	1.3
Galliformes birds (hazel hen, *Bonasa bonasia*, black grouse, *Lyrurus tetrix*, juv., capercaillie, *Tetrao urogallus*, juv.)	9.0
Jay (*Garrulus glandarius*)	2.5
Small passerines	24.5
Other birds (mallards, *Anas platyrhynchos*, curlew, *Numenius arquata*, cuckoo, *Cuculus canorus*, rook, *Corvus frugilegus*, and others)	11.4
Birds total	**48.7**
Fish (ide, *Leuciscus idus*)	5.2
Total *N* specimens of prey	78

birds (49%) (Table 4.39). Gavrin (1953) supposed that most of the large birds consumed by red kites were the leftovers from goshawks' prey, scavenged by kites. In the late 1930s, scavenging by red kites was frequently notified by Dackevich (1977) who regularly observed these raptors feeding on carcasses of domestic ungulates at a carrion discarding site in Białowieża Glade.

4.22 Hobbies – Aerial Hunting for Swallows, Larks and Dragonflies

Hobbies prey on small birds and large insects. Although their breeding territories mostly encompass forests, these birds hardly ever hunt in tree stands. Evident from their diet, hobbies' prime hunting habitats are open areas: marshy river valleys, meadows, farmlands, villages, large glades, and clear-cuts. An overwhelming majority of prey was captured by hobbies in flight; just a few voles, ungulate carcasses, flightless Carabidae beetles, totalling 3%, were undoubtedly picked up from the ground. Among over 30 species of birds that make up 95% of food biomass consumed by hobbies, swallows and larks are unchallenged dominants, followed by linnets and yellowhammers (Table 4.40). The hobby's specialisation on swallows is so great, that even in the diet of one pair that nested deep in the interior of Białowieża Forest, swallows were their dominant prey. These falcons must have exploited swallows nesting under the bridges on forest rivers.

Golodushko (1960) observed that in 1952–1958, the dates of spring arrivals of hobbies at their breeding quarters in Białowieża Forest matched exactly or were a little later than those of swallows. Also, hatching of young hobbies coincided with the mass appearance of swallow fledglings (5–12 July). Hobbies hunt on swallows in the late afternoon or at dusk. Golodushko (1960) observed these falcons flying very high and making rapid dives on swallows that foraged in the air. When hobbies hunt other birds (passerines living in meadows and forest ecotones), they fly lower and make short dives. To an occasional observer, they seem to suddenly appear from nowhere and disappear from view in a second.

On clear days, in full daylight, hobbies may be seen flying very high in the sky, hunting for some insects together with swallows and swifts to which they do not pay any attention at that time. From dusk until full darkness, over marshy valleys of the forest rivers, hobbies often pursue dragonflies, which they capture with talons and eat in flight. In drier habitats (forest-farmland ecotones), hobbies predominantly capture various Coleoptera bettles (predominantly Cerambycidae and Chrysomelidae) (Table 4.41). Although eaten in great numbers, insects made up only 1–5% of the total food biomass of hobbies (Table 4.40).

Evening hunts over rivers are also the very time and place where hobbies capture noctule bats. In 1952–1958, in the eastern part of the Forest, bats made up 22% of prey biomass of hobbies (Table 4.40). In no other species of predator did bats form such a notable constituent of the diet. Of the forest

Table 4.40. Diet composition of the hobby in BPF in spring-summer

Item	Eastern part		Western part		
	%Prey	%Bio	%Occ	%Prey	%Bio
Mole (*Talpa europaea*)	–	–	0.7	0.2	0.4
Parti-coloured bat (*Vespertilio murinus*)	0.9	1.8	–	–	–
Common noctule (*Nyctalus noctula*)	3.0	12.8	–	–	–
Lesser noctule (*Nyctalus leisleri*)	0.3	0.6	–	–	–
Undetermined bat	1.8	6.4	2.2	0.6	0.3
Chiroptera total	**6.0**	**21.6**	**2.2**	**0.6**	**0.3**
Red squirrel (*Sciurus vulgaris*)	–	–	1.4	0.4	0.7
Bank vole (*Clethrionomys glareolus*)	–	–	1.4	0.4	0.4
Common vole (*Microtus arvalis*)	–	–	0.7	0.2	0.2
Undetermined small rodent	2.7	6.9	–	–	–
Rodents total	**2.7**	**6.9**	**3.5**	**1.0**	**1.3**
Ungulate carcass	–	–	4.3	1.2	2.1
Pigeon (*Columba* sp.)	–	–	1.4	0.4	0.7
Turtle dove (*Streptopelia turtur*)	0.6	13.6	–	–	–
Great spotted woodpecker (*Dendrocopos major*)	–	–	0.7	0.2	0.4
Undet. woodpecker (*Dendrocopos* sp.)	–	–	0.7	0.2	0.4
Sky lark (*Alauda arvensis*)	1.5	7.9	16.5	9.8	22.7
Sand martin (*Riparia riparia*)	0.3	0.6	–	–	–
Swallow (*Hirundo rustica*)	–	–	59.7	22.8	25.8
Undetermined swallows and martins	8.7	22.2	–	–	–
Tree pipit (*Anthus trivialis*)	–	–	2.9	1.0	1.3
Meadow pipit (*Anthus pratensis*)	0.6	1.7	–	–	–
Pipit (*Anthus* sp.)	–	–	2.2	0.6	0.8
Lark (*Alauda* sp.) or pipit (*Anthus* sp.)	0.6	3.2	1.4	0.4	0.9
Yellow wagtail (*Motacilla flava*)	–	–	2.2	0.8	0.9
White wagtail (*Motacilla alba*)	–	–	3.6	1.2	1.8
Dunnock (*Prunella modularis*)	–	–	0.7	0.2	0.3
Robin (*Erithacus rubecula*)	–	–	0.7	0.2	0.2
Whinchat (*Saxicola rubetra*)	–	–	2.9	0.8	0.9
Blackbird (*Turdus merula*)	–	–	2.9	0.8	1.4
Song thrush (*Turdus philomelos*)	–	–	2.9	0.8	1.4
Great reed warbler (*Acrocephalus arundinaceus*)	0.3	1.6	–	–	–
Icterine warbler (*Hippolais icterina*)	–	–	0.7	0.2	0.2
Lesser whitethroat (*Sylvia curruca*)	–	–	0.7	0.2	0.1
Whitethroat (*Sylvia communis*)	–	–	1.4	0.4	0.3
Undetermined warbler (*Sylvia* sp.)	–	–	0.7	0.2	0.2
Great tit (*Parus major*)	–	–	1.4	0.4	0.5

Table 4.40 (continued)

Item	Eastern part		Western part		
	%Prey	%Bio	%Occ	%Prey	%Bio
Undetermined tits (*Parus* sp.)	–	–	2.2	0.6	0.7
Nuthatch (*Sitta europaea*)	–	–	1.4	0.4	0.6
Red-backed shrike (*Lanius collurio*)	0.3	1.3	–	–	–
Jay (*Garrulus glandarius*)	–	–	1.4	0.4	0.7
Chaffinch (*Fringilla coelebs*)	–	–	2.9	0.8	1.3
Greenfinch (*Carduelis chloris*)	–	–	3.6	1.2	2.1
Goldfinch (*Carduelis carduelis*)	–	–	5.0	1.8	1.6
Linnet (*Carduelis cannabina*)	–	–	10.1	4.1	4.6
Undetermined finch (*Carduelis* sp.)	–	–	1.4	0.4	0.5
Hawfinch (*Coccothraustes coccothraustes*)	–	–	4.3	1.4	2.5
Yellowhammer (*Emberiza citrinella*)	0.9	3.8	8.6	2.8	5.4
Undetermined small passerine	3.3	10.3	28.8	8.2	13.3
Birds total	**17.1**	**66.2**	**95.0**	**63.7**	**94.5**
Fish	–	–	0.7	0.2	0.4
Beetle (Coleoptera)	21.8	1.6	41.7	30.8	1.0
Dragonfly (Odonata)	52.4	3.7	1.4	0.4	+
Other and undetermined insects	–	–	12.9	1.9	+
Invertebrates total	**74.2**	**5.3**	**51.1**	**33.1**	**1.0**
Total *N* prey remains, pellets, or biomass (g)	334	2348 g	139	490	7827 g

Eastern part: data on prey brought to nests by adult birds, 1956–1958 (Golodushko 1960). Western part: analysis of pellets and prey remains, 1993–1994 (B. Jędrzejewska, W. Jędrzejewski, E. Pugacewicz, and K. Zub, unpubl.). Explanations as in Table 4.30.

community of bats which, according to Kurskov (1958), is dominated by the common pipistrelle (39% in forest colonies), noctule bat (38%) and lesser noctule (13%), hobbies most often captured the noctule (at least 50% of bats taken). In Białowieża Forest, noctules strongly prefer river valleys and forest-meadow ecotones, and their foraging activity peaks soon after sunset and, again, before sunrise (Rachwald 1992). In contrast, pipistrelles forage in dense woods, a habitat rather avoided by hobbies. The parti-coloured bat (at least 15% of all bats captured by hobbies) is a dominant species in rural colonies of bats (53% in the community; see Sect. 2.7).

The food requirements of hobbies' hatchlings (birds hand-reared by Golodushko 1960) increased from 10 g/day in 4-day-old eyases to 98 g/day in 1-month-old fledglings ready to leave nest (body mass 275 g). During its whole nest-life (1 month), a juvenile hobby consumed about 1500 g of prey.

Table 4.41. Insects captured by hobbies in Białowieża Forest. Sources as in Table 4.40. Summed figures for taxonomic orders of insects are in boldface type

Taxonomic group	Number of specimens			Percent
	Eastern part	Western part	Total	
Odonata (total)	175	2	177	**43.2**
Aeschnidae	87	–	87	21.2
Libellulidae	24	–	24	5.9
Undetermined Odonata[a]	64	2	66	16.1
Heteroptera	–	1	1	**0.2**
Hymenoptera[a]	–	1	1	**0.2**
Coleoptera (total)	73	151	224	**54.6**
Carabidae (total)	4	7	11	2.7
Carabus sp.	–	1	1	
Undetermined Carabidae	4	6	10	
Silphidae	–	3	3	0.7
Staphylinidae	6	–	6	1.5
Scarabaeidae (total)	1	22	23	5.6
Geotrupes sp.	–	1	1	
Undetermined Scarabaeidae	1	21	22	
Melolonthidae (total)	–	7	7	1.7
Melolontha sp.	–	6	6	
Cetonia aurata L.	–	1	1	
Dytiscidae	–	2	2	0.5
Coccinellidae (total)	–	9	9	2.2
Coccinella septempunctata L.	–	3	3	
Undetermined Coccinellidae	–	6	6	
Cerambycidae (total)	14	40	54	13.2
Spondylis buprestoides L.	–	21	21	
Callidium violaceum L.	–	17	17	
Undet. Cerambycidae	14	2	16	
Chrysomelidae (total)	25	12	37	9.0
Agelastica alni L.	–	1	1	
Undetermined Chrysomelidae	25	11	36	
Undetermined Coleoptera	23	49	72	17.6
Lepidoptera (total)	–	2	2	**0.5**
Sphingidae (total)	–	1	1	0.2
Sphinx sp.	–	1	1	
Pieridae (total)	–	1	1	0.2
Pieris rapae L.	–	1	1	
Undetermined insects	–	4	4	1.0
Total	248	162	410	100

[a] In addition to this material, Gavrin (1953) reported that stomachs of two shot hobbies contained 3 dragonflies (Odonata) and 59 winged ants (Formicidae).

4.23 Tawny Owls – Specialisation on Yellow-Necked Mice with Secondary Adaptations for Hunting Birds, Shrews, and Frogs

When we began investigations on feeding habits of tawny owls in the pristine forests of Białowieża National Park in the winter of 1985/1986, finding a single pellet on forest floor seemed a hopeless task. Owls could cast pellets from every old tree 20–40 m high, and pellets were soon covered by falling plant debris or snow and destroyed by rain, fungi and insects. However, once we focused on potential roost trees, mostly old limes with cavities or broken tops, our search became more fruitful and eventually yielded over 1000 pellets for analysis.

There were hardly any species of small vertebrates (rodents, insectivores, birds, amphibians, bats) living in the forest that were not found in tawny owl pellets. Among the diverse foods of owls, rodents were of prime importance, constituting, on average, 53% of the biomass consumed in spring–summer and nearly 80% of that taken in autumn–winter (Table 4.42). The yellow-necked mouse was the most important prey species, captured more often than expected from its share in forest rodent community.

Three other resources were significant buffer prey to owls: birds (4% of food biomass in autumn–winter, 16% in spring–summer), insectivores (7 and 16%, respectively), and amphibians (8 and 14%; Table 4.42). The list of birds captured by tawny owls includes some 20 species, of which the most common were thrushes: the song thrush in spring–summer, and migrating redwings in autumn–winter. Other frequently taken birds were tits, chaffinches, flycatchers, jays, and finches in the genus *Carduelis*. About 20–30% of birds (and probably more because of many unidentified passerines) captured by the owls in either season were resident species, a proportion similar to that in the breeding bird community. There was no particular selection of insectivores by tawny owls. Common shrews, the dominants in the insectivore community, were captured by owls most frequently, although in terms of biomass, moles (an order of magnitude heavier than shrews) contributed more to owls' food (Table 4.42). Among amphibians eaten by owls, the common frog dominated as in the living community, whereas the very numerous toads *Bufo bufo* (22% in the forest community of amphibians) were strongly avoided.

Month-to-month and year-to-year changes in owl diet reflected many factors: variation in numbers of prey, environmental conditions, and owls' ability to cope with these adversities. The seasonal pattern of an owl's consumption of seven important prey types, shown in Fig. 4.39, was generally shaped by the changing availability of these resources. Bimonthly changes in the proportion of yellow-necked mice in the tawny owl diet roughly followed the course of mouse abundance, with the exception of winter and early spring, when mice were rather scarce but owls preyed on them heavily. The bank vole's share in the owl diet was smaller than that of mice and its bimonthly changes did not reflect the seasonal variation in vole abundance. *Microtus* voles formed a stable addition to the owls' diet year round. Generally, tawny

Table 4.42. Diet composition of the tawny owl in spring-summer (16 April–30 Sept.) and autumn-winter (1 Oct. 1–15 April) seasons in Białowieża National Park (Polish part of BPF) in 1985/1986–1992 (Jędrzejewski et al. 1994a)

Item	Spring-summer			Autumn-winter		
	%Occ	%Prey	%Bio	%Occ	%Prey	%Bio
Mole (*Talpa europaea*)	7.4	2.1	10.1	2.1	0.7	2.9
Common shrew (*Sorex araneus*)	26.1	13.5	5.3	18.4	9.8	3.6
Lesser shrew (*Sorex minutus*)	10.0	3.9	0.6	6.2	2.5	0.4
Undetermined shrew (*Sorex* sp.)	1.0	0.3	0.1	0.5	0.1	0.1
Water shrew (*Neomys fodiens*)	0.8	0.2	0.1	1.8	0.6	0.3
Mediterranean water shrew (*Neomys anomalus*)	0.2	0.1	+	–	–	–
Undet. water shrew (*Neomys* sp.)	0.2	0.1	+	0.5	0.2	0.1
Insectivores total	**37.1**	**20.2**	**16.2**	**25.2**	**13.9**	**7.3**
Brown hare (*Lepus europaeus*)	0.2	0.1	0.7	0.2	0.1	0.6
Bank vole (*Clethrionomys glareolus*)	37.9	16.0	13.5	54.1	31.7	24.5
Water vole (*Arvicola terrestris*)	0.4	0.1	0.4	0.7	0.2	0.7
Pine vole (*Pitymys subterraneus*)	7.8	2.3	1.9	6.8	2.9	2.3
Root vole (*Microtus oeconomus*)	10.0	3.3	5.9	11.9	5.7	9.3
Short-tailed vole (*Microtus agrestis*)	2.2	0.6	0.6	2.3	0.8	0.8
Common vole (*Microtus arvalis*)	3.8	1.3	1.3	5.5	2.4	2.1
Undetermined vole (*Microtus* sp.)	9.0	2.6	3.9	7.0	2.5	3.3
House mouse (*Mus musculus*)	0.2	0.1	+	0.3	0.1	0.1
Common rat (*Rattus norvegicus*)	–	–	–	0.6	0.2	2.8
Harvest mouse (*Micromys minutus*)	1.6	0.5	0.2	1.9	0.7	0.2
Striped field mouse (*Apodemus agrarius*)	–	–	–	0.8	0.3	0.2
Yellow-necked mouse (*Apodemus flavicollis*)	41.9	15.9	22.4	49.9	22.7	31.7
Wood mouse (*Apodemus sylvaticus*)	0.6	0.2	0.2	2.1	0.8	0.7
Birch mouse (*Sicista betulina*)	0.4	0.1	0.1	0.2	0.1	+
Forest dormouse (*Dryomys nitedula*)	0.6	0.2	0.3	–	–	–
Fat dormouse (*Glis glis*)	1.6	0.5	1.4	–	–	–
Undetermined rodent	1.4	0.4	0.4	2.3	0.7	0.8
Rodents total	**78.0**	**44.1**	**52.5**	**89.6**	**71.8**	**79.5**
Weasel (*Mustela nivalis*)	–	–	–	0.3	0.1	0.3
Woodpecker (*Dendrocopos* sp.)	0.4	0.1	0.4	–	–	–
Pipit (*Anthus* sp.)	0.2	0.1	0.1	–	–	–
Wren (*Troglodytes troglodytes*)	0.4	0.1	0.1	–	–	–
Robin (*Erithacus rubecula*)	0.2	0.1	0.1	0.2	0.1	+
Blackbird (*Turdus merula*)	1.0	0.3	1.4	–	–	–
Song thrush (*Turdus philomelos*)	2.0	0.6	1.9	0.5	0.1	0.5
Redwing (*Turdus iliacus*)	0.6	0.2	0.6	1.1	0.4	1.1
Undetermined thrush (*Turdus* sp.)	2.2	0.6	2.2	0.6	0.2	0.7

Table 4.42 (continued)

Item	Spring–summer			Autumn–winter		
	%Occ	%Prey	%Bio	%Occ	%Prey	%Bio
Undet. warbler (*Acrocephalus* sp.)	0.2	0.1	0.1	–	–	–
Undet. warbler (*Sylvia* sp.)	0.6	0.2	0.1	–	–	–
Goldcrest (*Regulus regulus*)	–	–	–	0.2	0.1	+
Undet. flycatcher (*Ficedula* sp.)	0.8	0.2	0.1	–	–	–
Undet. tit (*Parus* sp.)	1.2	0.3	0.2	0.3	0.1	0.1
Nuthatch (*Sitta europaea*)	0.4	0.1	0.1	–	–	–
Jay (*Garrulus glandarius*)	1.2	0.3	2.9	0.2	0.1	0.4
Chaffinch (*Fringilla coelebs*)	1.6	0.5	0.6	–	–	–
Goldfinch (*Carduelis carduelis*)	0.6	0.2	0.1	0.3	0.1	0.1
Bullfinch (*Pyrrhula pyrrhula*)	–	–	–	0.2	0.1	0.1
Hawfinch (*Coccothraustes coccothraustes*)	0.4	0.1	0.3	0.3	0.1	0.3
Undetermined bird	10.8	3.0	4.9	1.3	0.4	0.5
Birds total	**22.0**	**7.1**	**16.2**	**4.9**	**1.6**	**3.8**
Common spadefoot (*Pelobates fuscus*)	0.2	0.1	0.1	0.2	0.1	0.1
Common frog (*Rana temporaria*)	22.0	9.5	6.7	10.6	4.4	4.0
Moor frog (*Rana arvalis*)	10.8	3.9	2.7	5.7	3.1	2.4
Undetermined frog (*Rana* sp.)	17.2	5.2	4.1	6.8	2.5	1.9
Amphibians total	**41.3**	**18.7**	**13.6**	**20.8**	**10.1**	**8.4**
Insects	20.0	9.6	0.7	5.9	2.2	0.1
Plant material	0.8	0.2	0.1	0.7	0.2	+
Total *N* pellets, prey specimens or biomass (g)	501	1773	35 893 g	615	1944	42 747 g

%Occ, percentage of occurrence of a given prey in pellets; %Prey, percentage of all prey specimens; %Bio, percentage of total biomass of consumed food.

owls relied on rodents most heavily in November–December (89% of consumed biomass) and least so in July–August (47% of consumed biomass).

The proportion of birds in the owls' diet was highest in May–June (on average 24% of consumed biomass) at the birds' peak breeding season, and lowest (1%) in November–December (Fig. 4.39). Shrews formed a fairly stable 3–7% of biomass taken by owls year-round and moles made a significant contribution in July–August (16%). Amphibians were preyed upon somewhat more intensively in March–April, during the frogs' spawning season, and in September–October, when they migrated to hibernation sites.

During our study on tawny owl feeding habits (1985/1986–1992), forest rodents underwent the entire wave of fluctuations typical of old growth deciduous forests: 5 years of moderate densities and 2 years of outbreak and crash

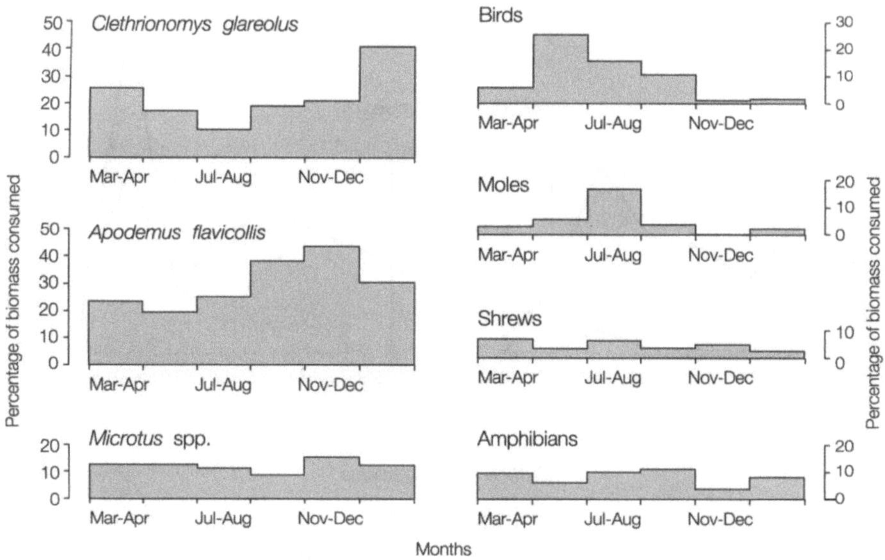

Fig. 4.39. Bimonthly changes in percent biomass of seven main groups of food resources in tawny owl diet in Białowieża National Park (data pooled for 1985/1986–1992). Jędrzejewski et al. (1994 a)

'cycle' triggered by a heavy crop of oak, hornbeam and maple seeds (see Sect. 2.4). How did tawny owls cope with a 30-fold variation in rodent densities? In their dietary response, owls followed the changes in abundance of yellow-necked mice (Fig. 4.40), but were able to hunt them intensively even when they were scarce. It was only the prolonged crash of mouse numbers in 1991–1992 that caused a decline of this species in owl diet to a mere 5%. Thus, the tawny owl clearly specialises on mice. The geographic range of *Strix aluco* is very similar to that of the genus *Apodemus*. The tawny owl's hunting behaviour is adapted to the behaviour of that prey and its activity pattern is the same as that of mice, with a peak around midnight (Mikkola 1983; Wójcik and Wołk 1985). Furthermore, microhabitat selection by mice, i.e. their equal utilisation of densely covered and almost bare ground under the forest canopy (Southern and Lowe 1968) and their routes on uprooted trees and fallen logs rather than beneath them (Olszewski 1968), facilitate owls' predation on them. Even during snowy winters, granivorous mice actively gather seeds on the snow.

The dependence of owls on bank voles is much weaker than that on mice. The share of bank voles in owl diet was shaped by two factors: the abundance of voles (positive relation) and the abundance of mice (negative relation). Moreover, in winter folivorous-granivorous bank voles forage in subnivean spaces, which are hardly penetrable to avian predators. In consequence, winter consumption of bank voles by owls in years of their moderate and high numbers was significantly affected by the depth of snow cover.

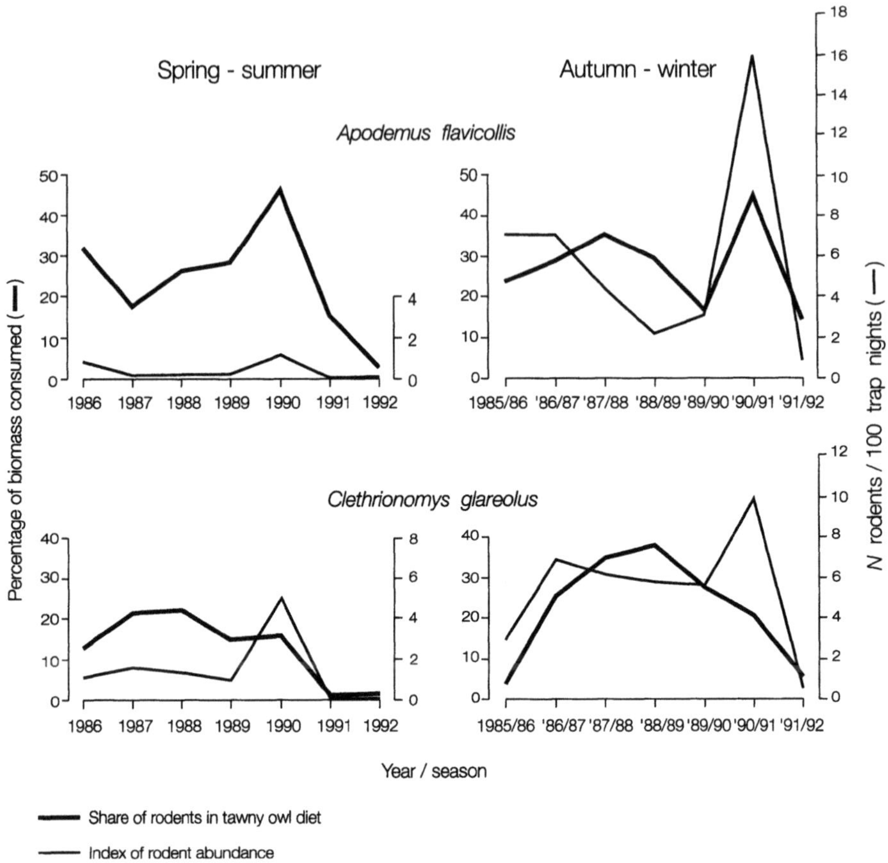

Fig. 4.40. Year to year variation in percent biomass of yellow-necked mouse *A. flavicollis* and bank vole *C. glareolus* in the diet of tawny owls in spring–summer and autumn–winter in relation to the indices of rodent abundance. Jędrzejewski et al. (1994a)

Mean percentage occurrence of bank vole remains in tawny owl pellets declined from 82% when there was no snow to 47% when snow was 30–40 cm deep (Jędrzejewski et al. 1994a). During spring and summer, vegetation on the forest floor helps the voles to avoid owl predation. The lowest proportion of bank voles in the tawny owl's diet occurred in summer, a period of relatively high vole numbers but also of lush vegetation. Finally, the crepuscular activity rhythm of voles (activity peaks at dawn and dusk, Wójcik and Wołk 1985) is quite divergent from that of tawny owls. Functional (dietary) responses of tawny owls to changes in mouse and vole numbers were best described by logarithmic curves (Fig. 4.41). Consumption of birds, shrews, and frogs was inversely proportional to the share of rodents in the tawny owl's diet (Fig. 4.42). Interestingly, the three buffer foods were used in parallel. Only in the case of frogs did cold winters cause their slightly lower consumption by owls.

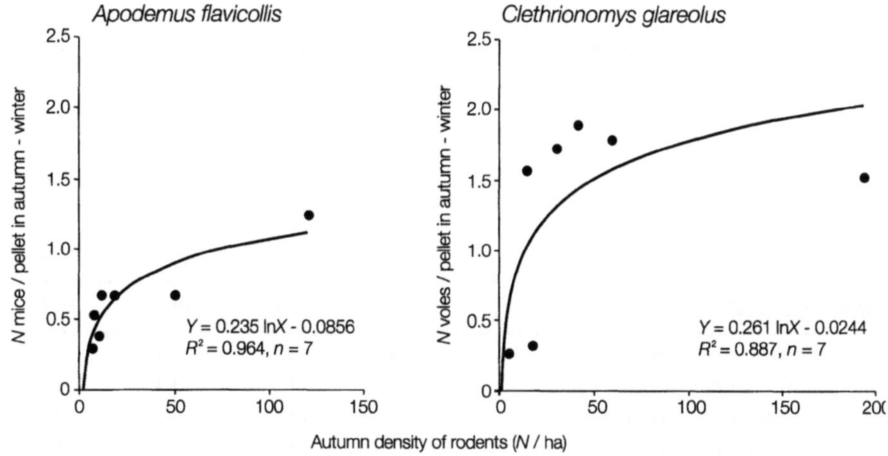

Fig. 4.41. Functional (dietary) responses by tawny owls in autumn–winter (1 Oct.–15 April) to variation in autumn (September/October) density of forest rodents in Białowieża National Park. Each point = one autumn–winter season (1985/1986–1991/1992). Jędrzejewski et al. (1996a)

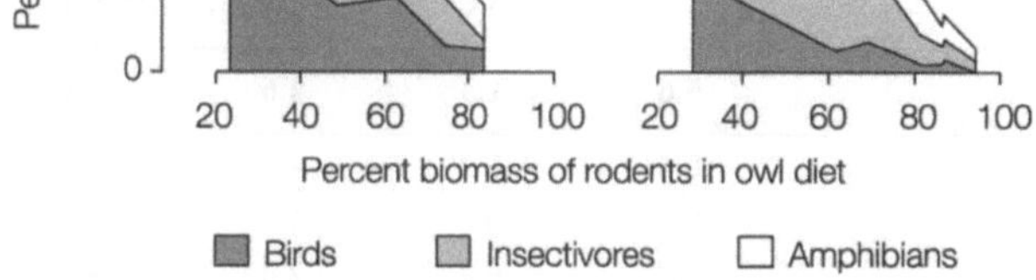

Fig. 4.42. Relationships between the percent biomass of rodents and percent biomass of buffer foods in the diet of tawny owls (1985/1986–1992, Białowieża National Park). Jędrzejewski et al. (1994a)

Thus, tawny owls' feeding strategy in the pristine deciduous and mixed forests was based on specialisation on forest rodents, particularly the yellow-necked mouse, and their ability to buffer the changes in rodent abundance by utilisation of three other food resources. How successful was this strategy in assuring the owls' survival of rodent crashes and environmental adversities? The average daily food consumption of tawny owls is about 60 g (Ceska 1980; Lowe 1980). In Białowieża Forest, in years of moderate rodent numbers, tawny owls consumed on average 69 g of prey daily (Jędrzejewski et al. 1996a). During outbreaks of rodents, owls enjoyed daily meals of 81 g, on

Table 4.43. Diet composition of the tawny owl in the Belarussian part of BPF in spring–summer and autumn–winter seasons in 1946–1947. Data from Gavrin (1953) on 33 stomachs and 113 pellets

Item	Spring–summer		Autumn–winter	
	%Prey	%Bio	%Prey	%Bio
Mole (*Talpa europaea*)	4.6	11.7	–	–
Undetermined Soricidae	16.1	5.4	39.1	15.4
Insectivores total	**20.7**	**17.1**	**39.1**	**15.4**
Brown hare (*Lepus europaeus*)	1.8	4.8	1.2	3.5
Bank vole (*Clethrionomys glareolus*)	1.4	1.0	3.4	2.9
Water vole (*Arvicola terrestris*)	9.7	25.1	2.3	7.0
Root vole (*Microtus oeconomus*)	1.4	1.5	3.4	4.4
Short-tailed vole (*Microtus agrestis*)	5.5	5.3	–	–
Common vole (*Microtus arvalis*)	19.3	15.4	10.3	9.7
Undetermined vole (*Microtus* sp.)	7.8	7.5	–	–
House mouse (*Mus musculus*)	0.5	0.3	9.2	7.7
Common rat (*Rattus norvegicus*)	–	–	1.2	3.5
Striped field mouse (*Apodemus agrarius*)	–	–	1.2	1.4
Yellow-necked mouse (*Apodemus flavicollis*)	4.6	6.0	5.8	8.8
Undetermined rodent	2.8	3.8	14.9	24.3
Rodents total	**53.0**	**65.9**	**51.7**	**69.7**
Hazel hen (*Bonasa bonasia*)	0.5	1.2	2.3	7.0
Thrush (*Turdus* sp.)	2.8	7.2	–	–
Undetermined passerine	2.3	2.9	2.3	3.4
Birds total	**5.6**	**11.3**	**4.6**	**10.4**
Lizard (*Lacerta* sp.)	0.5	0.2	–	–
Frog (*Rana* sp.)	0.4	0.3	1.1	1.0
Beetle (*Geotrupes* sp.)	9.2	0.2	2.3	+
Other and undetermined insects	8.8	0.2	–	–
Invertebrates total	**18.0**	**0.4**	**2.3**	**+**
Total *N* prey specimens or biomass (g)	217	5192 g	87	1766 g

average, but during a crash, they provided themselves with 55 g of food per day only. Therefore, at least some owls starved and died. In consequence, a 30% decline in tawny owl numbers was recorded (see Sect. 3.24).

Spatial variation in owl feeding habits was well evidenced by yet two other studies conducted in Białowieża Forest. Gavrin (1953) observed that at the south-eastern verges of the forest, near the meadows in river valleys, farmlands and villages, tawny owls hunted chiefly for rodents of the open landscape: voles in the genus *Microtus* and water voles (Table 4.43). In still more

Table 4.44. Diet composition of tawny owls inhabiting the Białowieża village in autumn-winter 1982/1983. (Ruprecht and Szwagrzak 1987)

Item	%Prey	%Bio
Common shrew (*Sorex araneus*)	0.5	0.2
Common long-eared bat (*Plecotus auritus*)	0.5	0.2
Bank vole (*Clethrionomys glareolus*)	3.4	3.4
Root vole (*Microtus oeconomus*)	3.4	5.1
Common vole (*Microtus arvalis*)	6.8	7.5
House mouse (*Mus musculus*)	14.6	14.4
Common rat (*Rattus norvegicus*)	0.5	1.8
Harvest mouse (*Micromys minutus*)	4.9	2.3
Striped field mouse (*Apodemus agrarius*)	1.5	2.0
Yellow-necked mouse (*Apodemus flavicollis*)	1.5	2.6
Undetermined mouse (*Apodemus* sp.)	2.4	4.0
Rodents total	**39.0**	**43.1**
Undetermined thrush (*Turdus* sp.)	0.5	1.8
Great tit (*Parus major*)	1.9	2.0
Marsh tit (*Parus palustris*)	0.5	0.3
Siskin (*Carduelis spinus*)	0.5	0.4
House sparrow (*Passer domesticus*)	3.4	5.7
Undetermined passerine	0.5	0.8
Birds total	**7.3**	**11.0**
Common spadefoot (*Pelobates fuscus*)	1.0	0.6
Undetermined frog (Anura)	51.7	44.9
Amphibians total	**52.7**	**45.5**
Total *N* prey specimens or biomass (g)	205	3541 g

transformed habitat, the village of Białowieża, Ruprecht and Szwagrzak (1987) found that the main food of rural tawny owls was house mice, *Microtus* voles, and house sparrows (Table 4.44). Therefore, living in various habitats, tawny owls had rather stable preferences for rodents, birds, shrews and amphibians and simply 'adjusted' to the local species composition of these groups.

4.24 Pygmy Owls - Tyrants of Passerine Birds

In spring and early summer when the whole forest sounds of birds singing, it is obvious that many predators enjoy meals of birds, their eggs, and hatchlings. However, it seems unlikely that any predator would rely on avian prey in winter, when the only birds seem to be redwings and waxwings coming and going in noisy flocks, and woodpeckers whose drumming echoes in a quiet forest. On windless days, tits and goldcrests bestir themselves and search for invertebrates in high canopies, their presence marked by low chirps and particles of lichen sprinkling down onto the snow. The pygmy owl, a predator not much bigger than many passerine birds, does persevere in bird hunting year-round.

In Białowieża Forest, birds made up to 42% of the food biomass of pygmy owls in autumn–winter and 59% in spring–summer (Table 4.45). In various years, the breeding season food of owls consisted of birds in quantities of 25 to 73% by biomass (Golodushko and Samusenko 1961; Jędrzejewska and Jędrzejewski 1993), although the lower figure is certainly an underestimate (see below). The list of birds falling as prey to pygmy owls includes 22 species, ranging in size from wrens and goldcrests (6 g) to hawfinches (55 g) and great spotted woodpeckers (80 g), the last ones bigger than pygmy owls (65 g). Besides birds, pygmy owls frequently captured small mammals, of which the bank vole was positively selected for.

While species of mammals taken by pygmy owls were the same in warm and cold seasons, the structure of bird prey clearly varied seasonally. In spring and summer, chaffinches, robins, flycatchers and warblers, all of them migratory species, were frequently captured, although not as much as would have been expected from their substantial shares in the bird community. Tits were the single most important group of birds in this owl's diet year-round, taken twice as often as could have been expected from their proportion in the total number of forest birds. In autumn and winter, pygmy owls often captured tits, goldcrests, and woodpeckers (Table 4.45). Also, *Carduelis* finches were the owls' favourite prey, taken far more commonly than their relative occurrence in the bird community (Jędrzejewska and Jędrzejewski 1993). This pattern of a pygmy owl feeding on birds (high shares of chaffinch, robin and tits in summer, and tits, goldcrests and *Carduelis* finches in winter) is very common geographically from taiga forests (Finland, Russia, Norway, Sweden), to temperate forests (Belarus, Poland, Germany, Austria) (review in Jędrzejewska and Jędrzejewski 1993).

The majority of birds are captured by pygmy owls in the canopy layer. Having caught a small passerine bird, the owl plucks it partly while sitting on branches or immediately takes it to a hole in a tree. Killing larger prey is a difficult task. Gavrin (1953) and Borowski (1961) observed owl attacks on a woodpecker and a hawfinch, respectively. In each case, the owl attacked a bird sitting in a tree from behind. The two, predator and prey, fell down and struggled for a long while. Locked together, they could have been approached as closely as 3 m (Borowski 1961) or even captured by hand (Gavrin 1953). Large prey are too heavy to be carried on wings, so the owls begin consumption on the ground and then take the rest to the cavity.

To pygmy owls, shrews are small enough to be swallowed whole but bigger mammals and all birds are plucked and partitioned. Owls frequently keep buffer stores of food in their cavities, especially in winter. In Białowieża

Table 4.45. Diet composition of the pygmy owl in Białowieża Forest in the spring-summer and autumn-winter seasons

Item	Spring-summer					Autumn-winter	
	Western part			Eastern part		Eastern part	
	%Occ	%Prey	%Bio	%Prey	%Bio	%Prey	%Bio
Common shrew (*Sorex araneus*)	–	–	–	5.5	3.6	11.1	8.1
Lesser shrew (*Sorex minutus*)	–	–	–	5.6	1.3	12.5	3.4
Undetermined shrew (*Sorex* sp.)	0.8	0.4	0.2	28.7	11.5	36.8	16.7
Insectivores total	**0.8**	**0.4**	**0.2**	**39.8**	**16.4**	**60.4**	**28.2**
Bank vole (*Clethrionomys glareolus*)	76.4	33.7	34.0	39.8	54.2	11.8	18.2
Water vole (*Arvicola terrestris*)	–	–	–	–	–	0.7	5.2
Pine vole (*Pitymys subterraneus*)	4.9	2.2	2.2	–	–	–	–
Undetermined vole (*Microtus* sp.)	1.6	0.7	0.8	–	–	2.1	4.3
Yellow-necked mouse (*Apodemus flavicollis*)	1.6	0.7	1.3	–	–	–	–
Undet. mouse (*Apodemus* sp.)	2.0	0.9	1.6	1.9	4.6	0.7	2.0
Birch mouse (*Sicista betulina*)	6.5	2.9	1.4	–	–	–	–
Rodents total	**85.8**	**41.1**	**41.3**	**41.7**	**58.8**	**15.3**	**29.7**
Great spotted woodpecker (*Dendrocopos major*)	–	–	–	–	–	0.7	5.1
Lesser spotted woodpecker (*Dendrocopos minor*)	–	–	–	–	–	1.4	2.6
Undetermined woodpecker (*Dendrocopos* sp.)	0.8	0.4	0.7	–	–	–	–
Dendrocopos or *Pyrrhula*	1.2	0.5	1.0	–	–	–	–
Wren (*Troglodytes troglodytes*)	3.7	1.6	0.8	–	–	–	–
Robin (*Erithacus rubecula*)	24.8	10.9	11.0	0.9	1.3	–	–
Redstart (*Phoenicurus phoenicurus*)	1.6	0.7	0.7	–	–	–	–
Undet. warbler (*Sylvia* sp.)	1.2	0.5	0.5	–	–	–	–
Undet. warbler (*Phylloscopus* sp.)	5.7	2.5	1.5	–	–	–	–
Goldcrest (*Regulus regulus*)	1.2	0.5	0.2	–	–	4.1	1.9
Collared flycatcher (*Ficedula albicollis*)	0.4	0.2	0.1	–	–	–	–
Pied flycatcher (*Ficedula hypoleuca*)	3.7	1.8	1.3	2.8	2.7	–	–
Undet. flycatcher (*Ficedula* sp.)	1.6	0.7	0.5	–	–	–	–
Coal tit (*Parus ater*)	–	–	–	0.9	0.6	–	–
Blue tit (*Parus caeruleus*)	4.5	2.0	1.3	–	–	–	–
Great tit (*Parus major*)	5.3	2.4	2.5	11.1	16.0	4.1	6.8
Undetermined tit (*Parus* sp.)	10.6	4.7	4.1	–	–	0.7	0.6

Table 4.45 (continued)

Item	Spring–summer					Autumn–winter	
	Western part			Eastern part		Eastern part	
	%Occ	%Prey	%Bio	%Prey	%Bio	%Prey	%Bio
Nuthatch (*Sitta europaea*)	1.6	0.7	1.0	–	–	–	–
Tree creeper (*Certhia familiaris*)	2.8	1.3	0.7	–	–	–	–
Red-backed shrike (*Lanius collurio*)	0.4	0.2	0.3	–	–	–	–
Chaffinch (*Fringilla coelebs*)	17.5	7.9	11.7	–	–	–	–
Serin (*Serinus serinus*)	0.4	0.2	0.1	–	–	–	–
Greenfinch (*Carduelis chloris*)	1.2	0.5	0.9	–	–	–	–
Goldfinch (*Carduelis carduelis*)	0.4	0.2	0.1	–	–	–	–
Siskin (*Carduelis spinus*)	1.6	0.7	0.6	–	–	–	–
Bullfinch (*Pyrrhula pyrrhula*)	1.2	0.5	0.8	–	–	–	–
Hawfinch (*Coccothraustes coccothraustes*)	2.0	0.9	1.6	–	–	0.7[a]	3.5[a]
Undetermined passerine	34.6	15.3	14.5	2.8	4.2	12.6	21.6
Birds total	**79.7**	**57.8**	**58.5**	**18.5**	**24.8**	**24.3**	**42.1**
Insects	1.6	0.7	+	–	–	–	–
Total *N* pellets, prey specimens or biomass (g)	246	553	9397 g	108	1348 g	144	1586 g

Western part of BPF: Jędrzejewska and Jędrzejewski (1993 and unpubl. data). Material included pellets collected in 1989, 1993 and 1994 in Białowieża National Park. Eastern part of BPF: Golodushko and Samusenko (1961), material included prey specimens found in stores in breeding and nesting holes ($N=13$ stores), and pellets ($N=110$), collected in 1956–1958. Explanations as in Table 4.42.

[a] Also, Borowski (1961) reported observations of three hawfinches captured by pygmy owls in Białowieża National Park in 1947–1957.

Forest, Golodushko and Samusenko (1961) found stores of one to eight small mammals in a cavity. They reported that owls more frequently stored mammals than birds which, as preferred prey, were often consumed immediately upon capture. Since more than half the data collected by Golodusho and Samusenko (1961) were on prey from stores, the share of mammals in owl diet (shown in Table 4.45) is overestimated.

Pygmy owl's specialisation on passerine birds in both summer and winter is a rather unusual evolutionary achievement in boreal and temperate forests. It is fully acknowledged in the Belarussian name of this predator, *veryabiny sych*, the sparrow owl.

4.25 Tengmalm's Owls – Boreal Menu of Bank Voles, Tits, and Goldcrests

Pellets of Tengmalm's owls collected in the breeding seasons of 1993 and 1994 in several territories of these birds in the western part of the forest predominantly contained forest rodents (78% of biomass consumed), of which the bank vole was by far the most important (Table 4.46). The second group of this owl's prey was small birds (22% of biomass). When hunting birds, Tengmalm's owls selected year-round residents, species of boreal or boreal-nemoral geographic ranges: tits (47% of all birds identified from pellets versus 9% in the breeding bird community) and goldcrests (20% of captured birds vs. 6% in the community). No data on the winter diet of Tengmalm's owl is so far available from Białowieża Forest.

4.26 Long-Eared Owls – Specialisation on *Microtus* Voles

Over the temperate and boreal zones of Europe, pellets of long-eared owls contain hardly anything else than *Microtus* voles (review in Mikkola 1983). Typically, in Białowieża Forest, owls fed on rodents, the majority of which were *Microtus* (84% of prey, 82% of food biomass). Ruprecht and Szwagrzak (1987), whose data are presented in Table 4.47, collected pellets of long-eared owls in three years (1967, 1969, and 1983). Between-year variation in the share of *Microtus* in owl diets was rather small, from 74 to 100% of prey specimens. Unfortunately, no data on vole abundance in those years are available.

4.27 Eagle Owls – Night Hunters of Large Birds

Fifteen prey specimens were recovered from eagle owl pellets and feeding remains collected in the regions just north of Białowieża Forest in 1993. Ten species of birds consituted a total of 93% of food biomass, the rest was a carcass of wild boar. The list of identified avian prey included: three mallards (36% of food biomass), a partridge, a hazel hen, a black headed gull, a wood pigeon, two tawny owls, a crow, a raven, a jay, and a blackbird. Although very small, this material seems representative when compared with more data for the whole north-eastern Podlasie. Pugacewicz (1995) reported that birds composed 46% of prey specimens captured and 78% of food biomass taken by eagle owls. Most of the birds (41% of total biomass consumed) were large birds of open marshes and wetlands (e.g. ducks, coots, lapwings, gulls, black grouse). Mammals (predominantly small rodents) made 54% of prey specimens and 22% of food biomass. Pugacewicz (1995) noted that proportions of birds and mammals in the eagle owl's diet varied between seasons: small mammals predominated in winter, whereas birds in spring–summer.

Table 4.46. Diet composition of the Tengmalm's owl in the Polish part of BPF in spring-summer 1993–1994 (W. Jędrzejewski, B. Jędrzejewska, E. Pugacewicz, and K. Zub, unpubl.). Explanations as in Table 4.42

Item	Spring–summer		
	%Occ	%Prey	%Bio
Common shrew (*Sorex araneus*)	0.7	0.4	0.2
Lesser shrew (*Sorex minutus*)	4.4	2.9	0.5
Insectivores total	**5.1**	**3.3**	**0.7**
Bank vole (*Clethrionomys glareolus*)	65.7	44.9	42.8
Pine vole (*Pitymys subterraneus*)	2.2	1.2	1.3
Short-tailed vole (*Microtus agrestis*)	8.0	4.9	6.4
Common vole (*Microtus arvalis*)	5.8	3.3	3.5
Undetermined vole (*Microtus* sp.)	8.8	4.9	6.1
Harvest mouse (*Micromys minutus*)	0.7	0.4	0.2
Yellow-necked mouse (*Apodemus flavicollis*)	5.8	3.3	5.8
Wood mouse (*Apodemus sylvaticus*)	3.6	2.5	2.8
Undetermined mouse (*Apodemus* sp.)	10.2	5.8	8.4
Birch mouse (*Sicista betulina*)	1.5	0.8	0.4
Rodents total	**93.4**	**72.0**	**77.7**
Middle spotted woodpecker (*Dendrocopos medius*)	0.7	0.4	0.7
Robin (*Erithacus rubecula*)	2.9	1.6	1.6
Song thrush (*Turdus philomelos*)	0.7	0.4	0.7
Undetermined thrush (*Turdus* sp.)	0.7	0.4	0.7
Undet. warbler (*Phylloscopus* sp.)	0.7	0.4	0.2
Goldcrest (*Regulus regulus*)	4.4	2.5	0.7
Blue tit (*Parus caeruleus*)	0.7	0.4	0.3
Great tit (*Parus major*)	2.9	1.7	1.7
Undetermined tit (*Parus* sp.)	6.6	3.7	3.6
Tree creeper (*Certhia familiaris*)	1.5	0.8	0.4
Undetermined passerine	18.2	10.3	11.0
Birds total	**31.4**	**22.6**	**21.6**
Insects	3.6	2.1	+
Total *N* pellets, prey specimens or biomass (g)	137	243	4324 g

Table 4.47. Diet composition of the long-eared owl in the Polish part of BPF in the spring-summer seasons of 1967, 1969, and 1983 (Ruprecht and Szwagrzak 1987; data based on analysis of pellets). Explanations as in Table 4.42

Item	%Prey	%Bio
Bank vole (*Clethrionomys glareolus*)	4.5	3.6
Water vole (*Arvicola terrestris*)	0.6	2.5
Root vole (*Microtus oeconomus*)	15.4	18.9
Short-tailed vole (*Microtus agrestis*)	0.6	0.7
Common vole (*Microtus arvalis*)	54.5	49.0
Undetermined vole (*Microtus* sp.)	13.4	13.4
House mouse (*Mus musculus*)	4.5	3.6
Yellow-necked mouse (*Apodemus flavicollis*)	2.6	3.8
Undetermined mouse (*Apodemus* sp.)	2.6	3.3
Rodents total	**98.7**	**98.8**
Undetermined passerine bird	1.3	1.2
Total *N* prey specimens or biomass (g)	156	3296 g

4.28 Peregrines - Sky Diving for Large Birds

Although the peregrine has not bred in BPF since the late 1950s, its feeding habits are worth presenting and comparing with other predators. In their large territories, encompassing clear pine forests as well as open marshes, meadows, and farmlands, peregrines seemed to know perfectly the daily schedules of bird passages. They hunted birds in flight, so dusk and dawn display flights of woodcocks, foraging and roosting passages of ducks and garganeys in marshes, and daylight flights of rooks to and from the farmland colony were the most frequently exploited opportunities (Gavrin 1953). No other predator fed so exclusively on 'flying objects' as peregrines (Table 4.48). In both studies by Gavrin (1953) and Duchic (1952), rooks were the most common prey of these falcons, followed by Charadriformes birds (woodcocks, lapwings, snipes), wild ducks, and forest Columbiformes. To capture a flying bird, peregrines dived from a height of few dozen metres and striked the prey with their talons. When the prey was large and heavy (e.g. a duck), both predator and prey fell to the ground. Smaller prey was hit in air and taken directly in flight to the perch in a tall tree or to a nest with eyases.

Table 4.48. Diet composition of the peregrines in the Eastern part of BPF in the spring–summer seasons of 1946–1951. Data for 1946–1947 from Gavrin (1953) and for 1950–1951 from Duchic (1952, cited after Fedyushin and Dolbik 1967), both based on prey brought to nests by adult birds. Dots denote prey with no quantitative information on their shares in the peregrine diet. Explanations as in Table 4.42

Item	1946–1947	1950–1951	
	%Prey	%Prey	%Bio
Noctule bat (*Nyctalus noctula*)	2.1	7.8	1.0
Mammals total	**2.1**	**7.8**	**1.0**
Little grebe (*Podiceps ruficollis*)	0.7	1.0	0.6
Garganey (*Anas querquedula*)	6.9	10.7	13.1
Quail (*Coturnix coturnix*)	–	1.0	0.4
Water rail (*Rallus aquaticus*)	–	1.0	0.5
Spotted crake (*Porzana porzana*)	0.7	1.9	0.7
Lapwing (*Vanellus vanellus*)	7.6	7.8	5.8
Common snipe (*Gallinago gallinago*)	5.5	1.9	0.7
Woodcock (*Scolopax rusticola*)	8.3	5.8	5.9
Black-tailed godwit (*Limosa limosa*)	0.7	1.0	0.9
Redshank (*Tringa totanus*)	0.7	–	–
Common sandpiper (*Tringa hypoleucos*)	–	1.0	0.2
Stock dove (*Columba oenas*)	.	4.8	4.9
Wood pigeon (*Columba palumbus*)	.	1.0	1.8
Turtle dove (*Streptopelia turtur*)	.	1.9	1.1
Wood pigeons and doves (total)	6.9		
Roller (*Coracias garrulus*)	0.7	1.9	1.0
Hoopoe (*Upupa epops*)	0.7	–	–
Black woodpecker (*Dryocopus martius*)	–	1.0	1.1
Whinchat (*Saxicola rubetra*)	0.7	–	–
Thrushes (*Turdus* sp.)	1.4	–	–
Great tit (*Parus major*)	0.7	–	–
Jay (*Garrulus glandarius*)	1.4	.	.
Magpie (*Pica pica*)	1.4	.	.
Jackdaw (*Corvus monedula*)	2.1	.	.
Rook (*Corvus frugilegus*)	31.0	39.7	54.6
Starling (*Sturnus vulgaris*)	2.7	1.0	0.3
Other birds	17.1	7.8 [a]	5.4
Birds total	**97.9**	**92.2**	**99.0**
Total *N* of prey specimens or biomass (g)	145	103	27 757 g

[a] Mainly jays, *Garrulus glandarius* and magpies, *Pica pica*.

4.29 Food of Predators Inhabiting Human Settlements: Barn Owls and Stray Dogs

Barn owls inhabiting Białowieża village hunt for small mammals: shrews and rodents. They may capture them around settlements but more often they do so in the fields and meadows. In spring and summer, shrews (especially the common shrew) were the dominant prey of owls, making 45% of their food biomass (Table 4.49). Swallows were quite often captured in spring and summer. Together with a few other rural and farmland birds, they contributed 15% to the total food eaten by owls. In autumn–winter, barn owls captured more rodents than shrews, and birds nearly disappeared from their diet. The few bats captured in the cold season must have been unfortunate hibernating co-habitants of owls' corners.

Ruprecht and Szwagrzak (1987) also studied the spring–summer food of barn owls in a small town (Hajnówka) and the open farmland environment at the western outskirts of Białowieża Forest. Owls living there took more rodents (83% of prey, mainly *Microtus* voles) than owls in Białowieża Glade (22%). Among birds captured there by barn owls, house sparrows were as numerous as swallows.

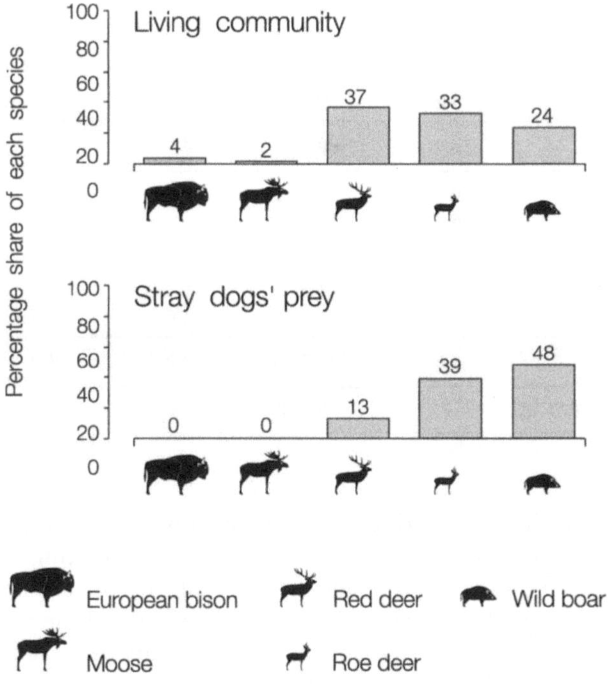

Fig. 4.43. Selection of prey species from the community of ungulates by stray dogs in the exploited forests of BPF in 1987/1988–1993/1994. Okarma et al. (1995)

Table 4.49. Diet composition of the barn owl in Białowieża village in spring–summer (1968 and 1971) and autumn–winter (1965 and 1979) (Ruprecht and Szwagrzak 1987; analysis of pellets). Explanations as in Table 4.42

Item	Spring–summer		Autumn–winter	
	%Prey	%Bio	%Prey	%Bio
Common shrew (*Sorex araneus*)	61.1	42.0	45.2	27.3
Lesser shrew (*Sorex minutus*)	6.3	1.6	5.8	1.3
Water shrew (*Neomys fodiens*)	1.7	1.6	1.9	1.6
Mediterranean water shrew (*Neomys anomalus*)	–	–	1.9	1.3
Insectivores total	**69.1**	**45.2**	**54.8**	**31.5**
Serotine bat (*Eptesicus serotinus*)	–	–	1.0	0.6
Common pipistrelle (*Pipistrellus pipistrellus*)	–	–	0.9	0.6
Chiroptera total	**–**	**–**	**1.9**	**1.2**
Bank vole (*Clethrionomys glareolus*)	1.7	2.5	–	–
Pine vole (*Pitymys subterraneus*)	–	–	1.0	1.2
Root vole (*Microtus oeconomus*)	4.5	10.2	21.1	41.5
Short-tailed vole (*Microtus agrestis*)	–	–	1.0	1.6
Common vole (*Microtus arvalis*)	10.3	16.8	9.6	13.8
House mouse (*Mus musculus*)	0.6	0.8	2.9	3.7
Harvest mouse (*Micromys minutus*)	0.6	0.4	6.7	4.1
Undet. mouse (*Apodemus* sp.)	4.6	9.4	–	–
Rodents total	**22.3**	**40.1**	**42.3**	**65.9**
Sky lark (*Alauda arvensis*)	0.6	1.8	–	–
Swallow (*Hirundo rustica*)	6.8	10.6	–	–
Black redstart (*Phoenicurus ochruros*)	0.6	0.9	–	–
House sparrow (*Passer domesticus*)	0.6	1.4	–	–
Undetermined passerine	–	–	1.0	1.4
Birds total	**8.6**	**14.7**	**1.0**	**1.4**
Total *N* prey specimens or biomass (g)	175	2039 g	104	1379 g

Feeding habits of domestic dogs wandering astray in the forest were of interest to game wardens as far as the dogs preyed on wild ungulates. By hunting in the forest, dogs supplemented their diets made up basically of household scraps. In 1984–1994, in the Polish part of BPF, game wardens found the remains of 56 ungulates mutilated by dogs (Okarma et al. 1995). This was a random sample of a much larger actual number of dogs' quarries. No such remains were found in Białowieża National Park. Stray dogs preferably hunted for wild boar and roe deer (Fig. 4.43). Dogs' specialisation on wild boar may be the result of an old local poaching habit: big dogs are trained to capture wild boar. Of 22 roe deer killed by dogs, 50% were fawns <1 year old, which constituted only 29% of the roe deer population. Selection of red deer fawns by dogs was probably still stronger; five of seven deer killed were juveniles (Okarma et al. 1995).

In winter, dogs also scavenged on carcasses of ungulates that had died or were killed by wolves and lynxes. In March 1994, we witnessed the interaction of a lynx family and a dog in Białowieża National Park. Upon returning to her killed deer, a mother lynx with kittens found a dog there helping itself. The dog retreated and then barked furiously at the lynxes from a safe distance. In such situations dogs, may fall prey to wolves and lynxes; dog hair was recovered a few times from scats of these predators.

4.30 Searching Modes of Carnivores in the Pristine Temperate Forest

When living in the same habitat or even hunting for the same prey species, various predators may utilise different microhabitat requisites, search for prey in a different way and at different times of a day. We investigated this problem by snow tracking eight carnivore species in the pristine forests of Białowieża National Park in 1985/1986–1991/1992. While following a trail and pacing its length, an observer noted all habitat attributes encountered by a tracked carnivore: fallen trees and logs, their root plates, standing trees and their bottoms, roads, and rivers. The mode of their utilisation (i.e. walking on a fallen log or under it, going on ice or swimming) were also recorded. Furthermore, to obtain a random sample of the occurrence of habitat attributes, we noted them along straight-line walks with a compass in a randomly picked direction. Data collected this way are a good enough representation of the relative occurrence of the above listed attributes in Białowieża National Park. However, the data collected by straight-line walks are not a random sample in terms of modes of utilisating the forest; the decision of whether to walk under or on an encountered fallen/hanging tree will basically depend on an observer. Some data from snow tracking weasels, stoats, polecats, American mink, pine marten, foxes, lynxes, and wolves – regarding their hunting activities – have already been presented in sections dealing with their feeding habits. Here, we will focus on the aspect of microhabitat use by these carnivores.

Graphic characteristics of habitat use by eight carnivore species and those described by random walks are presented in Fig. 4.44. Test *G* applied to pairwise comparisons of all species showed a unique pattern of habitat use for each of the eight carnivores (G from 36.9 to 1100.6, df=1, $p<0.001$). Then, utilisation of the main habitat attributes (fallen trees, their root plates, standing trees, rivers, and roads) by each predator species was compared with their availability in the studied pristine forests. In addition to dense stands of trees of all ages and sizes, virgin forests, as shown in Fig. 1.1. and Figs. 1.6 to 1.12, are rife with fallen trees, logs, and branches. To some carnivores, fallen trees were just obstacles to avoid (particularly by wolf, lynx and red fox), others used them more or less proportionally to their abundance, and for one species, the weasel, they were strongly preferred sites for searching and hunting for rodents (Fig. 4.44).

As regards the way fallen trees were used, the five mustelids form a heterogeneous assemblage. Taking the total length of their trails along, under and on fallen trees as 100% in each species, we calculated what proportion of the trail was *on* a lying tree versus *under* it. Weasels moved largely under the fallen logs and trees (67% of trail) and only sporadically on them (4%). Similar behaviour was found in the polecat (56% under versus 16% on fallen trees) and American mink (60 and 21%, respectively). In weasels and polecats, such penetration of all kinds of crevices and holes was their way to search for rodents. (In harsh winters, polecats leave streams and rivers and venture into forests to hunt rodents.) Stoat and particularly pine marten behaved the other way. The stoat chose the routes on and under fallen trees equally often (31 versus 30%), and the marten usually went on the fallen log (44%) and rarely under it (30%). In all species, the proportion of trail *along* a fallen tree (summing to 100%) was less variable (19–39%).

Root plates of fallen trees mattered to lynxes only, which visited them somewhat more often (not significantly though) than expected. Lynxes used them as marking places and - occasionally - as posts to climb and look around from. Root systems of standing trees, a place frequently inhabited by forest rodents, were checked proportionally to their availability by weasels, stoats, and polecats. Weasels simply entered rodent dens and burrows, whereas stoats and polecats very often hunted rodents by digging at the tree bottoms to enlarge den entrances. Other predators did not have any interest in walking from one tree to another, indeed they avoided them just as humans do while walking in woods.

Small forest rivers and brooks were singularly important to polecat and mink (and obviously to otters, which were not snow tracked). The polecat walked on ice and penetrated under the brows of banks but rarely did they wade into water. In contrast, the mink readily went into water and swam (Fig. 4.44). Also wolves, and less so lynxes, positively selected the rivers but exclusively for the comfort of fast and easy walking on their ice-covered surface.

Facilitation of movement was also the very reason for wolf's, lynx's and fox's strong preferences for forest roads and paths (Fig. 4.44). Although in all three carnivores, the proportion of roads in total the trail length (36% in wolves, 16% in lynxes, and 4% in foxes) may be somewhat overestimated (because human observers also tended to pick up the trail that crossed or fol-

Fig. 4.44. Modes of movement and area searching by eight species of carnivores in the pristine forests of Białowieża National Park, as revealed by snow tracking in 1985/1986–1993. Data collected by human's compass walks approximated the availability of various habitat attributes in the forest. *1* forest roads and paths, *2–9* attributes of the forest, *10–12* attributes of small forest rivers

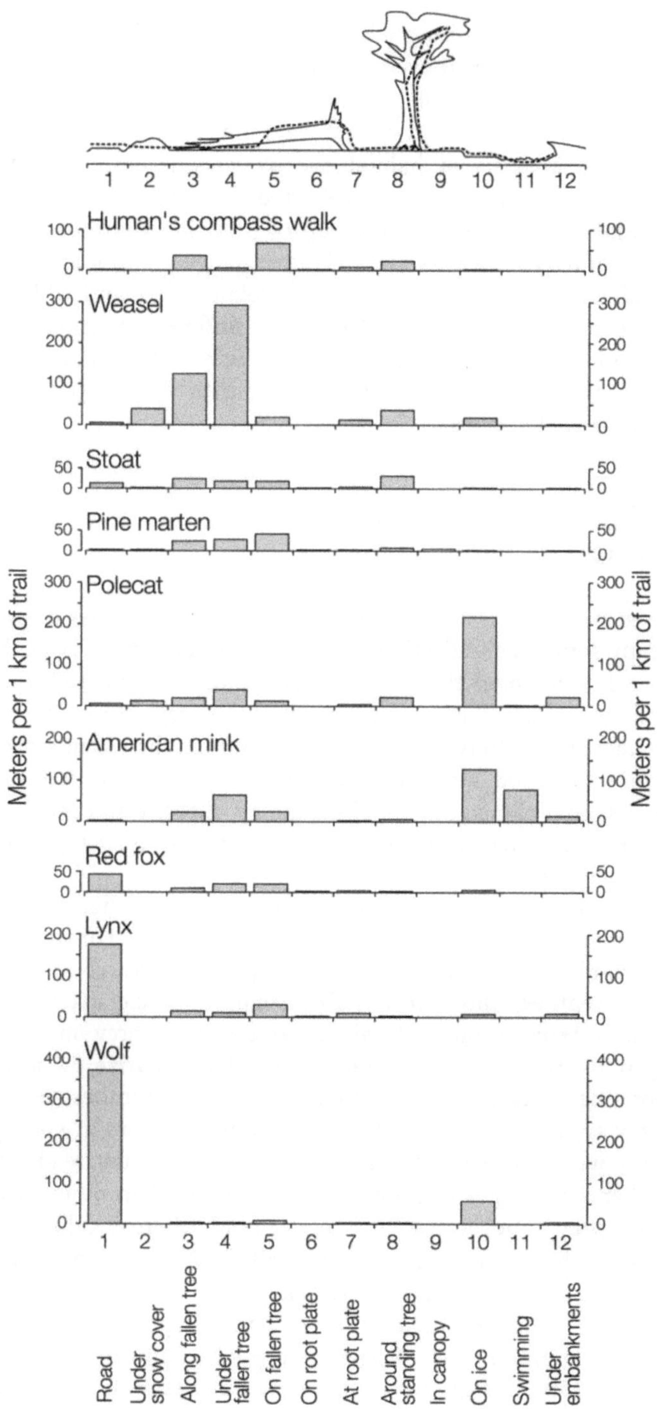

Human's compass walk
Weasel
Stoat
Pine marten
Polecat
American mink
Red fox
Lynx
Wolf
Meters per 1 km of trail
1 2 3 4 5 6 7 8 9 10 11 12
Road
Under snow cover
Along fallen tree
Under fallen tree
On fallen tree
On root plate
At root plate
Around standing tree
In canopy
On ice
Swimming
Under embankments

lowed the road), the inclination of the wolf and lynx to follow roads and paths was also confirmed by our radio-tracking and snow tracking studies in the whole Polish part of BPF. Foxes move from their dens and resting sites in forests to their main hunting habitat in the nearby open marshes and farmlands, so they often use the shortest and easiest way: the road or path.

To sum up, what this analysis of habitat use by various carnivores revealed is that searching modes of carnivores reflected their unique evolutionary adaptations to hunt a given species or group of prey. Furthermore, even when hunting on the same type of prey (weasel, stoat, polecat and marten on rodents, polecats and mink when hunting on amphibians, lynx and wolf on deer), each of the two or three species do it in partly different microhabitats of the same forest and in very different ways. Also, they may pursue their prey at various times of day and night as is the case with weasels (diurnal) and pine martens (nocturnal species).

4.31 Which Prey Groups Matter Most in the Diets of Temperate Forest Predators?

After being engrossed in the diversity and peculiarities of each species of predator, we will attempt some generalisations to find patterns and rules at the community level. Let us remember the cardinal features of the temperate zone that shape plant and animal life. First and foremost is the very strong seasonality, which divides each year into two parts: spring–summer, the warm season of growth and reproduction, and autumn–winter, the cold season of mere survival. Secondly, the essential climatic factors influencing, directly and indirectly, plant and animal performance are temperature and rainfall in spring–summer, and temperature and snow cover in autumn–winter.

In Chapter 2 we described the major taxonomic groups of herbivorous and insectivorous vertebrates inhabiting Białowieża Primeval Forest. The most common are ungulates, rodents, birds, and amphibians. Rarer in numbers are insectivores, reptiles, and medium-sized mammals (e.g. hare, beaver). By accepting these broad groups of animals as classes of prey and adding yet two other groups: fish and invertebrates, we will summarize the available information on the diets of all predators. Figure 4.45 synthesizes the roles of the nine prey types in the diets of 28 species of predators in spring–summer and autumn–winter seasons. Such presentation allows the reader to visually assess whether some types of prey are utilised more than others, and whether some of the prey types are sought year round whereas others are basically only sought in the warm season. First of all, it shows the outstanding role of small rodents, played year-round (Fig 4.45). Three quarters of predator species hunt for them deliberately. Rodents are the primary food for seven species in spring–summer (25% of species present) and for eight species in autumn–winter (50%). Not the commonest but certainly the most productive prey animals, accessible year round, small rodents are the mainstay of the predator-prey system in the temperate forests.

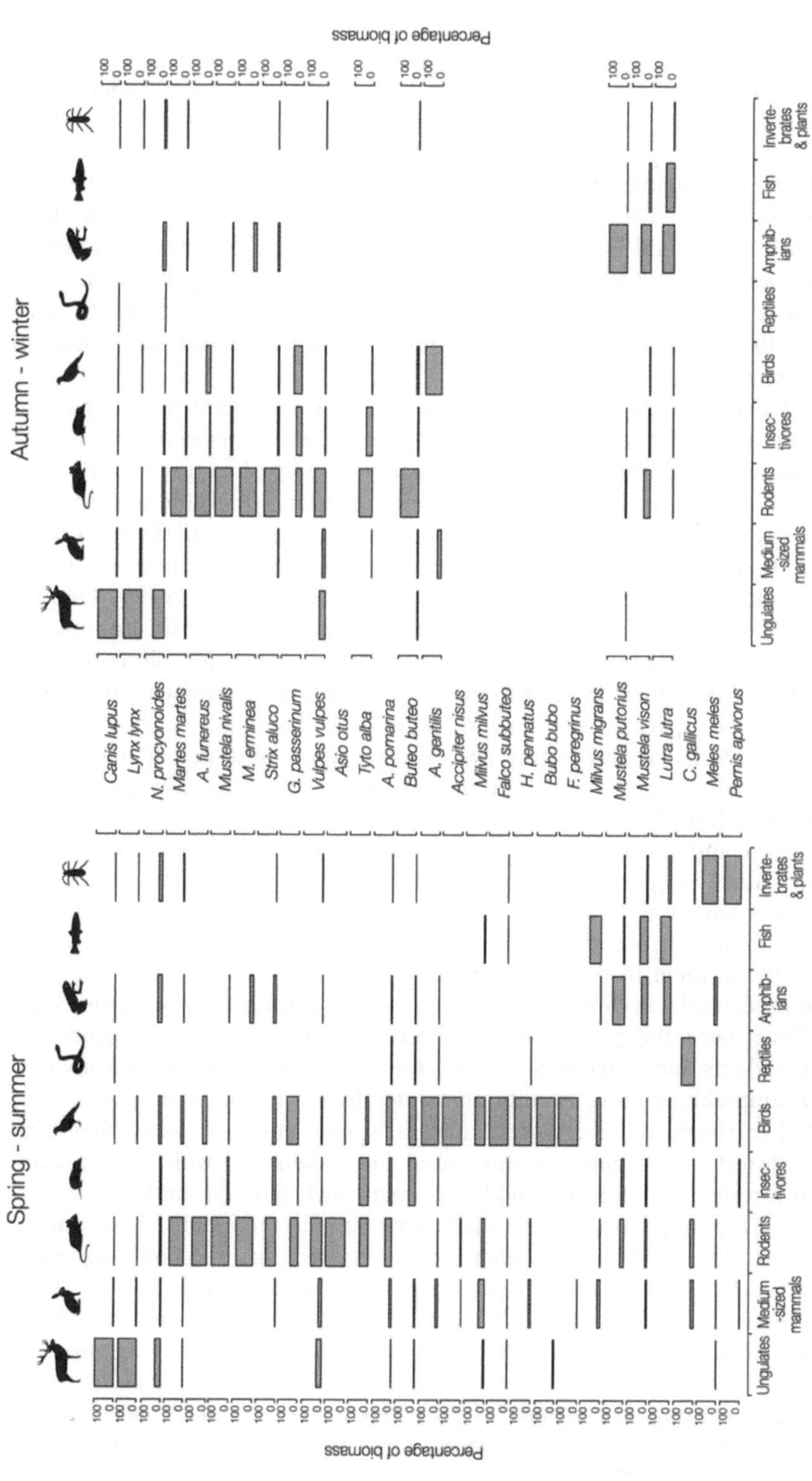

Fig. 4.45. Generalised picture of the spring–summer and autumn–winter diets of predators inhabiting Białowieża Primeval Forest (based on detailed data in Sects. 4.3–4.29). In stoat and Tengmalm's owl available data for one season were assumed to approximate their diets in the second season as well. In the case of eagle owl (feeding largely on migratory birds in spring–summer) such approximation was implausible

Main food as they are, rodents are seldom an exclusive prey to any predator. This results from their wide seasonal and multiannual variation in numbers. Only the long-eared owl, weasel, and stoat can be regarded rodent specialists, although the two carnivores, being resident, have at least some plasticity in their feeding habits (cf. Korpimäki et al. 1991). Many species relying on rodents are capable of supplementing their diets with one or more secondary prey type. The pine marten, tawny owl, Tengmalm's owl, red fox, common buzzard, and barn owl belong here. As a rule, the share of rodents in predators' diets is higher in autumn–winter than in spring–summer. This is also true when rodents are secondary prey (Fig. 4.45), because birds, amphibians, invertebrates and fish are less numerous and/or hardly accessible in the cold season.

To the whole predator community birds are as important as rodents. In spring–summer they constitute the main prey of seven species (25% of species present). In addition, birds are accessory or secondary prey to 17 species (61%; Fig. 4.45). All those 24 predators except one (the pine marten) are raptors. Predictable and not very variable from year to year, birds are a prey resource that gave rise to specialisation in several predators: hobbies, peregrines, booted eagles, sparrowhawks, and goshawks regularly cover >80% of their food demands by bird hunting. Of all raptors specialising on birds, only goshawks reside in their breeding quarters. Other species follow the migration patterns of their avian prey, be it short-distance movements to southwestern Europe (sparrowhawk) or long journeys to South Africa (hobby). Except for the goshawk and pygmy owl, all predators wintering in Białowieża Forest markedly 'lose interest' in hunting birds in autumn–winter, when birds are ten times less numerous than in spring–summer.

The third and fourth important prey groups are amphibians and ungulates, of rather similar importance to the predator community (Fig. 4.45). Amphibians have only one devoted predator, the polecat, to which they are the main food year-round, and two species, the otter and the mink which rely heavily on them in the cold season. Abundant though they are, frogs are very unpredictable in both autumn–winter and spring–summer (see Sect. 2.10). Thus, even the polecat must be adapted to survive long periods by hunting rodents and other small mammals. The three semiaquatic carnivores prey on amphibians more in autumn–winter (when common frogs are concentrated in rivers for hibernation) than in spring–summer (when they disperse in forests). The opposite was found in predators to which frogs, captured in forests, are secondary food (e.g. tawny owl. raccoon dog).

Hunting for ungulates, due to their large size, is monopolised by wolves and lynxes (Fig. 4.45). A few other predators (red fox, raccoon dog, pine marten, red kite, but also white-tailed eagle) deliberately search for and feed on carcasses of ungulates that either died from starvation, disease or cold or had been killed by wolves and lynxes. In the diets of all scavenging resident predators, the role of ungulate carcasses increases in the cold season.

Of the remaining five groups of prey (medium-sized mammals, insectivores, reptiles, fish, and invertebrates), three formed the main food of at

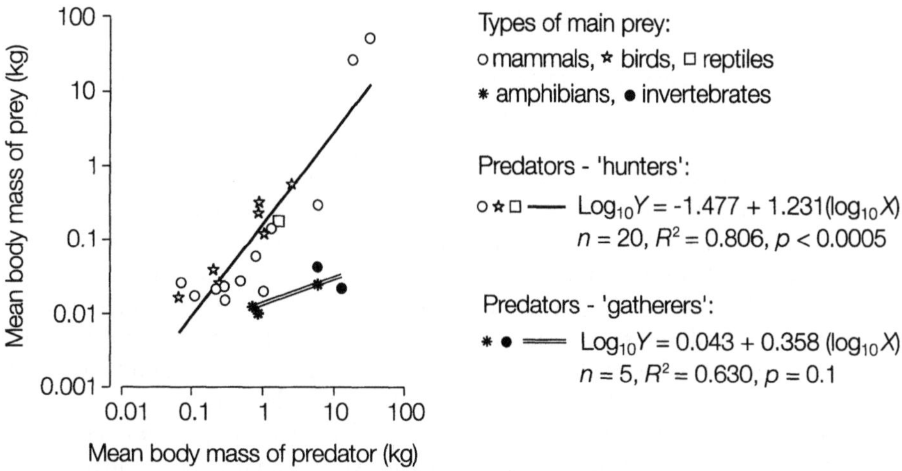

Fig. 4.46. Relationship between body mass of a predator and mean body mass of its prey. Each point = one predator species, $N=25$ species (black and red kites are excluded due to lack of detailed list of prey, and honey buzzard due to uncertainty as to what makes a 'prey' unit). In the case of raccoon dog, ungulate carcasses scavenged upon were not included in calculation of mean body mass of prey. Symbols denote the dominant prey type for each predator

least one predator species (reptiles for the short-toed eagle, fish for the otter, invertebrates for badgers and honey buzzards) though all in the warm season, only (Fig. 4.45). Insectivores and medium-sized mammals always played an inferior role in predators' diets.

The last thing we want to consider, based on grouping of prey into nine broad classes, is the body mass relationship between predators and their prey. In 25 species of predators we calculated the mean body mass of their prey (first, the mean mass of prey in each of the nine classes shown in Fig. 4.45, and then the mean of those nine classes weighted by the percentage biomass that each class contributed to the predator's diet). There was a clear positive relationship between body masses of the temperate forests' predators and the mean weights of their prey (Fig. 4.46). Furthermore, two groups of predators are distinguishable: one, which we called 'hunters' because their main prey was mammals, birds, or reptiles, and the second labelled as 'gatherers', comprising predators preying predominantly on less mobile prey (amphibians and invertebrates). Predators in the latter group are much larger than those hunting for equally small but more mobile prey. A case in point is the badger which feeds on earthworms although it is nearly as large as a lynx, which seizes ungulates.

4.32 The Pattern of Sharing Resources – Food Niche Overlaps

So far, we have discussed the roles of nine broad classes of prey in supporting the predator community. The next point to examine is how the predators share their resources. For this analysis we have grouped prey species into 20 ecological classes based on size of prey and the habitat it occupies: (1) ungulates; (2) brown hare; (3) squirrels and other arboreal rodents; (4) moles; (5) other medium-sized mammals (e.g. hedgehogs, mustelids, beavers); (6) shrews; (7) forest rodents; (8) open terrain rodents; (9) small forest birds (body mass up to 60 g); (10) medium-sized forest birds (61–300 g); (11) large forest birds (>300 g); (12) small birds of open terrain (body mass criteria as in forest birds); (13) medium-sized birds of open terrain; (14) large birds of open terrain; (15) reptiles; (16) amphibians; (17) fish; (18) insects; (19) other invertebrates, and (20) plant material.

Based on this classification, Pianka's index of niche overlap α was calculated for all possible pairs of predators (the barn owl, an inhabitant of villages, excluded). As shown in Figure 4.47, in spring–summer, pairs with little overlap (0–0.2) predominated (68%). Each consecutive class of α value was represented by still fewer pairs: class 0.21–0.4 by 16%, 0.41–0.6 by 9%, 0.61–0.8 by 3%, and 0.81–1 by 4%. Similar frequency distribution of the α index was found in autumn–winter with 57% of pairs in class 0–0.2, 15% in class 0.21–0.4, 10% in class 0.41–0.6, 3% in class 0.61–0.8, but – unlike in the warm season – 15% of pairs were in the highest class of overlap (0.81–1). The two frequency distributions significantly diverged from an even share of 20% in each class as expected if predators shared the resources indifferently ($G=120.7$, $df=4$, $p<0.001$ in spring–summer, $G=76.9$, $df=4$, $p<0.001$ in autumn–winter). This indicates avoidance of using the same resources by predators.

A visual examination of Fig. 4.47 ascertains that in most cases there is very distinct separation of predators' food niches. The only 'regions' where notably bigger overlaps occur are within a group of rodent-eaters (tawny owl, pine marten, weasel, Tengmalm's owl, stoat, red fox), a smaller group of amphibian eaters (polecat, mink, otter), and a pair of ungulate hunters (wolf, lynx). Interestingly, due to the high species and size diversity of birds, the overlap of diets of several bird-eating predators is markedly reduced. Generally, mean α's, for each species and the whole community, are higher in autumn–winter than in spring–summer, when available food resources are more abundant and far more diversified.

At the end of Chapter 2 we analysed the pattern of habitat niche overlap in Białowieża's community of predators. What we found was a random distribution of α indices, suggesting that predators do not strive to avoid using the same habitats. What they do avoid is using the same food resources. How is the food niche overlap between predators related to habitat niche overlaps in the same pairs? Although data points for 210 pairs are scatterred, there is a marked tendency that with habitat overlap growing from class 0–0.2 to class 0.61–0.8, the food niche overlap also increases (Fig. 4.48). However, with still more similar habitat niches ($\alpha=0.81$ to 1) most of the predator pairs tend to have rather dissimilar food spectra. Thus, most predators that share the same habitat hunt for entirely different prey.

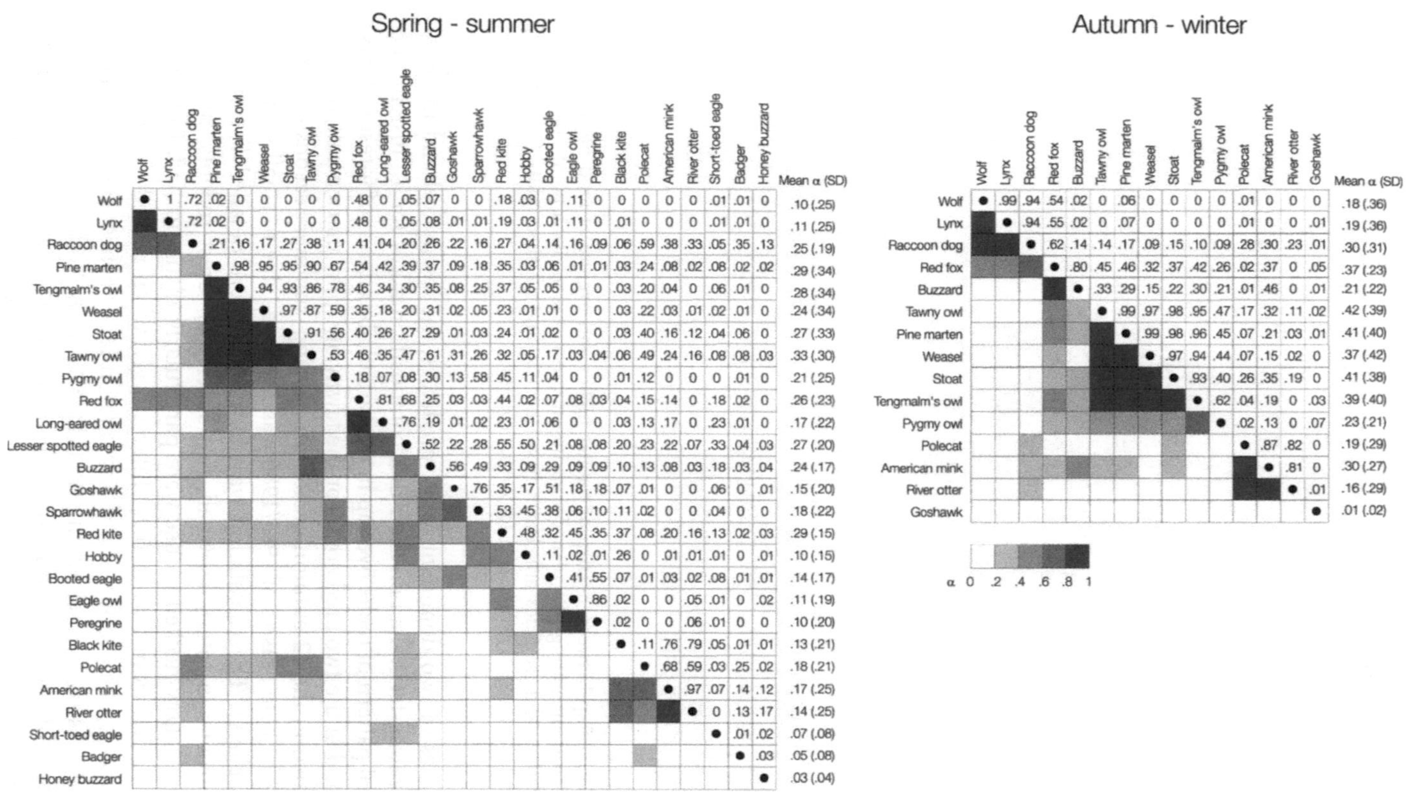

Fig. 4.47. Food niche overlaps in a predator community of BPF in spring–summer and autumn–winter. Index of overlap α (after Pianka 1973) varies from 0 (exclusive food niches) to 1 (identical niches). It was calculated based on 20 taxonomic-ecological groups of prey (see text), according to the formula: $\alpha_{ij} = (\sum p_{ia} \cdot p_{ja}) \cdot [(\sum p_{ia}^2) \cdot (\sum p_{ja}^2)]^{-1/2}$, where α_{ij} is the overlap between predators i and j, p_{ia} is the fraction of ath prey in the total biomass consumed by ith predator, p_{ja} is the fraction of ath prey in the total biomass consumed by jth predator

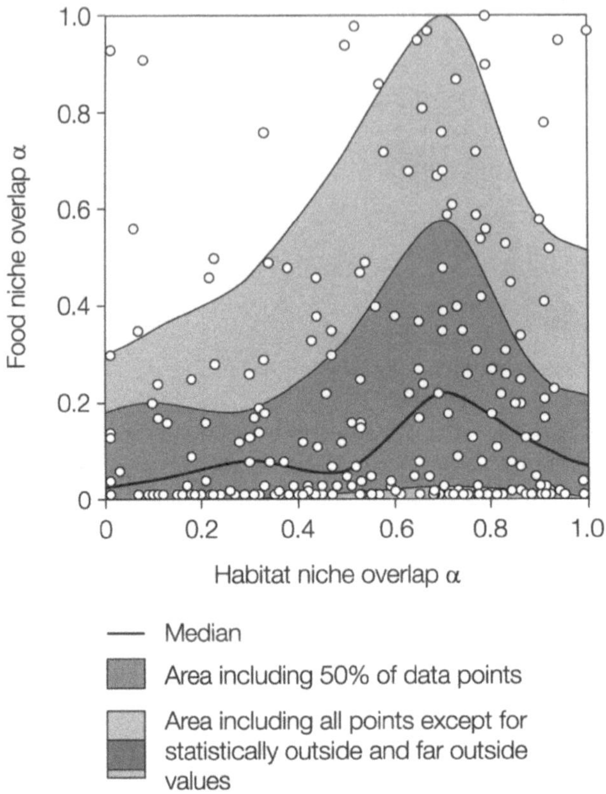

Fig. 4.48. Relationship between overlap of habitat niches and overlap of food niches in 210 pairs of predators inhabiting BPF; data for spring–summer season. Each point = one pair of predators. Median and isoclines are drawn based on box-and-whiskers plots (Tukey 1977, Velleman and Hoaglin 1981) computed for $\alpha = 0.2$ classes

In the cases of two dominant forest predators, the tawny owl and the pine marten, the long-term data on their diet composition allow for an insight into the temporal pattern of niche overlap in relation to abundance of forest rodents, the main food of the two predators. For the tawny owl-pine marten pair, index α was calculated in the 12 autumn–winter and spring–summer seasons of 1986–1991/1992 (the cold season of 1985/1986 was neglected due to small samples of scats and pellets). The calculation was done on a more detailed list of prey items (12 groups in total), above all, two species of forest rodents, the bank vole and the yellow-necked mouse were particularly distinguished. In various seasons, α varied from 0.61 to 0.95. Interestingly, when related to abundance of forest rodents, the overlap of owl and marten food niches appeared highest at moderate densities of their main prey (Fig. 4.49 A). At both very low and very high numbers of rodents, the food niches of the two predators became more disparate. In seasons of rodent shortage this was due to utilisation of buffer foods, partly different in each predator. Pine marten preyed on birds and con-

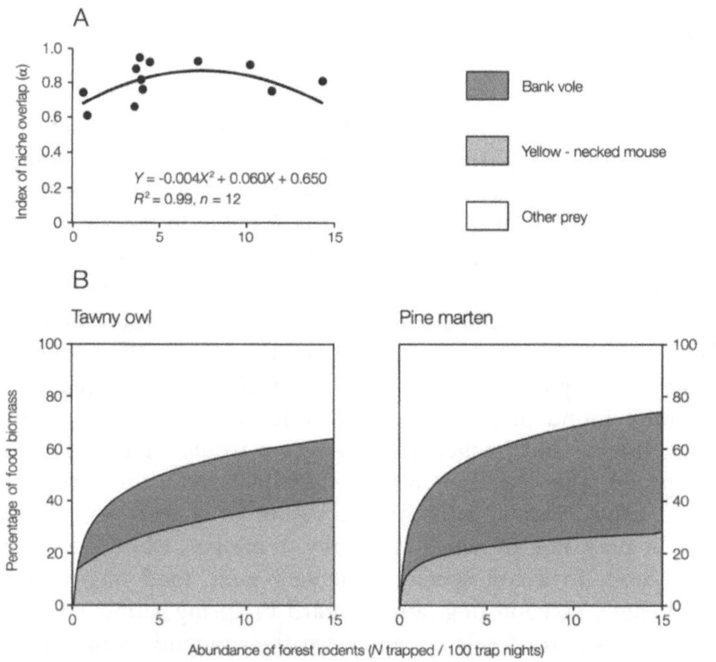

Fig. 4.49. **A** Food niche overlaps (Pianka's index α) between tawny owl and pine marten in Białowieża National Park in relation to the abundance of forest rodents. Each point = one season (spring–summer or autumn–winter) in 1986-1991/1992. **B** Generalised picture of diet composition of tawny owls and pine martens in relation to forest rodent abundance. With growing densities of forest rodents, tawny owls prey more on yellow-necked mouse, whereas pine martens prey more on bank voles

sumed much fruit in the warm seasons, and it hunted for insectivores and scavenged on ungulate carcasses in autumn–winter. Tawny owls compensated for the lack of rodents by more intense hunting on birds, amphibians, and shrews. When forest rodents were very abundant, the separation of food niches between the two predators was caused by the fact that each of them relished in hunting basically one preferred species: tawny owl fed mainly on yellow-necked mice, whereas martens captured mostly bank voles (Fig. 4.49 B).

Zalewski (1996) investigated the similarities of martens' and tawny owls' hunting for rodents at a still more detailed level: the choice of age classes of the bank vole population in those same years. He found that, in all seasons, tawny owls took heavier voles than pine martens, and also heavier than the mean in the population. In years of moderate densities of bank voles (1986–1989), the more numerous were rodents, the more similar was the age structure of voles captured by tawny owls and those killed by martens. The analysis by Zalewski (1996) as well as an earlier study by Goszczyński (1977) on six species of predators hunting for *Microtus arvalis*, documented that even when exploiting one population of prey, various species of predators harvested its different segments.

4.33 Breadth of Food Niches and Choice of Primary and Secondary Prey

Going back to the classification of all prey species found in the diets of Białowieża's predators into 20 taxonomic-ecological groups, we calculated the index of food niche breadth *B* (after Levins 1968), which in this case would vary from 1 (if a predator consumed one food type only) to 20 (if each of the 20 food types made up 5% in diet). In fact, the *B* index in the 28 species of predators varied from 1 to 11 (Table 4.50), with a mean value for the community of 3.1 in spring–summer and 2.3 in autumn–winter. In most species (71% in the warm season and 82% in the cold season), the index of food niche breadth was 1 to 3, which means that a predator seriously relied on one to three major food types. The narrowest food niches, composed of a single prey type were found in the wolf, lynx, long-eared owl, honey buzzard, weasel, stoat, hobby, and badger. Intuitively, we would include in this group two other species, although their *B*'s are slightly above 2: the peregrine and the short-toed eagle. Species with higher values of *B* feed on two or more major types of prey, one of which is usually dominant, while the other serves as a buffer food. In a few species with very wide food niches (red kite, lesser spotted eagle, raccoon dog, and buzzard in spring–summer), it is hard to say what their primary foods are and what their secondary ones are.

Table 4.50 also suggests that there is a clear seasonal change in niche breadth in resident species whose food niches are not particularly narrow. Predators that feed mainly on rodents (pine marten, tawny owl, red fox, buzzard, barn owl) or amphibians (polecat, otter, mink) have wider food spectra in spring–summer, when small vertebrates other than their main prey as well as invertebrates are abundant and easily available. In autumn–winter, when small vertebrates and invertebrates are less abundant, these predators concentrate on hunting their main prey despite the fact that it is also less numerous or less accessible than in spring–summer. The opposite trend is manifest in the two resident predators that make their living by bird hunting: the goshawk and the pygmy owl. In spring–summer, a season abounding in resident as well as migratory birds, food niches of these raptors are narrower than in autumn–winter, when shortage of birds forces them to supplement their diets with mammalian prey.

In the community of Białowieża's predators, the breadth of the food niche does not correlate with habitat niche breadth or species' dominance ranks. Therefore, neither wide nor narrow food spectra as such are keys to widespread occurrence or a high position in the community. In each predator species (perhaps with the exception of the raccoon dog) even a wide food niche does not mean indiscriminate feeding on any prey. Predators have their preferred species in a group of vertebrates constituting their main prey. For instance, roe deer is strongly selected from the ungulate community by lynx, and red deer by wolves. Yellow-necked mouse is preferred to other forest rodents by tawny owls, and bank voles by pine martens. What is more, the same applies to secondary, buffer prey. When hunting for birds, pine martens capture many more hole nesters than expected at random. Rodents and especially so birds, with their great diversity of species, nesting habits, and body size offer many possibilities for predators' specialisation.

In Tables 4.51 and 4.52 we compared species structures of rodent and bird communities with those captured by nearly a dozen species of carnivores

Table 4.50. Gradient of food niche breadth (index *B* after Levins 1968) in spring–summer and autumn–winter in the community of predators inhabiting BPF. *B* was calculated for 20 taxonomic-ecological groups of prey ($B = 1/\sum p_i^2$, where p_i is a fraction of a given prey group in the total biomass consumed) and thus would vary from 1 (the narrowest niche) to 20 (the widest niche). Figures in parentheses – due to lack of data for this season, the *B* value from the other season was used as rough approximation

Species	Index of food niche breadth *B*	
	Spring–summer	Autumn–winter
Wolf	1.07	1.01
Lynx	1.13	1.20
Long-eared owl	1.23	–
Honey buzzard	1.36	–
Weasel	1.49	1.49
Hobby	1.65	–
Stoat	(1.81)	1.81
Badger	1.92	–
Peregrine	2.03	–
Pygmy owl	2.03	3.72
Short-toed eagle	2.19	–
Tengmalm's owl	2.26	(2.26)
Goshawk	2.39	4.22
Polecat	2.55	1.12
River otter	2.59	2.04
Black kite	2.63	–
Eagle owl	2.75	(2.75)
Pine marten	2.85	2.07
Barn owl	2.86	1.92
Sparrowhawk	2.93	–
American mink	3.52	2.90
Red fox	3.77	3.56
Booted eagle	4.21	–
Tawny owl	4.66	2.52
Common buzzard	5.35	1.36
Raccoon dog	6.25	2.83
Lesser spotted eagle	6.47	–
Red kite	10.72	–
Mean *B* (SD)	3.10 (2.08)	2.28 (0.95)

Table 4.50 (continued)

Class of *B* index	Number of species (percentage) in a given class of niche breadth:	
	Spring–summer	Autumn–winter
1–2	8 (28.6%)	7 (41.2%)
2.01–3	12 (42.8%)	7 (41.2%)
3.01–5	4 (14.3%)	3 (17.6%)
5.01–7	3 (10.7%)	0
7.01–9	0	0
9.01–11	1 (3.6%)	0

and raptors. Rodent species are arranged in a forest-open field gradient and so are the predators' preferences. When hunting rodents, tawny, pygmy, and Tengmalm's owls, pine martens, and polecats rely on forest species, as might be expected from the habitat preferences of these predators. Also noteworthy are the shares of bank voles (a boreal-nemoral species) versus yellow-necked mice (a nemoral species) among all captured rodents. Boreal in their biogeographic ranges and preferring coniferous forests in their local occurrence in BPF, Tengmalm's and pygmy owls, strongly select for bank voles (Table 4.51). Pine marten, originally an inhabitant of all types of forests in Europe's boreal zone, is connected with the boreal-nemoral bank vole, too. By contrast, tawny owl and polecat, species typical of Europe's deciduous and mixed forests and preferring deciduous forests also at a local scale, hunt for yellow-necked mice more than expected from their share in the forest rodent community (Table 4.51).

Raccoon dogs and common buzzards utilise both forests and nearby open marshes and meadows and this is reflected by nearly equal shares of forest and 'field' rodents in their diets. Among rodents captured by mink, red fox, sparrowhawk, and buzzard, *Microtus* voles predominate, and in the lesser spotted eagle and long-eared owl, they form the majority of all rodents caught. Of three species of *Microtus*, only two, the root vole and the common vole, matter for predators and the role of one or the other reflects predators' adaptations to hunt in either tall vegetation reed and sedge marshes (where root vole abound) or drier fallows, meadows, and fields usually with shorter plant cover (where the common vole is dominant). Raccoon dogs and mink strongly prefer root voles, red foxes hunt intensively for both species, whereas buzzards, sparrowhawks, lesser spotted eagles and long-eared owls select for the common vole (Table 4.51).

Interesting and more complicated is the cross-species comparison of predators' preferences for birds (Table 4.52). Pine martens, Tengmalm's, pygmy and tawny owls, goshawks, and sparrowhawks hunted forest birds nearly exclusively. Surprisingly, lesser spotted eagles, and especially so common buzzards, also obtained most of their avian prey in forests. By contrast, hob-

Table 4.51. Selectivity of predators towards rodents available in forests and open lands of BPF. Figures are percentage shares of a given rodent species among all rodents identified from predator scats or pellets. Figures in italics – Ivlev's selectivity index *D* (after Jacobs 1974); *D* ranges from –1 (total avoidance) to 0 (predation proportional to occurrence) to 1 (the strongest positive selection). Selectivity indices are calculated for dominant rodent species only, to avoid errors resulting from small sample sizes. Important and positively selected rodent species by a given predator are in boldface type. Choice of rodents by ecotone predators (i.e. predators utilising both forests and open terrain) was compared with averaged measures of rodent abundance in forests and open areas. Stoat, an important rodent predator in the marshes of river valleys, is not included due to scarce data, representative of forests, only. Abbreviations are composed of Latin names of rodents. *(Mic)* – all *Microtus* combined

Predator	Percentage share of rodents in community or among all rodents identified from predators' diets									
	Cgl	*Afl*	*Psub*	*Sbet*	*Magr*	*Moe*	*Marv*	*(Mic)*	*Aag*	Other
Forest community	53.1	26.0	7.0	4.5	4.3	1.8	1.1	(7.2)	1.0	1.2
Pygmy owl	**82.1**	3.9	5.2	7.0	?	?	?	(1.8)	–	–
D	*0.60*	*–0.79*								
Tengmalm's owl	**62.3**	16.0	1.7	1.1	10.9	–	7.4	(18.3)	–	0.6
D	*0.19*	*–0.30*								
Pine marten	**54.6**	19.3	6.4	0.7	3.3	9.8	4.7	(17.8)	0.4	0.8
D	*0.03*	*–0.19*								
Tawny owl	41.7	**34.3**	4.6	0.1	1.7	10.6	4.5	(16.8)	0.4	2.1
D	*–0.23*	*0.20*								
Polecat	30.5	**31.5**	14.3	–	7.9	8.9	–	(16.8)	–	6.9
D	*–0.44*	*0.13*								
'Ecotone' rodents	33.2	13.2	3.5	2.4	3.1	33.5	5.9	(42.4)	4.3	1.1
Raccoon dog	**40.9**	8.3	–	–	–	**47.7**	–	(47.7)	–	3.1
D	*0.16*	*–0.25*				*0.29*	*–1*			
Common buzzard	32.8	6.3	3.3	0.2	3.1	18.5	**34.3**	(55.9)	–	1.5
D	*–0.01*	*–0.39*				*–0.38*	*0.79*			
Sparrowhawk	33.1	1.5	–	0.8	5.4	10.8	**43.8**	(60.0)	–	4.6
D	*0*	*–0.82*				*–0.61*	*0.85*			
Red fox	19.9	6.7	0.4	–	3.0	**36.8**	**31.4**	(71.2)	0.5	1.3
D	*–0.33*	*–0.36*				*0.07*	*0.76*			
American mink	21.6	–	1.8	–	–	**71.4**	–	(71.4)	–	5.2
D	*–0.29*	*–1*				*0.66*	*–1*			
Open area rodents	13.3	0.4	–	0.2	1.8	65.2	10.6	(77.6)	7.5	1.0
Lesser spotted eagle	6.1	3.1	5.2	–	3.9	17.5	**57.5**	(78.9)	0.2	6.5
D						*–0.80*	*0.84*			
Long-eared owl	4.5	5.2	–	–	0.7	18.8	**65.6**	(85.1)	–	5.2
D						*–0.78*	*0.88*			

Table 4.52. Selection of birds by nine species of predators compared with the availability of birds in forests, marshy river valleys, and settlement glades in the Polish part of BPF. Figures are percentage shares of all birds identified from predator scats or pellets. Selectivity index *D* as in Table 4.51. Choice of bird species by predators (forest or open area birds) was compared with respective communities of birds. In birds of open areas, numbers in parentheses are the percentage share of a given species in a bird community of settlement glades and villages (conservatively, Ivlev's *D* was calculated using the higher of the two measures of abundance). The three most important and positively selected bird species by each predator are in boldface type. Abbreviations consist of Latin names (species, genus, or family) of birds. *Al+An* denotes *Alauda* sp. and *Anthus* sp. combined

Predator		Percentage share of birds in a forest community or among all birds identified from predators' scats or pellets													
		Frin coele	*Phyllo*	*Erith rube*	*Parus*	*Turdus*	*Sylvia*	*Regul regul*	*Ficed*	*Picid*	*Cardu*	*Garr gland*	*Colum*	Other	Total forest birds
Bird community		18.3	13.5	10.7	9.5	9.4	6.3	3.9	3.4	1.8	0.9	0.8	0.6	20.8	100
Pine marten		–	–	3.1	3.1	**21.9**	–	–	**12.5**	**15.6**	–	6.3	3.1	34.4	100
	D	*–1*	*–1*	*–0.58*	*–0.53*	*0.46*	*–1*	*–1*	*0.60*	*0.82*	*–1*	*0.79*	*0.68*		
Tengmalm's owl		–	3.3	**13.3**	**46.7**	6.7	–	**20.0**	–	3.3	–	–	–	6.7	100
	D	*–1*	*–0.64*	*0.12*	*0.79*	*–0.18*	*–1*	*0.72*	*–1*	*0.30*	*–1*	*–1*	*–1*		
Pygmy owl		**18.6**	5.9	**25.7**	**21.1**	–	1.3	1.3	6.3	0.8	3.8	–	–	14.8	99.6
	D	*0.01*	*–0.43*	*0.49*	*0.44*	*–1*	*–0.67*	*–0.51*	*0.31*	*–0.39*	*0.63*	*–1*	*–1*		
Tawny owl		11.4	–	1.4	8.6	**41.4**	4.3	–	**5.7**	2.9	4.3	**8.9**	–	8.3	97.2
	D	*–0.27*	*–1*	*–0.79*	*–0.05*	*0.74*	*–0.20*	*–1*	*0.26*	*0.24*	*0.66*	*0.85*	*–1*		
Buzzard		4.2	2.1	2.6	4.2	**61.1**	–	–	–	**7.2**	–	**6.2**	2.6	8.2	98.4
	D	*–0.67*	*–0.76*	*–0.64*	*–0.41*	*0.88*	*–1*	*–1*	*–1*	*0.62*	*–1*	*0.78*	*0.63*		
Goshawk		2.0	0.2	0.2	0.9	**24.3**	–	–	0.2	**21.9**	0.2	**31.5**	4.6	11.8	97.8
	D	*–0.83*	*–0.97*	*–0.97*	*–0.84*	*0.51*	*–1*	*–1*	*–0.89*	*0.88*	*–0.64*	*0.97*	*0.78*		
Sparrowhawk		10.6	3.2	8.7	**22.5**	**16.6**	1.7	0.2	3.0	**10.2**	0.9	3.0	0.4	9.0	90.0
	D	*–0.31*	*–0.65*	*–0.11*	*0.47*	*0.31*	*–0.59*	*–0.91*	*–0.06*	*0.72*	*0*	*0.59*	*–0.20*		
Lesser spotted eagle		–	–	–	–	**41.2**	–	–	–	1.5	2.9	1.5	7.4	2.9	57.4
	D	*–1*	*–1*	*–1*	*–1*	*0.74*	*–1*	*–1*	*–1*	*–0.09*	*0.53*	*0.31*	*0.86*		

Table 4.52 (continued)

Predator		Percentage share of birds in an open area community or among all birds identified from predators' scats or pellets										
		Acroce	*Locust*	*Carpo eryth*	*Ember*	*Al+An*	*Motac*	*Hirun rust*	*Card cann*	*Perd perd*	Other	Total nonforest birds
Bird community		24.3	15.0	11.1	10.0	4.2(12.5)	0.9 (3.3)	0.1 (5.1)	0.1 (1.6)	+ (0.2)	34.3	100
Hobby		–	–	–	5.1	**21.2**	3.7	**40.9**	7.3	–	1.7	79.9
	D	*-1*	*-1*	*-1*	*-0.35*	*0.31*	*0.06*	*0.86*	*0.66*	*-1*		
Lesser spotted eagle		–	–	–	7.4	**19.1**	–	–	–	**11.8**	4.3	42.6
	D	*-1*	*-1*	*-1*	*-0.16*	*0.25*	*-1*	*-1*	*-1*	*0.97*		

bies, peregrines, booted eagles, and eagle owls predominantly pursued birds of open areas. (The last three species mentioned are not presented in Table 4.52 due either to small data samples or the incomparable nature of dietary information with the available information on living bird communities.)

The predators' selection of species from bird communities is peculiar. None of the dominants (chaffinch and *Phylloscopus* warblers in forests, *Acrocephalus* and *Locustella* warblers in open marshes) is positively selected by any of the nine predators presented in Table 4.52. Forest birds especially sought after are woodpeckers, thrushes, jays, pigeons, tits, and robins – residents or short-distance migrants. Woodpeckers and thrushes in particular are targeted by all predators except for the smallest ones. None of the predators hunting for forest birds is specialised on tropical migrants.

For a thorough analysis of selectivity of open area birds by predators, we lack data on the diets of harriers and information on eagle owls is too scanty. Material on booted eagles and peregrines from the 1950s in the Belarussian part of BPF cannot be confronted with living bird communities. Nonetheless, at least in the case of the booted eagle (see Table 4.37), a strong positive selection of lapwings is manifest. Table 4.52 shows that the lesser spotted eagle prey selectively on larks, pipits and partridges, and the hobby on swallows, larks, pipits, and linnets. Unlike in forests, birds selected by predators in open terrain include residents, short distance migrants and tropical migrants.

Based on the diets of all carnivores and raptors inhabiting BPF and the analysis of their preferences to various species of prey as compared with their availability, we put together the 'identification table' of predators and their primary and secondary prey (Table 4.53). In the pristine temperate forests of Białowieża, very few predators appeared true specialists on one prey type. Most species supplemented their diets with one, two or three secondary prey types. Which types of main and secondary prey are found in the diets of predators most frequently? The commonest 'mixture' is that of rodents, birds, and insectivorous mammals: it occurs in tawny owls, buzzards, pine martens, lesser spotted eagles, and pygmy owls. Mixed diets of rodents-frogs (or frogs-rodents) are characteristic of stoats, mink, and polecats (Table 4.53). Other (rare) combinations are those of frogs-fish (mink in the warm season and otters year-round), and invertebrates-amphibians (badgers).

Already, this crude generalisation suggests that the predation impact on some prey groups is partly determined by fluctuations of the other prey. Fluctuations of abundant yet highly variable prey, such as rodents, can drive a chain of effects into less common prey: birds and shrews in particular. This problem will be discussed in detail in Chapter 5.

Table 4.53. List of main and secondary prey of predators inhabiting Białowieża Primeval Forest. Names in parentheses are preferred prey species or genus

Predator species	Primary prey	Secondary prey (in order of decreasing importance)		
		1	2	3
Wolf	Red deer	Wild boar	Roe deer	Livestock
Lynx	Roe deer	Red deer	Brown hare	
Raccoon dog	Ungulate carcass	Amphibians (*Rana*)	Rodents (*Microtus*)	Invertebrates
Tengmalm's owl	Forest rodents (*Clethrionomys*)	Small forest birds (tits, *Regulus*)		
Pine marten	Forest rodents (*Clethrionomys*)	Birds (*Turdus*, hole nesters)	Insectivores	Ungulate carcass, fruit
Tawny owl	Forest rodents (*Apodemus*)	Forest birds (*Turdus*)	Shrews (*Sorex*)	Amphibians (*Rana*)
Weasel	Rodents	Insectivores		
Stoat	Rodents (*Microtus*)	Amphibians (*Rana*)		
Long-eared owl	*Microtus* voles			
Red fox	*Microtus* voles	Ungulate carcass	Brown hare	Forest rodents (*Clethrionomys*)
Common buzzard	*Microtus* voles	Forest birds (*Turdus*)	Moles	Forest rodents (*Clethrionomys*)
Lesser spotted eagle	*Microtus* voles	Birds (*Alauda*, *Anthus*, *Perdix*, *Turdus*)	Moles	Amphibians (*Rana*)
Pygmy owl	Small forest birds (*Erithacus*, tits)	Forest rodents (*Clethrionomys*)	Shrews	
Sparrowhawk	Small and medium-sized forest birds (tits, *Turdus*)	Rodents (*Microtus*)		
Goshawk	Medium-sized forest birds (jay, woodpeckers, *Turdus*)	Red squirrel	Small and large forest birds	
Booted eagle	Medium-sized birds of open areas (*Vanellus*)	Medium-sized forest birds		
Hobby	Small birds of open areas (*Hirundo rustica*)	Flying insects		
Peregrine	Large birds of open areas			
Polecat	Amphibians (*Rana*)	Forest rodents (*Apodemus*)	Shrews	
American mink	Amphibians (*Rana*)	Rodents (*Microtus*)	Fish	

Table 4.53 (continued)

Predator species	Primary prey	Secondary prey (in order of decreasing importance)		
		1	2	3
Otter	Amphibians (*Rana*)	Fish		
Badger	Earthworms	Amphibians (*Bufo bufo*)		
Honey buzzard	Wasp and bumblebee nests			
Short-toed eagle	Snakes and snake-like lizards			

4.34 Foraging Strategies and Functional Responses of Predators

Finally, we will focus on defining the foraging strategies exhibited by the temperate forest predators. In ecological literature, terms used to define predators' dietary plasticity are many: monophagous versus polyphagous predators, specialist versus opportunistic or facultative ones, and specialist versus generalist predators. Since the influential paper by Andersson and Erlinge (1977), this last distinction has been widely accepted as a terminological frame for empirical studies and as a basic assumption in modelling (e.g. Hansson 1987; Loman 1988; Hanski et al. 1991; Korpimäki 1993). According to Andersson and Erlinge (1977), predators specialising on rodents are so well adapted for hunting and killing rodents that they always constitute the bulk of their diet. Thus the numbers of specialist predators will be closely related to rodent density. Generalists, on the contrary, exploit a wide range of food items and feed on rodents when they are easily available. Otherwise they shift to other food. Generalist numbers will be little influenced by variation in rodent density.

Convenient though it is, this classification seems to miss the very mechanism of predators' foraging strategies. Furthermore, formulated for rodent predators, it may lead to ambiguous conclusions: a sparrowhawk, for instance, can be classified as specialist by an ecologist studying predation on birds and as generalist by someone investigating predation on rodents. The discussion is not only about terms. Far-reaching consequences of this classification come from prescribing qualitatively different numerical and functional responses to specialist and generalist predators. Their numerical responses were already discussed in Chapter 3, functional responses will be dealt with here.

To define foraging strategies of predators inhabiting the temperate forest zone, we need to recall that instead of a bimodal distribution of the 30 studied species into specialists and generalists, we found an uninterrupted gradient from very narrow food niches to fairly wide ones. The second essential point is that every species, irrespective of diet width, is specialised on some main prey, be it a taxonomic-ecological 'class' of animals, genus or species. This specialisation always means positive selection of a given prey out of the available community, and fairly strong spatial and temporal connections between predator and prey. These connections are manifested in continued hunting on that primary prey even when it becomes scarce. Such behaviour towards primary prey characterised predators with very narrow food niches (lynx, weasel, long-eared owl) as well as fairly wide ones (pine marten, tawny owl, red fox). In addition, many species with fairly wide food niches possess specialisations for some secondary prey.

When classifying predators into specialists and generalists, Andersson and Erlinge (1977) proposed that the essential difference in their predatory action resulted from different functional responses to changes in prey numbers. Referring to four types of functional responses proposed earlier by Holling (1959), they expected specialists to exhibit asymptotic (type II) functional response and generalists to have an S-shaped (type III) response. If true, this distinction would have profound consequences for a total predation impact on prey: destabilising or even leading to prey cycles in specialist predators and stabilising in generalist predators (Murdoch and Oaten 1975; Loman 1988; Hanski et al. 1991, 1993).

While portraying the predators of Białowieża Forest, particular attention was given to their functional (dietary) responses to varying availability of prey. Material on various species is far from being homogeneous: in some species it shows the share of a given prey in the diet, in others true consumption (killing) rates. Furthermore, variation in prey densities was shown in either temporal or spatial aspects. Nonetheless, the overall picture appeared rather unambiguous: functional (dietary) responses by predators to changes in abundance of their primary prey were always asymptotic (type II). This holds true for 'specialists' (e.g. lynx) and 'generalists' (e.g. tawny owl) alike (Fig. 4.50), and reflected the close connections between predators and their main prey. Data on none of the six investigated predator species showed any resemblance to an S-shaped functional response.

Predators' responses to variation in abundance of their secondary prey were usually shaped by two factors: the abundance of secondary and primary prey. With growing densities of secondary prey, their share in predators' diets grew slowly in a linear manner or increased exponentially only when prey was extremely abundant (Fig. 4.50). Usually, however, predators' consumption of secondary prey was shaped by variation in abundance of primary prey, a case in point being wolves' preying on wild boar and roe deer, which was determined by the density of red deer. Thus, predation on some animals (wild boar, birds, shrews) is linked to fluctuations of other animals (red deer, rodents). Defined as the alternative prey hypothesis (Angelstam et al. 1985), this phenomenon has received much attention in studies on predator-prey relationships in small vertebrate communities of the European boreal and arctic zones, where cycles of microtine rodents influence, possibly via predators' action, the population dynamics and reproductive success of sympatric small vertebrates (Angelstam et al. 1984; Korpimäki 1985; Järvinen 1985, 1990; Marcström et al. 1988).

Finally, we have to stress that all functional responses of predators analysed in this chapter had an important extrinsic component: weather condi-

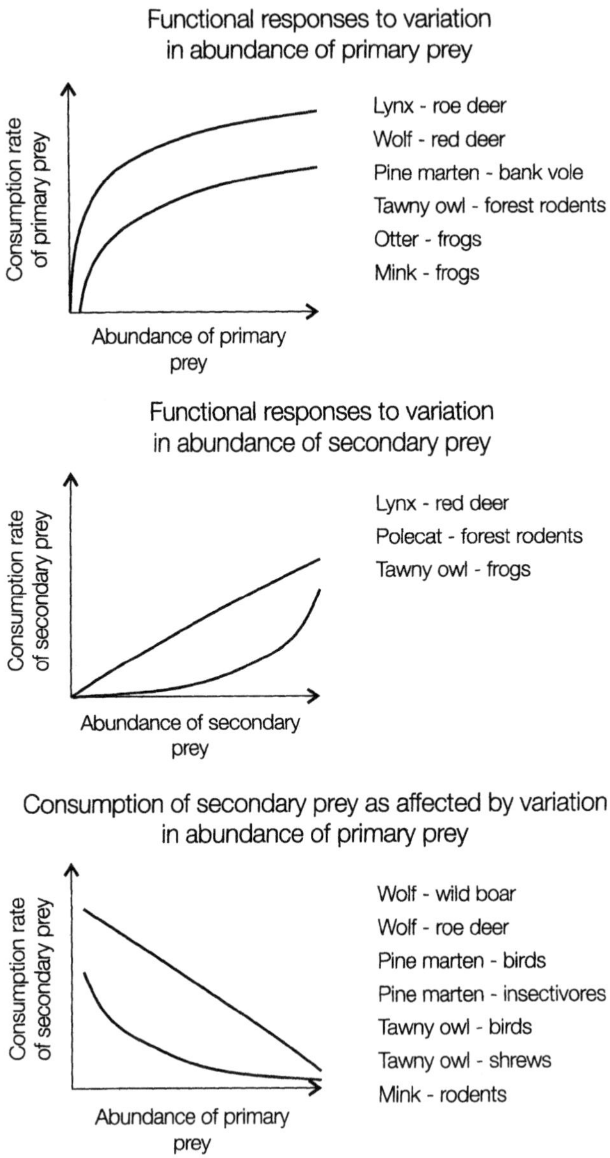

Fig. 4.50. Summary of functional (dietary) responses of predators inhabiting Białowieża Primeval Forest to variation in abundance of their primary and secondary prey resources. Prescribed to schematic figures are predator-prey pairs in which a given type of relationship was found

tions that strongly modified availability of some prey to predators. The role of snow cover as a barrier against successful rodent hunting for raptors and medium-sized carnivores has been long known and is again and again supported by new data from the boreal zone (Formozov 1946; Sonerud 1986; Nybo and Sonerud 1990). In this book, we have shown that snow cover and winter temperature play an important role in the temperate zone, too, and that they adversely affect predators' success in hunting rodents, amphibians, and positively affect their success with ungulates.

Having presented the densities and diets of predators and their reactions to variation in prey abundance and availability, in the next chapter we will analyse the magnitude of their predation on major groups of vertebrate prey. Our goal will be to compare the impact of predation with the effects of varying food supplies and weather conditions on prey populations.

5 Predation Impact and Its Importance to Prey Populations

Let us give thanks in our own and our ancestors' name
For oaks and their rough-barked dignity,
For pines, their trunks flaming in the sun,
For clear green clouds of vernal birch groves
And for the candlesticks of the autumnal wilderness, aspens.
Czesław Miłosz, *Lithuania, after fifty-two years*, 1992

5.1 Introduction

How many deer do wolves and lynxes kill annually and how does this amount compare with deer numbers and reproductive potential? Do predators rob enough nests and capture enough birds to keep the bird population lower in numbers than might be expected from food resources in the forest? If predators are responsible for the whole winter mortality of rodents, would all rodents survive winter when protected from predation? Questions of this kind are asked in this last chapter. Not all of them are given answers.

In our analysis of the magnitude of predation, which is a product of predator numbers and their killing rates, we shift from the widest, ecosystem perspective to a close look at single predator-prey pairs. We begin by estimating the average amounts of crude biomass of various prey taken by all vertebrate predators per unit area in the spring–summer and autumn–winter seasons. This calculation, based on mean densities of predators, their average reproductive parameters, and averaged diet compositions, is aimed at providing a general view of what the main predators and essential prey resources are in the biome of Europe's primeval temperate forests. These results also provide frames for further, more detailed analyses of predation on ungulates, rodents, insectivores, birds, and amphibians. In each of these groups, first, the average magnitude of predation in the two seasons (spring–summer and autumn–winter) is compared with mean densities of a given prey. Then, by calculating percentage contributions of various predator species to total predation, we identify chief enemies of each prey group, and – wherever there are enough data – attempt to show how predation impact varies with changing prey densities, as determined by predators' numerical and functional responses.

The predation impact, i.e. number of prey taken by predators, is most commonly expressed as percentage of prey densities. Holling (1959) was the first to propose various forms of percentage predation curves in relation to prey densities, as expected from predator's functional responses. Oversimplified though they were, his schemes were presented in ecology textbooks (e.g. Begon et al. 1986) and formed the foundation for theoretical and empirical work on the possible regulatory impact of predation on prey populations (e.g. Sinclair 1989; Sinclair et al. 1990; Messier 1991; Hanski et al. 1993). Recently, Messier (1995 a) has mathematically reanalysed the shapes of percentage predation curves expected from various functional and numerical responses by predators. Messier's work made it clear that the type of functional response per se is not decisive for the shape of percentage predation. Instead, it is a product of functional and numerical responses that should be analysed if one seeks the explanation as to whether or not predation regulates prey densities.

Regulation of prey populations by predation may occur when predators have a direct density-dependent effect over the whole range of densities at which their prey is found. Density dependence means that mortality caused by predation, as a percentage of the prey population, increases with growing density of prey (Fig. 5.1, graph I). Such regulation was suggested to have occurred in some small rodent populations in southern Scandinavia (Erlinge et al. 1984) but this was challenged by an alternative interpretation showing inverse density dependence in predation (Kidd and Lewis 1987).

Sinclair (1989) proposed that for regulation to occur, it is enough that percentage predation is density-dependent over most of the range of low densities of prey (graph II in Fig. 5.1). When a prey population erupts and forms an outbreak, the percentage predation becomes inversely density-dependent and predators cannot change the course of events. Such a model of population regulation was proposed for house mice predated by several raptor and carnivore species in Australia (Sinclair et al. 1990) and moose exploited by wolves in North America (Messier 1995b).

If predation is inversely density-dependent (depensatory predation according to Messier 1995a) over most of prey densities (graphs III and IV in Fig. 5.1), predators only limit and do not regulate their prey. Such patterns of predation have often been reported from field studies on rodents (Pearson 1966, 1971; Goszczyński 1977; Jędrzejewski et al. 1995, 1996a; Zalewski et al. 1995) and ungulates (caribou in Messier 1995b).

As exemplified by the cited works, the problem of regulation by predators is tackled mainly by mammalogists, and it arises from the puzzle of regular fluctuations of numbers recorded for many small and large mammals. The main question that appeared in the field of mammalian ecology is whether predation can change the pattern of prey population dynamics (Fig. 5.2A, B). On the one hand, specialist predators such as weasels and stoats, through time lags in their predation response, are suspected of driving the cycles of microtines in northern Fennoscandia (Korpimäki et al. 1991; Hanski et al. 1993; Norrdahl 1995). On the other hand, it has been proposed that predation by generalist predators, due to their S-shaped functional response, is the chief factor stabilising rodent fluctuations in the temperate zone of Europe and preventing the rodent population from cycling (Erlinge et al. 1983, 1991; Erlinge 1987; Hansson 1987; Loman 1988). Both of these hypotheses are undermined by the fact that microtines do cycle in the absence of mustelid predators (Chernyavskii 1979; Chernyavskii and Dorogoi 1981) and, secondly, that cycles occur also in the temperate zone of Europe in regions with abundant generalist predators (Goszczyński 1977; review in Jędrzejewski and Jędrzejewska 1996).

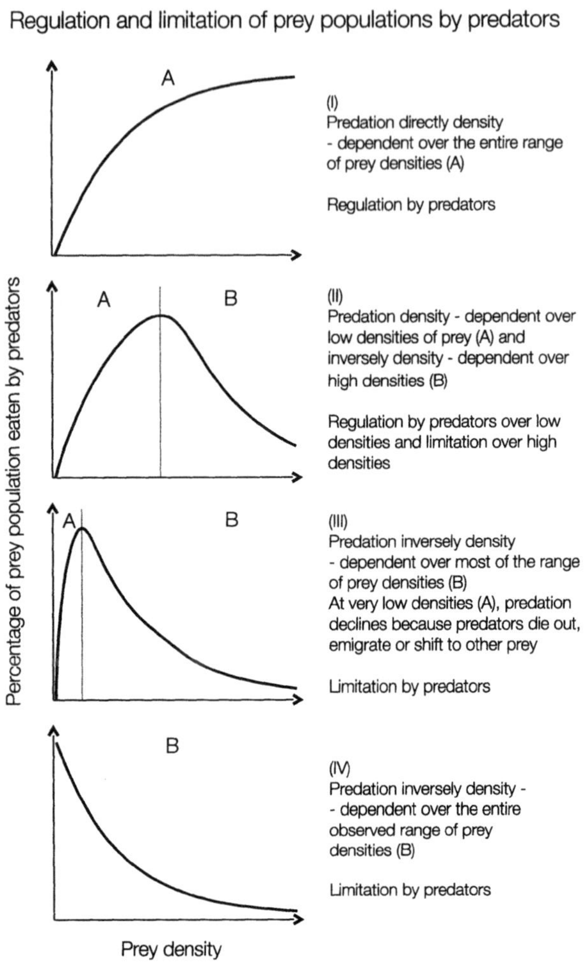

Fig. 5.1. Schemes of various curves of percentage predation in relation to prey densities proposed as occurring when predator impact on prey varies from both regulation and limitation of prey numbers *top* to limiting prey numbers only *bottom*. Compiled from theoretical and empirical works by Sinclair (1989), Sinclair et al. (1990), Messier (1995 a), Zalewski et al. (1995), and Jędrzejewski et al. (1995, 1996 a)

The question of how effectively predators limit or lower the densities of prey is often asked by ornithologists, whose study objects exhibit far more stable population dynamics than those of mammalogists. The problem discussed in studies on predation on birds is whether predators lower the productivity of a bird population, only, or if they also lower the density of the 'parental' breeding stock below the level set by food resources or nest site availability (reviews in Newton 1993, 1994) (Fig. 5.2 C, D).

Many studies that analysed the impact of predation tended to 'extract' the target predator-prey pairs from their abiotic and biotic setting, and to treat

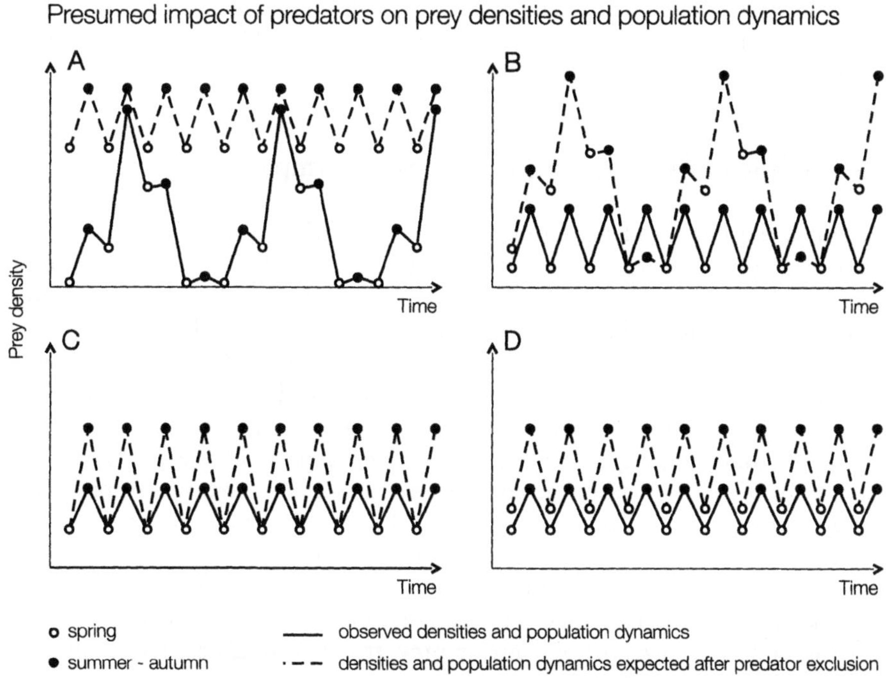

Fig. 5.2. Schematic representation of presumed impacts of predation on prey densities and population dynamics. **A** Predators drive cycles in population of otherwise noncyclic prey. **B** Predators stabilise the dynamics of otherwise cyclic prey. Problems A and B have often been addressed in studies on small rodents. **C** Predators decrease reproductive success of prey but not the density of the breeding population. **D** Predators decrease (limit) both reproductive success and density of breeding population. Problems C and D are often addressed in studies on birds

the predator-prey interrelationships as if no external factor modified or affected them. In a real world, however, foxes die of rabies, tawny owls starve to death in winters with very deep snow cover, wolves and lynxes are hunted or poached by humans, and prey individuals breed more successfully when food resources are abundant – to list just a few 'modifiers' of predator-prey interactions. We focused on such extrinsic factors and tried to quantify them wherever possible.

Our ultimate goal in this book is to assess the role of predation compared with other factors (mainly food and abiotic conditions) in shaping population dynamics and densities of ungulates, rodents, shrews, birds, and amphibians in the temperate forests of Europe. In some cases we did this by quantitative weighing of various factors, in others we verbally analysed the subject and tried to formulate conclusions.

We hope that in this chapter we have convincingly shown that there is no single 'correct' answer to the question: 'How much does predation matter to animal populations?' The answer is different for cold- and warm-blooded animals, herbivores and insectivores, large and small animals, common and rare vertebrates. What is more, the answer may be different for the same group of prey in various biomes.

5.2 Methods of Calculating the Predation Impact

To assess the impact of predation on prey populations, we need to know the number of prey specimens removed by a population of predators from a unit area per unit time. This is a product of two variables: (1) the number of predators, and (2) their killing rates. In practice, the latter variable is rarely estimated directly due to obvious technical problems. An indirect way to assess predators' killing rates is based on their daily food intake (known from enclosure experiments) and diet composition (studied in the field).

On the following pages we will present two aspects of the predation impact, both based on data collected between 1985–1996. First, the average estimate of pressure by all predators living in Białowieża Primeval Forest will be presented for the spring–summer and autumn–winter seasons. Secondly, between-year variation in predation impact is shown for several species of the most numerous forest predators (wolf, lynx, tawny owl, pine marten, weasel, polecat, common buzzard) and major prey groups (ungulates, forest rodents, shrews, birds, and amphibians).

Based on the same rationale, our calculations of predation were done somewhat differently for each of the following groups of predators: (1) wolf and lynx, (2) other carnivores, and (3) birds of prey. In the case of wolf and lynx, data on their actual killing rates (i.e. number of prey killed by one predator per month) obtained by radio-tracking and intense search for prey remains were multiplied by densities of these predators: number of packs in wolves and number of individuals in lynxes. Killing rates by lynxes were estimated as a sum of the mean consumption time of one prey and the mean searching time, and then multiplied by proportions of a given ungulate species among all prey found. Killing rates were calculated separately for adult males, subadults, and females with kittens. Details on methods are provided by Okarma et al. (1997) and Jędrzejewski et al. (in press).

In all other carnivores, we used the estimates of their daily food intakes (from the literature, see Table 5.1 for sources) and data on their diet composition in BPF. The product of these two measures was used as an approximation of predators' daily killing rates. Thus the magnitude of their predation in spring–summer and autumn–winter, expressed as the number of prey specimens removed from an average hectare (N_{prey}) was calculated as:

$$N_{prey} = D \times (DFI \times B_{prey}) / BM_{prey} \times n_{days},$$

where D is the density of a predator (number per hectare); DFI is the daily food intake (in grams), B_{prey} is the fraction of a given prey biomass in the predator's diet, BM_{prey} is the mean body mass of prey, and n_{days} is the

Table 5.1. Parameters of predators' daily food intake, time of their presence in the study area, number of pellets casted per day, and coefficients to correct for loss of prey specimens in digestion, used in calculation of predation impact in Białowieża Primeval Forest

Species of predator	Daily food intake (g/day)		*N* days present in study area per year			*N* pellets/day		Correction coefficient	Source of data on daily food intake
	Adult	Juvenile	Adult AW	Adult SS	Juv SS	SS	AW		
Lynx	2200	2200	197	168	–	–	–	–	Okarma et al. (1997)
Wolf	5348	5348	197	168	–	–	–	–	Jędrzejewski et al. (in press)
Red fox	470	447	197	168	84 [a]	–	–	–	Ryszkowski et al. (1973)
Raccoon dog	300	240	127 [b]	168	84 [a]	–	–	–	Korhonen (1988 a, b)
Badger	600	500	75 [b]	168	107 [c]	–	–	–	Goszczyński (1985)
Pine marten	189.6	151	197	168	137 [d]	–	–	–	Ryszkowski et al. (1973)
Weasel	36	36	197	168	84 [a]	–	–	–	Jędrzejewska and Jędrzejewski (1989)
Stoat	51	41	197	168	84 [a]	–	–	–	King (1989)
Polecat	175	140	197	168	84 [a]	–	–	–	Erlinge et al. (1983)
American mink	150	120	197	168	84 [a]	–	–	–	Dunstone (1993)
River otter	1100	880	197	168	168	–	–	–	Kruuk (1995)
Honey buzzard	99.5	99.5	–	168	84 [e]	–	–	–	As for common buzzard
Short-toed eagle	150	150	–	140	84 [e]	1 [f]	–	1.19 [g]	This book (mean biomass per pellet)
Goshawk	161	119	197	168	84 [e]	1 [f]	1 [f]	1.19 [g]	Kenward et al. (1981)
Sparrowhawk	78.5	78.5	–	168	56 [e]	1 [f]	–	1.05 [g]	This book (mean biomass per pellet)
Lesser spotted eagle	101.4	101.4	–	168	90 [e]	1 [f]	–	1.19 [g]	This book (mean biomass per pellet)
Common buzzard	99.5	99.5	98	168	110 [e]	1 [f]	1 [f]	1.19 [g]	This book (mean biomass per pellet)
Hobby	56.3	56.3	–	124	50 [e]	1 [f]	–	1.05 [h]	This book (mean biomass per pellet)
Pygmy owl	38.2	38.2	197	168	60 [i]	1.1 [j]	1.25 [j]	1.05 [h]	This book (mean biomass per pellet)
Tawny owl	71.6	71.6	197	168	138 [k]	1.1 [j]	1.25 [j]	1.19 [g]	This book (mean biomass per pellet)

Table 5.1 (continued)

Species of predator	Daily food intake (g/day)		*N* days present in study area per year			*N* pellets/day		Correction coefficient	Source of data on daily food intake
	Adult	Juvenile	Adult AW	Adult SS	Juv SS	SS	AW		
Tengmalm's owl	31.6	31.6	197	168	30[k]	1.1[j]	1.25[j]	1.05[h]	This book (mean biomass per pellet)
Long-eared owl	63	63	45	168	84[k]	1.1[j]	1.25[j]	1.05[h]	Ceska (1980)

AW, Autumn–winter.
SS, Spring–summer.

[a] It was assumed that young are present till mid-July, i.e. the seasonal peak of carnivores' numbers; in weasels calculations were based on densities censused in summer.
[b] Based on snow tracking of raccoon dogs (W. Jędrzejewski, unpubl.) and observations of badgers (A.N. Bunevich, unpubl.).
[c] After Sumiński (1989).
[d] After Goszczyński (1985).
[e] After Pugacewicz (1996).
[f] According to data on North American diurnal raptors obtained by Duke et al. (1976).
[g] Data for *Strix aluco* (from Raczyński and Ruprecht 1974) used for all medium-sized owls;
[h] data for *Tyto alba* (Goszczyński 1976) accepted for all small owls; also, Duke et al. (1976) reported that disappearance of prey specimens in the process of digestion increased with growing body size of owls.
[i] After Domaszewicz (1993) and Mikkola (1983).
[j] Data on *Strix aluco* (Lowe 1980) used for all owls.
[k] After Mikkola (1983), Domaszewicz (1993), and Dackevich (1997).

number of days per season a predator was present or active. The duration of stay was different in juvenile and adult predators (Table 5.1). It was also a relevant variable in badgers and raccoon dogs, which sleep in winter, and in migratory raptors.

In calculating the between-year variation in magnitude of weasel predation, we made a simplifying assumption that they feed exclusively on forest rodents, and we included surplus killing as governed by ambient temperature (see details in Jędrzejewski et al. 1995). More detailed descriptions of methods on calculating predation by medium-sized and small predators can be found in Jędrzejewski and Jędrzejewska (1993) and Zalewski et al. (1995).

In birds of prey, calculation of predation was based on the mean number of prey individuals recovered from pellets (see Sect. 4.2). From enclosure experiments it was known what the rate of casting pellets per day was, as well as what percentage of prey individuals disappeared in the process of digestion (Table 5.1). Thus, in birds of prey, the magnitude of their predation, i.e. the number of prey specimens (N_{prey}) removed from an average hectare during spring–summer or autumn–winter season was calculated as:

$$N_{prey} = D \times N \times CC \times n_{pellet} \times n_{days},$$

where D is the density of predators (numbers per hectare), N is the mean number of prey specimens in one pellet, CC is the coefficient to correct for disappearance of prey individuals during digestion, n_{pellet} is the mean number of pellets cast per day, and n_{days} is the number of days per season a raptor was present in the study area.

The honey buzzard and the goshawk were exceptions: their total consumption of prey was calculated based on approximate daily food intakes (see Table 5.1). Further details on methods of estimating the magnitude of predation by raptors are given in Jędrzejewski and Jędrzejewska (1993) and Jędrzejewski et al. (1996a). All cited papers also provide thorough discussions of the reliability and possible errors of the presented methods of calculating predation. Here, suffice it to say that all sets of data (densities of predators, rates of prey consumption, diet composition) are possible sources of bias. Through the series of calculations needed to estimate the magnitude of predation these errors may 'neutralise' each other but there again, they may increase at compound interest. There is no question that some of the results presented on the following pages are fairly rough, preliminary estimates. Still, if they are treated with caution, these data can shed much light on the fascinating subject of predation.

5.3 Total Consumption of Prey Biomass by All Predators

From an average square kilometre of Białowieża Forest, all vertebrate predators from weasels and pygmy owls to eagles and wolves harvested 713 kg of crude biomass of prey per year (i.e. 7129 g/ha; Fig. 5.3). The larger portion

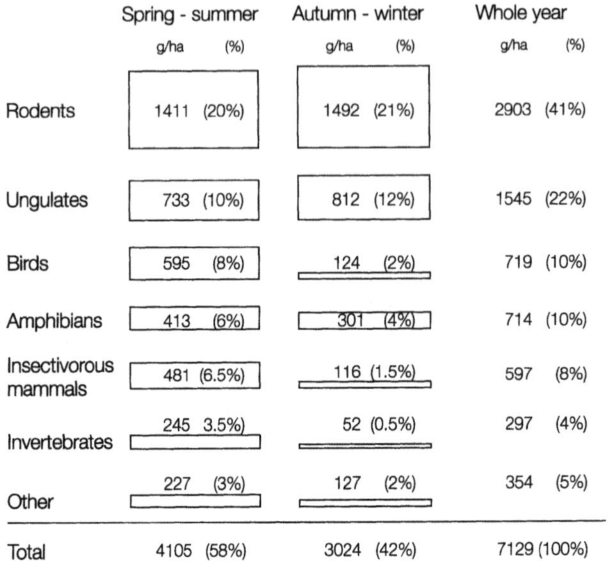

Fig. 5.3. Estimates of mean consumption of various groups of prey by all carnivores and birds of prey in spring–summer (May–September), autumn–winter (October–April), and in whole year in Białowieża Primeval Forest. Consumption of a given prey resource is shown in absolute quantities (g/ha) and as percentage of total prey biomass taken by all predators

(58%) of that amount is consumed during spring–summer rather than in autumn–winter (42%). The relative contributions of various prey groups in the total biomass taken by predators are striking. Rodents are the unchallenged dominants (41%), followed by ungulates (22%), birds (10%), and amphibians (10%). Rodents, ungulates, and frogs are harvested by predators in similar amounts in spring–summer and in autumn–winter, whereas birds, insectivorous mammals, and invertebrates bear the main brunt of predation in spring–summer. In the temperate forests, fish, reptiles, and medium-sized mammals matter very little in the total biomass of consumed prey; together with plant material, they are listed as 'other' food in Fig. 5.3.

The whole predation impact was calculated for 22 species of carnivores and raptors. Their percentage contributions to the total consumed biomass are listed in Table 5.2. The tawny owl appeared as the single most important predator of the pristine temperate forest; it consumed one quarter of all the prey biomass eaten by predators in spring–summer, and one third in autumn–winter. Co-dominant predators, taking ≥5% of the prey biomass each, were the common buzzard, fox, wolf, lynx, pine marten, and polecat. Generally, the roles of various predators corresponded with their ranks in the predator community (cf. Sect. 3.31). Based on this relationship, we estimated that the exclusion of seven rare species from this analysis (due to very little data on their ecology) underestimated the total consumption of prey biomass by some 2% only.

Table 5.2. Percentage contributions by various species of predators to the total consumption of prey (in terms of biomass) in spring-summer, autumn-winter, and whole year combined. Dominants are in boldface type (≥5% contribution)

Species of predator	Percentage contributions to the total consumption of prey		
	Spring-summer	Autumn-winter	Whole year
Tawny owl	**26.5**	**30.7**	**28.3**
Common buzzard	**13.7**	**8.3**	**11.4**
Red fox	**8.6**	**11.1**	**9.7**
Wolf	**7.5**	**12.2**	**9.5**
Lynx	**6.4**	**10.4**	**8.1**
Pine marten	**7.8**	**7.0**	**7.5**
Polecat	**5.7**	**5.0**	**5.4**
Weasel	**5.2**	4.3	4.8
Raccoon dog	4.9	2.3	3.8
Badger	4.0	1.9	3.1
Goshawk	1.6	2.1	1.8
River otter	1.3	2.2	1.7
Honey buzzard	2.1	–	1.2
Pygmy owl	0.7	0.8	0.8
Sparrowhawk	1.2	–	0.7
American mink	0.7	0.7	0.7
Stoat	0.6	0.7	0.6
Lesser spotted eagle	0.9	–	0.5
Long-eared owl	0.3	0.1	0.2
Tengmalm's owl	0.1	0.2	0.1
Hobby	0.1	–	0.1
Short-toed eagle	0.1	–	+
Total biomass of prey consumed (g/ha)	4105 (100%)	3024 (100%)	7129 (100%)

Admittedly, there are several other species that act as predators of vertebrates in Białowieża Forest: black stork *Ciconia nigra*, white stork *C. ciconia*, snakes and vipers, shrikes, to say nothing of predaceous fish such as pikes. The fact that they are neglected in this analysis led to an underestimation of the total consumption of aquatic amphibians and fish (e.g. Gavrin 1953). Nevertheless, as regards predation on terrestrial and semiaquatic vertebrates, we believe we cover the large part of predation on them by vertebrate predators.

The ranking of predators' roles in the total consumption of prey (shown in Table 5.2) demonstrated some intriguing features of predator-prey ecology in the primeval forests of temperate Europe. Of eight important predators, only four (the tawny owl, the common buzzard, the red fox, and the weasel) are doing fairly well in contemporary temperate woodlands of our continent.

The other four are either extinct from most of their former range (lynx, wolf; e.g. Kratochvil 1968; Promberger and Schröder 1992) or rare (pine marten, polecat; Lockie 1964; Walton 1968; Weber 1988a). Particularly, wolf and lynx have been eradicated in many regions for so long that their lack is hardly noted or accounted for in ecological studies on ungulate-vegetation relations (e.g. Clutton-Brock et al. 1985).

In Białowieża Forest, except for the high rank of the common buzzard, all migratory raptors play minute roles in total predation and contribute altogether <5% to the total consumed biomass. This is not to say that they are unimportant and their extinction would not mean a great loss to the forests of Europe. On the contrary, it is the presence of many rare species that gives primeval habitats their hallmark.

Out of 713 kg/km^2 of prey biomass consumed annually by all predators, 680 kg is comprised of terrestrial and semiaquatic vertebrates. How does this amount compare with the standing crop of biomass and production of prey populations? Rough though it is, such comparison may put our exploration on the right path. Let us begin by analysing rodents, which appeared the mainstay of predator-prey relationships. Predators annually harvest a biomass which exceeds by three times its mean standing crop (Table 5.3). Such heavy predation is made possible by the very high productivity of rodents. Even so, rodents seem to be exploited by predators to the extreme: half of their annual production is taken by predators in spring–summer and the other half in autumn–winter. Predators hardly give other mortality factors an opportunity to act.

As regards shrews, they are under fairly heavy predator pressure, too, but factors other than predation seem to play decisive roles in the autumn–winter mortality of shrews (Table 5.3). The situation with birds is similar, but the low importance of predation in birds' mortality in the cold season is obviously due to the fact that migratory birds are then beyond the reach of local predators. Of all small vertebrates, amphibians seem to be least harvested by predators (Table 5.3). Ungulates are characterised by a very large standing crop of biomass and rather low productivity. Predators annually harvest very little (10%) of ungulates' standing crop of biomass but this is quite a lot when compared with their annual production. Also, predators seem to be an important factor in the autumn–winter mortality of large hooved mammals.

It must be borne in mind, however, that the intensity of a predator's exploitation of a given prey does not simply translate into the importance of predation in shaping the densities and population dynamics of that prey. Only when rates of prey reproduction at various levels of food resources are compared with rates of predation can that problem be elucidated. Moreover, a crucial question is whether or not in the absence of predators other natural enemies (such as disease, starvation) take over their role. The detailed analysis of predation patterns and mechanisms will begin with a close look at ungulate-large predator relationships.

Table 5.3. Total biomass of main groups of prey taken by all predators in spring–summer and autumn–winter seasons in BPF compared with the average standing crop of biomass, production and mortality of a given prey group. Data on prey biomass from Chapter 2

Parameter	Biomass of prey (kg/km^2)				
	Rodents	Shrews	Birds	Amphibians	Ungulates
Standing crop of biomass in spring	27	6	47	130	994
Standing crop of biomass in summer/autumn	167	26	114	260	1157
Spring–summer predation	141	14	60	41	60
Autumn–winter predation	149	7	12	30	71
Annual predation as percent of mean standing crop of biomass[a]	300%	130%	90%	35%	10%
Spring–summer predation as percent of total production[b]	50%	40%	45%	25%	25%
Autumn–winter predation as percent of winter mortality[c]	100%	35%	20%	25%	45%

[a] Mean of spring and summer/autumn estimates of prey biomass.
[b] Total production assessed roughly as a sum of observed increase in prey from spring to summer/autumn and the amount taken by predators during that period.
[c] Winter mortality estimated as difference between summer/autumn and the following spring biomass of prey.

5.4 Predation on Ungulates

5.4.1 Species Structure of Total Predation on Ungulates

Annually, wolves and lynxes killed an average of 3.47 ungulates/km^2, which constituted 130.4 kg/km^2 of crude biomass. Of the total number of preyed ungulates, lynx took 63% and wolf 37%. But since lynxes' mean prey was under half the size of wolves', the respective percentage contributions in terms of biomass were 46% by lynx and 54% by wolf (Table 5.4). Messier and Crete (1985) estimated that 75% of an ungulate body is consumable. Indeed, thick bones were utilised neither by host predators nor by scavengers. Old bones, jaws, and antlers of ungulates still lie about in the forest months after an animal was killed by wolves or lynxes. The actual consumption of total crude biomass of all killed ungulates was at least 65%, including 60% by wolf and lynx combined and 5% (maximum 15%) by scavengers (Table 5.4). Scavenging will be a subject of Section 5.10; here, we only mention the minor roles of foxes, raccoon dogs, buzzards and several other species in utilising ungulates killed by large predators.

As the brown bear, the third native predator of ungulates in temperate Europe, has been eradicated from Białowieża Forest, the question arises as to how much it would have contributed to total predation on ungulates. In cen-

Table 5.4. Percentage shares of wolf and lynx in the total predation on ungulates, and percentage contributions of various predators to the consumption of total biomass of ungulates killed by wolves and lynx

Species	Spring–summer	Autumn–winter	Whole year
Percentage contribution to the total biomass of killed ungulates			
Wolf	**53.8**	**53.9**	**53.9**
Lynx	**46.2**	**46.1**	**46.1**
Percentage contribution to the consumption of ungulates killed by wolf and lynx			
Wolf	**35.9**	**33.4**	**34.6**
Lynx	**26.3**	**25.4**	**25.8**
Red fox	0.9	2.6	1.8
Raccoon dog	1.0	0.7	0.8
Common buzzard	1.0	0.6	0.8
Pine marten	0.5	0.5	0.5
Polecat	–	0.2	0.1
Badger	+	+	+
Other scavengers and unused[a]	9.4	11.6	10.6
Assumed non-consumable portion[b]	25.0	25.0	25.0
Total biomass (g/ha)	599	705	1304

[a] Among other scavengers, the most important were ravens and wild boar (see Sect. 5.10).
[b] After Messier and Crete (1985).

tral Europe, bears are largely herbivorous (Frąckowiak and Gula 1992), but they also feed on carrion and hunt ungulates. A review of data from nature reserves and national parks of Central Europe and the former Soviet Union showed that the average role of brown bear in total predation on ungulates was from 1% in roe deer to 40% in moose (B. Jędrzejewska and W. Jędrzejewski, in prep.). Based on these mean values, we would guess that the presence of brown bear in Białowieża Primeval Forest could have increased the total predation on ungulates by only 5–10%.

5.4.2 Magnitude and Mechanisms of Wolf and Lynx Predation on Ungulates

Predation on the ungulates of Białowieża Forest (Fig. 5.4) was analysed in two time scales. First, absolute measures of predation (numbers of ungulates killed annually by wolves and lynxes) were obtained and compared with prey densities. Secondly, 100-year data on the densities of the two predators and ungulates were mathematically analysed to show ungulate-predator relationships based on relative measures of predation rates. These complementary sets of data revealed the patterns and mechanisms of wolves' and lynxes' impact on large mammals.

Fig. 5.4. A wolf feeding on a female red deer. Wolves are the most important predators of red deer. Photo: W. Jędrzejewski

In 1991–1995, wolves annually took 57–105 red deer/100 km^2 and lynxes killed 42–70 red deer/100 km^2 (Table 5.5). Predation by wolves constituted merely 9–13% of spring-summer densities of red deer, and predation by lynxes 6–13%. When compared with the annual increase in deer due to breeding, both wolf and lynx predation appeared significant: 32–47% by wolves and 21–43% by lynxes. Still bigger were the quotas of the hunting harvest, which in 1991–1995 increased two to three times compared with earlier years (see Sect. 2.3). With an annual harvest of 131–140 ind/100 km^2, the red deer population lost 16–25% of its spring-summer densities and as much as 53–85% of its yearly production. The hunting season lasts from October through February. During that time, wolves and lynxes cannot lower their predation, which is already predetermined by the existing density of adult predators and the number of their half-grown young. Thus, the hunting harvest and predation were additive and far exceeded the annual increase in the red deer population due to reproduction (altogether 121–168%, mean 145%). After such heavy exploitation by humans and predators (not to mention other factors of mortality such as poaching, disease and traffic accidents) there was a steep decline in the number of red deer in 1991–1996 (Fig. 5.5).

The situation and fate of the roe deer population were similar, only the relative contributions of lynx, wolf, and human harvests were different than those in red deer. Lynxes annually killed 110–181 roe deer/100 km^2, which made up 21–36% of deer densities in spring-summer and 52–85% of its yearly increase due to breeding (Table 5.5). Wolves, to which roe deer is a secondary prey, took 19–25 roe deer/100 km^2 per year, that is 3–4% of deer densities and 8–10% of their annual production. The hunting harvest, with annual quotas of 85–114 deer shot per 100 km^2, took 10–20% of densities and 26–48% of yearly production of roe deer. Again, additive effects of the hunting harvest and predation, together amounted to 86–143% of roe deer's

Table 5.5. Predation impacts by wolves, lynx and hunting harvest on populations of red deer, roe deer, moose, and wild boar in the Polish part of Białowieża Primeval Forest in 1991/1992–1993/1994 and 1995/1996. (Okarma et al. 1997; Jędrzejewski et al., in press)

Parameter	Year				Mean (SD)
	1991/92	1992/93	1993/94	1995/96	
Wolf population dynamics					
Number of wolf packs (n/100 km^2)	0.58	0.43	0.43	0.43	0.47 (0.1)
Density of wolves (n/100 km^2)	3.3	2.1	2.1	2.3	2.45 (0.6)
Lynx population dynamics					
Density of lynx (n/100 km^2)	5.1	4.9	5.2	3.0	4.44 (1.0)
Mean number of kittens per mother	2.0	1.4	1.8	1.0	1.6 (0.4)
Red deer (*Cervus elaphus*) Population dynamics					
Density in March (n/100 km^2)	607	416	359	463	461 (106)
Percentage of adult females	46	44	48	47	46 (2)
Number of juveniles per adult female	0.9	0.9	0.9	0.9	0.9
Juveniles born in spring (n/100 km^2)	251	165	155	196	192 (43)
Density in May/June (n/100 km^2)	858	581	514	659	653 (149)
Wolf predation					
Total yearly predation (n/100 km^2)	105	78	57	62	76 (22)
Predation as percentage of:					
spring/summer density of red deer	12	13	11	9	11 (2)
red deer increase due to breeding	42	47	37	32	39.5 (6)
Lynx predation					
Total yearly predation (n/100 km^2)	65	70	66	42	61 (13)
Predation as percentage of:					
spring/summer density of red deer	8	12	13	6	10 (3)
red deer increase due to breeding	26	42	43	21	33 (11)
Hunting harvest					
Annual hunting harvest (n/100 km^2)	133	131	131	140	134 (4)
Hunting harvest as percentage of:					
spring/summer density of red deer	16	23	25	21	21 (4)
red deer increase due to breeding	53	79	85	71	72 (14)
Roe deer (*Capreolus capreolus*) Population dynamics					
Density in March (n/100 km^2)	492	425	288	324	382 (93)
Percentage of adult females	46	45	51	41	46 (4)
Number of juveniles per adult female	1.45	1.45	1.45	1.45	1.45
Juveniles born in spring (n/100 km^2)	328	277	213	193	253 (62)
Density in May/June (n/100 km^2)	820	702	501	517	635 (153)

Table 5.5 (continued)

Parameter	Year				Mean (SD)
	1991/92	1992/93	1993/94	1995/96	
	Wolf predation				
Total yearly predation (n/100 km^2)	25	21	21	19	22 (2)
Predation as percentage of:					
spring/summer density of roe deer	3	3	4	4	3.5 (1)
roe deer increase due to breeding	8	8	10	10	9 (1)
	Lynx predation				
Total yearly predation (n/100 km^2)	169	169	181	110	157 (32)
Predation as percentage of:					
spring/summer density of roe deer	21	24	36	21	26 (7)
roe deer increase due to breeding	52	61	85	57	64 (15)
	Hunting harvest				
Annual hunting harvest (n/100 km^2)	85	114	102	86	97 (14)
Hunting harvest as percentage of:					
spring/summer density of roe deer	10	16	20	17	16 (4)
roe deer increase due to breeding	26	41	48	45	40 (10)
	Moose (*Alces alces*) Population dynamics				
Density in March (n/100 km^2)	37	12	14	2.5	16 (15)
Percentage of adult females	42	39	50	44	44 (6)
Number of juveniles per adult female	1.0	1.0	1.0	1.0	1.0
Juveniles born in spring (n/100 km^2)	15	5	7	1	7 (6)
Density in May/June (n/100 km^2)	52	17	21	3.5	23 (20)
	Wolf predation				
Total yearly predation (n/100 km^2)	2	0	1	1	1 (1)
Predation as percentage of:					
spring/summer density of moose	4	0	5	29	10 (13)
moose increase due to breeding	13	0	14	100	32 (46)
	Hunting harvest				
Annual hunting harvest (n/100 km^2)	7	4.5	5.5	1	4.5 (2)
Hunting harvest as percentage of:					
spring/summer density of moose	13	26	26	29	24 (7)
moose increase due to breeding	47	90	79	100	79 (23)
	Wild boar (*Sus scrofa*) Population dynamics				
Density in March (n/100 km^2)	458	209	308	225	300 (114)
Percentage of adult females	15	16	16	16	16 (1)
Number of juveniles per adult female	3.2	5.9	5.9	5.9	5.2 (1.4)
Juveniles born in spring (n/100 km^2)	220	197	291	212	230 (42)
Density in May/June (n/100 km^2)	678	406	599	437	530 (130)

Table 5.5 (continued)

Parameter	Year				Mean (SD)
	1991/92	1992/93	1993/94	1995/96	
	Wolf predation				
Total yearly predation (n/100 km^2)	28	19	38	36	30 (9)
Predation as percentage of:					
spring/summer density of wild boar	4	5	6	8	6 (2)
wild boar increase due to breeding	13	10	13	17	13 (3)
	Hunting harvest				
Annual hunting harvest (n/100 km^2)	142	90	45	77	81 (45)
Hunting harvest as percentage of:					
spring/summer density of wild boar	21	22	8	18	17 (6)
wild boar increase due to breeding	65	46	15	36	40 (21)

annual production (and was obviously reinforced by poaching, disease, traffic accidents etc.), and caused a serious decline in deer numbers (Fig. 5.5).

The other three ungulates (moose, European bison, wild boar) differed from the two deer species as regards predation pressure. European bison are largely free of predation (Fig. 5.6). Lynxes cannot threaten them at all. Of the few bison carcasses fed upon by wolves (registered in 1985–1996), only one could have been an actual kill. Moose, the least numerous ungulates of Białowieża Forest, are killed by wolves regularly though in very low numbers. The annual predation rate of 0–2 moose/100 km^2 mattered very little, either to density or to the yearly increase in the moose population, except in 1995, when moose became extremely rare, predominantly due to a heavy hunting harvest (Table 5.5). In BPF, lynxes have not recently hunted moose. Wild boar fell as prey to both wolf and lynx, but the latter kills boar so sporadically that we neglected it in calculating predation impacts. Wolves annually took 19–38 wild boar/100 km^2 and hunters 45–142 boar. Together, wolf predation and the hunting harvest amounted to 28–78% of wild boar's yearly increase due to breeding. Thus, there must have been other factors of mortality acting on the wild boar population as well, the more so as its population rapidly declined in 1991–1996 (Fig. 5.5).

The presented data, albeit covering 4 years only, revealed several essential points. First, predators are far more important for roe and red deer than for moose, bison and wild boar. Secondly, the roe deer suffers predominantly from lynx predation whereas red deer suffers from the combined pressure of wolf and lynx, the first one being a more important enemy to red deer. Thirdly, predation and harvesting by humans have additive effects on ungulate populations. And, finally, during a steep decline of deer numbers in 1991–1996, wolf densities remained fairly stable and those of lynx declined by 40% (with 1-year lag) due to the markedly lower reproductive success of lynx. Percentage predation

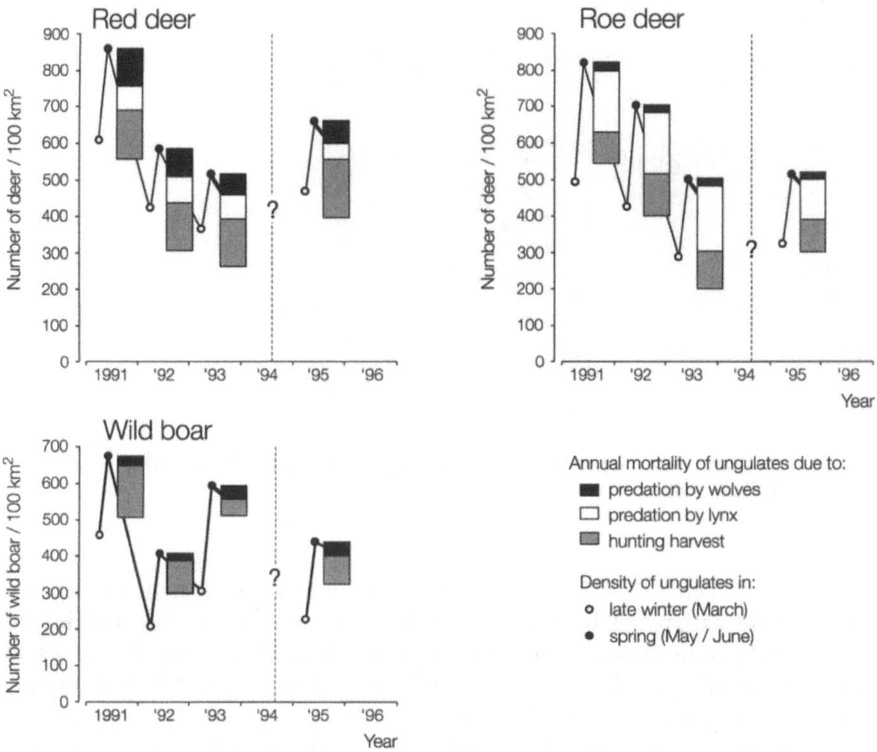

Fig. 5.5. Annual predation by wolf and lynx and annual hunting harvest compared with mean densities of red deer, roe deer, and wild boar in the Polish part of BPF in 1991–1995/1996. *Vertical lines* separate results of ungulate censuses obtained by somewhat different methods. The decline of ungulate numbers accelerated in 1996–1997. Jędrzejewski et al. (in press)

by wolves and lynxes was fairly stable from year to year or showed a tendency to increase with declining density of prey (Table 5.5).

Now we will change focus and look at ungulate-predator relationships in the last 100 or more years. This set of data does not provide absolute measures of predation (i.e. number of ungulates killed by predators) but it gives indices of predation impact, all the more interesting as the available information covers unintended (and replicated) 'experiments' on predator extermination.

In Fig. 5.7 population dynamics of wolves are laid on that of red deer. It appears that dramatic growths and declines of red deer numbers in 1890–1993 alternated with those of wolf density. Over most of that time, however, human action was impelling the fluctuations. Extermination of wolves in 1870–1890 and persistent shooting of all immigrant wolves until 1915 had 'cleared the ground' for reintroduction of red deer and allowed for their exponential increase. Wool-sorter's disease, that broke out in the red deer population in 1912, had a fairly weak effect on deer numbers (Sablina 1955). It

Fig. 5.6. European bison, due to its large body size, is not affected by predation. Photo: W. Jędrzejewski

was only between 1915–1921, during a few years of warfare, that red deer became nearly extirpated by humans. In those very years, wolves quickly resettled Białowieża Forest, now relieved from predator control. In the 1930s, wolves declined again in consequence of resumed control but they flourished again in the years of World War II (Fig. 5.7). For about 30 years of high numbers of wolves (in 1920–1950), red deer densities remained very low. Their fast increase occurred only after the extermination of wolves that

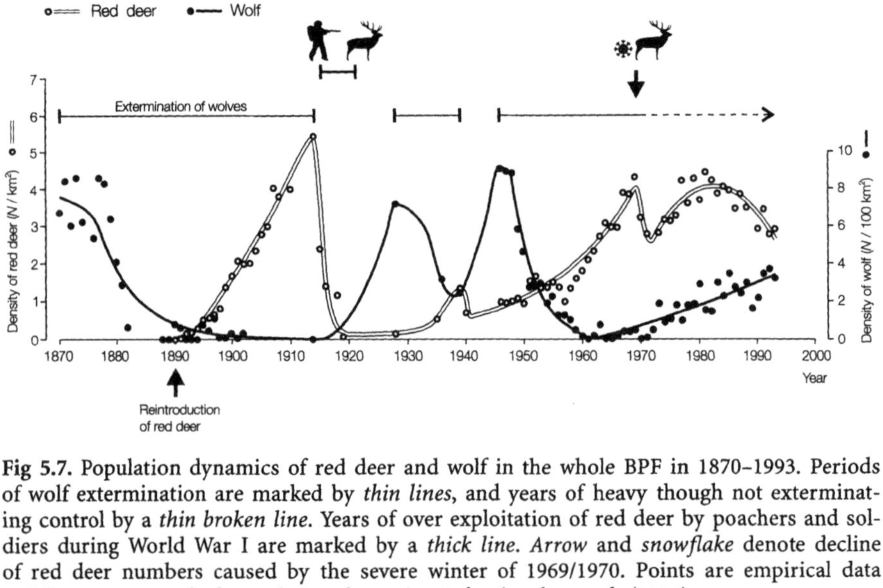

Fig 5.7. Population dynamics of red deer and wolf in the whole BPF in 1870–1993. Periods of wolf extermination are marked by *thin lines*, and years of heavy though not exterminating control by a *thin broken line*. Years of over exploitation of red deer by poachers and soldiers during World War I are marked by a *thick line*. *Arrow* and *snowflake* denote decline of red deer numbers caused by the severe winter of 1969/1970. Points are empirical data and lines are smoothed population dynamics. Jędrzejewska et. al. (1997)

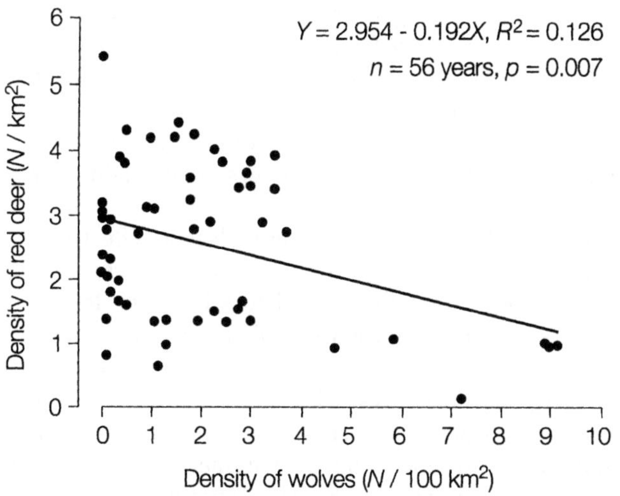

Fig. 5.8. Density of red deer in relation to wolf density in BPF in 1897–1993. Each point denotes 1 year. Jędrzejewska et al. (1997)

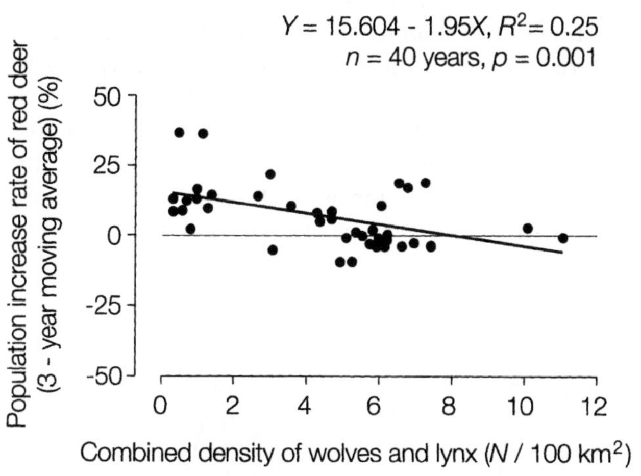

Fig. 5.9. Population increase rate of red deer in relation to combined density of large predators (wolves and lynx) in BPF. Each point denotes 1 year. Jędrzejewska et al. (1997)

began in 1946. In recent years (1980–1993), a heavy though not exterminating harvest of wolves in the Belarussian part of BPF and their protection in the Polish part allowed survival, indeed a steady increase in the wolf population. Plotted against each other, red deer and wolf densities appeared negatively correlated (Fig. 5.8), with changes in wolf numbers explaining 13% of the total variation in deer numbers. Red deer numbers did not correlate with densities of lynx.

Another parameter of red deer population analysed was percentage population increase from year to year. To avoid possible sampling errors, it was smoothed by calculating a 3-year moving average. As defined, the rate of population increase in red deer was negatively correlated with combined densities of wolves and lynxes (Fig. 5.9). The regression equation predicts that the increase in red deer population turns negative when combined densities of wolves and lynxes exceed 8 ind/100 km^2. In fact, even lower densities of predators together with additive effects of other factors of mortality often caused a decline in the red deer population.

Does this all mean that hunters are right in seeing wolves as effective executioners of red deer populations? If we recall wolves' dietary response to changes in deer densities it appears that those predators kill very few deer when the densities of deer are low (Fig. 5.10, left panel). However, wolves did not decline when their favourite prey was scarce because they were able to exploit buffer prey (wild boar, roe deer, livestock). In fact, wolves were numerous in Białowieża Forest even when red deer were rare or absent. Admittedly, due to constant human disturbance, the true numerical response of wolves to changes in deer numbers remains unknown. Nonetheless, it is certain that in the multispecies community of ungulates in temperate Europe wolf densities do not strongly depend on red deer numbers. In the last 100 years, the numerical relationship between wolves and deer densities in Białowieża Forest has been largely shaped by humans (see Fig. 5.10, middle panel). With a logarithmic functional response and a negative numerical relationship, the percentage predation by wolves was inversely density-dependent over most observed densities (Fig. 5.10, right panel). Although the Y-scale is arbitrary and says nothing about the magnitude of percentage predation, the curve of the predation rate in Fig. 5.10 shows some concrete points. Firstly, red deer suffered the heaviest predation when their average densities were about 1 ind/km^2. Secondly, when deer densities were lower than that,

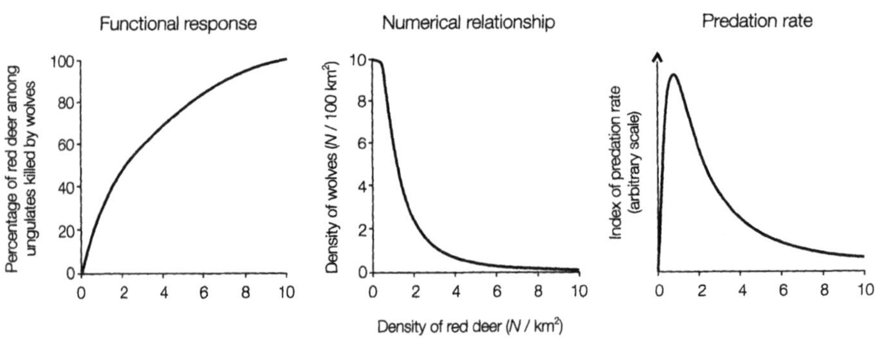

Fig. 5.10. *Left panel* curve of wolf functional (dietary) response to varying density of red deer (derived from empirical studies, see Figs. 4.11 and 4.12). *Middle panel* curve of numerical relationship between wolf and red deer numbers in 1897–1993 (cf. Figs. 5.7 and 5.8). *Right panel* curve of total predation response of wolves, calculated as a product of functional and numerical responses, to changes in red deer densities in BPF. This curve does not show real values of percentage predation (note the lack of numerical *Y*-scale), but it correctly shows the shape of the predation curve

percentage predation by wolves declined rapidly so that predators would not cause extermination of their prey. At deer densities higher than 1 ind/km^2, wolf predation declined again but in this case due to control of wolves by humans. All in all, the hunters are right in the sense that if red deer numbers are scarce, wolves can hold them low for years.

In Fig. 5.11, the population dynamics of roe deer in the last 130 years is shown together with that of its main predator, the lynx. Roe deer numbers were fairly stable during that whole period though they underwent two irruptions, both of them coinciding with the extermination of lynx. However, the rapid declines in roe deer numbers, starting in 1915 and 1969/1970, were not triggered by changes in predation rates but by human exploitation and a harsh winter, respectively. Interestingly enough, both declines coincided with the recovery of a lynx population saved from persecution. In 1897–1993, when data on both roe deer and lynxes were available, densities of roe deer and those of lynxes appeared negatively correlated, with changes in lynx numbers explaining 37% of a total variation in roe deer numbers (Fig. 5.12). Similarly, the population increase rate of roe deer was significantly affected by lynx density (Fig. 5.13). The regression equation predicts that if lynx densities exceed 4.4 ind/100 km^2, the population increase in roe deer will turn negative.

As was the case with wolf predation on red deer, we calculated the relative index of lynx predation rate on roe deer, based on the lynx dietary response to changes in deer density and the actual numerical relationship between the prey and the predator (Fig. 5.14). Lynxes are persevering hunters of roe deer and they shift to other prey only when deer densities are very low (<0.5 ind/km^2). In Białowieża Forest, lynxes have access to two buffer foods, red deer and brown hares, so their numbers do not critically depend on roe deer densities. Indeed, the highest numbers of lynxes occurred when roe deer was

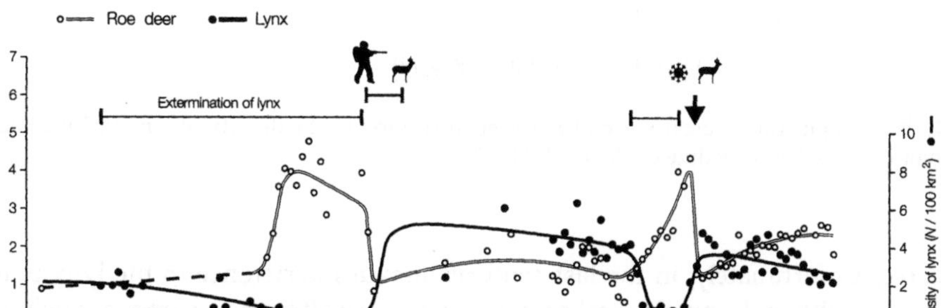

Fig. 5.11. Population dynamics of roe deer and lynx in the whole BPF in 1860–1993. Periods of lynx extermination marked by *thin lines*. Years of over exploitation of roe deer by poachers and soldiers during World War I are marked by a *thick line*. *Arrow* and *snowflake* denote decline of roe deer numbers caused by the severe winter of 1969/1970. Points are empirical data on roe deer and lynx densities, lines are smoothed population dynamics. Jędrzejewska et al. (1997)

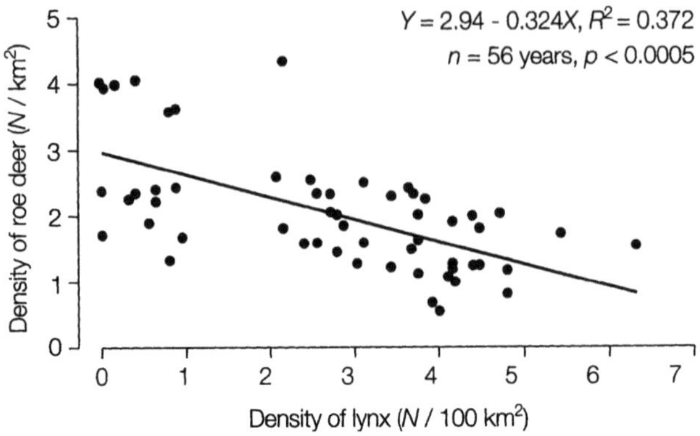

Fig. 5.12. Density of roe deer in relation to lynx density in BPF in 1896–1993. Each point denotes 1 year. Jędrzejewska et al. (1997)

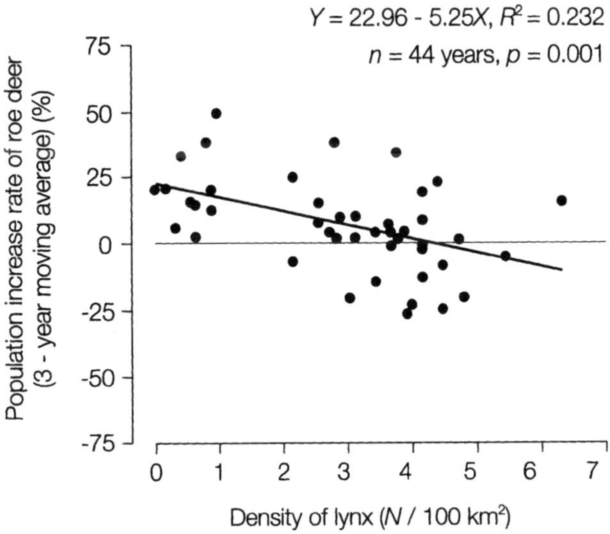

Fig. 5.13. Population increase rate of roe deer in relation to the density of lynx in BPF. Each point denotes 1 year. Jędrzejewska et al. (1997)

scarce. Unfortunately, in the last century, human interference in the lynx population (although much weaker than that in wolf) was strong enough to make the numerical relationship between the lynx and its main prey negative (Fig. 5.14, middle panel). Consequently, the index of predation rate (calculated as a product of lynx numerical relationship and functional response to changes in roe deer densities) was greatest at very low and moderate abundances of roe deer (0.5–1.5 ind/km^2) and declined quickly with increasing abundance of deer (Fig. 5.14, right panel). We emphasize that in this series of data, low predation rates at high densities of deer was not least because of

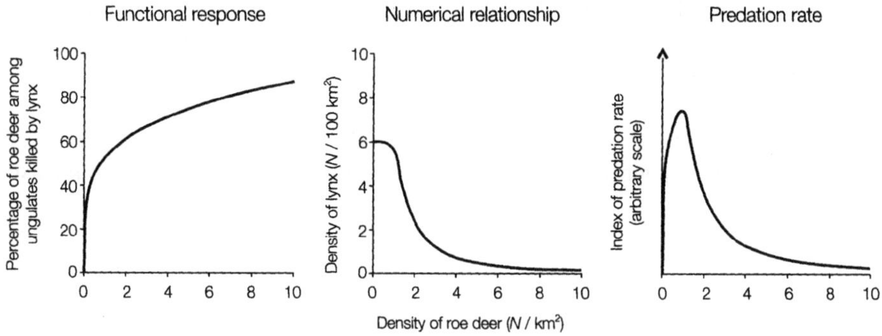

Fig. 5.14. *Left panel* curve of lynx functional (dietary) response to varying density of roe deer (derived from empirical study, see Fig. 4.4). *Middle panel* curve of numerical relationship between lynx and roe deer numbers in 1896–1993 (cf. Figs. 5.11 and 5.12). *Right panel* curve of total predation response of lynx, calculated as a product of functional and numerical responses, to changes in roe deer densities in BPF. This curve does not show real values of percentage predation (note the lack of numerical *Y*-scale) but it correctly shows the shape of predation curve

human exploitation of the lynx population. However, the fact that the lynx was able to maintain high densities in years of low densities of roe deer indicates that, in an undisturbed predator-prey relationship, predation by lynxes on a low density population of roe deer would be rather severe. However, if roe deer densities are very low, percentage predation by lynxes declines steeply (Fig. 5.14).

The 100-year long data on population dynamics of moose and wild boar as well as the 200-year series on European bison numbers are very weakly, if at all, related to densities of large predators. Only for moose and wolf numbers was the negative relationship statistically significant ($R^2=0.102$, $p=0.01$, $n=62$ years; Jędrzejewska et al. 1997).

5.4.3 Relative Roles of Predation and Other Factors in Shaping the Densities and Population Dynamics of Ungulates

In his inspiring paper on population regulation, Holmes (1995) stressed that populations are affected by interactions of many factors (and not "weather *or* predation *or* food *or* disease"). He also acknowledges the trickiness of the real world where "conditions are continually changing, so that the 'equilibrium' tracked by density-dependent mechanisms constitutes a moving target" (Holmes 1995, p. 11). The remark about the interactive influence of many factors applies to Białowieża Forest's ungulates, the more so as their populations are affected by natural as well as man-related factors. In the decade of 1984–1994, game wardens registered the carcasses of ungulates in the exploited forests of BPF. In 1986–1993, we did that in the pristine forests of Białowieża National Park. Based on that material (a total of nearly 2500 carcasses), Okarma et al. (1995) analysed the relative roles of natural causes of mortality in the five species of ungulates (Fig. 5.15). In 1984–1993, both species of native predators occurred in the forest and winter conditions varied from very mild to severe. Predation played the most important role in roe deer deaths (86% of the total mortality from natural causes, mainly by lynx) and red deer deaths (82% of mortality, mainly by wolves). Only a quarter of the natural mortality of moose

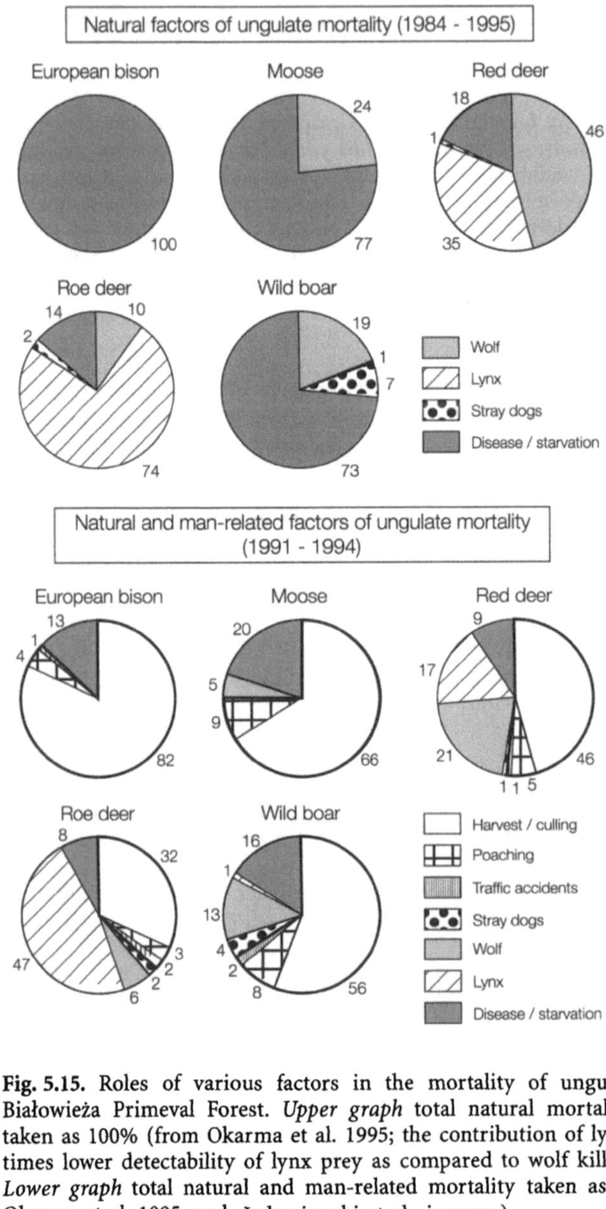

Fig. 5.15. Roles of various factors in the mortality of ungulates in the Polish part of Białowieża Primeval Forest. *Upper graph* total natural mortality (mean yearly estimates) taken as 100% (from Okarma et al. 1995; the contribution of lynx predation corrected for 6 times lower detectability of lynx prey as compared to wolf kills, W. Jędrzejewski, unpubl.). *Lower graph* total natural and man-related mortality taken as 100% (based on data from Okarma et al. 1995, and Jędrzejewski et al., in press)

and wild boar, and none in bison, was due to predation (Fig. 5.15). In contrast to the data on hunting harvests, the numbers of carcasses found are only small samples of the absolute (unknown) numbers. Thus they were used as relative indices. In 1991–1994, when both the indices of various mortality factors and absolute numbers of ungulates killed by wolves, lynxes, and hunters were available, we assessed the roles of both natural and man-related factors in the total mortality of ungulates (Fig. 5.15). During this period, the hunting harvest (or culling in the case of bison) appeared as the primary cause of deaths in all species except for roe deer. But, as already mentioned, the years of 1991–1994 were characterised by an elevated hunting harvest, compared with earlier years.

The hunting harvest as well as natural mortality of ungulates varied from year to year. Suffice it to say that when recording ungulate carcasses in the winter of 1969/1970, the harshest on the record, game wardens found 2452 roe deer, red deer and wild boar that had died from cold, starvation and disease (Okarma et al. 1995). In the series of 'normal' winters (1984/1985–1993/1994), deaths from starvation and/or disease in bison, moose, roe deer, and red deer showed no correlation with snow depth. In wild boar, however, mortality from starvation/disease was principally influenced by the acorn crop of the previous year, and snow cover. Increase in the values of both variables caused an increase in the number of wild boar carcasses found (Okarma et al. 1995). The mechanisms of these relationships have already been mentioned in Section 2.3: oak produces heavy seed crops at 6 to 9-year intervals (most recently in 1982, 1989 and 1996). After the fall of mast, wild boar survive well, and next spring, they breed early and intensively (Fig. 5.16). However, the nil acorn crop that follows the mast year causes very high mortality in both adult and young boar. Snow cover, if deep enough, and frozen soil make it difficult for wild boar to root and get food. During the periods with

Fig. 5.16. A herd of wild boar with several small piglets in spring. Animals of high reproductive potential, wild boar quickly respond numerically to heavy crops of acorns. Photo: S. Wąsik

deep snow cover and nil or poor acorn crop, boars starve, are victims of disease, and die en masse (Fig. 5.17; cf. Caboń 1958).

If 10 years was time enough to determine the major factors of ungulate deaths and their average roles in each species' mortality, then 100-year data allowed a close look at how those factors work. By multiple regression analysis, we determined which environmental and intrapopulation factors explained the variation in numbers and rate of population increase of ungulates. Furthermore, for all independent variables, we calculated semipartial correlations squared (sr^2) which show how much a given factor contributed to the total variation explained.

In European bison, the mean annual temperature and political instability in the region appeared the most important factors shaping densities between 1802–1993 (Table 5.6). However, they explained only 9% of the total variation in bison numbers, thus the factors underlying the major part of that variation were not revealed. Over most of the 19th century and until 1915, the prime goal of game management of BPF was to increase the number of bison by various protection measures. The same approach was pursued in the Polish and Belarussian parts of BPF after 1952, during the early decades of the restitution of a free-living population. After 1980, the numbers of European bison inhabiting the Polish part of BPF have been subject to more intense culling in order to stabilise them at an administratively acceptable level of about 250 head (Krasiński et al. 1994). Thus, during the last two centuries, the level of bison density in BPF has been determined by humans in one way or another.

The population increase rate of European bison has been largely explained by political stability, density of bison, and total biomass of other wild ungulates (Table 5.7). In other words, changes in the numbers of bison from year to year have been predominantly shaped by exploitation by humans (that, in

Fig. 5.17. Wild boar suffer mass mortality after poor crops of acorns and during severe winters. A lair with four dead wild boar; April 1987, Białowieża National Park. Photo: K. Zub

Table 5.6. Summary of multiple regression analysis of the impact of various factors on the numbers of ungulates in BPF. Series of data cover years 1802-1993 in bison, and 1860, 1890–1993 in other ungulates. For wild boar, two analyses were conducted: for 1860, 1890–1993 (all material on boar numbers but no full data on snow cover and acorn crop), and for 1960–1993 (information on acorn crop and snow cover available). Semipartial correlation squared (sr^2, after Tabachnick and Fidell 1983) expresses the unique contribution of a given factor (independent variable) as a proportion of the total variance of the ungulate numbers (dependent variable). Thus, the importance of each factor in shaping the numbers of ungulates decreases with lower sr^2 values. In each species of ungulate, the most important factor is in boldface type. (Jędrzejewska et al. 1997)

Parameter	Bison	Moose	Red deer	Roe deer	Wild boar	
Number of years	135	62	56	56	57	24
Percent of variation explained (R^2)	9%	27%	20%	40%	34%	27%
Significance (p)	0.003	<0.0005	0.009	<0.0005	<0.0005	0.039
Contributions by various factors to the total variation explained (sr^2):						
Mean annual temperature (5-year moving average)	(+) **7.5%**	(+) **16.8%**	(+) 6.5%	(+) 1.9%	(+) **25.8%**	(+) **20.2%**
Wolf density		(–) 9.6%	(–) **9.7%**		(–) 6.5%	
Lynx density				(–) **36.4%**		
Acorn crop (in year n-2)						(+) 11.7%
Political instability	(–) 1.5%	(–) 0.5%	(–) 1.9%	(–) 1.7%	(–) 0.8%	

(+) Positive correlation; (–) negative correlation.

years of political chaos, went out of control) and to a lesser extent by intra- and interspecific competition (Jędrzejewska et al. 1997).

Between 1798–1993, 27% of variation in moose densities was largely explained by temperature and wolf densities (Table 5.6). The rate of annual increase in moose numbers was influenced by the biomass of all other wild ungulates and moose density (Table 5.7). The role of interspecific competition was most important. Variation in wolf densities had no discernible impact on moose population growth.

Wolf density was the strongest determinant of red deer numbers (Table 5.6). Other significantly contributing factors were mean annual temperature (positive effect) and political instability (negative effect). Variation in lynx numbers had no effect on red deer densities. Four factors explained a total of 60% variation in the population increase rate of red deer in the entire series of data (Table 5.7). Most important were intraspecific competition and the combined density of wolves and lynxes, both exerting a negative effect on deer population growth. Next in importance, the mean annual temperature had a notable positive influence. Political instability was least important since it acted severely during some years only.

Lynx numbers appeared to be the essential factor shaping the densities of roe deer (Table 5.6). Variation in the mean annual temperature (positive ef-

Table 5.7. Summary of multiple regression analysis of the impact of various factors on the population increase rate (3-year moving average in all series of data except for the 23-year series in wild boar) of ungulates in BPF (1890–1993). For bison two analyses were done: for 1802–1993 (all material on bison numbers but no full data on the biomass of other ungulates) and 1890–1993 (information on numbers of other ungulates available). Also, for wild boar two analyses were conducted: for 1890–1993 and for 1960–1993 (see explanation in Table 5.6). (Jędrzejewska et al. 1997)

Parameter	Bison		Moose	Red deer	Roe deer	Wild boar	
Number of years	97	49	52	40	44	44	23
Percent of variation explained (R^2)	42%	77%	34%	60%	45%	33%	75%
Significance (p)	<0.0005	<0.0005	<0.0005	<0.0005	<0.0005	0.003	<0.0005
Contributions by various factors to the total variation explained (sr^2):							
Political instability	(–) **35.7%**	(–) **61.9%**	(–) 0.5%	(–) 2.2%	(–) 5.0%	(–) 2.4%	
Density of own species	(–) 4.2%	(–) 11.4%	(–) 7.6%	(–) **29.0%**	(–) 11.2%	(–) **22.2%**	(–) **13.2%**
Biomass of all other wild ungulates		(–) 2.8%	(–) **12.9%**				
Mean annual temperature (5-year moving average)				(+) 14.6%	(+) 7.8%	(+) **19.9%**	
Acorn crop in the preceding year							(+) **27.2%**
Combined density of wolf and lynx				(–) **27.5%**			
Lynx density					(–) **32.6%**		
Wolf density						(–) 4.7%	
Mean daily snow cover in November–March							(–) 2.7%

(+) Positive correlation; (–) negative correlation.
See Table 5.6 for explanation of statistical denotations.

fect) and political instability (negative effect) added very little to the total variation explained. Similarly, the population increase rate of roe deer was shaped predominantly by lynx density. Secondary factors were intraspecific competition, mean annual temperature, and political instability (Table 5.7). In total, various factors explained 45% of the observed variation in roe deer population growth from year to year.

The numbers of wild boar were primarily shaped by mean annual temperature; this is evident in the whole series of data on boar numbers (1860, 1890–1993) as well as in the more comprehensive material from recent decades (1960–1993) (Table 5.6). Other factors were the acorn crop of the preceding autumn–winter season (positive effect) and density of wolves (negative effect). The latter factor only showed a discernible influence on wild boar numbers in the long series of data.

Multiple regression analysis conducted for the whole series of data on wild boar numbers indicated that the rate of population increase of wild boar was shaped by the mean annual temperature, wild boar density, wolf numbers, and political instability. These factors explained 33% of the total variation (Table 5.7). In the shorter series (1960–1993), data on acorn crops and snow cover were available, and these two factors together with boar density, explained 75% of the total variation in the population increase rate of wild boar. The acorn crop was definitely the single most important determinant of wild boar increase rate from year to year.

The five species of ungulates inhabiting Europe's temperate lowland forests differ in the factors which shape their densities and population dynamics. We think that there are three essential features underlying these differences: food resources utilised, susceptibility to predation, and intrinsic rate of reproduction. Ungulates of the temperate forests feed on two main resources: (1) browse and ground vegetation, and (2) acorn crop. Browse and ground vegetation exhibit no rapid year-to-year variation in abundance, hence they are predictable to consumers. Furthermore, the relationship between browse and ground vegetation and ungulates is interactive (see Caughley and Sinclair 1994), i.e. ungulate dynamics reacts to the level of these resources, but also the rate of growth of browse and ground vegetation responds to the number of animals using them. This mutual relationship leads to some kind of equilibrium density in herbivores. In contrast, oak seed crops are extremely variable from year to year, with masts occuring at 6–9-year intervals and being followed by 1 or 2 barren years. Consumers of acorns have no influence on the rate of their renewal (reactive relationship), and hence do not influence the amount of food available to subsequent generations. These two kinds of resource lie at the base of the different population dynamics of ungulates in Białowieża Forest. Most variable from year to year, numbers of wild boar followed the variation in acorn crop, whereas all other species, feeding on more stable resources, exhibited more stable population dynamics.

Density dependent decline of ungulate population growth is assumed to reflect food competition (cf. Messier 1991). Also, we think that the positive influence of temperature, noticeable in all ungulates, acted predominantly through improving the food supply and feeding conditions. In 1948–1993, annual periods of vegetative growth in BPF (i.e. number of days with mean daily temperature >5°C) lasted for 207.6 days, on average (SD 10.8), and varied from 188 days in 1956 to 237 days in 1990. Longer vegetative periods and a shorter duration of snow cover directly translate to better feeding conditions for all ungulates, particularly those which obtain part or all of their food from the ground layer. Moreover, variation in annual temperature causes long-term changes in food supply. The multiannual pattern of acorn crops is shaped by temperature, with mast years occurring at shorter intervals during decades of warming (Pucek et al. 1993). Most likely long-term trends in regeneration of deciduous trees and shrubs, including species of ungulates' preferred browse, depend on climatic trends. This phenomenon, noted by forestry inventories and observations, needs thorough investigation. Furthermore, a warming climate may also mean higher productivity of the whole forest biome. The direct impact of adverse weather conditions (especially severe winters with deep snow) on ungulate survival was also noted, but even then it acts together with forage deprivation (when snow is deep) and/or deterioration (during severe frost).

The major natural factors shaping bison and moose populations were mean annual temperature and competition. Thus, food was the prime natural factor limiting population growth and densities of bison and moose in the temperate forests. These large ungulates have very high daily demands for food: 19–32 kg fresh weight in bison (Gębczyńska and Krasińska 1972) and 16–20 kg in moose (Morow 1976). Interspecific (with other ungulates) and intraspecific competition for food, acting through the density-dependent decline of population increase, is the most reliable regulatory mechanism in bison and moose. The density at which the population increase rate declined to zero may indicate the carrying capacity of Białowieża Forest for a given species of ungulate. If this is so, the densities of bison determined by the food carrying capacity would be about 0.5 ind/km^2 (see Fig. 2.13 in Sect. 2.3). Maximum recorded densities of bison were about 1.5 ind/km^2 in the absence of red deer (the 19th century), and 0.4 ind/km^2 when red deer was present and when deliberate culling was conducted (the 20th century). The carrying capacity of BPF for moose, as estimated from the regression equation (population increase rate versus density) would be about 0.3 ind/km^2 (see Fig 2.13). The highest recorded densities of moose were 0.6 ind/km^2 (with no red deer in the community), and 0.2 ind/km^2 (with red deer present). Interestingly, after attaining densities near or above the carrying capacity of habitat, bison and moose populations began to decline.

Predation does not limit the densities of bison and only weakly those of moose. This results from the fact that in a multispecies community, bison and moose are ungulates of the largest body size. As such they form a negligible contribution to predators' diets. Exploitation by humans affected their population growth and limited densities. Bison especially with the slowest reproduction rate among native wild ungulates was seriously affected by humans, both by deliberate culling or management and by uncontrolled exploitation in years of political instability.

Red deer and roe deer are animals whose densities and population increase rates are strongly negatively correlated with densities of predators. The cause-effect interpretation of those correlations is corroborated by the fact that the long-term data represent a series of unintended experiments. First, each extermination or substantial reduction of predators was followed by an increase in prey numbers. More importantly, an 'experiment' long postulated, but so far never reported to have been done (Skogland 1991; Boutin 1992; Van Ballenberghe and Ballard 1994), took place in BPF: reduction of prey numbers in a period of high numbers of predators. In 1915–1920, when wolves and lynxes quickly recolonised the forests of Białowieża, roe deer and red deer numbers had already been reduced due to uncontrolled exploitation by humans. Both wolf and lynx kept the populations of their main prey (red deer and roe deer, respectively) at low densities for 30–40 years, i.e. till the next temporary extermination of predators. In the roe deer-lynx system, the same 'experiment' was 'replicated' when the severe winter of 1969/1970 reduced the numbers of roe deer by 70%. Again, high numbers of lynxes hampered the recovery of the roe deer population so much that 25 years past that catastrophic winter decline, roe deer density was still only half of that recorded in 1969. Results of these unintended human-caused and natural 'experiments' are strong arguments for the limiting role of predation on red and roe deer. Despite being nonregulatory sensu Sinclair (1989), predation by wolves and lynxes effectively kept deer populations at low densities for decades.

The relationship between the increase rate (year-to-year changes) and the population density of red deer was examined separately for the years with wolves and lynxes present in BPF and those when the two predators were exterminated (Fig. 5.18). When wolves and lynxes occurred in the forest, red deer population increase was very low or negative. In years when both predators were extirpated (1897–1914 and 1960–1969), the population increase rate of red deer was markedly higher. The regression equations let us predict that, regardless of the predators' status, red deer population growth will be halted at densities of about 4–5 ind/km^2. Thus, densities of red deer attained around 1914, 1967–1969, and in 1975–1985 approached the level set by food limitation.

Density-dependent decline in the rate of population growth (year-to-year changes) was manifest in the whole series of data on roe deer. However, in years when large predators occurred in the forest, roe deer population increase was much lower than in years when both wolves and lynxes were exterminated from BPF (1897–1914 and 1960–1969) (Fig. 5.18). The regression equations allowed us to predict that, when not limited by predation, roe deer population growth would be halted by competition for food at mean densities of about 4 ind/km^2. Thus, densities of roe deer attained in 1910–1914 and 1967–1969 were near the upper level set by food availability. With predators present, roe deer population stabilised at a mean density of 2 ind/km^2.

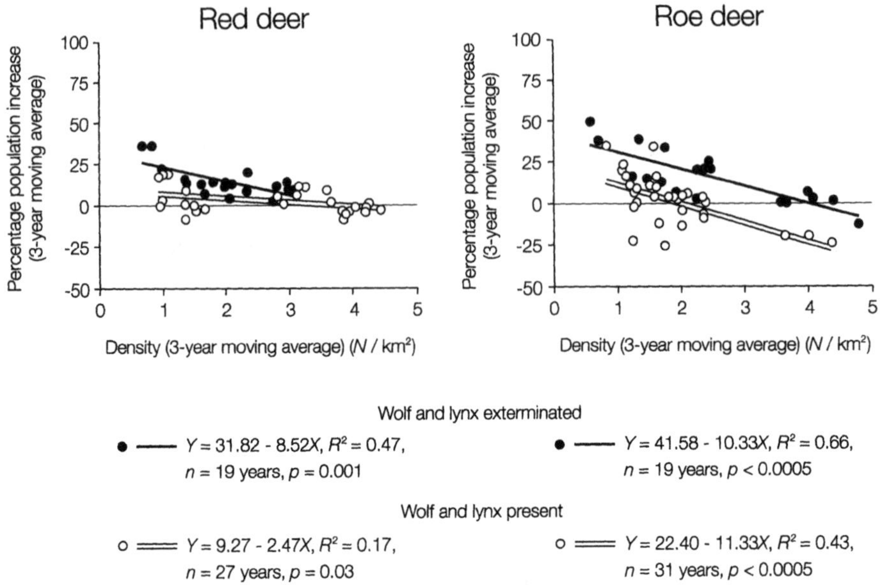

Fig. 5.18. Rates of annual population increase of red deer and roe deer in relation to their densities (3-year moving average of increase rate and density) in years when the two large predators were exterminated (1897–1914 and 1960–1969) and when they occurred in Białowieża Primeval Forest

Based on the two correlations between deer density and population increase rate (as differentiated by predators' presence, see Fig. 5.18), and the shape of the total predation response to changes in deer density (see Figs. 5.10 and 5.14), we reconstructed the plausible mechanisms of wolf and lynx impact on red and roe deer and showed it in the following graphical model (Fig. 5.19). In the absence of large predators, percentage population increase of deer from year to year declined linearly with growing densities (Fig. 5.19, top panel). Predation rates varied with deer densities in a non-linear way according to Figs. 5.10 and 5.14. For the functioning of prey populations, the points of intersection of the two functions: predation curve and population increase line were critical. At these points (which can be read from Fig. 5.18), the whole annual increase in deer populations is harvested by predators. In the case of red deer, the predation curve and population increase lines intersect at $y=0$, and predators only slowed down the rate of prey growth (Fig. 5.19, middle panel). Even when hunted by wolves and lynxes red deer increased but reached the level set by food availability after a much longer time than they did when free of predation. Admittedly, such a modest role for predation was largely due to heavy and prolonged persecution of wolves (cf. Fig. 5.7). Another important feature of predator impact on red deer was a rapid decline in the predation rate at very low densities of deer (<1 ind/km^2). This resulted from wolves' dietary response (shifting to buffer prey), and shows that predators would not exterminate their prey. The absolute year-to-year increase in deer numbers in relation to density (Fig. 5.19, bottom panel) was much lower with predators present than with no predators in the forest.

In roe deer, the curve of predation rate intersected the line of inversely density-dependent population growth at a point $y>0$ (Fig. 5.19, top panel).

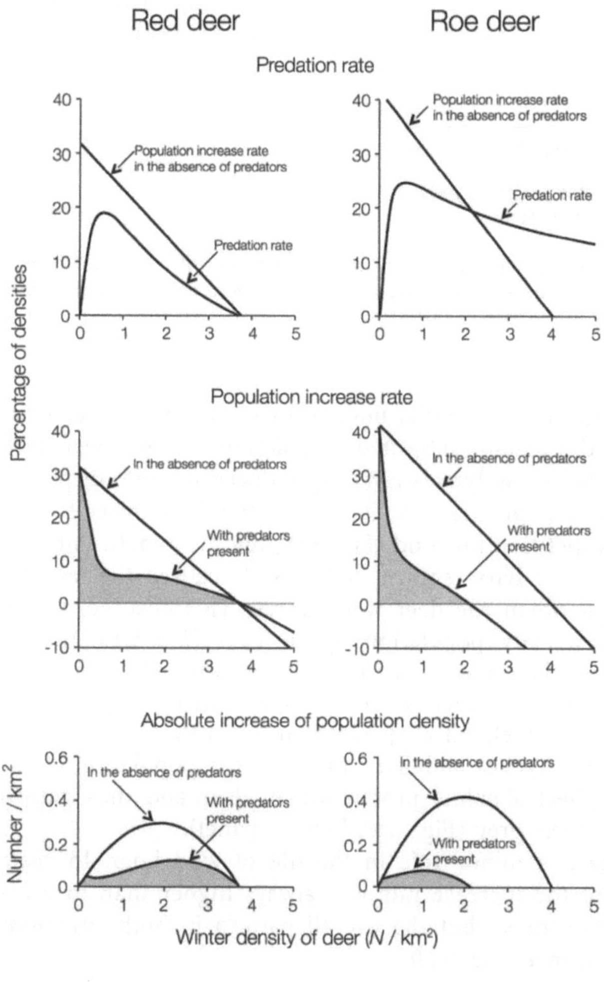

Fig. 5.19. Graphical models of mechanisms of predation impact on red deer and roe deer populations in Białowieża Primeval Forest in 1890–1993. *Upper panel* presumed predation rates (shapes of curves from Figs. 5.10 and 5.14) and potential increase rates of deer populations from year to year (from Fig. 5.18) in relation to density of deer. *Middle panel* potential population increase rates of deer (observed in years with no predators) and increase rates of deer populations harvested by predators, both in relation to deer density. *Shaded* is the area and magnitude of deer population growth when harvested by wolves and lynxes. *Bottom panel* absolute measures of year to year increase of deer numbers at a given density in the absence of large predators and in years when predators were present. *Shaded* is the increase of deer when exploited by wolves and lynxes. See text for further explanations

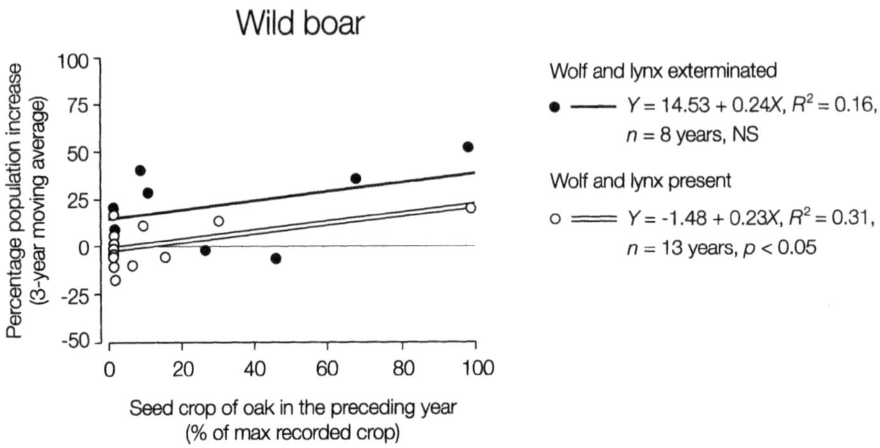

Fig. 5.20. Rates of annual population increase (3-year moving average) of wild boar in relation to the crop of oak seeds in years when wolves and lynxes were exterminated (1960–1969) and when they occurred in the forest (1970–1993)

This may be deduced from nearly parallel lines of increase rates versus density in predator-free and predator-subjected population of roe deer (see Fig. 5.18). Therefore, predation by lynxes not only reduced the rate of potential increase in the roe deer population but it also limited the density to well below the level set by competition for food (Fig. 5.19, middle panel). Furthermore, the dietary response of lynx (shown in Fig. 5.14) caused the fast decline in the predation rate when roe deer were scarce. This ensured a high increase rate in the low density population of roe deer (Fig. 5.19, middle panel). Since the lynx population was decidedly less disturbed by humans than wolves, we may accept this model of roe deer-lynx functioning as being close to the natural system which once operated in Europe's temperate forests. It is worth noting that lynxes stabilised the roe deer population at the density of potentially highest absolute production of deer and thus gained the maximum yield from their prey (Fig. 5.19, bottom panel).

Finally, predation played a minor role in the life of wild boar. In years with no predators present the increase rate was always higher than in years when they occurred in the forest, but the overall pattern in both situations was shaped by food abundance (Fig. 5.20).

5.5 Predation on Rodents

5.5.1 Species Structure of Total Predation on Rodents

Rodents appear to be the mainstay of predator-prey relationships in temperate forests, especially when small and medium-sized carnivores and birds of prey are considered. Although all species of predators capture rodents now and then, five species monopolised the consumption of rodents: tawny owls (45%), pine marten, red fox, common buzzard, and weasel. Together, they consumed 94% of the whole rodent biomass eaten by predators (Table 5.8).

There are some differences in the species structure of predation on forest rodents, compared with rodents of open marshes in the river valleys (Table 5.9). In both spring–summer and autumn–winter, over 90% of all forest rodents taken by predators fell as prey to resident predators. There is not a single migratory raptor that comes to breed in the temperate forests which seriously relies on forest rodents as food. Furthermore, although nearly all carnivores and birds of prey now and then capture forest rodents, 84% of total predation is exerted by three species only, the pine marten, the weasel, and above all the tawny owl, which takes every second forest rodent captured by predators (Table 5.9). To red foxes and buzzards, which contribute about 5% each to the total predation, forest rodents are a secondary (if not tertiary) prey resource.

In Białowieża Forest as a whole (and not only in open marshes, voles' prime habitat), *Microtus* voles were hunted principally by red foxes, tawny owls and buzzards, the last being particularly important in autumn–winter (Table 5.9). The large shares by tawny owls and pine martens (typical forest predators) in the total predation impact on *Microtus* stem from the fact that the fairly narrow, open valleys of forest rivers are penetrated by forest preda-

Table 5.8. Percentage contributions by various predators to the total consumption of rodent biomass in Białowieża Primeval Forest. Spring–summer is 16 April–30 September, Autumn–winter is 1 October–15 April. Dominating species are in boldface type

Predator species	Spring–summer	Autumn–winter	Whole year
Tawny owl	**40.3**	**49.5**	**45.0**
Pine marten	**16.7**	**11.9**	**14.2**
Red fox	**13.8**	**12.2**	**13.0**
Common buzzard	7.2	**15.3**	**11.4**
Weasel	**13.2**	7.7	**10.4**
Polecat	3.0	0.4	1.7
Stoat	1.4	1.1	1.2
Raccoon dog	1.0	0.5	0.7
Pygmy owl	0.8	0.5	0.7
Long-eared owl	0.8	0.1	0.4
American mink	0.2	0.4	0.3
Tengmalm's owl	0.2	0.2	0.2
Badger	0.2	0.1	0.1
Sparrowhawk	0.1	+	+
Goshawk	0.1	+	+
Other species total	+	+	+
Total rodent biomass (g/ha)	1411	1492	2903

Table 5.9. Percentage contributions of various predators to the total predation (in terms of biomass and numbers of rodents taken by predators) of forest rodents (combined numbers of bank voles, yellow-necked mice and pine vole) and open marsh rodents (mainly *Microtus* spp.). Mean values for BPF in 1985–1995

Species of predator	Forest rodents			Open marsh rodents		
	Spring–summer	Autumn–winter	Whole year	Spring–summer	Autumn–winter	Whole year
Tawny owl	**42.7**	**61.0**	**51.6**	**29.5**	**25.6**	**27.3**
Pine marten	**19.8**	**17.6**	**18.7**	13.3	5.7	9.0
Weasel	**16.1**	**10.5**	**13.4**	2.9	1.4	2.0
Red fox	5.7	5.7	5.7	**32.3**	**24.0**	**27.6**
Common buzzard	**8.2**	1.4	4.9	8.4	**38.0**	**25.2**
Other predators total	7.5	3.8	5.7	13.6	5.3	8.9
Total biomass consumed per hectare (g)	918	880	1798	430	561	992
Total number of rodents consumed per hectare	41.4	39.5	80.9	16.4	23.6	40.0

tors, and that *Microtus* voles venture to the forests, too. If only the open marshes had been considered, the shares of foxes, buzzards, weasels and stoats (and possibly long-eared owls and harriers) in the total amount of predation would have been larger.

5.5.2 Magnitude and Mechanisms of Predation on Forest Rodents

As might be expected, species of rodents dominating by numbers in the living community are also those paying the heaviest tolls to predators: the bank vole and the yellow-necked mouse (61% of total numbers of rodents taken by predators), and *Microtus* voles (31%). Pine voles constituted a mere 4% of all rodents preyed upon (Table 5.10). The numbers shown in Table 5.10 are average figures for the Białowieża Primeval Forest as a whole, in the fairly broad time frame of 1985/1986–1995. The estimate of predation on forest rodents was compared with their densities averaged for over 20 years (Fig. 5.21). If we assume that the observed population increase from spring until autumn (64.5 ind/ha) plus the number taken by predators during that season (42 ind/ha) equal the total yearly production of rodents, then predators annually harvest about 80% of rodent production: half of that amount being taken in spring–summer and another half in autumn–winter, when rodents usually do not breed. In the cold season, predators are the most important cause of rodent mortality; they take 62% of all rodents that disappear between autumn and spring (Fig. 5.21). Data on densities of *Microtus* spp. in Białowieża Primeval Forest are not yet available, so the mean magnitude of predation on

Table 5.10. Total predation on rodents (expressed as numbers of rodents taken from an average hectare by all predators) in Białowieża Primeval Forest (data pooled for 1985–1995)

Rodent species	Spring–summer		Autumn–winter		Whole year	
	N/ha	%	*N*/ha	%	*N*/ha	%
Bank vole	23.4	38	22.7	36	46.1	37
Yellow-necked mouse	15.2	25	14.9	23	30.1	24
Microtus voles total	16.4	27	21.9	34	38.3	31
Pine vole	2.8	5	1.9	3	4.7	4
Other rodents total	3.1	5	2.7	4	5.8	4
Forest rodents total	41.4	68	39.6	62	81.0	65
Open marsh rodents	19.5	32	24.5	38	44.0	35
Small rodents total	60.9	100	64.1	100	125.0	100

them (averaged for the whole forest and not only for open marshes in river valleys) cannot be compared with population densities.

The pattern of predation on forest rodents (bank vole and yellow-necked mouse) was analysed based on the data from the pristine deciduous and mixed forests of the Białowieża National Park from 1985/1986 to 1991/1992. Since the three species of predators (tawny owl, weasel, pine marten) are responsible for, on average, 84% of the total predation, we focused on them (Fig. 5.22). As shown in Section 2.4, the population dynamics of forest rodents is characterised by several (4 to 7) years of moderate densities followed by 2 years of outbreak and crash triggered by a heavy crop of tree seeds occurring in synchrony in oak, hornbeam and maple. In years of moderate densities of forest rodents (1985–1989) predators harvested a fairly stable number of rodents from year to year and from season to season. In autumn–winter, tawny owls took 12–29 rodents/ha, weasels 4–10 rodents/ha, and martens 2–8 rodents/ha. Their predation in warm seasons was of similar magnitude despite the fact that the numbers of predators were considerably higher then (juveniles present). However, in spring and summer, the proportion of rodents in the diets of tawny owls and martens substantially declined and gave way to seasonally abundant birds, frogs and insectivores (see Chap. 4). Thus, in the spring–summer seasons of moderate densities of forest rodents, tawny owls took 15–25 rodents/ha and weasels 8–10 rodents/ha. (No data on between-year variation in marten reproduction were available, so marten's predation could not be calculated.)

The outbreak of forest rodents that happened in 1990 after the heavy crop of tree seeds in the previous autumn evoked changes in predation which, though discernible, could hardly keep up with the rate of rodent increase. While rodent numbers in autumn 1990 were on average 6 times higher compared with those of 1985–1989, predation by tawny owls increased merely 2.4 times in spring–summer and 1.6 times in autumn–winter. The growths in weasel predation oc-

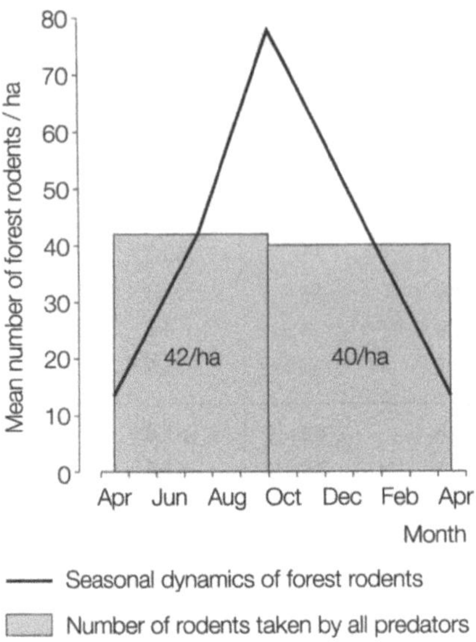

Fig. 5.21. Average predation impact on forest rodents (*N* rodents taken by all predators from 1 ha) in spring–summer (16 April–30 September) and autumn–winter (1 October–15 April) compared with their mean densities (bank voles and yellow-necked mice combined) in 1985–1992

curred in spring–summer (1.7 times) but not in autumn–winter. Predation by pine marten did not increase either (Fig. 5.22). During the consequent crash of rodent numbers in 1991, total predation on them declined, too, but not as sharply as rodents. In 1991, rodent densities were on average five times lower than in 1985–1989. Tawny owls harvested 3–5 times fewer rodents, weasels 3–4 times and martens 2 times fewer rodents than in moderate years (Fig. 5.22).

When plotted against autumn densities of rodents, predation by the three species of predators in autumn–winter appeared to grow logarithmically (tawny owl, weasel) or hardly at all (marten) with increasing rodent densities (Fig. 5.23). In tawny owls, the shape of predation response is determined by functional (dietary) response as well as numerical response (determined by highly variable breeding success and little variation in the size of the resident population). The two responses to changing prey densities were best described by logarithmic curves (see Chaps. 3 and 4), thus the total winter predation response by tawny owls was logarithmic, too (Fig. 5.23). Weasels, as rodent specialists, responded to variation in rodent numbers mainly by numerical changes, but – as in tawny owls – their positive reaction to growth of rodent numbers was restricted to the breeding season (spring–summer) only. In the winter following the summer–autumn of rodent outbreak, densities of weasels – and thus their predation effect as well – were already as low as in moderate years. Due to delayed implantation, martens responded numerically to rodent outbreak with 1-year lag. Therefore, high densities of martens coincided with the rodent crash. Despite the fact that consumption of rodents by individual martens declined markedly during the rodent crash, winter predation by pine martens was rather stable (Fig. 5.22 and 5.23). Autumn-winter predation by the three species expressed as percentage of autumn densities of rodents was

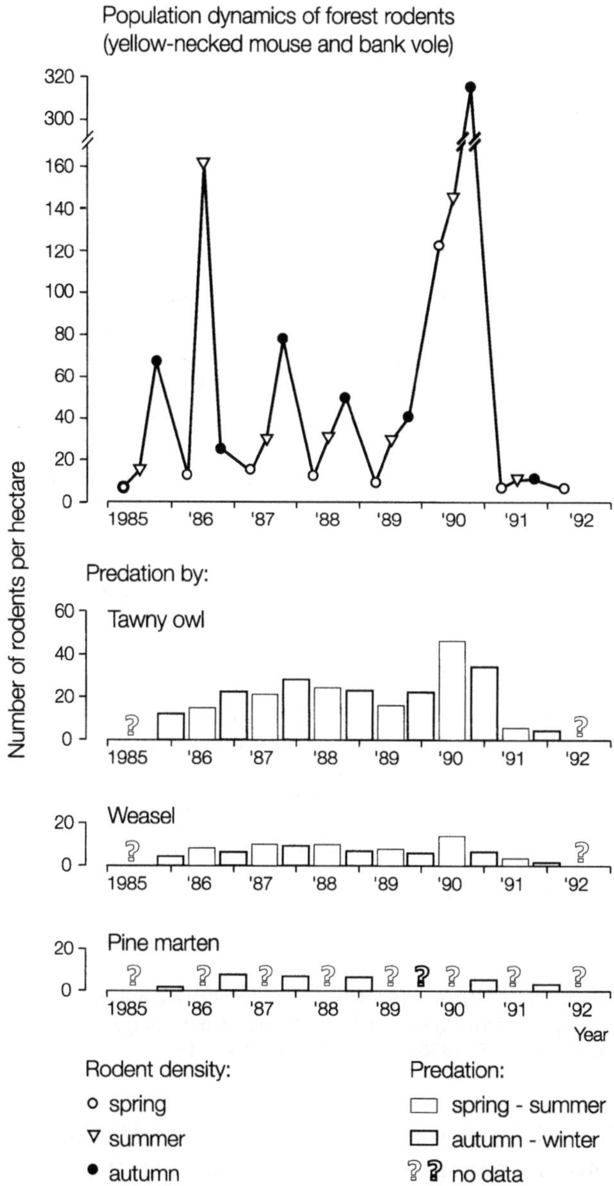

Fig. 5.22. Population dynamics of forest rodents and predation impact on rodents by the three most important predators: tawny owls, weasels and pine martens in the pristine forests of Białowieża National Park. Seasons as in Fig. 5.21

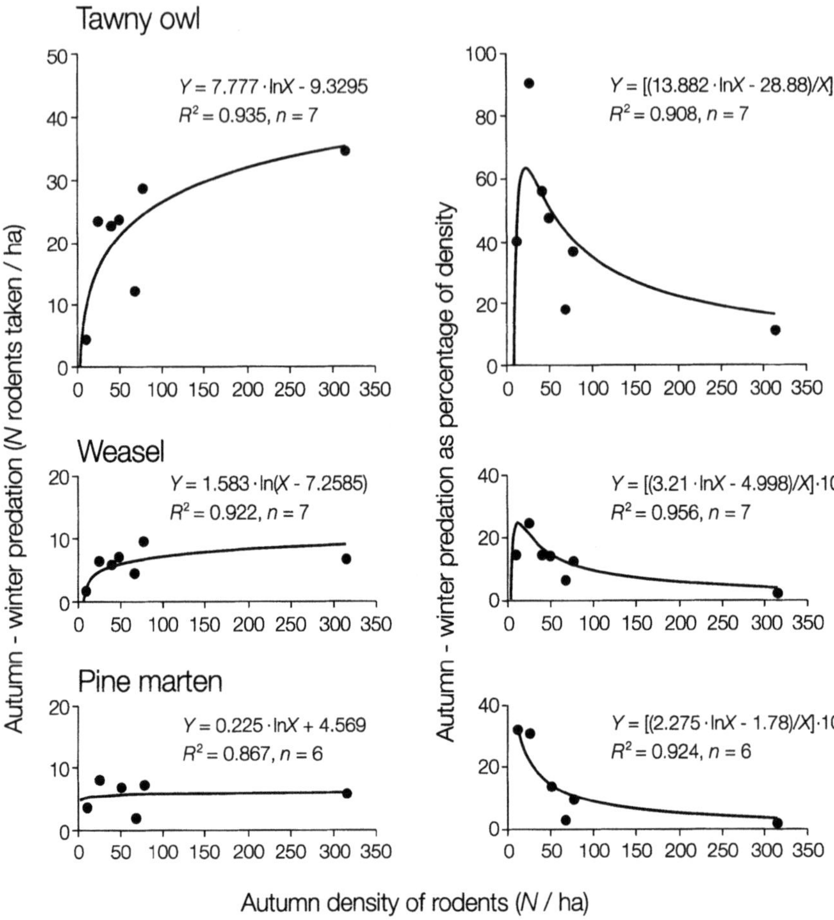

Fig. 5.23. Autumn–winter predation on forest rodents (bank voles and yellow-necked mice combined) by tawny owls, weasels and pine martens in relation to autumn densities of rodents in Białowieża National Park, in 1985/1986 – 1991/1992. *Left panels* absolute magnitude of predation (*N* rodents removed per 1 ha). *Right panels* predation as percentage of autumn densities. Sources: Jędrzejewski et al. (1995, 1996a) and Zalewski et al. (1995)

inversely density-dependent: high at low and moderate densities and unimportant when rodents irrupted in response to a heavy seed crop (Fig. 5.23).

Important though they are, fluctuations in prey densities are not the sole factor of variation in predators' numbers and predation rates. The weasel is a case in point; these animals are no less prey than they are predators. Predation on tawny owls and pine martens was also occasionally recorded (see Sect. 5.9). However, the most dangerous and unpredictable 'killer' of tawny owls is the deep and long-lasting snow cover that makes it difficult to hunt for rodents and buffer foods, such as frogs and insectivores. Tawny owl populations were reported to have been decimated after winters as severe as that of 1995/1996 (see Sect. 3.24). Also, during 'normal' winters, the depth of snow cover affected the martens', tawny owls' and foxes' predation rates on rodents, particularly on bank voles. The deeper the snow, the safer bank voles are from predation. As a result, in the snowy winters of 1985/1986 and 1986/1987, total predation on bank voles was lower than expected from rodent densities alone (Jędrzejewski and Jędrzejewska 1993).

In spring and summer, predators harvested rodents during their intense breeding season when numbers of rodents grew. Thus, predation can be compared with the production rate of rodent populations based on a simplifying assumption that the total production equals the observed increase plus the amount eaten by predators throughout the warm season. In tawny owls and weasels, spring–summer predation (number of rodents taken from an average hectare) grew logarithmically with the increasing production rate of the rodent population (Fig. 5.24). Thus, the higher the rate of rodent increase due to reproduction, the smaller the importance of predation. With such an inverse relationship, birds of prey and carnivores exerted a strong impact on summer numbers of rodents during the crash year (by taking about 80% of total production), and a negligible impact during the outbreak year, when rodents - stimulated by the extra food supply - indulged in intense breeding (Fig. 5.24). The same applies to the pre-outbreak winter of 1989/1990 when, due to the extra food supply, rodents reproduced, the more so as none of the predatory species was capable of winter breeding to keep pace with increasing rodent populations.

How do those findings pertain to the hypothesis of the essential role of predation in shaping population dynamics of rodents (mentioned in Sect. 5.1)? The hypothesis is based on the assumption of qualitatively different impact by specialist predators (weasels and stoats that are supposed to dominate among resident predators in the north) and generalist predators (foxes, buzzards, tawny owls, and pine martens that dominate at southern latitudes) (Hansson 1987; Hansson and Henttonen 1988; Hanski et al. 1991). The crucial assumptions of the predator regulation hypothesis are: (1) specialist predators have destabilizing effects on rodent numbers through continued effective hunting of the few remaining rodents after the decline, and may drive rodent cycles by a substantial time lag in their numerical response (Andersson and Erlinge 1977; Hanski et al. 1993; Norrdahl 1995); and (2) generalist predators in the presence of abundant alternative prey have a stabilising effect on rodent numbers because of an S-shaped functional response and fairly stable numbers, which in consequence are expected to produce density-dependent percentage predation on rodents (Andersson and Erlinge 1977; Erlinge et al. 1984, 1988; Hansson 1987).

In the deciduous and mixed forests of Białowieża, we found no qualitative difference between impact by generalist (tawny owls, pine martens) and specialist predators (weasels). Each of the three species exerted the heaviest impact at low and/or moderate densities of rodents. According to the predator regulation hypothesis, this is expected of specialists (weasels), only. Moreover, at very low densities of rodents, predation by weasels was small when compared with that by tawny owls and pine martens, because weasels died

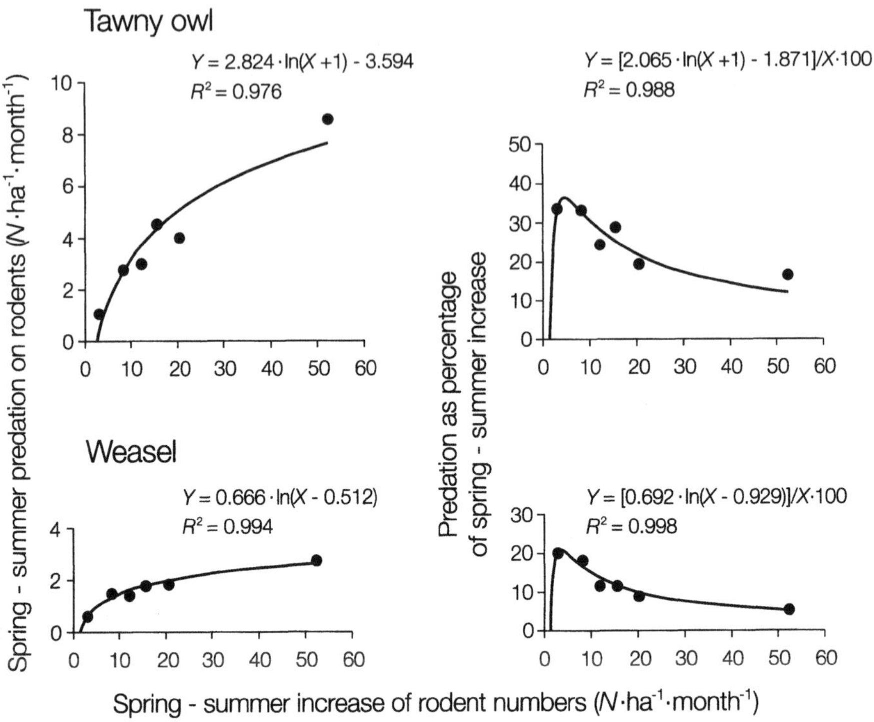

Fig. 5.24. Predation rates on forest rodents (bank voles and yellow-necked mice combined) by tawny owls and weasels in spring–summer (*N* rodents taken per ha per month) in relation to the rate of rodent population increase from spring till late summer in BNP, in 1986–1991. Calculation of summer increase rate of rodent numbers was based on a simplified assumption that total increase of rodents was the sum of the observed increase of density (recorded by trapping) and the amount taken by the three predators (owl, weasel, marten) in the same spring–summer season. In the case of pine marten (no specific data for various seasons) a mean value for six spring–summer seasons was used in each year

out or emigrated, whereas generalist predators, due to their ability to supplement diet with alternative prey, maintained fairly high numbers and continued to hunt rodents. Both the tawny owl and pine marten had functional responses described by a logarithmic function and not an S-shaped one. A 1-year lag time in numerical response has only been observed in a carnivore with delayed implantation (marten) and not in weasels. Thus, the role of generalists was quite the reverse to that proposed by the predator regulation hypothesis (Table 5.11). All predators limited rodent numbers and possibly deepened the crash (that occurred every 6–9 years) but did not regulate (stabilise) the densities.

Table 5.11. Results of the study on predation on rodents in Białowieża Primeval Forest (E Poland) compared with the predictions of the predation regulation hypothesis (PRH). (Jędrzejewski and Jędrzejewska 1996)

Parameter	Generalist predators		Specialist predators	
	Predicted by PRH	Observed in Białowieża Forest	Predicted by PRH	Observed in Białowieża Forest
Dietary response to rodent fluctuations	Shift to alternative prey, when rodents scarce; S-shaped functional response	Continued hunting, when rodents scarce; functional response of logarithmic type (–)	Continued hunting when rodents scarce	Continued hunting when rodents scarce (+)
Numbers	Stable	Fairly stable numbers of tawny owls; delayed numerical response (1-year lag) by martens (±)	Delayed numerical response to rodent fluctuations	Numerical response with no delay (–)
Predation impact (% predation)	Density-dependent, heaviest at high rodent densities	Inversely density-dependent, heavy at low numbers of rodents (heaviest at the most frequent level of rodent densities) (–)	Inversely density-dependent, heaviest at declining numbers of rodents	Inversely density-dependent, heaviest at low numbers of rodents (+)
Consequences for rodent populations	Regulation (stabilisation) of rodent numbers	Limitation of rodent numbers, deepening the crash (–)	Destabilisation, driving the cycles	Limitation of rodent numbers, deepening the crash (–)

(+) Confirmation of hypothesis; (–) results contradictory to the hypothesis.

5.5.3 Relative Roles of Predation and Other Factors in Shaping the Densities and Population Dynamics of Rodents

In this Section we endeavour to find out (1) why woodland rodents are noncyclic in most years whereas rodents inhabiting open sedge marshes in the forest exhibit regular 3–4-year cycles, and (2) what determines the levels of rodent densities. Rodents inhabiting temperate forests are small and very productive animals that forage on high quality parts of green plants and on seeds. Abundance of their food varies greatly between warm and cold seasons (forest floor vegetation) and from year to year (tree seeds). All vertebrate predators of rodents are much bigger than their prey and thus have slower reproduction rates. Glaring as they are, these facts must be a starting point for our trying to define the role of predation and other factors in determining the densities and population dynamics of rodents in Białowieża Primeval Forest.

Bank voles and yellow-necked mice reproduced in spring through early autumn and their densities attained at the end of the breeding season depended on foód abundance (Fig. 5.25 A; see also Fig. 2.40 in Chap. 2). Regression lines shown in Fig. 5.25 predict that when the seed crop fails and rodents rely predominantly on green plants as food, their autumn densities are about 25 ind/ha, only. Crops of tree seeds act as natural 'food addition experiments': with a linear increase in seed fall, densities of rodents rise to the second power.

In spring and summer of various years predators harvest a fairly stable and small number of rodents. Most likely, reproduction compensates for losses from predators. But even if we make a conservative assumption that predation impact is additive and compare its magnitude with presumed total densities of rodents in autumn, it emerges that predation plays an inferior role in shaping the post-breeding densities of forest rodents (Fig. 5.25 B). The overall pattern of density fluctuations was determined by food abundance. At their best, predators might have limited the densities; more so when food resources were poor (by over 50%) than in years of extra abundant food (by <20% of densities reached in such years). Greatest percentage predation happens at the densities that occurred most frequently, in nearly 70% of years (Fig. 5.25 C). Thus, predators maximally exploit the most frequent densities of forest rodents.

In autumn–winter, rodents did not breed and declined (except for cold seasons following heavy seed crops). The higher the autumn densities, the more of them died before spring. Percentage winter mortality of rodents was density-dependent (cf. Fig. 2.24 in Chap. 2) while percentage predation was not. Predators took fairly stable numbers of rodents in years of their moderate and low densities and notably fewer rodents during a crash (Fig. 5.26 A). Also, the consequences of the decline in predation (primarily due to decline in predator numbers) carried on to a year following a crash, vide winter 1985/1986. Predation was heaviest at 10–60 rodents/ha, i.e. the density that occurred most often, on about 60% of years (Fig. 5.26 B). Again, predators maximally exploit the most frequent densities of rodents. Compared with total winter mortality for rodents, predation appeared a crucial factor of mortality in years of low and moderate abundance of forest rodents but played a minor role (<20%) in rodent decline after an outbreak (Fig. 5.26 C).

Autumn–winter decline of rodents living in temperate forests seems an essential feature of their population dynamics, decisive for its non-cyclicity. What is the cause of that regular decline? Its prominent characteristic is density dependence: the more numerous the rodents become in autumn, the higher proportion of them disappears through winter (see Fig. 2.24). This indicates that food shortage is a critical factor underlying the autumn–winter decline. On the other hand, we said that autumn–winter predation is usually so heavy that, in all years except for outbreaks, it can be responsible for most of the rodent deaths. However, predation does not act in a density-de-

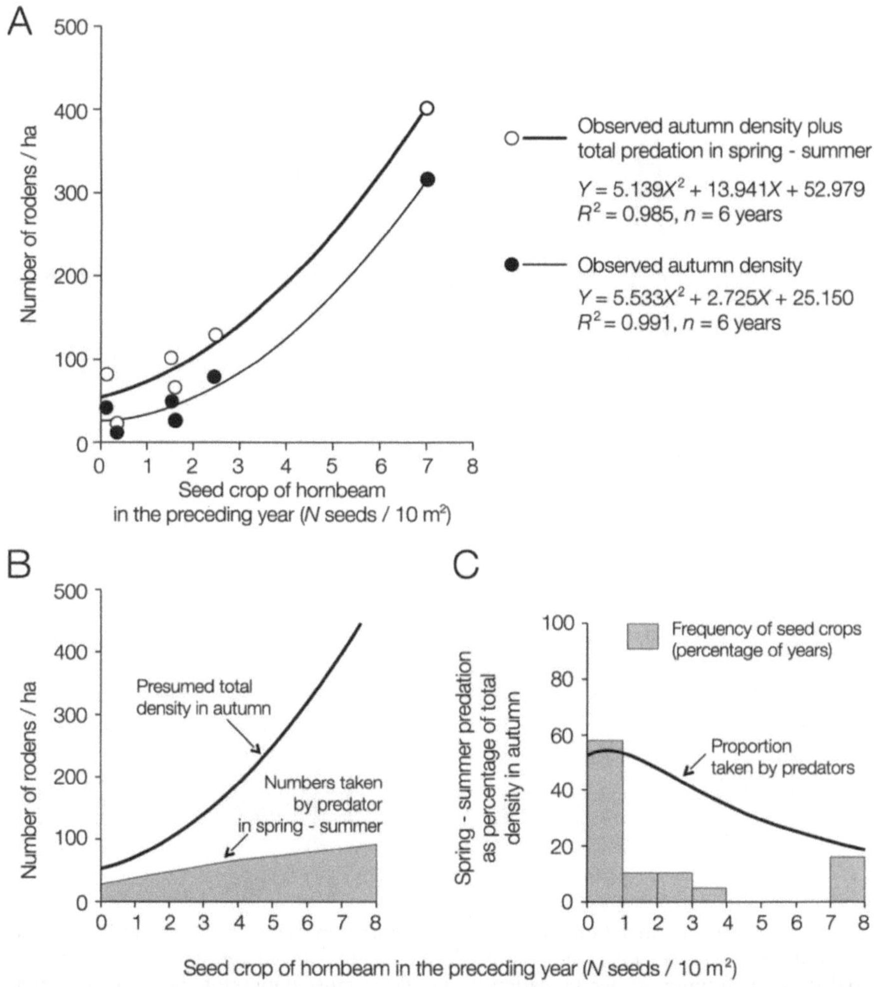

Fig. 5.25. The roles of food (seed crop of hornbeam) and predation in shaping autumn densities of forest rodents (bank voles and yellow-necked mice) in BNP in 1986–1992 (for population dynamics of rodents see Fig. 5.22). **A** Autumn densities of rodents in 1986–1991 in relation to hornbeam seed crop in the preceding year (*solid points*). Total number of rodents taken by tawny owls, weasels, and martens in spring–summer were added to autumn densities in each year (*open points*) to compare the magnitude of predation with presumed total densities. **B** and **C** Graphical model of the roles of food and predation in shaping autumn densities of forest rodents (based on empirical data from graph **A** and Fig. 5.22)

pendent way (see Fig. 5.26), so it cannot be a regulatory factor. Therefore, we propose that an ultimate cause of the winter mortality of forest rodents is shortage of food in winter. In most years, a proximate factor of rodent death is predation: predators act as executioners of rodents that, due to lack of food, have poor prospects of surviving in any case.

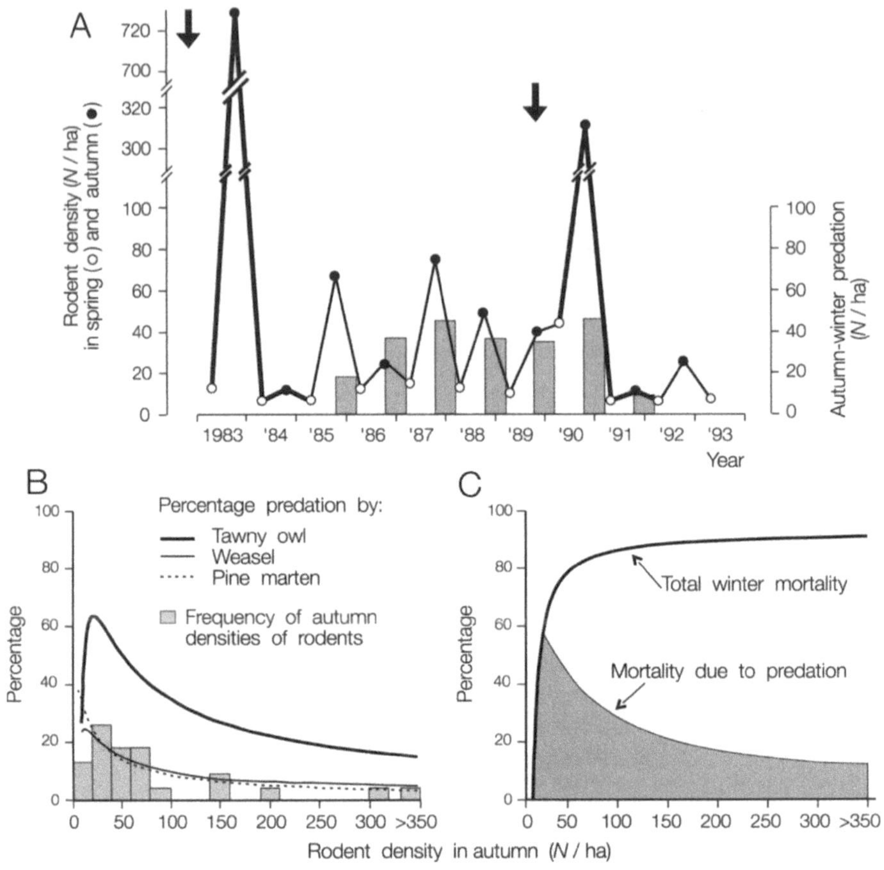

Fig. 5.26. The role of predation in the autumn–winter mortality of forest rodents (bank voles and yellow-necked mice) in pristine deciduous and mixed forests of Białowieża National Park. **A** Autumn–winter predation by the three most important species of predators in relation to rodent densities. In years of low densities, rodents still bred in autumn so the number of rodents removed by predators throughout the cold season was somewhat higher than the density of rodents recorded in September/October. *Arrows* denote heavy crops of tree seeds. **B** Percentage predation by the three predators in autumn–winter in relation to autumn densities of rodents. Predation rates by all species compared to the frequency distribution of autumn densities of rodents recorded during 23 years *bars* appeared to peak at the most frequent densities. **C** Percentage winter mortality of rodents in relation to their autumn densities and the role of predation (combined effect of the three predators) in the total winter mortality. Percentage predation *shaded area* was calculated from empirical data (graph **A**) and corrected for overestimates of predation rates in years of low rodent densities. Jędrzejewski and Jędrzejewska (1996)

Although a quantitative analysis of rodent mortality versus food abundance is not possible (no long-term data on winter abundance of ground vegetation), there is much indirect evidence supporting this explanation. First and foremost, heavy seed crops acting as natural 'food addition experiments' caused better survival and winter breeding of rodents, thus indicating that in non-mast years, reproduction was halted by food shortage. Furthermore, during warm winters (in years with small crops of seeds), when the herb layer persisted longer and plants were not severely deteriorated by frost, yellow-necked mice survived somewhat better and bank voles began breeding about 1 month earlier than in other years (Pucek et al. 1993).

Interpretation of non-cyclicity of rodents in Białowieża Forest as due to winter shortage of food is consistent with the extensive data on rodent dynamics patterns in seven Palaearctic biomes (Jędrzejewski and Jędrzejewska 1996). Rodents exhibit 3–5-year cycles in habitats where winter food is abundant enough to regularly allow for winter breeding. The standing crop of ground vegetation (i.e. all the aboveground vegetation in tundra, grassland, steppe, desert, and farmland, but only the forest floor vegetation in forests) can be regarded as a rough indicator of food abundance for rodents. From tundra to desert, the mean index of rodent cyclicity was significantly positively correlated with the mean standing crop of ground vegetation (Fig. 5.27). In 44 long-term series of rodent trapping used for this analysis, there was a continuum from highly cyclic to non-cyclic populations and not the dichotomy of those two types. In this context, it is clear why *Microtus* voles living in Białowieża's river valleys were cyclic: the standing crop of ground vegetation in those open sedge marshes was about 4 tons dry wt./ha.

Mean densities of rodents in the analysed biomes and habitats in the Palaearctic were not related to the standing crop of plant biomass, but they correlated with productivity of ground vegetation, i.e. amount of new biomass produced during the course of a growing season. The better the productivity the higher the densities of rodents (Fig. 5.28). Such a correlation also exists for the productivity of plants and the maximum recorded densities of rodents (Jędrzejewski and Jędrzejewska 1996). Thus, from both a local and a biogeographic perspective, there is evidence that food regulates densities and population dynamics of small rodents. Importantly, a distinction should be made between two ways in which food resources act. First, a standing crop of vegetation is decisive for good survival of rodents and, if high enough, for their winter breeding. This leads to cycles of rodents. Second, spring–summer productivity of vegetation translates into amount of high quality food in the warm season and determines densities that rodents can reach in one season.

A graphic scheme of that mechanism as pertaining to woodland and marshland rodents in Białowieża Forest is proposed in Fig. 5.29. Both in forests and in open terrain, the increase phase of rodents takes place in spring–summer, during a period of intense production of fresh plant biomass, which provides high quality food for rodents. In forests, the standing crop of biomass of ground vegetation (overshadowed by tree canopies) is rather low, and in winter it is truly scarce (see Chap. 1). It allows for neither good

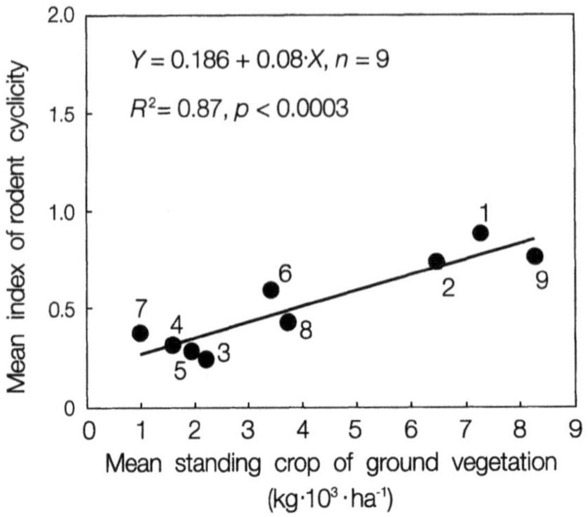

1 tundra
2 taiga forests
3 mixed forests
4 deciduous forests
5 steppe woodland
6 steppe
7 desert
8 grasslands in the zone of deciduous forests
9 farmlands in the zone of deciduous forests

Fig. 5.27. Relationship between mean index of rodent cyclicity and the mean standing crop of biomass of the ground vegetation in summer assumed to represent food availability to rodents in both spring–summer and autumn–winter seasons. Summary of data from 9 biomes in the Palaearctic region. Cyclicity index (after Hansson and Henttonen 1985) was calculated for each biome and type of habitat for the studies covering ≥5 years. The higher the index, the bigger the amplitude of rodent multiannual fluctuations. In deciduous forests, the average crop of tree seeds (an important food for rodents) was added to the mean standing crop of ground vegetation. Jędrzejewski and Jędrzejewska (1996)

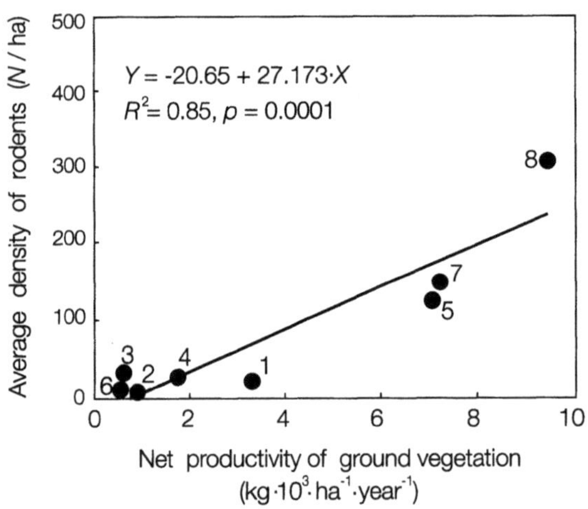

1 tundra
2 taiga forests
3 mixed forests
4 deciduous forests
5 steppe
6 desert
7 grasslands in the zone of deciduous forests
8 farmlands in the zone of deciduous forests

Fig. 5.28. Relationship between mean productivity of ground vegetation and mean density of rodents in the Palaearctic biomes. In the case of deciduous forests, the average crop of tree seeds was included. Mean densities of rodents were calculated for available spring and autumn data points combined from various studies. Only studies on the whole community of rodents or one heavily dominating species were considered. Jędrzejewski and Jędrzejewska (1996)

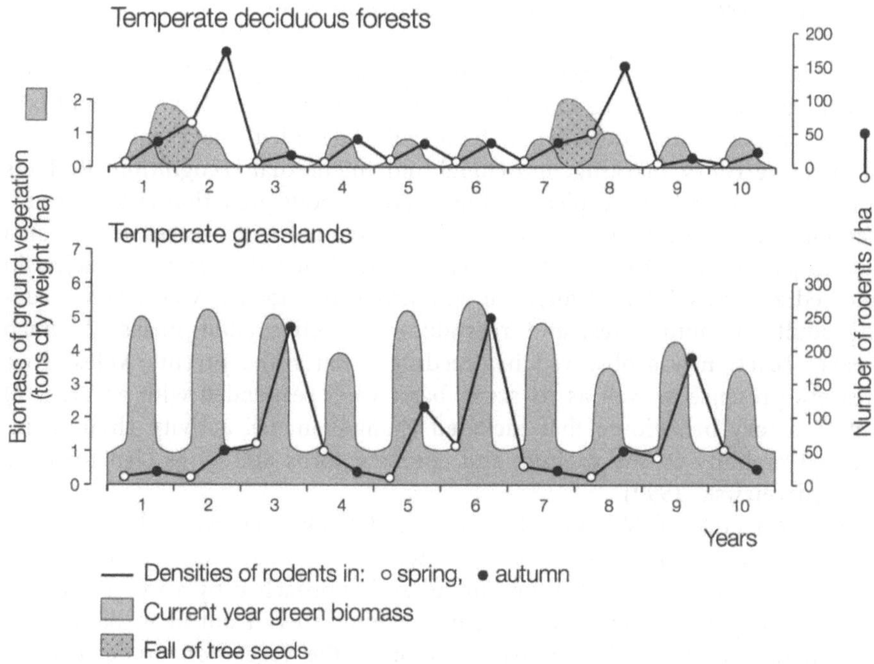

Fig. 5.29. Schemes of the cyclic and non-cyclic population dynamics of rodents as determined by food resources: standing crop of ground vegetation, its productivity, and seasonal changes in abundance. Graphs are schemes based on average measures of rodent and vegetation abundance (data from Jędrzejewski and Jędrzejewska 1996) in the presented habitats

survival nor winter breeding of rodents. Opportunity for winter reproduction and outbreak arises only after a heavy crop of tree seeds (Fig. 5.29). In the riverside meadows and marshes, the standing crop of ground vegetation in winter is fair enough to allow for better survival and even winter breeding of voles. In consequence, exponential population growth practically continues over two or three seasons. The final outbreak of rodents is made possible by high production of vegetation in spring-summer, but in autumn-winter rodent density 'overshoots' the food limit. Having grown very numerous on highly productive spring-summer vegetation, rodents face the rapid deterioration and vanishing of plant biomass in autumn, so a steep decline of their population begins (Fig. 5.29).

Having investigated predation on rodents for over 10 years, we became convinced that predators play an inferior role in governing the populations of such productive animals. The principal role in driving rodent fluctuations is taken over by food – its quantity and quality.

5.5.4 Antipredatory Behaviours of Forest Rodents

Incapable of radically changing rodents' population dynamics, predators are nonetheless a strong force in evolution of their behaviour. Enclosure and laboratory experiments showed that bank voles 'decoded' the carnivore scents used for territory marking, individual and intersexual recognition. Rodents use these as an advanced piece of information about predation risk. In semi-natural enclosures, bank voles avoided pens tainted with weasel odour for a few to as long as 10 days after a weasel had visited the place (Jędrzejewski and Jędrzejewska 1990). Avoidance was strong in subadult voles, non-breeding adults of both sexes, and reproductively active adult males. A much weaker reaction was observed in breeding females and juvenile voles. To a weasel's presence as well as its scent, bank voles responded with an array of antipredatory behaviours that included changes in diel activity rhythm, increased mobility on the ground, and climbing forbs and twigs (Jędrzejewska and Jędrzejewski 1990).

Escape tactics of yellow-necked mice and bank voles were also observed in the wild, during radio-tracking of weasels in Białowieża National Park (Jędrzejewski et al. 1992b). When found and approached by a weasel in their den or cavity, the rodents most often attempted to run out of it. Bank voles ran on the ground and disappeared amongst the vegetation or dead leaves and sometimes were observed to stay immobile or to climb a tall forb. Yellow-necked mice never stayed calm and immobile. Instead, they often climbed trees or shrubs after having run out of their den or cavity (Fig. 5.30).

Antipredatory behaviour in bank voles did improve their survival. Older voles that showed a whole array of escape tactics had markedly better chances of surviving weasel's attacks than young voles (Jędrzejewska and Jędrzejewski 1990). However, severe risk of weasel predation has a negative effect on voles' performance, too. Experimental exposure of bank vole pairs to mustelid odour suppressed their reproduction, probably due to avoiding copulation by female voles (Ylönen 1989; Ylönen et al. 1992; Ronkainen and Ylönen 1994).

Perhaps the most interesting feature of bank voles' antipredatory adaptation is their inherent ability to precisely recognise the scent of at least six species of carnivores: weasel, stoat, marten, polecat, fox, and raccoon dog (Jędrzejewski et al. 1993d). What is more, voles reacted specifically to the odour of each of them. For instance, they escaped out of burrows and climbed twigs in response to weasel scent but hid in burrows and stayed immobile when perceiving fox scent. A significant positive correlation was found between the number of qualitatively different antipredatory behaviours of voles and the degree of the carnivore's specialisation in that prey: the higher the proportion of bank voles in a carnivore diet, the more diverse the voles' antipredatory behaviour against that predator (Jędrzejewski et al. 1993d).

Fig. 5.30. Yellow-necked mouse climbing an aspen sapling when escaping a weasel's attack. Photo: S. Wąsik

Experimental exposure to the odour of tawny owl evoked no reaction from bank voles. Due to the aerial hunting of raptors, rodents lack efficient, direct recognition of their risk of predation and rely on indirect cues such as light intensity and the presence of overhead cover (e.g. Price et al. 1984; Merkens et al. 1991).

5.6 Predation on Shrews

5.6.1 Species Structure of Total Predation on Insectivores

Mammals of the order Insectivora have received far more attention as predators than as prey. One need only think of Holling's classical work (1959) or the recent review by Hanski (1992). In the temperate forests of Europe, insectivorous mammals are rather slender resources for predators: they constituted 8% of total prey biomass taken annually by all carnivores and raptors (see Fig. 5.3). Insectivores come in three distinct classes of body size: hedgehogs (mean body mass of adult 670 g), moles (60 g), and shrews (3–14 g). Of these three, moles provide the amplest resource for predators in terms of biomass (60%, Table 5.12). However, moles and hedgehogs are not very active and thus hardly available to predators in the cold season; 90% of their biomass eaten by predators is taken in spring and summer. In shrews, that seasonal difference is two-fold, only.

Out of the total predatory impact on Soricidae, tawny owls hold sway by taking 52% of all consumed shrews (Table 5.13). Polecats take 13%, and raccoon dogs, weasels and pine martens, 7–8% each. All these predators, except for the polecat, captured the common shrew *Sorex araneus* (a dominant in the soricid guild) definitely more often than other species of shrews. Polecats live along rivers, brooks, and in bog alderwoods and catch water shrews *Neomys* at least as frequently as *Sorex* (see Sect. 4.11).

Some carnivores, especially so pine martens and red foxes, kill shrews and leave them uneaten. During snow tracking of martens in Białowieża National Park, we found on average 0.3 abondoned shrews per 10 km of trail. If we take this into account, the total predation of shrews by martens was, on average, 25% higher than that calculated from consumed shrews only (Zalewski et al. 1995). Red foxes left as many as 0.6 killed and uneaten shrews per 10 km of trail. We estimated that foxes consumed every second shrew they killed. Thus, their total predation on shrews was twice that estimated from consumption, only. With the correction for prey discarded by foxes and martens, the overall impact of carnivores and raptors on Soricidae would be 5% higher.

Predation on moles is monopolized by common buzzards and tawny owls (Table 5.13). Hedgehogs are sporadic (and certainly not easy!) meals for predators. Of 4.5 hedgehogs harvested annually by predators from an average square kilometre, common buzzards took 77%, badgers 18%, pine martens and lesser spotted eagles 2% each, and short-toed eagles 1%.

Table 5.12. Mean biomass of shrews, moles, hedgehogs, and all insectivores eaten by all predators from an average hectare in Białowieża Primeval Forest (averaged data from 1985–1995)

Prey group	Spring–summer		Autumn–winter		Whole year	
	Biomass (g/ha)	%	Biomass (g/ha)	%	Biomass (g/ha)	%
Shrews Soricidae	137	28.5	69	59.5	206	34.5
Mole *Talpa europaea*	316	65.7	45	38.7	361	60.5
Eastern hedgehog *Erinaceus concolor*	28	5.8	2	1.8	30	5.0
Insectivores total	481	100	116	100	597	100

Table 5.13. Percentage contributions by various predators to the total biomass of shrews (mainly the common shrew *Sorex araneus*) and moles consumed by all predators from an average hectare in BPF, 1985–1995. Dominants are in boldface type (≥5% of total predation)

Species of predator	Shrews (Soricidae)			Moles (*Talpa europaea*)		
	Spring–summer	Autumn–winter	Whole year	Spring–summer	Autumn–winter	Whole year
Tawny owl	**48.2**	**60.5**	**52.4**	**34.7**	**60.0**	**37.8**
Polecat	**19.1**	0.9	**13.0**	–	0.3	+
Raccoon dog	**9.7**	3.9	**7.8**	0.1	1.3	0.3
Weasel	**6.5**	**7.9**	**7.0**	2.6	**11.3**	3.7
Pine marten	**6.5**	**7.4**	**6.8**	1.5	**9.4**	2.5
Pygmy owl	0.1	**10.2**	3.4	–	–	–
Red fox	1.3	**7.2**	3.3	0.5	**6.7**	1.2
Buzzard	4.1	0.7	3.0	**57.5**	**8.4**	**51.4**
Honey buzzard	2.2	–	1.5	1.0	–	0.8
American mink	1.1	0.6	0.9	–	1.1	0.1
Badger	0.6	0.5	0.5	0.6	1.5	0.7
Lesser spotted eagle	0.5	–	0.4	1.4	–	1.3
River otter	–	0.1	+	–	–	–
Tengmalm's owl	+	+	+	–	–	–
Short-toed eagle	+	–	+	–	–	–
Wolf	–	+	+	–	–	–
Lynx	–	+	+	–	–	–
Goshawk	–	–	–	0.1	–	0.1
Hobby	–	–	–	+	–	+
Total biomass consumed (g/ha)	137	69	206	315	45	360
Mean *N* prey (ind/ha)	17.1	8.6	25.7	5.2	0.7	5.9

5.6.2 Magnitude and Mechanisms of Predation on Shrews

In Białowieża Forest, none of the woodland predators specialises on shrews. To most predators, they are occasional prey. To a few, shrews are secondary prey, captured deliberately when the supply of primary prey is poor. Shrews are buffer prey to rodent-eaters: tawny owls, pine martens, and weasels. In summer, the polecat, a carnivore specialised on frogs, catches a fair number of shrews. Consequently, insectivores are not in such interactive relationships with their enemies as are rodents. Most probably, multiannual fluctuations in shrew numbers generally evoke no numerical or functional responses in any predator. We must admit, however, that our study on predation did not cover an unusual irruption of soricids in the early 1970s (see Sect. 2.6), which might have been reflected in the diets of predators.

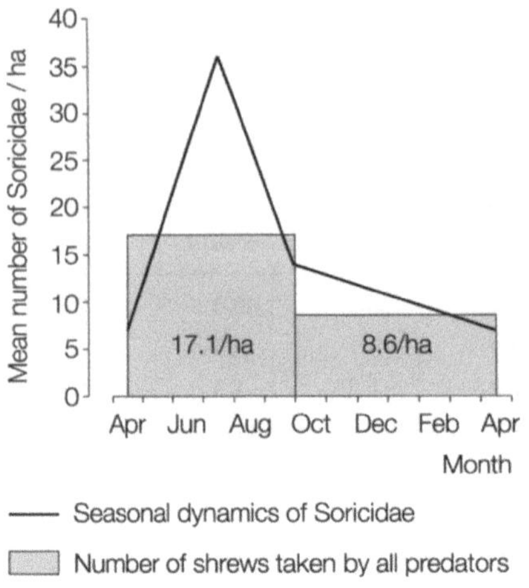

Fig. 5.31. Average predation impact on Soricidae (*N* shrews taken by all predators from 1 ha) in spring–summer (16 April–30 September) and autumn–winter (1 October–15 April) compared with the average level of their densities in BPF

The heaviest absolute predation on shrews (on average 17 ind/ha taken by all predators) occurs in spring–summer, i.e. when shrews reproduce, reach their seasonally highest numbers, and then decline (Fig. 5.31). If we assume that predators harvest shrews at a constant rate throughout the warm season, we arrive at rough estimates of 20% of spring production and 40% of summer numbers being lost to predators. In autumn–winter, when shrews always decline but at a lower rate than in summer-autumn, predators are responsible for a major part, indeed all of the winter mortality of shrews.

Long-term trapping of shrews in Białowieża National Park (Pucek et al., in prep.) allowed us to look for the patterns of predation on shrews by the tawny owl and the pine marten. These two species were responsible for 55% of total predation in spring–summer and 68% in autumn–winter. In 1985–1992, tawny owls took from 3.2 to 12.9 shrews/ha in spring–summer (mean 8.4). In autumn–winter, owls' predation was, on average, one-third lower: from 2.1 to 7.1 shrews/ha (mean 5.4) (Fig. 5.32). Martens took from 0.1 to 2 shrews/ha in the cold season. In neither tawny owls nor martens was the predation correlated with shrew densities (Figs. 5.32 and 5.33). The relative predation rate by tawny owls (i.e. total predation divided by shrew abundance index) was inversely density dependent: it was highest at low numbers of shrews and declined when shrews grew more numerous (Fig. 5.33).

For tawny owls, shrews are one of the three resources used for buffering shortages of rodents (the other two being birds and frogs). But while the negative correlation between rodents and shrews was prominent in the diets of owls, it was only weakly (and not significantly) marked in total predation by the tawny owl population. When forest rodents were scarce (in 1991–1992), owls' predation on shrews was heavier than the average level. However, during the outbreak of rodents in summer 1990, when the role of shrews in owl diet was very small, total predation was also quite heavy, because owls had significantly more young to feed than in other years. Conse-

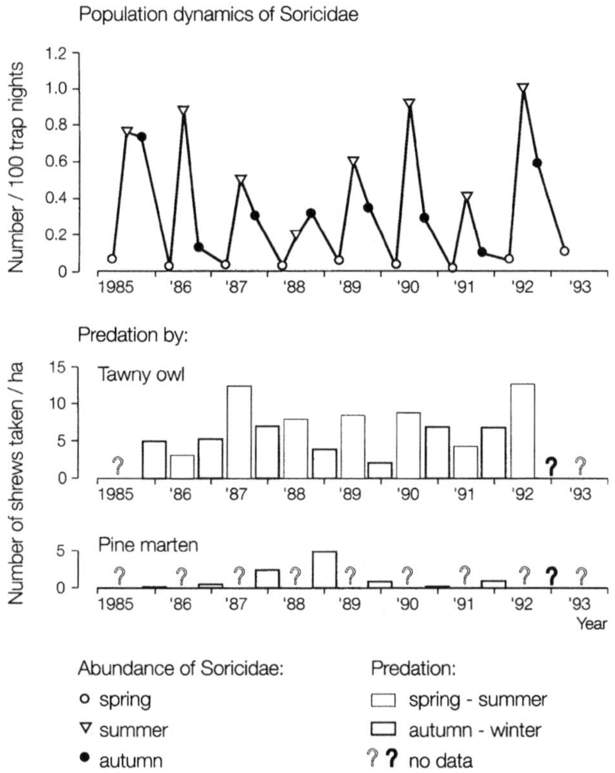

Fig. 5.32. Population dynamics of shrews (combined numbers of the dominant common shrew *Sorex araneus* and *S. minutus, S. caecutiens, Neomys fodiens, N. anomalus*) and predation impact on shrews by tawny owls and pine martens in the pristine forests of Białowieża National Park. Seasons as in Fig. 5.31

quently, in temperate forests, large amplitude fluctuations of rodent numbers do not exert a strong predator-linked impact on shrew numbers.

Is predation a potent factor shaping the numbers and population dynamics of shrews in Białowieża Forest? We have too limited data at our disposal to solve this problem. Direct estimates of shrew densities are scarce and come from years other than those of our study on predation (see Sect. 2.6). The long-term (23-year) trapping of shrews (Pucek et al., in prep.) yielded indices of *N* shrews/100 trap nights, which cannot be translated to densities. More centrally, there are no data on year-to-year changes in abundance of soil invertebrates, the basic food resource of shrews.

The available, indirect evidence suggests that both seasonal and multiannual fluctuations of shrew populations are most likely driven by abundance of food: soil invertebrates and leaf-eating caterpillars that descend to the soil for pupation (see Chap. 2). The lack of density dependence in the predation impact on shrews is a persuasive argument for a limiting but not regulating role of predation in population dynamics of shrews.

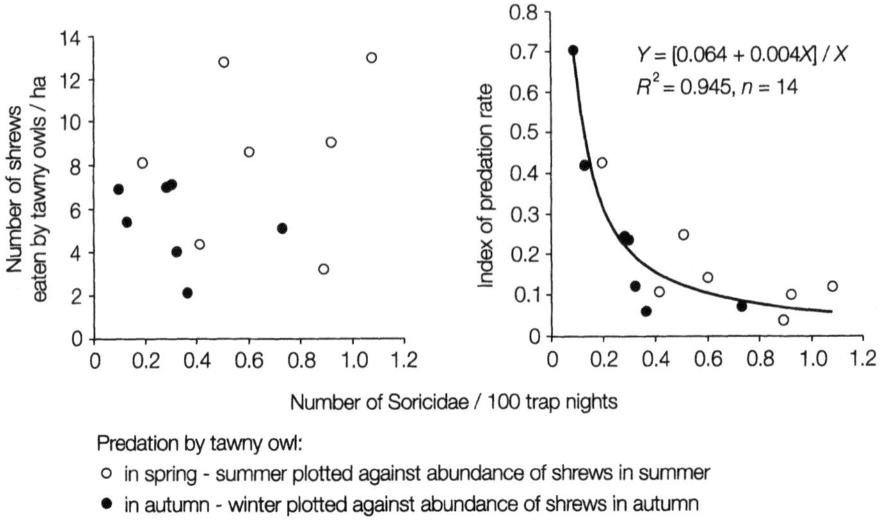

Fig. 5.33. Autumn–winter and spring–summer predation on shrews by tawny owls in relation to indices of shrew abundance in Białowieża National Park, in 1985/1986–1992. *Left panel* absolute magnitude of predation (*N* shrews removed per 1 ha). *Right panel* relative index of predation rate (magnitude of predation divided by 100 times index of shrew abundance)

The non-regulatory impact of predation does not mean it is trivial for shrews. Even if the ultimate cause of population decline is food shortage, probably the commonest death for shrews is due to predation. The link between these two factors was convincingly demonstrated in laboratory experiments by Saariko (1992). Common shrews recognised the scent of a weasel and, generally, decreased foraging activity for 5–15 min after a weasel's visit to the feeders. However, individual variation in shrews' responses was large: heavier individuals could afford to reduce their foraging activity, whereas small individuals with poor energy reserves continued foraging even when the perceived risk of predation was high.

5.7 Predation on Birds

5.7.1 Species Structure of Total Predation on Birds

In spring and summer, birds are the third most important source of food (after ungulates and rodents) to predators. In the cold season, as a result of migration of both songbirds and several raptors, avian prey matters very little in the total biomass consumed by carnivores and birds of prey (see Fig. 5.3). The dominant enemies of birds are raptors, contributing 85% to the total predation in spring–summer and 88% in autumn–winter (Table 5.14). Of those, buzzards, tawny owls, and goshawks are of paramount importance.

Table 5.14. Percentage contributions by various predators to the total biomass of birds (fledglings and adults) consumed by all predators. Spring–summer is 16 April–30 September, autumn–winter is 1 October–15 April. Dominant predators are in boldface type (≥5% of total consumption)

Predator	Spring–summer	Autumn–winter	Whole year
Buzzard	**31.5**	7.7	**27.4**
Tawny owl	**29.5**	**28.6**	**29.3**
Goshawk	**9.3**	**42.4**	**15.0**
Sparrowhawk	**8.2**	–	**6.8**
Pine marten	**7.0**	2.7	**6.2**
Pygmy owl	2.8	**8.5**	3.8
Raccoon dog[a]	4.4	0.9	3.8
Red fox[a]	1.8	4.7	2.4
Lesser spotted eagle	1.9	–	1.6
Weasel	0.6	1.7	0.8
Hobby	0.7	–	0.6
Polecat	0.6	+	0.5
Honey buzzard	0.5	–	0.4
Lynx	0.4	0.8	0.4
Otter	0.3	0.5	0.4
Tengmalm's owl	0.2	0.8	0.3
Badger	0.1	0.2	0.1
American mink	+	0.4	0.1
Wolf	+	+	+
Long-eared owl	+	+	+
Short-toed eagle	+	–	+
Total biomass consumed (g/ha)	595	124	719

[a] Foxes and raccoon dogs often scavenge on carcasses of dead birds (especially in winter) and remnants of birds killed by goshawks and other raptors.

Mammalian predators add little to the total predation, the only exception being the pine marten that takes 6% of all avian biomass eaten by predators.

An essential feature of the species structure of predation on birds is that most predation is by non-specialists, namely buzzards and tawny owls. Small and medium-sized birds are important prey of these two raptors in spring and summer, but the buzzard is a generalist that feeds on rodents, moles, birds and amphibians, and the tawny owl captures primarily forest rodents (see Chap. 4). The key roles of buzzards and tawny owls in predation on birds are a consequence of their high densities in the temperate forests. Predators that are specialised on birds contribute 18% to the total predation on birds in spring–summer (goshawk, sparrowhawk, hobby) and 42% in autumn–winter (all by goshawk, a year round resident).

Table 5.15. Percentage contributions by various predators to the total annual consumption of fledglings and adults of some common birds of Białowieża Forest. The three most important predators in each case are in boldface type

Predator	Thrushes (*Turdus* spp.)	Woodpeckers (Picidae)	Tits (*Parus* spp.)	Chaffinch (*Fringilla coelebs*)
Buzzard	**45.6**	**40.9**	13.5	13.6
Tawny owl	**39.3**	9.3	**18.7**	**39.6**
Sparrowhawk	4.8	**13.1**	**29.8**	**22.2**
Pygmy owl	0.1	4.9	**24.8**	**20.5**
Goshawk	**5.3**	**23.2**	0.4	1.6
Pine marten	1.8	5.5	9.3	–
Other predators total	3.1	3.1	3.5	2.5
Total biomass consumed (g/ha)	222.9	46.6	16.6	16.4

Species structure of predation on all birds is the result of predators' preferences for bird species (see Chap. 4). Thus, predation on various birds may differ in relative amounts by various predators. We compared the importance of several raptors and one carnivore in the total predation on thrushes, woodpeckers, tits, and chaffinches, which together constituted about 40% of all avian biomass consumed by predators (Table 5.15). Invariably, the same guild of six 'bird-eaters' was responsible for nearly all predation in each case, but contributions by given predators were related to body sizes of predators and prey. Bigger birds (woodpeckers, thrushes) suffered most from predation by bigger raptors (buzzards, goshawks, tawny owls), whereas the majority of smaller birds (chaffinches, tits) fell as prey to smaller raptors (sparrowhawk, pygmy owl). Pine martens' shares in total predation weighed more in cavity nesters than in other birds.

5.7.2 Magnitude and Mechanisms of Predation on Birds

We have discussed prey in a sequence which began with mammals (eutherian reproduction), proceeded to birds (external eggs, parental care), and ended with a discussion of amphibians (external eggs, metamorphosis, no parental care). Along that gradient, the reader's attention shifts from homeothermic vertebrates that are not seriously affected by abiotic conditions but depend on frequent feeding, to heterothermic animals that can survive months without food but are critically dependent on ambient conditions such as humidity and temperature. Presumably, with increasing independence on abiotic factors, small animals will be more affected by predation because they have to spend much time on daily foraging, which makes them a predictable prey for predators.

Birds seem to be a case in point. If any predator's actions are likely to be observed directly, they are aerial hunts of sparrowhawks or hobbies on songbirds. By no other method can predation be studied as tangibly as by checking the robbed nests of birds. However, despite these apparent 'conveniences' of their life habits for investigating predation, birds are actually one of the most difficult study objects in this field. What should a researcher know to reliably conclude whether or not a given bird species is affected by predation? Let us consider two examples. In wood warblers, one should estimate predation rates on eggs, nestlings, and adult birds in their breeding sites in the temperate forests of Europe, then predation rates on migratory birds on their way to and from winter quarters, and finally, predation in their winter ranges in Africa. The known effects of predation should ideally be compared with those of food availability and adverse weather during breeding, migration, and wintering. In resident birds, woodpeckers for instance, one would need to know the impact of the temperate forests' predators on eggs, nestlings, dispersing juveniles, and resident adults, and be able to compare this with the interacting effects of food abundance and weather. Needless to say, there is not a single study that has done all this for either a migratory, or a resident bird.

In our attempt to quantify the predation impact on birds in the primeval forests of Białowieża, we combined data from two sources: the results of our studies on consumption of adult and juvenile birds (at least half-fledged so the feather remains are recognizable in scats and pellets) by carnivores and raptors in 1985/1986–1995, and results of several studies on bird breeding densities and nest predation in the 1980s–1990s (e.g. Tomiałojć et al. 1984; Wesołowski 1985; Walankiewicz 1991).

The mean biomass of birds consumed by all predators in the warm and the cold seasons from an average hectare of BPF was shown against the background of seasonal dynamics of bird abundance, expressed in terms of biomass (Fig. 5.34). (Converting the biomass into number of birds consumed by predators could not have been done with reasonable accuracy, since every species of predator took birds of somewhat different mean body mass. Worse still, for many species of predators, the sample of birds identified to species was too small to evaluate the average weight of captured birds.) Compared with the estimated mean standing crop of birds, total predation on them seems heavy, especially in spring–summer. If we assume that predation during the birds' breeding season was additive and was not compensated by increased production, we arrive at a maximum estimate of 45% of total annual production being lost to predators. Decline in bird numbers from summer to autumn is caused by migration, predation, and death from other causes. The role of predation in changes of bird numbers in autumn–winter cannot be assessed; it is only certain that autumn and spring migrations are processes of the utmost importance in causing the observed overall pattern of bird dynamics. In winter, local predators may play a substantial role only among the resident species of birds.

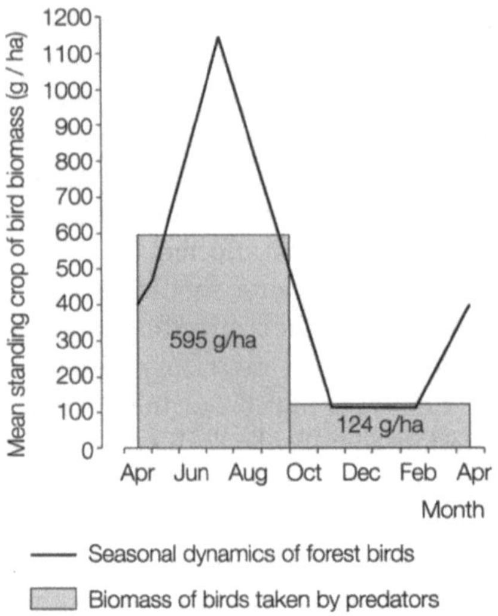

Fig. 5.34. Average predation impact on birds (biomass of birds taken by all predators from 1 ha) in spring–summer (16 April–30 September) and autumn–winter (1 October–15 April) compared with the average level of bird abundance (see Table 2.11 for sources of data)

The magnitude of predation on particular species of birds (Fig. 5.35) is very difficult to assess, not least because, in pellets and scats of predators, feathers of small and dull-coloured birds had much poorer chances of being identified to species than those of big and brightly coloured ones. To overcome this problem, we analysed predation on broad ecological groups of birds: residents, short-distance migrants, and tropical migrants, each of them including dull-coloured and brightly coloured birds, small and large species. Of the whole biomass of birds consumed by predators year-round, 53% was constituted by short-distance migrants, 44% by residents, and only 3% by tropical migrants (Table 5.16). Nearly the same proportions were found in the breeding community of birds, as weighted by the duration of their stay in BPF. Predators exploited migrants and residents proportionally to their availability, perhaps with the exception of tropical migrants that seem to suffer from predation less than expected from their share in the bird community. In Chapter 4 we showed that hardly any predator breeding in the temperate forests of Europe is specialised on tropical migrants or selects them positively out of the bird community. Warblers in the genus *Phylloscopus* seem particularly prone to predation. Their small size (adult body mass 7–9 g) makes them attractive only to the smallest raptors (pygmy owl, Tengmalm's owl, sparrowhawk), although they, too, capture very few *Phylloscopus* warblers. This is in agreement with data obtained by Wesołowski (1985), who noted that, in the breeding season, the predation rate on breeding female wood warblers was 4.3% per month.

It is also worth noting that of all short-distance migrants taken by predators, thrushes (the song thrush, the blackbirds, and – in autumn – migrating redwings) are the chief resource of predators. Found in the diets of nearly all predators, ranging in size from lynx to pygmy owl, thrushes contributed 59% of the biomass of short-distance migrants consumed by predators and 31% of the total bird biomass eaten by them annually. Among resident birds, the most important to predators were woodpeckers, constituting 15% of resident bird biomass and 6.5% of all bird biomass taken by predators (cf. Table 5.15).

Between-year variation in the magnitude of predation on birds was studied in 1985/1986–1992 in Białowieża National Park (Zalewski et al. 1995;

Fig. 5.35. Wood pigeon *Columba palumbus* killed and partly consumed by a predator. Photo: S. Wąsik

Table 5.16. Predation (in biomass taken per hectare by all predators during a given season or a whole year) on three groups of birds differing in migratory habits, compared with the spring standing crop of bird biomass in Białowieża National Park. Calculation of bird biomass based on data from Tomiałojć et al. (1984) and Tomiałojć and Wesołowski (1994, 1996). Weighted percentage of biomass of each migratory group of birds was calculated based on duration of their stay in Białowieża Forest: residents, 365 days; short-distance migrants, 183 days (April–September); tropical migrants, 123 days (May–August)

Parameter	Residents	Short-distance migrants	Tropical migrants	All birds
Biomass of birds in spring (g/ha)	122	295	54	471
(%)	26	63	11	100
Weighted percentage (%)	42	52	6	100
Biomass consumed by predators in spring–summer (g/ha)	227	344	25	595
(%)	38	58	4	100
Biomass consumed by predators in autumn–winter (g/ha)	88	35	1	124
(%)	71	28	1	100
Biomass consumed by predators in a whole year (g/ha)	315	379	26	719
(%)	44	53	3	100

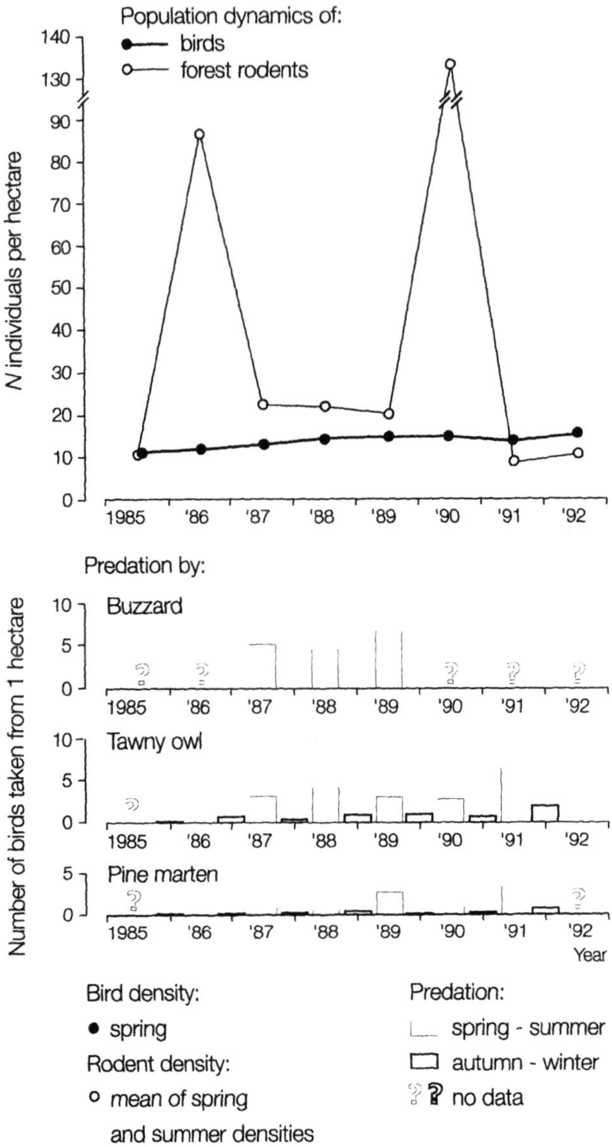

Fig. 5.36. Spring densities of breeding birds, spring–summer densities of forest rodents (combined numbers of bank voles and yellow-necked mice) and predation impact on birds by common buzzards, tawny owls, and pine martens in the pristine forests of Białowieża National Park, 1985–1992. Seasons as in Fig. 5.34. Data on bird densities from Tomiałojć et al. (1984), Tomiałojć and Wesołowski (1994, 1996)

Jędrzejewski et al. 1996a). Tawny owls and pine martens feed predominantly on forest rodents, the abundance of which varied 30-fold from year to year. Birds were an important alternative prey, buffering for the temporal shortages of rodents. In the spring–summer seasons of 1986–1992, tawny owls captured from 1.6 to 6.4 birds/ha (Fig. 5.36). In autumn–winter, they took from 0 to 1.85 birds/ha, that is, on average, five times fewer birds than in the warm season. Pine martens captured 0.5–3.4 birds/ha in spring–summer and 0.02–0.7 birds/ha (on average seven times fewer) in autumn–winter. Both predators exerted the strongest impact on birds in 1991–1992, when numbers of forest rodents were very low (Fig. 5.36). In 1992, hunting on birds was facilitated by the fact that numbers of birds were highest then. In the cold seasons, severity of winter also contributed to between-year variation in predation on birds. During harsh winters, fewer birds reside in the forest than in mild winters. In the cold seasons with similar (moderate) densities of rodents but differing in mean ambient temperature, tawny owls captured on average 0.3 birds/ha in severe winters (1985/1986–1987/1988) and 0.8 birds/ha in mild winters (1988/1989–1989/1990). However, in the whole period 1985/1986–1992, tawny owls' and martens' predation on birds was primarily shaped by combined densities of bank voles and yellow-necked mice; the fewer the rodents, the heavier the toll of birds taken by predators. In non-linear regressions, changes in rodent densities explained from 76–98% of variation in predation on birds (Fig. 5.37).

Common buzzards are the most important predators of birds in spring and summer but they play an insignificant role in autumn–winter. An essential feature of buzzards' predation is the fact that they hunt for birds in June–August and predominantly capture fledglings (Fig. 5.38). In the spring–summer food of buzzards, birds made on average 34% of prey biomass, with

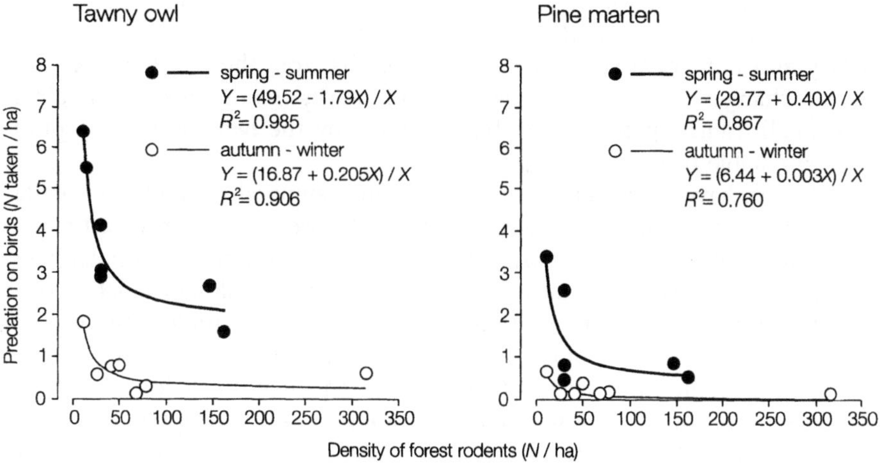

Fig. 5.37. Tawny owl and pine marten predation on birds as shaped by densities of forest rodents in Białowieża National Park in 1985/1986–1992. Each point denotes one season

Fig. 5.38. Chaffinch *Fringilla coelebs* fledgling, prematurely out of nest. Such fledglings are particularly susceptible to predation. Photo: S. Wąsik

negligible multiannual variation (22–36%), even in years of very low densities of rodents. Therefore, it seems that buzzards' predation on birds (although calculated for 3 years only; see Fig. 5.36) is generally less variable than that of tawny owls, the more so as densities of buzzards are rather stable from year to year (Pugacewicz 1996).

The goshawk and the sparrowhawk specialise on birds. In spring–summer, goshawks captured on average 0.4 birds (of mean body mass 155 g) per ha and sparrowhawks 1.1 birds (of mean body mass 46 g). In autumn–winter, resident goshawks took 0.3 bird/ha. The proportions of bird biomass in the diets of these two raptors varied little between years (81–86% in goshawk, 86–96% in sparrowhawk). Moreover, the densities and breeding successes of goshawks and sparrowhawks were rather stable (Pugacewicz 1996). Such dietary and numerical stability obviously results from the fact that the year-to-year changes in the spring densities of hawks' main prey, forest birds, were small. Hence, predation on birds exerted by the two species was probably very stable from year to year.

So far, we have presented the magnitude and temporal variability of predation on adult birds and fledglings. The estimates were based on investigation of predators' diets and densities. When it comes to bird eggs and hatchlings, this approach is useless because neither egg content nor downy chicks would leave recognisable remnants in predators' scats, pellets or under the plucking posts. Occasional finding of egg shells in a scat or a pellet merely reveals the fact of nest robbing but says little about its frequency. In Białowieża Forest, bird eggs were found in scats, pellets or stomachs of lynxes, red foxes, raccoon dogs, badgers, weasels, pine martens and buzzards. Furthermore, direct observations and traces left at robbed nests have shown that yellow-necked mice, red squirrels, woodpeckers, forest dormice, jays and other corvids are

Table 5.17. Role of predation in nesting losses (nests that failed to rear any young as percentage of all nests studied) in seven species of birds in Białowieża National Park

Species	Mean nesting losses (%)	Role of predation in total losses (%)	Source
Wren	50	92	Wesołowski (1983)
Wood warbler	70	>80	Wesołowski (1985)
Chiffchaff	40	70	Piotrowska and Wesołowski (1989)
Nuthatch	35	75	Wesołowski and Stawarczyk (1991)
Collared flycatcher	43	95	Walankiewicz (1991; unpubl. data)
Blackbird	67	98	Tomiałojć (1994)
White-backed woodpecker	40	73	Wesołowski (1995)

predators of eggs and the small chicks of songbirds. Woodpeckers in particular seem to be notorious enemies of cavity nesters such as flycatchers, nuthatches, and other woodpeckers (Walankiewicz 1991; Wesołowski and Stawarczyk 1991; Wesołowski 1995).

In Białowieża Primeval Forest, nesting losses were studied directly in seven species of birds (Table 5.17). Mean losses varied from 35% in nuthatches to 70% in wood warblers and blackbirds. In all species studied, losses due to predation dominated. The authors of all works cited in Table 5.17 emphasised that heavy predation on nests distinguished bird populations of Białowieża Forest from those living in man-transformed habitats in Europe. Furthermore, Tomiałojć (1994) and Wesołowski (1995) observed that for blackbirds and white-backed woodpeckers, respectively, the lower the numbers of rodents, the higher the rates of birds' breeding losses due to predation.

5.7.3 Relative Roles of Predation and Other Factors in Shaping the Densities and Population Dynamics of Bird

To tackle the problem of the role of predation in a forest bird community, we approached it differently from the case of rodents but in an analogous way to the analysis of ungulates. Available data on birds show only spring densities of the breeding community. Predation on birds was calculated for spring–summer and autumn–winter seasons. It is, therefore, not possible to calculate percentage predation and examine the density-dependence of predation. Instead, we carefully analysed the data on birds between 1976 and 1994 to detect whether years of outbreaks and crashes of forest rodents (occurring at 6–9-year intervals) had left discernible traces in bird densities and breeding success due to changes in otherwise 'normal' or average predation rates.

First, we synthesised the data on nesting losses in migratory birds feeding on leaf-eating caterpillars. The level of predation on nests appeared inversely related to the abundance of caterpillars, the main food of these birds, in the breeding season (Fig. 5.39). A generalised model based on empirical data shows that predation on nests of migratory birds increases from 5% at outbreaks of caterpillars to over 70% when caterpillars are very scarce. Losses from other factors also grow with decreasing food abundance, but they add no more than 10% to the losses from predation (Fig. 5.39). This relationship suggests that food supply is the ultimate factor of nest losses, whereas predation is a proximate factor. What is the mechanism of predation acting as executor of nesting failure at a poor supply of food? Having reviewed a large number of bird studies, Martin (1992) concluded that parent birds reduce the probability of predation on their young by perching near them to guard and defend them. However, time spent perching competes with time needed for foraging, so the better the food resources the more time parent birds can spend at their nests. Another factor contributing to lower predation rates on nests of migratory birds during outbreaks of winter moths is the fact that several predators of eggs and hatchlings (woodpeckers, jays, rodents, squirrels) may shift to feeding on caterpillars as well.

We have shown that the nesting success of residents and short-distance migrants not relying heavily on caterpillars as food (wren, nuthatch, white-backed woodpecker) was not related to caterpillar abundance (see Sect. 2.8). Studies conducted in tundra regions (Ryabicev et al. 1976; Järvinen 1985) have shown that predation rates on bird nests rose in years of low densities of rodents. In the temperate forests of Białowieża, elevated predation on

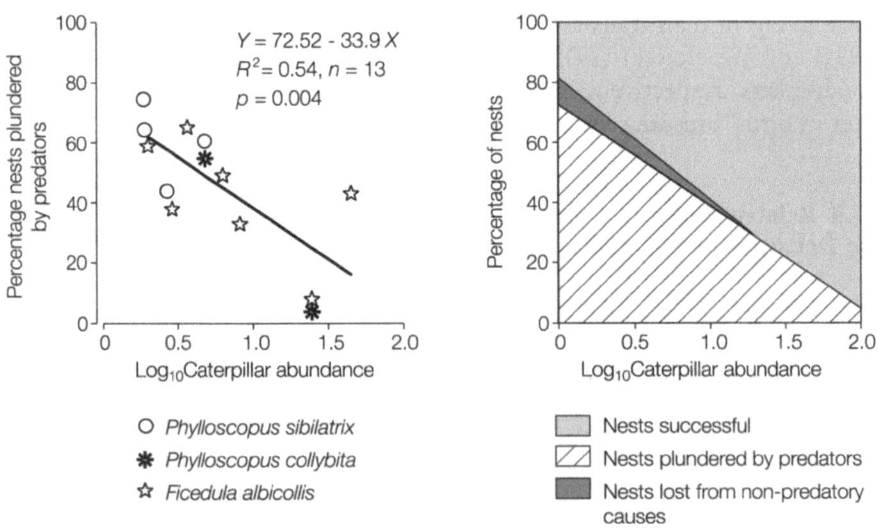

Fig. 5.39. Predation on nests of migratory birds in relation to abundance of leaf-eating caterpillars in birds' breeding season in Białowieża National Park. *Left panel* data from studies by Wesołowski (1985), Piotrowska and Wesołowski (1989), and Walankiewicz (1991, and unpubl. data). Each point represents one year. *Right panel* generalised model based on empirical data

adults and fledglings also occurs during rodent crashes but the picture of predation on nests is far more complicated. First, rodents themselves are mighty predators of bird eggs, especially in small songbirds that nest on the ground, in shrubs, and in tree cavities - places easily accessible to the yellow-necked mouse. The breeding success of five species of birds was studied for long enough to cover the years of outbreak, crash, and moderate densities of rodents (Table 5.18). Small birds (wren, wood warbler, flycatcher) suffered the heaviest losses in years of very high densities of rodents: in the three species, on average, only 20% of nests succeeded in rearing young during outbreaks, while their mean breeding success was 60% in years of moderate densities of rodents and 43% in years of rodent crashes. This indicates that carnivores and raptors worsen the breeding performance of small birds (during crashes of rodents) but not as heavily as do rodents rocketing in numbers after a heavy crop of tree seeds.

In the 5-year series of data on the fates of collared flycatchers' nests (1988–1992), food abundance, rodent density, and predation rate by 'true' predators explained, in a multiple regression analysis, a total of 98.4% of variation in flycatchers' breeding success (but due to the small sample, $n=5$, the relationship was statistically not significant: $p=0.16$). Of these three factors, most important was the positive role of food, i.e. abundance of caterpillars (semipartial correlation squared $sr^2=74\%$), followed by negative influences of rodent density ($sr^2=22\%$) and predation by 'true' predators, approximated by known spring–summer predation on all birds by tawny owls ($sr^2=18\%$).

Although rodents were also reported as predators of eggs of medium-sized birds (e.g. blackbirds, Tomiałojć 1994), they play a decidedly smaller role in their nesting losses than do carnivores and raptors. Breeding performance of blackbirds and white-backed woodpeckers was very poor (about 30%) in years of rodent crashes, when predators such as martens and tawny owls

Table 5.18. Influence of densities of forest rodents on breeding success of small birds (eggs presumably subject to heavy predation by rodents) and medium-sized birds (eggs and nestlings subject to heavy predation by carnivores and raptors) in Białowieża Primeval Forest, 1976–1995. Each figure stands for one breeding season. Sources as in Table 5.17

Species	Breeding success (% nests successful) in years of:		
	Outbreak of rodents	Moderate densities of rodents	Crash of rodents
		Small birds	
Wren	**12.5**	100, 50	50
Wood warbler	**13**	44, 34	28
Collared flycatcher	**34**	30, 61, 92, 56, 62, 70	50
		Medium-sized birds	
Blackbird	?	33, 53, 31, 29, 32	**27**
White-backed woodpecker	83	?	**36**

shifted to bird hunting (Table 5.18). In 1984–1989, the breeding success of blackbirds was strongly determined by rodent density ($R^2 = 94.5\%$, $p = 0.001$); the higher the rodent numbers, the better the breeding success of blackbirds, a relation obviously linked by carnivores' and raptors' shifting to birds (as alternative prey) in years of rodent scarcity.

Hence, if predation on eggs and chicks in Białowieża Forest is particularly heavy during two consecutive years, the outbreak and the crash of forest rodents, it is likely to reduce the breeding numbers of birds in the following seasons: in crash and post-crash years. Furthermore, carnivores and raptors which prey on rodents exert markedly heavier predation on adult birds in years of rodent crashes. Thus, spring densities of birds should be lower in years of rodent crashes and in the post-crash years than in any other years. For 44 species of forest birds, we calculated mean densities in 6 years covering crash and post-crash years of forest rodents (1978–1979, 1984–1985, 1991–1992) and compared them with mean densities in all other years (13 years between 1976–1994) (Table 5.19). In 15 species of birds, the mean densities reported during rodent crashes and 1 year later were over 20% lower than in the years of moderate and high densities of rodents. In 19 species, that decline was –1 to –20%, and 10 species did not show any decline or were even more abundant in years when rodents were scarce. The magnitude of decline in the analysed 44 species of birds was related to their nesting habits. Densities of ground nesters were, on average, 28% lower in rodent crash and post-crash years, and those of shrub nesters 23% lower (Fig. 5.40). Obviously, birds nesting on the ground and in shrubs are easier targets to all predators of rodents than birds nesting in trees. Hole-nesters' losses were higher in species breeding in large holes (more accessible to pine marten), than those nesting in small holes. Crown nesters suffered a small decline (by 2.5%, on average), although some common species were quite seriously affected (e.g. hawfinch, chaffinch, and song thrush).

Two species of those shown in Table 5.19 deserve comment: the wood warbler and the red-breasted flycatcher. Wood warblers were just shown to be severely affected by rodent predation on nests in years of rodent outbreaks (see Table 5.18), but they markedly increased in number soon afterwards (Table 5.19). This seeming paradox is explained by the fact that, in spring, migrating wood warblers settle in Białowieża Forest in numbers inversely proportional to spring numbers of forest rodents (Fig. 5.41). Considering the unusually poor breeding success of wood warblers in years of rodent outbreaks, their ability to assess rodent abundance upon making a decision where to settle for breeding seems an efficient antipredatory behaviour.

The densities of another species, the red-breasted flycatcher, were negatively correlated with those of its numerically superior and larger congeneric, the collared flycatcher (Fig. 5.41). In their detailed analysis of the whole community of birds in BNP, Wesołowski and Tomiałojć (1997) found that this was the only pair of congeners that showed such a negative correlation of numbers. Most probably, densities of red-breasted flycatcher in BNP are shaped by competition with the collared flycatcher.

Table 5.19. Mean densities of 44 common species of forest birds in crash and post-crash years of forest rodents (1978–1979, 1984–1985, 1991–1992) compared with mean densities in years of moderate and high densities of rodents (other years between 1976 and 1994). Data from studies in Białowieża National Park by Tomiałojć et al. (1984) and Tomiałojć and Wesołowski (1994, 1996)

Species	Mean density, N ind/km^2 (SD)		Percent change
	Moderate and outbreak densities of rodents ($n=13$ years)	Crash and post-crash years of forest rodents ($n=6$ years)	
	Ground nesters		
Tree pipit (*Anthus trivialis*)	10.7 (3.6)	7.1 (3.0)	-33
Hazel hen (*Bonasa bonasia*)	2.3 (1.6)	1.7 (1.2)	-26
Scarlet rosefinch (*Carpodacus erythrinus*)	3.2 (1.5)	2.3 (1.5)	-26
Chiffchaff (*Phylloscopus collybita*)	37.8 (6.0)	27.7 (5.0)	-27
Wood warbler (*Phylloscopus sibilatrix*)	92.5 (37.2)	100.8 (19.3)	+9
Woodcock (*Scolopax rusticola*)	1.2 (1.2)	0.9 (1.4)	-26
	Shrub nesters		
Long-tailed tit (*Aegithalos caudatus*)	3.9 (2.4)	2.7 (1.7)	-30
Robin (*Erithacus rubecula*)	136.3 (14.0)	102.8 (8.9)	-25
Dunnock (*Prunella modularis*)	35.6 (8.7)	24.9 (7.4)	-30
Blackcap (*Sylvia atricapilla*)	49.4 (6.9)	38.2 (8.3)	-23
Garden warbler (*Sylvia borin*)	8.9 (3.1)	5.8 (3.8)	-35
Wren (*Troglodytes troglodytes*)	39.4 (7.0)	33.6 (9.3)	-15
Blackbird (*Turdus merula*)	36.2 (4.4)	34.5 (10.1)	-5
	Nesters of large holes		
Stock dove (*Columba oenas*)	1.9 (1.1)	1.6 (0.5)	-15
White-backed woodpecker (*Dendrocopos leucotos*)	3.0 (1.0)	2.5 (1.3)	-17
Great spotted woodpecker (*D. major*)	16.1 (5.1)	9.7 (3.9)	-40
Middle spotted woodpecker (*D. medius*)	12.4 (2.0)	10.2 (4.1)	-17
Black woodpecker (*Dryocopus martius*)	2.1 (1.2)	2.0 (1.7)	-6
Three-toed woodpecker (*Picoides tridactylus*)	2.5 (1.5)	1.9 (1.8)	-26
Starling (*Sturnus vulgaris*)	29.0 (11.9)	25.4 (4.9)	-12
	Nesters of small holes		
Tree creeper (*Certhia familiaris*)	37.7 (7.8)	35.2 (8.4)	-7
Lesser spotted woodpecker (*Dendrocopos minor*)	4.3 (1.5)	3.9 (2.2)	-8
Collared flycatcher (*Ficedula albicollis*)	81.3 (25.7)	62.5 (14.6)	-23
Pied flycatcher (*Ficedula hypoleuca*)	15.7 (4.7)	11.6 (7.2)	-26
Coal tit (*Parus ater*)	9.3 (3.6)	9.2 (4.7)	-1
Blue tit (*Parus caeruleus*)	37.9 (11.1)	44.9 (27.7)	+18
Crested tit (*Parus cristatus*)	9.1 (2.1)	9.5 (3.7)	+4

Table 5.19 (continued)

Species	Mean density, N ind/km^2 (SD)		Percent change
	Moderate and outbreak densities of rodents ($n=13$ years)	Crash and post-crash years of forest rodents ($n=6$ years)	
Great tit (*Parus major*)	44.9 (6.8)	41.3 (16.0)	-8
Marsh tit (*Parus palustris*)	20.8 (5.9)	17.8 (7.2)	-15
Nuthatch (*Sitta europaea*)	40.1 (11.3)	37.2 (17.6)	-7
Crown nesters			
Goldfinch (*Carduelis carduelis*)	1.9 (1.2)	2.2 (0.9)	+14
Greenfinch (*Carduelis chloris*)	2.3 (1.6)	2.6 (1.8)	+10
Siskin (*Carduelis spinus*)	4.6 (4.3)	5.4 (3.0)	+17
Hawfinch (*Coccothraustes coccothraustes*)	49.2 (12.7)	44.7 (10.1)	-9
Wood pigeon (*Columba palumbus*)	12.0 (2.9)	12.6 (4.4)	+5
Red-breasted flycatcher (*Ficedula parva*)	10.7 (3.1)	12.1 (1.4)	+14
Chaffinch (*Fringilla coelebs*)	303.3 (39.5)	267.8 (43.9)	-12
Jay (*Garrulus glandarius*)	9.3 (1.4)	9.3 (5.0)	0
Icterine warbler (*Hippolais icterina*)	3.9 (2.8)	4.5 (2.0)	+16
Spotted flycatcher (*Muscicapa striata*)	21.6 (5.6)	18.3 (6.2)	-16
Golden oriole (*Oriolus oriolus*)	8.1 (2.9)	7.7 (3.0)	-5
Bullfinch (*Pyrrhula pyrrhula*)	5.5 (3.2)	4.6 (2.9)	-16
Goldcrest (*Regulus regulus*)	43.9 (15.8)	33.7 (16.1)	-23
Song thrush (*Turdus philomelos*)	68.9 (14.6)	59.8 (13.4)	-13

Lower densities of many birds observed during a crash of forest rodents and 1 year later are a very suggestive argument for predation limiting the numbers of birds breeding in the pristine temperate forest. However, predation must have interacted with other factors such as food supply in the breeding season (abundance of leaf-eating caterpillars in the preceding summer) and in winter (snow cover affecting food accessibility). In Section 2.8, we showed that an index of the combined influence of snow cover and caterpillar abundance explained 59% of the total year-to-year variation in densities of all birds. Spring densities of rodents (as an indirect index of predation rates on birds) added as another variable to the multiple regression analysis improved the total variation explained by a mere 2% ($R^2=61\%$, $n=19$ years, $p=0.001$). Thus, predation plays a minor role in the direct limitation of bird breeding numbers. The following examples will show how predation operated within the frames set by food conditions.

In Białowieża Forest, spring densities of woodpeckers and tits are determined chiefly by snow cover in the preceding winter; the higher the snow cover the fewer birds survive until spring. In woodpeckers, however, 6 years

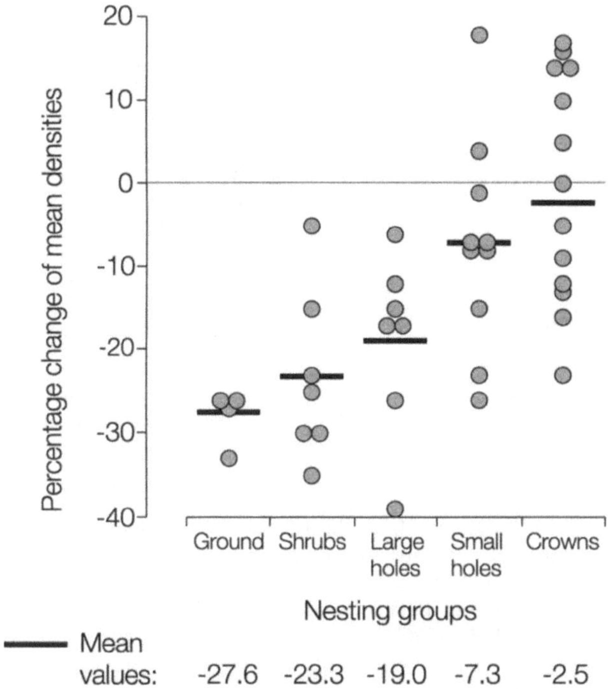

Fig. 5.40. Percentage change of mean densities of birds in years of low densities of forest rodents (crash and post-crash years, presumed heavy pressure of predators on birds, $n=6$ years), as compared with years of moderate and high densities of forest rodents (presumed average pressure of predators on birds, $n=13$ years) in relation to the height and type of nest location. Each point = one bird species. Data on bird densities in 1976–1994 from Tomiałojć et al. (1984), and Tomiałojć and Wesołowski (1994, 1996). Bird species and mean densities as in Table 5.19 with exceptions of *Phylloscopus sibilatrix* and *Ficedula parva*, which are shown separately in Fig. 5.41

of elevated predation rates caused about 30% lower densities than in years with 'normal' predation rates (Fig. 5.42). Interestingly, predators were equally effective in limiting woodpecker numbers in years of good as well as poor wintering conditions. In the 6 years, predation was additive to other factors of woodpeckers' deaths. In contrast, tits did not show markedly lower densities in the 6 years of presumably heavier predation (Fig. 5.42). This suggests that either predation on tits in those years did not increase markedly or that it did but predation was compensatory, not additive to other mortality factors.

The densities of chiffchaff and collared flycatcher strongly depended on the abundance of leaf-eating caterpillars in the preceding summer (that acted through the greater breeding success and the higher density of breeding birds in the following year). Within the greater influence of food supply, densities in the 6 years of increased predation were discernibly lower than in years of 'normal' predation rates. It is worth noting that the lowering of bird

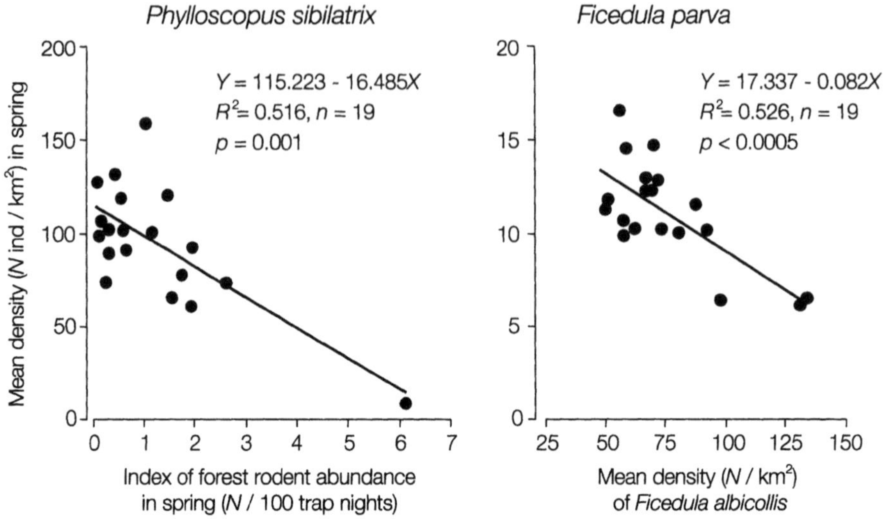

Fig. 5.41. Densities of two species of migratory forest birds: the wood warbler (a ground nesting species) shaped by avoidance of nest predation by forest rodents (bank voles and yellow-necked mice combined), and the red-breasted flycatcher shaped by competition with the collared flycatcher. Each point = one year. Data on bird densities from Tomiałojć et al. (1984) and Tomiałojć and Wesołowski (1994, 1996)

density by predation was much stronger when the food supply was poor than in years of abundant food (Fig. 5.42).

The above examples of bird species, genera, or families, albeit representative to their broad ecological groups, do not show the whole picture. Wesołowski and Tomiałojć (1997), who aimed at a detailed analysis of the 26 most common species of forest birds, have documented considerable variation in the responses of various bird species to changes in winter conditions, caterpillar abundance, and densities of forest rodents. Nonetheless, on the level of broad migratory groups, the overall pattern was manifest: densities of residents were shaped predominantly by winter conditions (mean winter temperature in the analysis by Wesołowski and Tomiałojć 1997), those of short-distance migrants by caterpillar abundance, rodent densities (as an indirect index of predation impact), and winter conditions; and densities of tropical migrants by caterpillar abundance.

Guided by the consistent results of the two analyses, by Wesołowski and Tomiałojć (1997) from the standpoint of avian ecology, and ours from the predation perspective, we propose a general model of the role of food and predation in shaping the densities and population dynamics of birds in Białowieża Forest. In 1976–1994, total breeding densities of resident birds varied about 2.5 times (Figs. 5.43 and 5.44), primarily in response to snow cover in the preceding winter (Table 5.20). In years of elevated predation impact, numbers of birds are not much lower than expected from winter conditions. Percentage increase in breeding densities of residents from year to year was shaped mainly by their own densities and snow cover (negative correla-

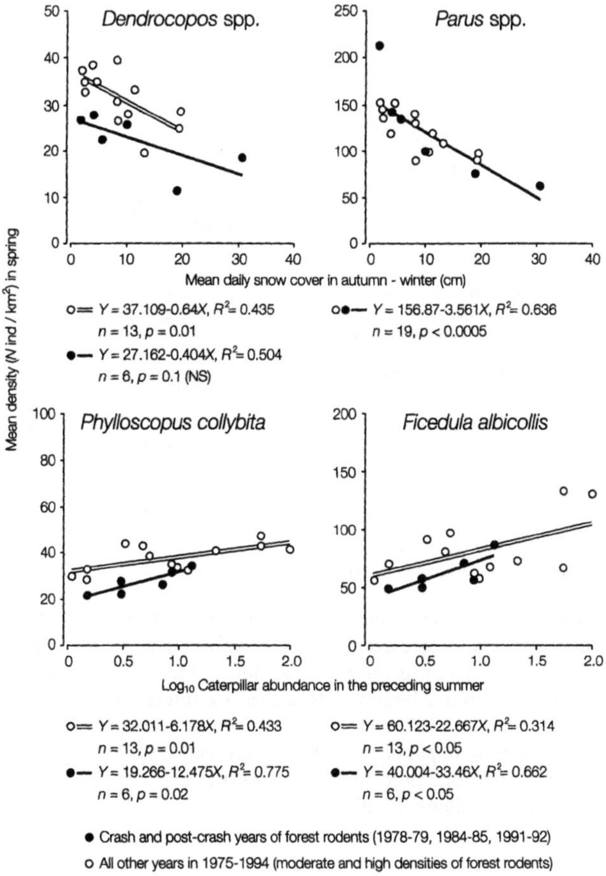

Fig. 5.42. Densities of some birds in BNP, in relation to food availability and predation. *Upper panels* spring densities of woodpeckers and tits (resident birds) in relation to depth of snow cover in the preceding winter, a factor decreasing access to food resources. *Lower panels* spring densities of chiffchaff and collared flycatcher (migratory insectivorous birds) in relation to abundance of leaf-eating caterpillars in the preceding summer. Each point = one year. *Open points* years with average predation pressure on birds. *Solid points* years with heavy predation on birds. Data on bird densities and caterpillar abundance from Tomiałojć et al. (1984), Tomiałojć and Wesołowski (1994, 1996), Wesołowski and Tomiałojć (1997)

tion in both cases; Table 5.20). Arguably, food availability in winter provides the ultimate check on residents' breeding numbers. Mortality inflicted by predators, however large, must be compensatory. We may predict that with no predators, numbers of resident birds in Białowieża National Park would not have been markedly higher. Also, a strong negative relationship between density of residents and the percentage increase in their numbers from year to year indicates that resident birds are regulated by competition for some resources, most likely the winter supply of food. It is worth noting that, in the pristine temperate forests of Białowieża, potential nesting sites (cavities,

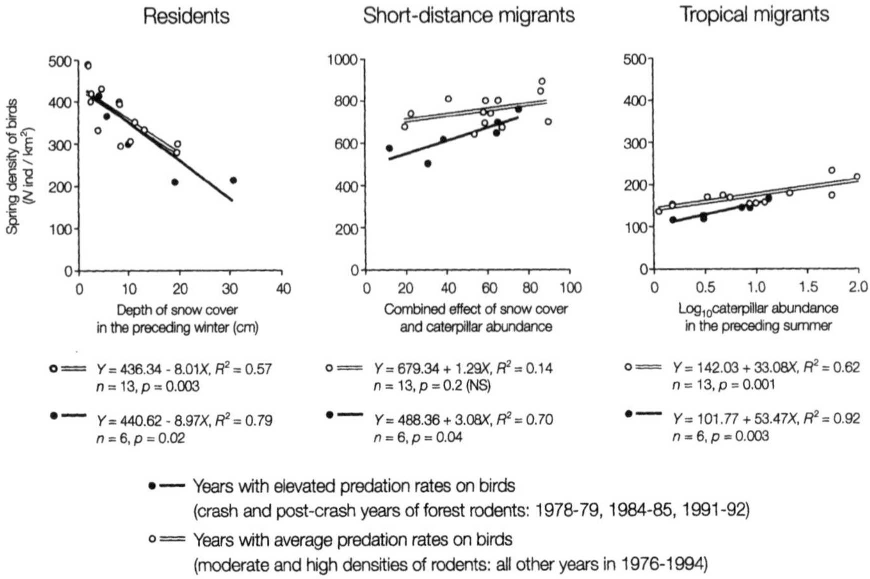

Fig. 5.43. Total densities of resident birds, short-distance migrants and tropical migrants in relation to food availability (snow cover and/or caterpillar abundance) in years of average and elevated predation pressure on birds. Tropical migrants do not include *Phylloscopus sibilatrix* and *Ficedula parva* whose densities are largely determined by interactions with rodents and interspecific competition, respectively (see Fig. 5.41). Index of combined effect of snow cover and caterpillar abundance *middle panel* as in Fig. 2.42 (Sect. 2.8). Data on bird densities from Tomiałojć et al. (1984) and Tomiałojć and Wesołowski (1994, 1996), on caterpillar abundance from Wesołowski and Tomiałojć (1997)

holes etc.) are abundant and are not regarded as a resource which would limit bird densities (Walankiewicz 1991; Wesołowski and Tomiałojć 1997).

In Białowieża Forest, breeding densities of tropical migrants (with notable exceptions of the wood warbler and the red-breasted flycatcher) are determined by food abundance, i.e. mainly caterpillars of winter moths (Table 5.20). It must be emphasised that the food supply acted most strongly via a time-lagged demographic effect: caterpillar abundance in year *n–1* caused higher production of young in that year and higher breeding density in year *n*. In 1976–1994, numbers of tropical migrants varied twofold in response to over 1500-fold fluctuations of caterpillar abundance. Breeding numbers of tropical migrants are largely determined by caterpillar supply. The percentage change of their numbers from year to year was influenced by their own density (negative effect) and caterpillar abundance (positive effect) (Table 5.20). A negative relationship between breeding densities and percentage increase in their numbers from year to year suggests that combined numbers of tropical migrants are regulated by some kind of competition. Furthermore, in years of poor food supply, elevated predation reduced breeding numbers of tropical migrants by about one third below the level that would otherwise occur, whereas with superabundant food, the reduction did not occur (Figs.

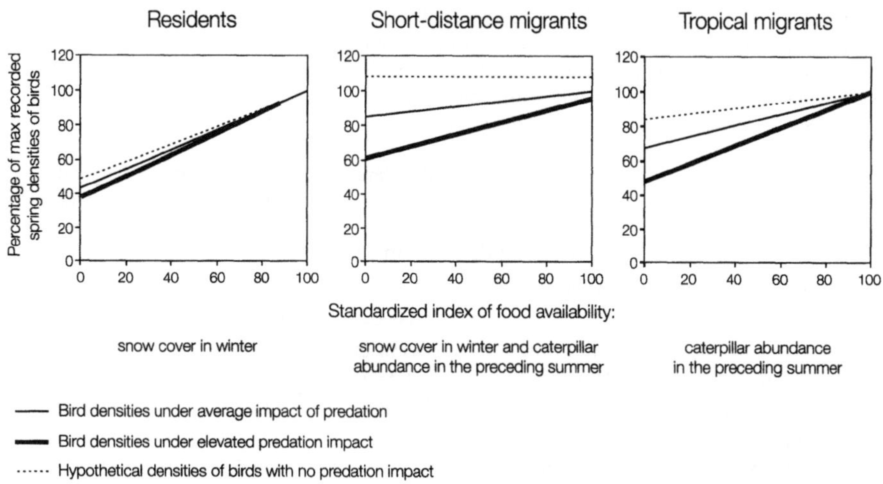

Fig. 5.44. Generalised model (based on empirical data shown in Fig. 5.43) of interactions of food availability and predation in shaping spring densities of birds breeding in pristine temperate forests. To enable comparison of the three ecological groups of birds, the best feeding conditions reported in 1976–1994 (the least snow cover and the most numerous caterpillars) were taken as 100%. Similarly, the highest densities of birds read from regression lines at average predation impact (see Fig. 5.43) were taken as 100%. Elevated predation impacts occur at 6–9 year intervals and lasts for 2 consecutive years (crash and post-crash of forest rodents)

5.43 and 5.44). In other words, the impact of predation decelerates with growing food abundance. Hypothetically, with no predation on tropical migrants, their maximum densities would have remained as observed but the minimum densities would have been about 1.7-fold higher than those observed. Predation increases the amplitude of tropical migrants' fluctuations but the fluctuations as such are driven by caterpillar outbreaks.

The situation of short-distance migrants is far more complicated. This is a very heterogeneous assemblage of birds. Some species rely heavily on leaf-eating caterpillars (e.g. the chiffchaff), others feed on soil invertebrates and supplement their diets with winter moth larvae (e.g. the blackbird; Tomiałojć 1994). Still others are more seriously affected by winter conditions than by food abundance during the breeding season (e.g. the wren; Wesołowski and Tomiałojć 1997). All this contributed to the fact that the bigger part of year-to-year variation in the numbers of short-distance migrants and their percentage increase remains unexplained (Table 5.20). Secondly, in the explained variation predation seems to contribute relatively more than in the cases of residents and tropical migrants.

Indeed, in years of elevated predation pressure, densities of short-distance migrants were a third lower when trophic conditions were poor (Figs. 5.43 and 5.44). Though the reduction of numbers by predators decelerated with improvement of food availability, it was still discernible during the best recorded trophic conditions. Thus, short-distance migrants, a group that in-

Table 5.20. Summary of multiple regression analysis of the role of various factors in shaping the combined numbers ($n = 19$ years) and percentage increase in numbers from year to year ($n = 18$) of three migratory groups of forest birds in Białowieża National Park, 1976–1994. Data on bird densities from Tomiałojć et al. (1984) and Tomiałojć and Wesołowski (1994, 1996), on caterpillar abundance from Wesołowski and Tomiałojć (1997)

Parameter	Resident birds	Short-distance migrants	Tropical migrants
Breeding densities in year *n*			
Percentage of total variation explained (R^2)	76%	37%	65%
Significance (*p*)	<0.0005	0.066 (NS)	<0.0005
Contributions by various factors to the total variation explained (sr^2):			
Snow cover in the preceding winter	(–) **39%**	(–) 9%	(–) 1%
Caterpillar abundance in year *n–1*	(+) 2%	(+) 8%	(+) **42%**
Spring density of rodents in year *n* (indirect index of predation on birds)	(+) 5%	(+) 5%	(+) 4%
Percentage change in breeding densities from year *n* to *n+1*			
Percentage of total variation explained (R^2)	59%	30%	60%
Significance (*p*)	0.005	0.15 (NS)	0.004
Contributions by various factors to the total variation explained (sr^2):			
Breeding density in year *n*	(–) **38%**	(–) **17%**	(–) **41%**
Snow cover in winter	(–) **34%**		
Caterpillar abundance in year *n*		(+) 6%	(+) **32%**
Change in rodent density from spring *n* to *n+1* (indirect index of predation on birds)	(+) 3%	(+) **14%**	(+) 7%

(+) Positive correlation; (–) negative correlation.
Other denotations as in Table 5.6.

cludes species preferred by predators (e.g. thrushes, robins, hawfinches, and woodpigeons), are most likely depressed by predation below what the pristine forests of Białowieża could otherwise sustain. This is corroborated by a rather weak relationship between combined densities of short-distance migrants and percentage increase of their numbers from year to year.

In years of rodent crashes, total predation on birds increased by one third, as compared with other years. Based on this information, we may estimate that breeding densities of short-distance migrants with no predators could have been maximally 20% higher than the average recorded in 1976–1994. Moreover, the importance of food availability (combined index of snow cover and caterpillar abundance) in shaping numbers of those birds waned with decreasing predation pressure (Figs. 5.43 and 5.44). When wholly free of predation, short-distance migrants are likely to maintain even more stable dynamics from year to year.

5.8 Predation on Amphibians

5.8.1 Species Structure of Total Predation on Amphibians

The life cycle of amphibians includes the stages of spawn and tadpoles that develop in water during a few months of spring and summer, and the stage of imagines which may live for several years, indeed over a decade, and lead more 'terrestrial' lives. The embryonic and larval forms of amphibians are largely beyond the feeding interest of carnivores and raptors. Fish, non-raptorial birds (e.g. ducks), and invertebrates may intensively harvest tadpoles. Once metamorphosed, amphibians begin a terrestrial life, and they become potential food to many predaceous animals, again not only carnivores and raptors but also other vertebrates such as water shrews (Buchalczyk and Pucek 1963), grass snakes and vipers (Pikulik et al. 1988), and black storks (Gavrin 1953). All these consumers of amphibians occur in Białowieża Primeval Forest. Thus, our analysis of amphibian consumption by raptors and carnivores covered only a part of the problem, undoubtedly a major part, because the 'true' predators are relatively large, year-round active homeotherms with relatively high food demands. Nonetheless, from an amphibian point of view they are not the sole enemy.

In the woodland landscape of Białowieża Primeval Forest, amphibians are distributed extremely non-homogeneously: in large numbers in wet habitats and negligible amounts in dry ones (see Sect. 2.10). Our estimates of mean annual predation on amphibians (as those on other groups of vertebrate prey) are averaged values for the whole Polish part of BPF that includes woodlands varying in humidity from very wet to dry, and open marshes in river valleys. Needless to say, such averaged estimates of predation say nothing about spatial variation in the role of amphibians in predator-prey relationships. That variation is obviously much bigger in frogs than in mammals or birds - animals of more homogeneous distribution in the temperate forests of Białowieża.

On average, carvivores and raptors annually harvest 714 g of amphibian per ha, 58% of this amount being consumed in spring–summer, and - interestingly enough - as much as 42% in autumn–winter when amphibians are hibernating. The most important amphibian-eaters were polecats (39%), tawny owls (32%), otters (8%), and raccoon dogs (8%; Table 5.21). A closer look at the contributions by various species reveals that the total predation on amphibians is exerted by two ecological groups of predators. First and foremost are semiaquatic mustelids: polecat, otter, and American mink. They heavily or even predominantly feed on amphibians and live near water, i.e. in habitats rich in amphibians. These predators possess adaptations to hunt for hibernating frogs. In the diets of polecat, otter, and mink, the share of amphibians is markedly bigger in the cold season than in the warm one. Thus, despite higher densities of predators in spring–summer (due to reproduction), their absolute predation on amphibians does not vary seasonally or is even somewhat greater in the cold season. Equally important is a group of

Table 5.21. Percentage contributions by various species to the total consumption of amphibian biomass by carnivores and raptors in spring–summer (16 April–30 Sept.), autumn–winter (1 Oct.–15 April), and the whole year. To calculate mean number of individuals eaten by predators from 1 ha, biomass was divided by mean body masses of frogs taken by various species. For the following species, these mean masses (estimated by measuring ilium width of frog remains recovered from scats or pellets) were: American mink 6.5 g, polecat 9 g, pine marten 10 g, otter 15 g, tawny owl 18.1 g. For other predators the following estimates were accepted arbitrarily: carnivores smaller than mink 6.5 g, terrestrial carnivores larger than pine marten 10 g, all raptors 18.1 g. Dominant predators are in boldface type (≥5% contribution)

Predator species	Spring–summer	Autumn–winter	Whole year
Polecat	**33.6**	**47.3**	**39.4**
Tawny owl	**35.7**	**25.9**	**31.6**
River otter	4.5	**12.9**	**8.0**
Raccoon dog	**10.7**	3.8	**7.8**
Badger	**6.1**	3.1	4.8
American mink	2.1	3.3	2.6
Common buzzard	3.5	–	2.0
Pine marten	1.2	1.4	1.3
Stoat	0.9	1.1	1.0
Weasel	1.1	0.9	1.0
Lesser spotted eagle	0.4	–	0.3
Red fox	0.2	0.2	0.2
Total biomass consumed (g/ha)	413	301	714
N individuals taken/ha	36	28	64

predators to which amphibians are secondary or accessory food. This group includes the tawny owl, raccoon dog, buzzard, badger, and several other species (Table 5.21). They all prey on amphibians more intensively in spring–summer than in the cold season since none of them is capable of digging for hibernating frogs or regular hunting in water. Consequently, their absolute predation on amphibians is much heavier in spring–summer than in autumn–winter. Furthermore, the spatial pattern of predation by these two groups differs. Semiaquatic mustelids take most of their prey from wet habitats (rivers, marshy river valleys, riverside ash-alder forests, and bog alderwoods), whereas exploitation of amphibians by all other predators is more evenly distributed over all woodland habitats wherever amphibians occur during their non-hibernatory phase of life.

In the total amount of amphibians eaten by predators, frogs (*Rana* spp.) strongly dominated (95% of biomass). Toads (*Bufo* spp.) made up 5%, only. Interestingly, in the consumption of toads, badgers played the chief role (48%), and were followed by common buzzards (22%), river otters (17%), and raccoon dogs (13%).

5.8.2 Magnitude and Mechanisms of Predation on Amphibians

Total consumption of frogs and toads by predators was compared with fairly rough (due to few data available) and conservative estimates of mean seasonal dynamics of amphibians in Białowieża Forest (Fig. 5.45). Carnivores and raptors appeared to make little use of available amphibian prey: they took about 30% of the standing crop in spring–summer and 10–15% in autumn-winter. In the cold season, losses inflicted by predation contributed, on average, 20% to the observed winter mortality of amphibians.

Analysis of the multiannual pattern of predation on amphibians was done for three species: the polecat that is specialised on frogs; the tawny owl which uses frogs as an important alternative prey, buffering for the declines of rodents; and the pine marten, which uses frogs as an accessory prey taken in small quantities (Fig. 5.46). Because no estimates of frog abundance were available for 1986–1991, predation impact by each predator was related to weather variables as indirect indices of amphibian availability. These variables were (1) the sum of rainfall in April–June that directly affects the activity of amphibians and improves their breeding success, and (2) mean temperature of autumn-winter which, we believe, represents the extent of ground activity of frogs in the cold season and their overwinter survival. If we consider both factors, the years of 1988 and 1989 (very or moderately rainy spring and mild winters afterwards) must have been most favourable for predators' exploitation of frogs (Fig. 5.46).

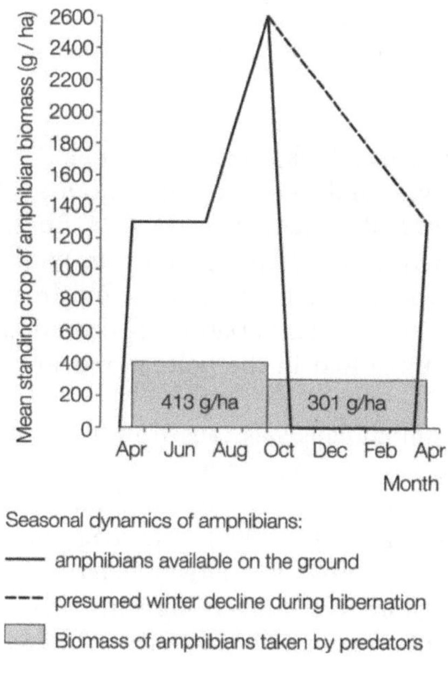

Fig. 5.45. Average predation impact on amphibians (biomass of amphibians taken by all carnivores and raptors from 1 ha) in the Polish part of Białowieża Primeval Forest, in spring–summer (16 April–30 September) and autumn-winter (1 October–15 April) compared with the average level of their abundance. See Section 2.10 for sources of data on amphibian abundance

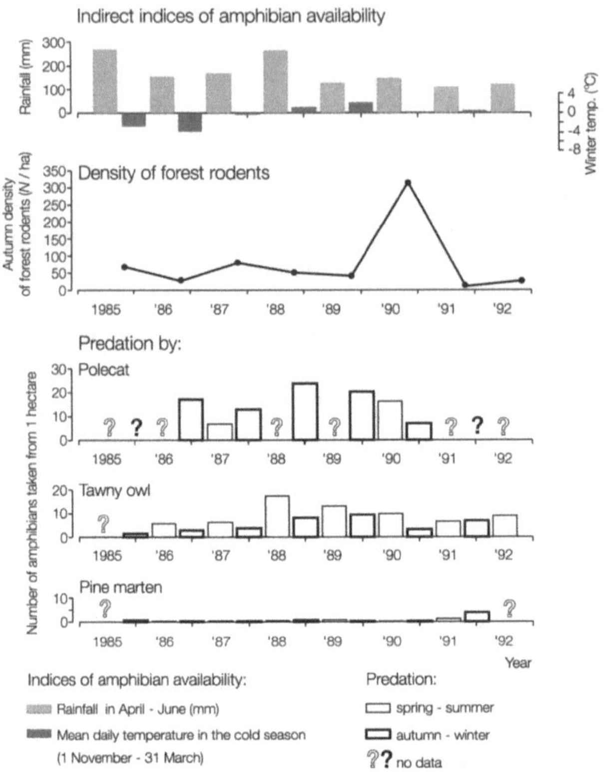

Fig. 5.46. Predation impact on amphibians by polecats, tawny owls, and pine martens in Białowieża National Park (1985/1986–1992) in relation to indirect indices of amphibian abundance and densities of forest rodents (bank voles and yellow-necked mice). Numbers and/or availability of amphibians increase with greater rainfall (in spring–summer) and warmer mean temperature (in autumn–winter)

In the pristine forests of Białowieża National Park, tawny owls took 2–9 frogs/ha (mean 5 frogs/ha) in autumn–winter and 6–17 frogs/ha (mean 9.5) in the spring–summer seasons of 1985–1992. Polecats harvested 9–24 frogs/ha (mean 16) in the cold season and 7–16 frogs/ha in spring–summer. While summer numbers of amphibians in Białowieża Forest can vary 20-fold from year to year, there was remarkably little interannual variation in predation rates by the chief frog-eaters: maximally three-fold in the polecat and four-fold in the tawny owl (Fig. 5.46).

Multiannual variation in predation on amphibians was caused by two factors: availability of amphibians and densities of forest rodents. Figure 5.46 shows that tawny owls' predation on frogs was highest in the wet summer of 1988 and in the mildest winters (1988/1989 and 1989/1990). The magnitude of spring–summer predation of tawny owls on frogs was largely determined by rainfall in April–June and mean temperature of the preceding winter, both factors having a positive effect (multiple regression analysis: $R^2 = 0.84$,

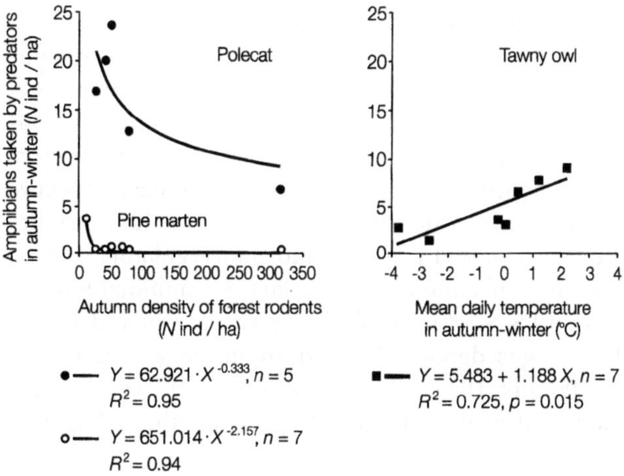

Fig. 5.47. *Left panel* polecat and pine marten predation on amphibians in autumn–winter seasons as shaped by autumn densities of forest rodents. *Right panel* tawny owl predation on amphibians in relation to mean daily temperature in the cold season (1 October–15 April). Each point denotes one season. Data on martens from Zalewski et al. (1995), on tawny owls from Jędrzejewski et al. (1996a)

$p=0.03$, $n=7$ years). Obviously, humid weather in spring–summer and a mild winter are conducive to more intense activity of amphibians. Furthermore, severe and long winters may take heavy tolls of hibernating amphibians and lower their numbers in the following breeding season.

Autumn–winter predation on amphibians by polecats, tawny owls, and pine martens was shaped by winter severity (acting through lowering the accessibility of frogs to predators) and density of forest rodents. The first factor was essential in tawny owls' impact on frogs, whereas the latter one was important for polecats and martens (Fig. 5.47). Years 1985–1989 were characterised by moderate (and not very variable between years) densities of forest rodents (see Fig. 5.46). In those years, rodent fluctuations had minor, if any, influence on predation on amphibians. During the rodent outbreak in 1990, polecats, carnivores specialised on amphibians, were engrossed in hunting rodents, which markedly relieved amphibians from predation. Pine martens captured very few frogs in years of high and moderate densities of rodents. It was only the prolonged crash of rodents in 1991/1992 that caused an increase in martens' impact on amphibians.

5.8.3 Roles of Predation and Other Factors in Shaping the Densities and Population Dynamics of Amphibians

Our attempt to answer the question of the role of predation in shaping the fluctuations and densities of amphibians will be rather preliminary. What we present below should be treated as a proposed hypothesis rather than a conclusive summary of fairly few data.

Factors that affect survival of amphibians from the egg stage to metamorphosis include chemistry of the water (especially acidity), drying of temporary ponds with spawn and tadpoles, harmful changes in water temperature, shortage of food, and last but not least, pollution of water. Severcov and Surova (1979) and Surova (1990) studied the role of predation in the total mortality of embryonic and larval common frogs and moor frogs near Moscow, Russia. They found that the greater the density, the higher was the mortality rate of amphibians, but predation played a minor and non-regulatory role in that process. Altogether, survival of embryonic and larval amphibians up to metamorphosis was very low (2.5–8.7%). Similarly, in various inland waters of Belarus, from 1 to 3.9% of eggs deposited by common frogs gave rise to froglets emerging on land (Pikulik 1985). Beattie et al. (1993), who studied the embryonic and larval survival of the common frogs in northern England, reported a similar figure of 2.1%.

Frog numbers in Białowieża Forest varied at least 20-fold between years. At the same time, the combined predatory impact by polecats, tawny owls and martens varied only three-fold between the minimum (in 1990/1991) and maximum (in 1988/1989) recorded figures. This alone suggests that predators have no biological capacities to keep pace with growing numbers of frogs increasing under favourable environmental conditions. On the other hand, predators that feed largely on amphibians (otter, polecat, mink) are 'conservative' in a sense that they continue to search and hunt for amphibians even during droughts or long and harsh winters.

In Chapter 2, we showed that summer densities of amphibians grew linearly with increasing amounts of rainfall in April–June of the preceding year, i.e. during the period of hatching of the youngest overwintered generation, which is the most numerous in the population. Surprisingly enough, the magnitude of total spring–summer predation on amphibians was not at all related to the previous year's rainfall, presumed to reflect amphibian abundance. In 1987–1991, predation plotted against rainfall in the preceding spring yielded a nearly constant relationship: $Y=29.1+0.004\,X$, which means that with rainfall varying from 125 to 263 mm, and presumed summer densities of amphibians ranging from 40 to 260 ind/ha, total predation on them in spring–summer was invariably about 30 ind/ha. These relationships are generalised in Fig. 5.48.

Assuming that total predation on amphibians in spring–summer is additive, we may estimate that their densities with no predators would be 30 ind/ha higher than those observed. It would mean that predation as a percentage of amphibian densities is strongly depensatory (i.e. inversely density-dependent): very heavy in bad years, when amphibians are scarce due to drought, and negligible in wet years, when amphibians flourish (Fig. 5.48). Heavy predation (>50% of summer densities of frogs taken by predators) occurred after years with less than 125 mm of rainfall in April–June. Throughout 48 years (1948–1995), such dry springs happened in 7 years (15%). Negligible predation (<15% of frogs taken by predators) occurred after wet springs (>200 mm rainfall), which happened in 19 years (40%; Fig. 5.48).

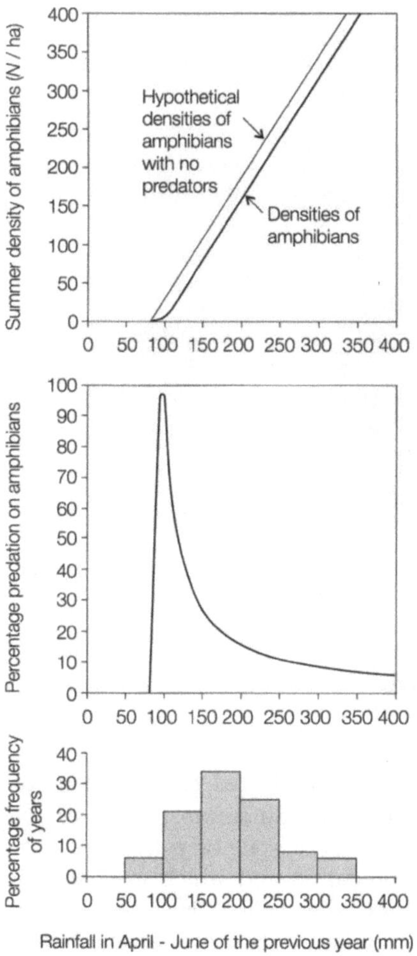

Fig. 5.48. Model of the role of predation in shaping the numbers of amphibians in Białowieża Primeval Forest. *Upper panel* summer densities of amphibians in relation to rainfall in April–June of the previous year (*thick line*, based on empirical data shown in Fig. 2.50) and densities enlarged by addition of the mean spring–summer predation on metamorphosed amphibians by all carnivores and raptors (*thin line*, based on empirical data shown in Fig. 5.46, corrected for contributions of all other predators). Note that the magnitude of predation did not vary with growing densities of amphibians. *Middle panel* total spring–summer predation on amphibians as percentage of their summer densities in relation to rainfall (April–June of the previous year) as the main factor shaping amphibian abundance. *Lower panel* percentage frequency of years with a given amount of rainfall in 1948–1995 in Białowieża Forest

In conclusion, we propose that abiotic conditions decisive to the breeding success of amphibians (and most likely to their winter survival, too) are prime factors shaping both densities and population dynamics of amphibians in the temperate forests of Białowieża (Fig. 5.49). Predation, which seems to be inversely density-dependent, only changes things from bad to worse during years of spring droughts, but it has an insignificant effect when the weather is favourable for amphibian reproduction.

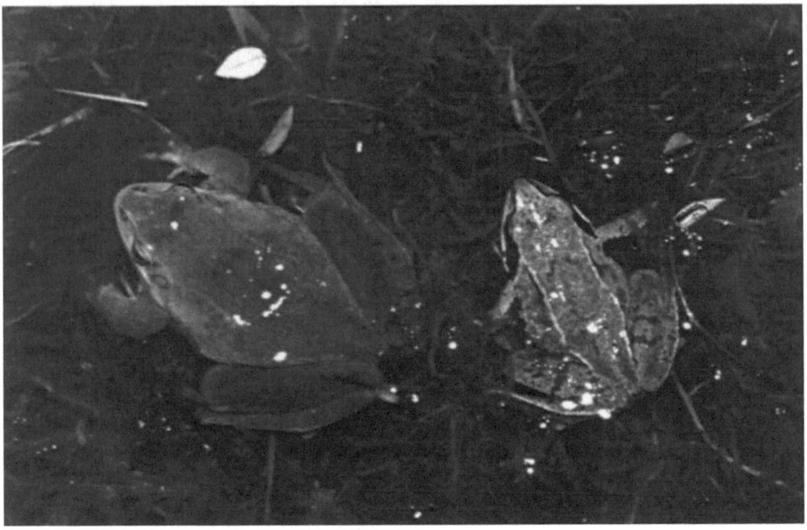

Fig. 5.49. Moor frogs in the spawning pond in spring. Numbers of frogs critically depend on rainfall in April–June, during egg and tadpole development. Photo: S. Wąsik

5.9 Intraguild Predation in Carnivores and Raptors of Białowieża Forest

Thirteen species of predators occasionally or regularly fed on specimens of one to six other species of predators (Table 5.22). Since most of cases reported in Table 5.22 are derived from the analysis of scats or pellets, we cannot say which of them represented actual predation and which were simply scavenging on a predator that had died from other causes. Nonetheless, the body size proportions (predators always being larger than prey species) suggest that in most cases it was actually an act of killing. Undoubtedly, scavenging (not listed in Table 5.22) involved a radio-collared lynx feeding on a carcass of a poached (and left in a snare) wolf and a raccoon dog feeding on a dead badger (observations of W. Jędrzejewski, K. Schmidt, and R. Kowalczyk). Instances of predation or attempted but unsuccessful predation that we recorded during our fieldwork in Białowieża Forest were more numerous. They included attacks by wolves and lynxes on raccoon dogs, killing of pine martens by lynxes, wolves' inspecting occupied badger setts, attacks by tawny owls on weasels, marten predation on raptor nests, and others (see Chap. 4).

The available data, including 164 cases of intraguild predation, allow the distinction of two different phenomena. First and most common are the cases when some smaller predators regularly fell as prey to larger ones. The main reasons for this is that the 'prey' species match in body size and habitat preferences the respective needs of the predator. The most striking examples are the lesser spotted eagle capturing weasels and stoats in the open

Table 5.22. Cases of intraguild predation in carnivores and raptors of Białowieża Primeval Forest. Information about predators being eaten by other predators is derived from analysis of scats, pellets, and prey remains on trails (data in Sections 4.3–4.23)

Predator	Species found as prey (number of cases in parentheses)
Wolf	Raccoon dog (5), lynx (2), red fox (2), badger (2), pine marten (1), tawny owl (1)
Lynx	Pine marten (3), raccoon dog (2), red fox (1)[a], long-eared owl (1), undetermined owl (1)
Red fox	Weasel (2), tawny owl (2)
Raccoon dog	Badger cubs (2), weasel (1)
White-tailed eagle	Raccoon dog (1)
Eagle owl	Tawny owl (2)
Lesser spotted eagle	Weasel (16), stoat (14), stoat or weasel (22), polecat (2), tawny owl (1)
Pine marten	Weasel (3), polecat (1), common buzzard (1), tawny owl (1)
Goshawk	Sparrowhawk (22), tawny owl (13), undetermined *Accipiter* (7), goshawk (4), long-eared owl (2), undetermined owl (2), polecat (1), undetermined marten (1)
Common buzzard	Weasel (11), tawny owl (8), stoat (1), little owl (1), Tengmalm's owl (1), goshawk or buzzard chick (1)
Honey buzzard	Polecat (1)
American mink	American mink (3)
Tawny owl	Weasel (2)

[a] After Kossak (1988).

marshes in river valleys, and wolves and lynxes deliberately hunting for raccoon dogs or badgers. Pine marten also belongs to this group, with its predation on eggs, hatchlings, and adults of raptors (see Chap. 3). We are also inclined to interpret the unusually high incidence of sparrowhawks and tawny owls falling as prey to goshawk as predator-prey interactions. Goshawks are specialised on hunting medium-sized forest birds, and the two smaller raptors may well fit the size class of goshawk's preferred prey.

The second phenomenon may be the result of interspecific competition and aggressive interactions, usually over food resources, that may end with smaller predators being killed by larger ones. Interestingly, the victim may be left uneaten (Fig. 5.50). Undoubtedly, cases of such interactions are very rare. They may include lynx and wolf killing smaller carnivores (namely pine martens and red foxes) that scavenge on their kills.

Generally, the larger the predator, the more species of other predators were found as its prey (Table 5.22). This trend is manifest in carnivores and diurnal raptors. It is obviously valid for owls as well, but our data on food of the largest species, the eagle owl, are too scarce to document it. Pugacewicz (1995) studied the diet of eagle owls in the whole north Podlasie region, and found goshawks (four individuals), common buzzards (three), tawny owls

Fig. 5.50. A young pine marten killed by lynx (a radio-collared female 'Sonia' with three kittens); Białowieża National Park, August 1994. A lynx began to eat the prey (note the marten's feet lying in front of it) but regurgitated it. Photo: W. Jędrzejewski

(three), long-eared owls (three), and a sparrowhawk among a total of 239 prey animals captured by eagle owls.

Mikkola (1983), who reviewed much literature on 13 species of European owls, found a positive correlation between body size of an owl species and its predation on other owls. He interprets this in terms of competition that often leads to fatal consequences for the weaker of the two conflicting individuals of different species. Having collated the data from Białowieża Primeval Forest, we are inclined towards the explanation that when larger predators kill and feed on smaller ones, it is most often a process of procuring food. Its 'side effect' may be annihilation of a competitor, if competitor it was.

The relative frequency with which a given species of predator fell as prey to other predators was related not so much to body size as to the commonness of a given species of 'prey' carnivore or raptor (Table 5.23). Two numerically dominant species of predators, the tawny owl and the weasel, were most often eaten by other predators. Indeed, predation on small mustelids has been so common (see Korpimäki and Norrdahl 1989) that some mustelid species possess well developed antipredatory traits and behaviours (Powell 1982; Weber 1988b).

Table 5.23. Percentage frequency of a given species of predator among all cases of predators found as prey of other predators ($N=164$)

Predator species	Percentage of all cases of predators being eaten by other predators
Weasel	28.7
Tawny owl	17.7
Stoat	15.3
Sparrowhawk	13.4
Raccoon dog	4.9
Pine marten	3.1
Polecat	3.1
Badger	2.4
Goshawk	2.4
Red fox	1.8
American mink	1.8
Long-eared owl	1.8
Lynx	1.2
Common buzzard	1.2
Tengmalm's owl	0.6
Little owl	0.6

5.10 Scavenging – Seemingly a 'Free Lunch', Sometimes at High Cost

5.10.1 Sources and Supply of Ungulate Carcasses

Predaceous mammals and birds which work hard for their food do not sneeze at carrion. Indeed, many carnivores and raptors scavenge more or less frequently. As an evolutionary adaptation of predators and an important process of functioning of animal communities, scavenging seems typical of open landscapes in Africa (e.g. Brown 1984, 1990; Henschel and Skinner 1990). Originally, however, scavenging was also widespread in the temperate forests of Europe, where it was based on a constant and predictable supply of ungulate carcasses. In those woodlands, now exemplified by Białowieża Forest, there were two major factors of ungulate mortality that made carcasses available to scavengers. First, large predators (wolves, lynxes, and in the past also the brown bear) kill ungulates but then eat most of the consumable tissue, usually leaving bones with little flesh to scavengers. Secondly, ungulates die from starvation, disease, and cold. In such cases, intact corpses can be found. Contemporarily, besides these natural factors, some man-caused mortality is also recorded (e.g. ungulates that have been gun-wounded, poached and left in snares, or killed in traffic accidents).

In Białowieża Forest, the roles of these factors were determined, based on the impressive sample of over 2400 ungulate carcasses found by game war-

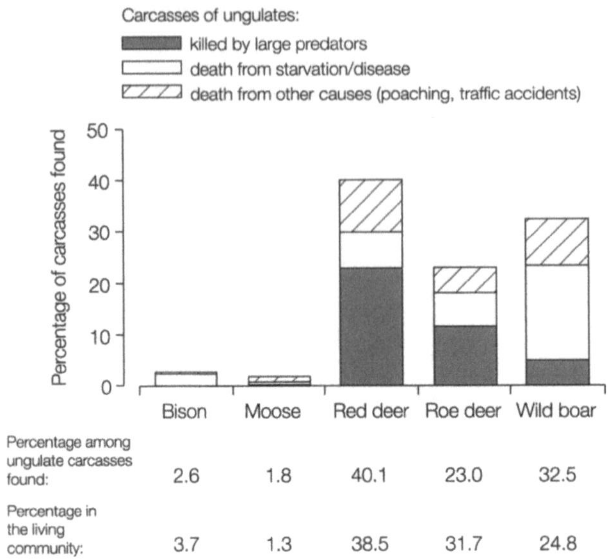

Fig. 5.51. Species structure and causes of death of ungulates whose carcasses were recorded in the Polish part of Białowieża Primeval Forest in 1984/1985–1993/1994 ($N=2411$ carcasses), compared with structure of the living community of ungulates in that period. Data from Okarma et al. (1995). Shares of each species in the total number of carcasses were directly proportional to their shares in the living community of ungulates ($r=0.94$, $p=0.02$, $n=5$)

dens and researchers in 1984/1985–1993/1994 (Okarma et al. 1995). As was to be expected, the proportions of carcasses of the five ungulate species reflected their shares in the living community of ungulates (Fig. 5.51). Red and roe deer carcasses were predominantly the leftovers of wolf and lynx meals, whereas wild boar were usually found as carrion of animals that had died from starvation and disease.

The number of wolf and lynx kills depend primarily on the population density of these predators. In 1985–1996, when wolf and lynx numbers were stable (see Chap. 3), there was a fairly constant, both in seasons and years, supply of ungulate remains for scavengers. The situation was different with ungulates that died from starvation, disease, and severe winter conditions. Their deaths were definitely more likely in winter; 13% of dead ungulates found throughout the year were recorded in December–January, 71% in February–March, 9% in April–May, and very few from June through November (Fig. 5.52). Interestingly, from the predators' perspective, the seasonal peak of carcass abundance (late winter) coincides with the lowest numbers and accessibility of rodents, birds, and amphibians.

Multiannual variation in the number of ungulate carcasses available to scavengers is very large. Suffice it to say that game wardens of the Białowieża, Hajnówka, and Browsk State Forestry Districts (eight people working with similar intensity year after year, with some help from other

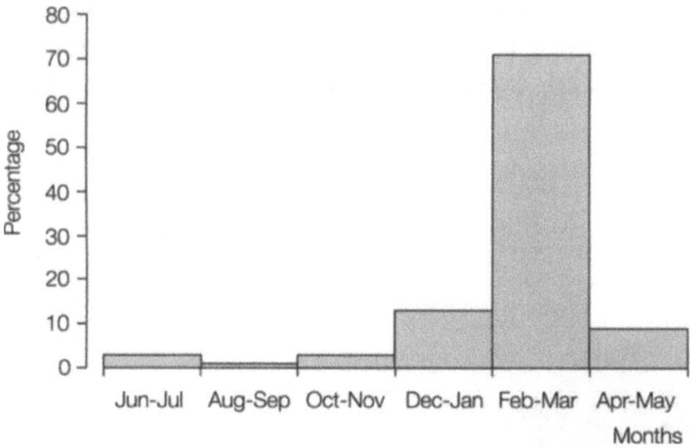

Fig. 5.52. Seasonal variation in 'supply' of carcasses of ungulates that die from diseases/starvation. *Bars* show number of carcasses found in a given bimonthly period as percentage of all carcasses recorded throughout a year. $N = 179$ fresh carcasses recovered in Białowieża National Park in 1986/1987–1995/1996

forestry services) annually registered from 94 carcasses in 1989/1990 (September through August) to 2452 in 1969/1970 from about 510 km^2. There are four main factors of this variation: density of ungulates, density of large predators, severity of winter, and oak seed crops (the last factor principally affects survival of wild boar). Harshness of winter is responsible for the extreme values in the whole range of variation. With the above-mentioned smallest and largest numbers of registered carcasses, one should know that the winter of 1989/1990 was the warmest one on record, with no snow cover and mean daily temperature from November through March of 2.2°C, whereas that of 1969/1970 was characterised by a mean temperature of −4.1°C, and snow cover that reached 95 cm and persisted for 4 months. Wild boar, which root for food in the humus and upper layer of the soil, are more susceptible than other ungulates to death with increasing snow cover (Okarma et al. 1995). Multiannual variation in acorn crops affected the death rates of wild boar, too (Fig. 5.53). Their survival was very good in winters after heavy crops (e.g. in 1989/1990) and poor in the following year when oak trees were barren. Consequently, between 1986 and 1993, the supply of wild boar carcasses varied enormously from year to year, as shown by records from the pristine forests of Białowieża National Park (Fig. 5.54).

Although we know many parameters of seasonal and multiannual variation in abundance of ungulate carcasses, it is very difficult to give precise estimates of their absolute numbers available per unit area as well as per unit time. Rough estimates may be derived from the following data. During the decade of 1986–1996, registering of ungulate carcasses showed that the three main sources of carcasses (wolf predation, lynx predation, and starvation or disease) 'supplied' comparable numbers of them (32, 34, and 34%). Based on these proportions, we calculate that, if lynxes annually killed 220 ungulates/100 km^2 and wolves 130 ungulates/100 km^2, then most likely about 150–200 ungulates/100 km^2 die annually from other causes. (Note that mortality of ungulates due to the hunting harvest is ignored here, because shot ungulates are taken away from the forest and, except for discarded guts, are not available to scavengers.)

Fig. 5.53. Carcass of a young boar (dead from starvation during severe winter) fed upon by scavengers; Białowieża National Park, winter 1986/1987. Photo: K. Zub

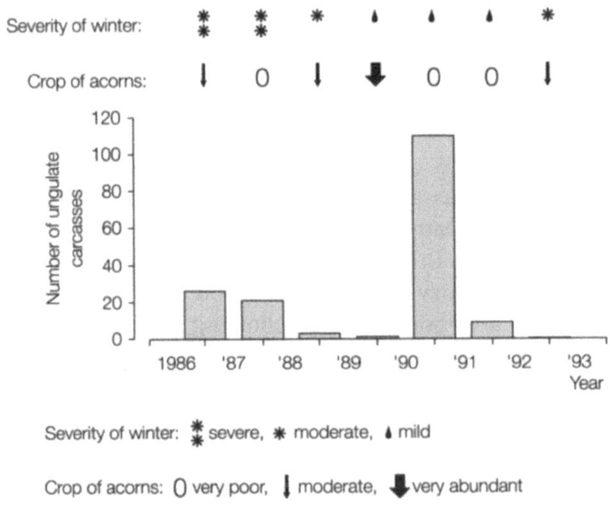

Fig. 5.54. Interannual variation in the number of carcasses of wild boar that had died from disease/starvation (recorded in Białowieża National Park in the cold seasons of 1986/1987–1992/1993). Acorn crop and severity of winters were the most important factors shaping this variation. Based on data from: Okarma et al. (1995)

5.10.2 Utilisation of Carcasses by Scavengers

Analyses of diets of over 30 species of Białowieża's predators (Chap. 4) have shown that the following species scavenged on ungulates (listed in order of decreasing shares of carcasses in their diets): raccoon dog, red fox, common buzzard, red kite, pine marten, hobby, lesser spotted eagle, polecat, and badger. Moreover, snow and radio-tracking and visual observations have demonstrated that wolf, lynx, white-tailed eagle, goshawk, and tawny owl sporadically scavenged, too.

Medium-sized and small predators consumed annually, on average, 316 g of ungulate carcasses per hectare (Table 5.24). Since the majority of that biomass was eaten in winter and early spring, little seasonal variation was detected if the total consumption was shown in two seasons (spring–summer and autumn–winter). Red foxes appeared to be the most significant scavengers, followed by raccoon dogs, buzzards and pine martens. Only one of these species, the raccoon dog, feeds on ungulate carcasses as a staple food. But even in this case it is a local rather than general feeding strategy of this carnivore. It results from the abundance of ungulate carcasses in the pristine forests of Białowieża. In other regions of Europe, where raccoon dog has spread, it is omnivorous and does not obligatorily rely on ungulate carcasses (e.g. Kritskaya 1961; Samusenko and Golodushko 1961; Włodek and Krzywiński 1986; Kauhala et al. 1993).

None of the other (native) predators relies on ungulate carcasses as primary food. To red foxes and pine martens carcasses are a secondary resource, utilised predominantly in the cold season. Consumption of ungulate carcasses by foxes and martens is governed by changes in availability and numbers of their primary prey (*Microtus* voles in the case of fox, and forest rodents in the case of marten) and not the availability of carcasses

Table 5.24. Percentage contributions by various species to the total consumption of ungulate carcasses (from an average hectare) by carnivores and raptors in spring–summer (16 April–30 Sept.), autumn–winter (1 Oct.–15 April), and the whole year. Dominants are in boldface type (≥5% contribution)

Species	Spring–summer	Autumn–winter	Whole year
Red fox	**52.0**	**63.5**	**57.4**
Raccoon dog	**34.6**	**26.4**	**30.8**
Common buzzard	**10.1**	2.8	**6.7**
Pine marten	2.5	**6.4**	4.3
Polecat	–	0.8	0.4
Badger	0.3	0.1	0.2
Lesser spotted eagle	0.4	–	0.2
Hobby	0.1	–	+
Total biomass consumed (g/ha)	166	150	316

(Jędrzejewski and Jędrzejewska 1992; Jędrzejewski et al. 1993c). Obviously, to some predators an opportunity to feed on carrion means a chance to survive critical periods.

Complementary data were collected by registration of signs (tracks, scats) and visual observations of scavengers at ungulate carcasses of various origin. The list of recorded scavengers includes at least 18 species, of which 12 were 'true' predators: carnivores and raptors (Table 5.25). Red foxes were most frequently noted (at 44% of all scavenged carcasses), followed by ravens, wild boar, and raccoon dogs. Ravens and boar usually came to carcasses in groups, so in terms of number of scavenging individuals they probably outnumbered foxes. In winters it was not uncommon to watch ravens flying from various directions, attracted by the croaks of those which had already spotted a carcass. We once counted up to 40 ravens excitedly flying around a carcass and trying to feed on it. Similarly, wild boars, which live in herds, were frequently commensals while feeding on carrion. Raccoon dogs live in pairs and often roam and forage in pairs. Other species, perhaps with the exception of small birds such as tits, live solitary lives and feed on carcasses solitarily.

Animals, for which scavenging is important or – in some periods – even a crucial food, possess adaptations to find dead ungulates and usually show non-random use of carcasses of various origin. To understand the prefer-

Table 5.25. Frequency of scavenging by various animal species at ungulate carcasses, as noted by signs (tracks, scats) left at carcasses and by visual observations; Polish part of BPF, 1985/1986–1995/1996. Figures are number of carcasses fed upon by a given scavenger as a percentage of all carcasses utilised by at least one scavenger. In each column, the sum of percentages is >100%, because many carcasses were fed upon by >1 species of scavenger. The dominant in each column is in boldface type

Species of scavenger	Percentage of the total number of carcasses utilised by scavengers			
	All carcasses[a] (N=231)	Ungulates killed by wolves (N=69)	Ungulates killed by lynx (N=55)	Dead ungulates (N=67)
Red fox	**44.2**	49.3	38.2	**52.2**
Raven	36.8	**53.6**	49.1	25.4
Wild boar	32.5	39.1	**58.2**	13.4
Raccoon dog	16.0	7.2	9.1	28.4
Pine marten	15.1	17.4	10.9	13.4
Jay	5.6	5.8	7.3	7.5
Common buzzard	4.8	8.9	3.6	3.0
White-tailed eagle	3.0	8.9	1.8	–

[a] This group also includes ungulates with an undetermined cause of death. Beside the listed species, the following scavengers were recorded: wolf 3.0%, lynx 1.3%, polecat, badger, goshawk, tawny owl, stray dog, woodpecker, tit, red squirrel 0.4% each, undetermined bird 1.3%, and undetermined scavenger 7.8%.

ences of scavengers, we have to look more closely at the types of available carcasses. Ungulates killed by wolves are usually big animals (mostly red deer), and are quickly and completely eaten by a pack within a few hours. A little flesh and soft tissue, most of the bones (especially the head, spinal cord, and legs), hide, some intestines and stomach content are left. Remnants of wolf kills are relatively easy to locate due to the vocal activity of wolves (howling) and signs of their 'gluttonous' feeding: pieces of prey are torn apart, blood (especially on snow) is seen in a radius of few metres, and soon the scent is discernible even to a less sensitive human nose. Wolves sometimes return to places of previous kills, if the prey is large or if their next hunting attempt fails, but they do not guard the kill for days. Remains of wolf kills had the highest frequency of scavengers recorded at them (63%). Here, the most common scavengers were ravens (Table 5.25), which were found at wolf kills more often than would have been expected at random. Indeed, in the primeval temperate forests, ravens seem to be evolutionarily adapted to search for and feed on wolves' kills. This fascinating subject needs a thorough investigation. So far, our observations indicate that ravens are able to recognise and decode wolves' behaviour and follow them by air (as foxes do on the ground). There is much rivalry between wolves and the harassing ravens.

Ungulates killed by lynxes are not easy targets for scavengers. Except for mother and kitten groups, lynxes live solitary lives. A successful hunt on roe or red deer provides a lynx with food for a few to several days, if a carcass is successfully guarded. The sophisticated methods of lynx's hiding, covering, and protecting the prey are described in Chapter 4. They all serve one purpose: to conceal the kill from scavengers searching by visual and olfactory cues (Fig. 5.55). Carcasses of roe deer, which are small and can be easily transported by a lynx to concealment, had the lowest frequency of scavengers' visits (28%). Interestingly, red deer are too heavy for a lynx to move to concealment and properly cover (Jędrzejewski et al. 1993b). Despite a lynx's efforts to guard them, as many as 77% of red deer killed were fed upon by scavengers. The most notorious scavenger at lynx kills was the wild boar (Table 5.25). This omnivorous animal possesses an excellent sense of smell, sensitive to the slightest odour of decaying meat. This enables wild boars to find even the well hidden kills. Worse still, herds of wild boar and single adult males were reported to take over the kill from a lynx and to force the host to abandon it. Scavengers were reported at 39% of all ungulates killed by lynxes.

Dead ungulates (mostly wild boar) are always intact carcasses, neither concealed nor exposed. They offer much more biomass of food than do leftovers of wolf and lynx kills. The majority of them (58%) were utilised by scavengers, predominantly by red foxes and raccoon dogs (Table 5.25). Foxes did not particularly select dead ungulates, but raccoon dogs strongly preferred this source of carcass (Fig. 5.56). In contrast, the wild boar very rarely scavenged on dead ungulates, still less on dead conspecifics. The risk of disease or parasite transmission may be high for animals scavenging on ungu-

Fig. 5.55. Roe deer killed by a lynx (a radio-collared male 'Makary') and completely covered with litter to protect it from scavengers. Białowieża State Forestry District, early spring 1991. Photo: D. Haycock

lates that had died from disease. Perhaps this factor explains rare cases of cannibalism in wild boar, and their frequent scavenging on wolf and lynx kills, despite the fact that these two predators (especially the wolf) are effective enemies of boar.

Not only were adaptations to find carrion decisive for the choice of ungulate carcasses by scavengers; also important seemed the potential danger of being eaten by the host predator. 'Free lunch' at a carcass may actually be inflicted with deadly risk and most animals scavenge only if faced with a serious shortage of food. The cases of raccoon dog and pine marten illustrate the choice of carcasses affected by predation risk. The raccoon dog, a slow clumsy animal, falls as prey to wolves and lynxes. Not surprisingly, it strongly avoids feeding on the remains of their kills (Fig. 5.56). Furthermore, other scavengers may be dangerous to this species. In the winter of 1995/1996, a raccoon dog that came to feed on remnants of red deer killed by wolves was captured by a white-tailed eagle using the same carcass. Pine martens that may escape into trees are rather safe from wolf attacks (so they frequently feed on the leftovers of their kills) but not so safe from lynx attacks. Three martens killed by predators in 1985–1996 fell as prey to lynxes. No wonder that martens scavenged at remains of lynx kills less often than expected at random (Fig. 5.56).

Ungulate carcasses often became crowded 'fast food bars' with up to six commensal species of scavengers recorded. Their interactions at a carcass, temporal sequence of feeding, possible competition and tolerance remain a really nice (though not easy) task for future researchers. Based on post factum recording of signs and tracks, we counted, on average, 1.8 species of scavengers having fed on a carcass utilised by any scavengers (range 1–6); 54% of them were fed upon by one species, 24% by two, 15% by three, and 7% by four to six species. Interestingly, remnants of wolf and lynx kills were

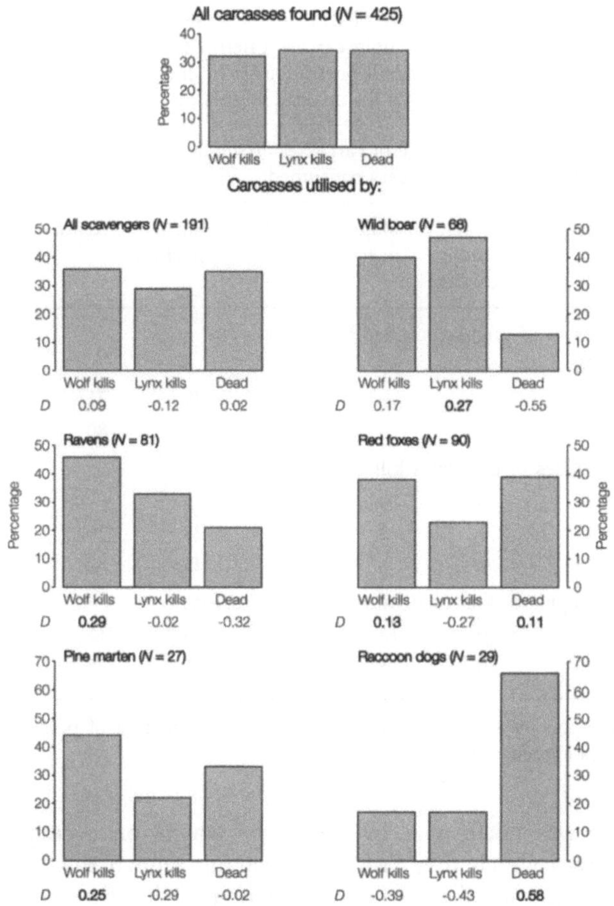

Fig. 5.56. Selectivity of various species of scavengers towards carcasses of ungulates killed by wolf, lynx and those that had died from disease and/or starvation. Selectivity index *D* (see explanation to Fig. 4.2) ranges from –1 (complete avoidance) to 0 (scavenging proportional to occurrence) to 1 (the strongest positive selection)

usually utilised by more species (on average 2, range 1–6; 47% by one species, 23% by two, 20% by three, and 10% by four to six species), than carcasses of dead ungulates (on average 1.5 species at each carcass visited by scavengers, range 1–4; 67% fed upon by one species, 21% by two, 7% by three, and 5% by four species of scavengers).

Based on a large enough sample of carcasses, we were able to visually assess the degree of their utilisation by the host predator (wolf or lynx), and scavengers pooled into two groups: wild boar and all other scavengers. Wolves consumed on average, 88% of their kills, and both wild boar and other scavengers were always finishing up the leftovers (Table 5.26). Lynxes also utilised a major part of their prey (on average 75%), but in some cases

Table 5.26. Utilisation of ungulate carcasses by host predators and scavengers (data from registration of ungulate carcasses in the Polish part of BPF in 1985/1986–1995/1996). Degrees of consumption (assessed visually): nearly whole carcass eaten, consumption of 76–100%; major part, 51–75%; minor part, 26–50%; small portion, 5–25%; and remnants, <5% eaten (scavengers feeding on remnants of wolf/lynx kills or dead ungulates)

Predator or scavenger feeding on a carcass	Sample size (*N*)	Mean percent use of a carcass [a]	Percentage of carcasses at which predators or scavengers consumed:				
			Nearly whole carcass	Major part	Minor part	Small portion	Remnants
			Carcasses of ungulates killed by wolves				
Host (wolf)	135	88	69	19	9	3	–
Wild boar	24	10	–	–	–	29	71
Other scavengers	60	9	–	–	–	20	80
			Carcasses of ungulates killed by lynx				
Host (lynx)	147	75	48	14	29	9	–
Wild boar	32	43	–	9	38	28	25
Other scavengers	42	13	–	–	–	43	57
			Carcasses of ungulates that died from starvation/disease				
Wild boar	15	64	40	13	20	13	13
Other scavengers	61	40	10	3	38	41	8

[a] Only carcasses that were fed upon by a given species are included, so the sum of percentages in each class of carcasses exceeds 100%.

prey not completely eaten was taken over by wild boar which, on average, utilised 43% of the carcasses which they fed upon. Having known the frequency of wild boar scavenging at all lynx kills, we estimated that they consumed roughly 10% of biomass of all lynx prey. Other scavengers (medium-sized and small carnivores and birds) contented themselves with leftovers of lynx meals. It was at the carcasses of ungulates dead from starvation and disease where wild boar and other scavengers could really gorge (Table 5.26). Wild boar consumed on average 64% of carcasses they fed upon (roughly 5% of all available biomass of dead ungulates), and other scavengers utilised 40% of biomass of visited carcasses (and 15% of all carcasses).

Undoubtedly, in the primeval temperate forests, scavenging was a widespread ecological phenomenon (Fig. 5.57). All species which intensively utilised carrion (foxes, ravens, raccoon dogs) are those which – facing the disappearance of their ancestral habitats – easily adapt to scavenging on garbage dumps.

Fig. 5.57. Raven *Corvus corax* scavenging on a carcass of dead bison; Czerlon Marsh, Białowieża State Forestry District, late winter 1995/1996. Photo: J. Chyra

5.11 Predation in Vertebrate Animal Communities of the European Temperate Forests – An Attempt at Synthesis

Our aim in this final chapter is to summarise the main features of predation on vertebrates in Europe's temperate woodlands, as revealed by studies in Białowieża Primeval Forest. Focused on predator ecology and predator-prey relationships, the subject is also embedded in the lower trophic levels (relationships between prey animals and their food resources: plants or invertebrates) and in the habitat and abiotic settings. One must bear in mind that all data presented in this book, albeit large and long-term, came from observational, descriptive studies and were usually subjected to regression analysis. All that such analysis can show is correlations and a *potential* explanation of variation, but not a proof of cause-and-effect relationships. However obvious the results may seem (e.g. correlation of rodent numbers with abundance of their food), we have to be aware that when attributing causality to correlations, we proceed from facts to interpretation. In the field studies conducted on animal ecology in Białowieża Forest, an experimental approach was very rarely applied. Apart from unintended manipulative 'experiments' with predator extermination that were done a few times in the 19th and 20th centuries, scientific experiments on predator impact were not performed.

Herbivorous and insectivorous vertebrates of Białowieża Forest fluctuated widely in numbers. The abundance of their food (plant material and invertebrates) showed very pronounced seasonal changes and manifold multiannual variation, particularly the summer (seasonally highest) abundance. The range of variation in food supply predetermined the potential variation of consumer densities. From that standpoint, we compared the results of studies on ungulates, rodents, shrews, birds, and amphibians in order to draw tentative conclusions as to: (1) whether fluctuations in food resources were the sole important factor or whether other factors changed the events in prey population dynamics as well, (2) whether any factors were effective enough to limit the numbers of those animals when, in conditions of good food supply, they approached the upper limit of recorded densities, (3) what factors acted on prey populations most seriously when their densities were low, and (4) what factors affected the breeding success and rate of population increase in various prey groups.

In all listed groups of vertebrates, predators exerted the heaviest impact and mattered most in situations of low densities of their prey (Table 5.27). In all studied predator-prey relationships, percentage predation compared with prey density was inversely density-dependent over the whole range of prey densities except very low ones, when the predation rate declined rapidly (models III and IV in Fig. 5.1). Such shapes of predation curves resulted from a notable 'conservatism' of predators in their dietary and/or numerical responses to quick changes in prey numbers (see syntheses to Chaps. 3 and 4). But even a very heavy predation impact on low densities of prey could not prevent the increase of prey numbers when food or abiotic conditions improved.

Such a significant role of predation in populations of temporarily low densities explains the antipredatory behaviours gained in evolution by many prey species. Percent mortality due to predation is very high in such bottleneck situations, thus individuals with efficient antipredatory behaviour will be strongly selected for.

Table 5.27. Summary of the roles of food resources (F), predation (**P**), abiotic weather factors (A), and human impact (H) in shaping the population dynamics, densities, and breeding success of the main species or groups of vertebrate prey in Białowieża Primeval Forest. Based on results presented in Sections 5.4–5.8

Prey	Factors important in limiting low densities of prey (below the level set by food abundance)	Factors important in limiting high densities of prey	Factors affecting breeding success and rate of population increase	Factors shaping seasonal and multiannual fluctuations of prey numbers
Roe deer	**P** H	F **P** A H	F **P** H	F **P** A H
Red deer	**P** H	F A H	F **P** H	F **P** A H
Bison, moose	H	F H	F H	F H
Wild boar	**P** A H	F A H	F **P** A	F A
Forest rodents	**P**	F	F **P**	F
Shrews	**P**	F	F **P**	F
Resident birds	**P** A	F A	F **P**	F **P** A
Short-distance migrants	**P**	F **P**	F **P**	F **P** A
Tropical migrants	**P**	F	F **P**	F **P**
Frogs	**P** A	A	A	A

Besides predation, abiotic factors (inclement weather) and human exploitation were also capable of lowering the already low densities of prey below the level expected from food supply alone. In resident birds and wild boar, weather interacted with food: deep snow cover obstructed the animals' access to food resources. In frogs, water is sine qua non for reproduction, thus rainfall in spring–summer probably always limits frog densities below the level potentially allowed by food supply (invertebrates).

All species of ungulates have at some time been limited by humans to below the level expected from food supply, be it poor or abundant. Indeed, exploitation by humans was the only type of 'predation' that could have led to extermination of 'prey' (bison and moose).

Another situation when predation appeared rather important was reproductive success of prey (Table 5.27). Young of many animals are particularly susceptible and attractive to predators. In European temperate forests, reproductive success of bison and moose, the largest land animals, is not really threatened by predation. Amphibians' breeding success (always low, averaging perhaps 2%, see Chaps. 2 and 5) does not seem to be much affected by predation, either. Most probably abiotic factors impose such a serious check on numbers of embryonic and larval amphibians that predation on them is insignificant. All other species and groups of vertebrates listed in Table 5.27 were found to suffer heavy predation compared with their reproductive potential or high losses due to predation during the breeding season. However, small vertebrates seem to produce excessive numbers of offspring, most of which have no chance of being recruited into the breeding population anyway. This is a 'doomed surplus' (Errington 1946) that predators may take with small consequences to prey populations.

Nonetheless, at least in deer (roe and red deer) and many birds, predators lowered the population growth or breeding success so much that the effects of their actions were lower densities of prey in the following years and/or prolonged limiting of prey numbers on a level markedly below that determined by food resources. Definitely, in prey with a slow rate of reproduction (ungulates and birds, see Table 2.14 in Chap. 2), temporarily heavier predation will take not only the 'doomed surplus' but some of the 'iron reserve' as well.

In conditions of excellent food supply, when prey populations flourished and attained high densities, the role of predation was insignificant (Table 5.27). In two cases only (roe deer and birds of short-distance winter migrations) were the predators capable of limiting prey numbers notably below the level set by food resources. The cases of red deer and wolf are uncertain, because prolonged extermination of wolves by humans only rarely allowed them to build up high densities. In most other vertebrates, in periods when prey was numerous, predators harvested bigger numbers of prey than ever, but compared to densities, percentage predation was very small. Moreover, good reproduction compensated for the losses from predation. In effect, ungulates (with the exception of roe deer), rodents, shrews, and many birds increased in densities until limited by availability of food resources. We suppose that amphibians rarely if ever reach the level of densities limited by food supply. As cold-blooded animals, they are resistant to temporary shortages of food but are rather susceptible to dry and cold weather. Thus, they seem to be limited by abiotic conditions at low as well as high densities.

Abiotic conditions, namely deep snow and hard frosts, could also reduce the numbers of deer, wild boar, and resident birds, and they acted in a density-independent way. Mortality during harsh winters occasionally took the form of mass deaths, when weather (cold) interacted with food accessibility (deep snow cover).

The overall role of predation in population dynamics of prey was usually subordinate: seasonal and multiannual fluctuations of prey numbers were driven by varying availability and abundance of food or by interactions of abiotic conditions and food (Table 5.27). Predators were not able to regulate prey densities (sensu Sinclair 1989), neither could they essentially change the pattern of prey fluctuations. In addition to deepening the phases of low densities in probably all vertebrate prey, predators markedly modified the course of fluctuations in two prey groups only, deer and birds, both characterised by fairly low reproductive rates. Large predators slow down the rate of deer population increase and may even limit their densities below the carrying capacity of the habitat (roe deer-lynx). Interestingly, in this case, limitation occurs at a prey density yielding the biggest annual production (in absolute measures). In this sense, lynx are prudent predators; they can stabilise their prey population around the level of the maximum sustained yield, to use the terminology of wildlife harvesting (Caughley and Sinclair 1994).

The modifying impact of predation on bird fluctuations was exerted in years of very low densities of forest rodents (in 1–2 of every 6–9 years) and was based on predators' dietary shift from rodents to birds. As proposed by an alternative prey hypothesis (Angelstam et al. 1985), cycles or any regular fluctuations of rodents in the northern hemisphere have consequences for other small terrestrial vertebrates, also exploited by rodent predators.

'Predation' by humans (in the form of a hunting harvest or uncontrolled exploitation in times of political chaos) significantly affected the dynamics of ungulate populations (Table 5.27).

Our analysis revealed a tight interaction between food resources and predation in affecting the densities of prey, particularly birds and rodents. In both those groups, effects of predators were stronger in conditions of poor supplies of food. This is consistent with results of manipulative experiments conducted on small rodents (Desy and Batzli 1989) and snowshoe hare (Krebs et al. 1995).

As regards the role of food in population dynamics of rodents, new and noteworthy is the differentiation of its two components, each operating in a different way: the standing crop of plant biomass (especially the winter crop) which shapes the multiannual pattern of dynamics, and annual productivity which determines densities (Sect. 5.5). Recognising these two characteristics of vegetation on both the local and the biogeographic scale may improve our understanding of the diverse patterns of animal fluctuations. It may be a pro-

Fig. 5.58. Lynx roaming in the forest by night. Mysterious and elusive, it is a symbol of all predators. Whether supremely important or incidental for other animal populations, they will always remain fascinating. Photo: W. Jędrzejewski

spective approach also in investigating other animals (e.g. insectivores, birds, ungulates).

Varying effectiveness of predation stems from the biological properties of animals that are preyed upon. Material presented in this book allows the following broad generalisations. Heterothermic animals (amphibians) critically depend on abiotic conditions and thus are very unpredictable prey (see Table 2.15 in Chap. 2). They are little affected by food and still less so by predation. Homeothermic animals are very susceptible to changes in food supply because they have to spend much time feeding to maintain a constantly high body temperature. Arguably, the need to 'go out' for foraging every day or night have made them a predictable prey for carnivores and raptors. Among warm-blooded animals, the importance of predation grows with decreasing reproductive capacity of the prey. Very high reproduction rates, such as those of forest rodents, shrews, and wild boar - each compared with other animals of its size - means a superior capacity to compensate for losses due to predation. Animals of lower reproductive rates (ungulates and birds) are potentially more affected by predators.

In this context, an essential feature is the ratio of predator and prey reproductive capacities. Predators of rodents, shrews, birds, and amphibians are larger than their prey and with no exception have slower intrinsic rates of reproduction. This makes them 'conservative' in their numerical responses to prey fluctuations. Being unable to keep pace with changes in numbers of their prey, predators of rodents and probably other small vertebrates are adapted to maximally exploit the densities that occur most frequently.

Ungulates are of the same size or larger than their predators (lynx and wolf). Thus, the predators may have a higher breeding capacity than their prey (ruminants), they can quickly react numerically to an increase in prey numbers, and are potentially very effective enemies of ungulates (Fig. 5.58). In predator-ungulate relationships, where predators are smaller and breed faster than their prey, the importance of predation on various prey species seems to be based on predators' behavioural ability to seize their prey. Roe and red deer sizes optimally match the hunting abilities of their predators. Moose and bison are too large for lynxes and big enough to be strongly avoided by wolves. Therefore, they are little affected by predation.

References

Alerstam T (1990) Bird migration. Cambridge University Press, Cambridge

Andersson M, Erlinge S (1977) Influence of predation on rodent populations. Oikos 29:591–597

Andrewartha HG, Birch LC (1954) The distribution and abundance of animals. University of Chicago Press, Chicago

Angelstam P, Lindström E, Widén P (1984) Role of predation in short-term population fluctuations of some birds and mammals in Fennoscandia. Oecologia 62:199–208

Angelstam P, Lindström E, Widén P (1985) Synchronous short-term population fluctuations of some birds and mammals in Fennoscandia - occurrence and distribution. Holarc Ecol 8:285–298

Aulak W (1967) Estimation of small mammal density in three forest biotopes. Ekol Pol 15(39):755–778

Aulak W (1970) Small mammal communities of the Białowieża National Park. Acta Theriol 15:465–515

Aulak W (1975) A comparison of the production of the ground flora and of the rate of organic fall in two areas of the association *Circaeo-Alnetum* Oberd., 1953, in the Białowieża National Park. Ekol Pol 23:65–81

Aulak W (1976) Development and production of ground flora in *Tilio-Carpinetum* Tracz. 1962 association as one of the elements of basic trophic level in the forest ecosystems. Zesz Nauk SGGW-AR w Warszawie, Rozprawy Naukowe 60:1–151 (in Polish with English Summary)

Aulak W, Goszczyński J (1986) An inventory of hoofed populations driven through entire forest complexes. Wiad Ekol 32:403–413 (in Polish with English Summary)

Bakharev VA (1986) Sovremennoe sostoyanie gerpetofauny Belovezhskoi Pushchi. [The present state of herpetological fauna of Białowieża Primeval Forest.] Zapovedniki Belorussii 10:83–88 (in Russian)

Bannikov AG, Belova ZV (1956) Materialy k izucheniyu zemnovodnykh i presmykayushchikhsya Belovezhskoi Pushchi. [Materials on amphibians and reptiles of Białowieża Primeval Forest.] Uch Zap Mosk Gorodskogo Pedagog Inst Im VP Potemkina 61:385–402 (in Russian with English Summary)

Beattie RC, Aston RJ, Milner AG (1993) Embryonic and larval survival of the common frog (*Rana temporaria* L.) in acidic and limed ponds. Herpetol J 3:43–48

Begon M, Harper JL, Townsend CR (1986) Ecology. Individuals, populations and communities. Blackwell Scientific, Oxford

Böhme (1977) Zur Bestimmung quartarer Anuren Europas an Hand von Skelettelementen. Wiss Z der Humboldt-Univ Berl Math-Naturwiss Reihe 26(3):283–300

Borowski S (1960) *Geotrupes stercorosus* (Sc.) (Coleoptera, Scarabaeidae) in the Białowieża National Park. Fragm Faun 8(23):337–365 (in Polish with English Summary)

Borowski S (1961) Some observations on pygmy owl, *Glaucidium passerinum* (L.) and on great grey owl, *Strix nebulosa* Forst. in National Park of Białowieża. Przegl Zool 5(1):59–60 (in Polish with English Summary)

Borowski S, Dehnel A (1952) Angaben zur Biologie der Soricidae. Ann Univ Mariae Curie - Skłodowska, Lublin - Polonia, Sectio C 7:305–448 (in Polish with German Summary)

Borowski S, Dzięciołowski R (1980) Browse supply in lowland forests of eastern Poland. Holarct Ecol 3:201–213

Borowski S, Okołów C (1988) The birds of the Białowieża Forest. Acta Zool Cracov 31:65–114
Boutin S (1992) Predation and moose population dynamics: a critique. J Wildl Manage 56:116–127
Boutin S (1995) Testing predator-prey theory by studying fluctuating populations of small mammals. Wildl Res 22:89–100
Brown CJ (1984) The behaviour of a tawny eagle at carrion. Madoqua 14(1):95–97
Brown CJ (1990) An evaluation of supplementary feeding for bearded vultures and other avian scavengers in the Natal Drakensberg. Lammergeyer 41:30–36
Brylińska M (ed) (1991) Ryby słodkowodne Polski. [Freshwater fish of Poland.] Państwowe Wydawnictwo Naukowe, Warszawa
Brzeziński M, Jędrzejewski W, Jędrzejewska B (1992) Winter home ranges and movements of polecats *Mustela putorius* in Białowieża Primeval Forest, Poland. Acta Theriol 37:181–191
Brzeziński M, Jędrzejewski W, Jędrzejewska B (1993) Diet of otters (*Lutra lutra*) inhabiting small rivers in the Białowieża National Park, eastern Poland. J Zool (Lond) 230:495–501
Buchalczyk T (1980) The brown bear in Poland. In: Martinka CJ, McArthur KL (eds) Bears – their biology and management. Bear Biol Assoc Conf 3:229–232
Buchalczyk T, Pucek Z (1963) Food storage of European water shrew, *Neomys fodiens* (Pennant, 1771). Acta Theriol 7:376–379
Buchalczyk T, Pucek Z (1968) Estimation of the numbers of *Microtus oeconomus* using the standard minimum method. Acta Theriol 13:461–482
Buchalczyk T, Gębczyńska Z, Pucek Z (1970) Numbers of *Microtus oeconomus* (Pallas, 1776) and its noxiousness in forest plantations. EPPO Publ Ser A 58:95–99
Bunevich AN (1983a) Ekologiya i troficheskie svyazi khishchnikov Belovezhskoi Pushchi (lisica, volk, enotovidnaya sobaka). [Ecology and trophic relationships of predators in Białowieża Forest (red fox, wolf, raccoon dog).] Unpublished report, Gosudarstvennoe zapovedno-okhotnichie khozyaistvo Belovezhskaya Pushcha, Kamenyuki (in Russian)
Bunevich AN (1983b) K razmnozheniyu lisicy v Belovezhskoi Pushche. [Reproduction of the red fox in Białowieża Forest.] Zapovedniki Belorussii 7:59–62 (in Russian)
Bunevich AN (1984) Chislennost i nekotorye osobennosti ekologii barsuka v Belovezhskoi Pushche. [Numbers and some features of badger ecology in Białowieża Forest.] Zapovedniki Belorussii 8:108–113 (in Russian)
Bunevich AN (1986a) Vliyanie volka na pogolove kopytnykh v Belovezhskoi Pushche. [The impact of wolf on ungulate numbers in Białowieża Forest.] IV Sezd Vsesoyuznogo Teriologicheskogo Obshchestva, Tezisi dokladov, AN SSSR, Moskva 3:40–41 (in Russian)
Bunevich AN (1986b) Pitanie lisicy v Belovezhskoi Pushche i ee okhrannoi zone. [Feeding by red fox in Białowieża Forest and its buffer zone.] Zapovedniki Belorussii 10:94–102 (in Russian)
Bunevich AN (1986c) Sravnitelnaya kharakteristika pitaniya nornykh khishchnikov Belovezhskoi Pushchi. [Comparative characteristics of food of denning predators in Białowieża Forest.] IV Sezd Vsesoyuznogo Teriologicheskogo Obshchestva, Tezisi dokladov, AN SSSR, Moskva 3:41–42 (in Russian)
Bunevich AN (1988) Dinamika chislennosti i pitanie volka v Belovezhskoi Pushche. [Population dynamics and food of the wolf in Białowieża Primeval Forest.] Zapovedniki Belorussii 12:108–113 (in Russian)
Bunevich AN, Dackevich VA (1985) Stacialnoe razmeshchenie i pitanie enotovidnoi sobaki v Belovezhskoi Pushche. [Spatial distribution and food of the raccoon dog in Białowieża Forest.] Zapovedniki Belorussii 9:114–120 (in Russian)
Butet A, Leroux ABA (1993) Effect of prey on a predator's breeding success. A 7-year study on common vole (*Microtus arvalis*) and Montagu's harrier (*Circus pygargus*) in a west France marsh. Acta Oecol 14:857–865
Caboń K (1958) Das Massensterben von Wildschweinen im Naturstaatspark von Bialowieza im Winter 1955/56. Acta Theriol 2:71–82
Caughley G, Sinclair ARE (1994) Wildlife ecology and management. Blackwell Scientific, Boston
Ceska V (1980) Untersuchungen zu Nahrungsverbrauch, Nahrungsnutzung und Energiehaushalt bei Eulen. J Ornithol 121:186–199

Chernyavskii FB (1979) Population dynamics and population indices of *Lemmus sibiricus* and *Dicrostonyx torquatus* on the Vrangel Island. Zool Zhurnal 58:553–563 (in Russian with English Summary)

Chernyavskii FB, Dorogoi VI (1981) O roli khishchnikov v dinamike chislennosti lemmingov (na primere Ostrova Vrangelya). [On the role of predators in the dynamics of lemming numbers (on the example of the Wrangel Island).] In: Ekologiya mlekopitayushchikh severo-vostochnoi Sibiri. [Ecology of mammals of north-east Siberia.] Izdatelstvo Nauka, Moskva, pp 32–50 (in Russian)

Churchfield S (1990) The natural history of shrews. Christopher Helm, A & C Black, London

Chyrvonaya Kniga Respubliki Belarus. [Red Data Book of the Belarus Republic.] (1993) Belaruskaya Encyklapedyya Im P Brouki, Minsk (in Belarussian)

Clutton-Brock TH, Major M, Guinness FE (1985) Population regulation in male and female red deer. J Anim Ecol 54:831–846

Dackevich VA (1977) Sezonnoe razvitie yavlenii prirody v Belovezhskoi Pushche (1946–1969 gg.). [Seasonal development of nature in Białowieża Forest (1946–1969).] Zapovedniki Belorussii 1:5–23 (in Russian)

Dackevich VA (1979) Pitanie lesnoi kunicy v Belovezhskoi Pushche. [Food of the pine marten in Białowieża Forest.] Zapovedniki Belorussii 3:67–70 (in Russian)

Dackevich VA (1983) Vidovoi sostav ryb vodoemov Belovezhskoi Pushchi. [Species composition of fish of Białowieża Forest waterbodies.] Zapovedniki Belorussii 7:76–81 (in Russian)

Dackevich VA (1997) Istoricheskii ocherk i nekotorye itogi ornitologicheskikh issledovanii v Belovezhskoi pushche (1945–1985 gg.). [Historical outline and some results of ornithological research in Białowieża Forest (1945–1985).] Unpublished report, Gosudarstvennyi Nacionalnyi Park Belovezhskaya Pushcha, Kamenyuki (in Russian)

Dackevich VA, Popenko VM (1987) Ornitofauna opushek Belovezhskoi Pushchi. [Avian fauna of forest edges of Białowieża Primeval Forest.] Zapovedniki Belorussii 11:120–129 (in Russian)

Dackevich VA, Vakula VA (1980) Chislennost teterevidnykh ptic v Belovezhskoi Pushche i faktory vliyayushchie na ee izmenenie. [Numbers of tetraonid birds in Białowieża Primeval Forest and factors affecting their variability.] Zapovedniki Belorussii 4:91–100 (in Russian)

Dackevich VA, Popenko VM, Kolosei LK (1985) Sostav, chislennost i biotopicheskoe raspredelenie sov Belovezhskoi Pushchi. [Species structure, numbers and habitat distribution of owls of Białowieża Primeval Forest.] Zapovedniki Belorussii 9:106–114 (in Russian)

Dąbrowski MJ (1953) The studies on the biomass of field-stratum carried on by the branch of IBL at Białowieża. Ekol Pol 1:45–56 (in Polish with English Summary)

Debrot S (1983) Fluctuations de populations chez l'Hermine (*Mustela erminea*). Mammalia 47:323–332

Debrot S, Mermod C, Fivaz G, Weber J-M (1982) Atlas des poils de mammiferes d'Europe. Universite de Neuchatel, Switzerland

Desy EA, Batzli GO (1989) Effects of food availability and predation on prairie vole demography: a field experiment. Ecology 70:411–421

Dinesman LG, Kaleckaya ML (1952) Metody kolichestvennogo ucheta amfibii i reptilii. [Methods of quantitative survey of amphibians and reptiles.] In: Metody ucheta chislennosti i geograficheskogo rasprostranieniya nazemnykh pozvonochnykh. [Methods of survey of numbers and geographical distribution of terrestrial vertebrates.] Moskva, pp 329–341 (in Russian with English Summary)

Domaszewicz A (1993) Sowy Puszczy Białowieskiej. [Owls of Białowieża Primeval Forest.] Unpublished report, Białowieża (in Polish)

Domaszewicz A (1997) Pygmy owl, *Glaucidium passerinum* in the Białowieża National Park – habitats, distribution, and numbers. Notatki Orn 38:43–50 (in Polish with English Summary)

Duchic VI (1952) Rol sokola-sapsana i zmeeyada v okhotnichem khozyaistve zapovednika Belovezhskaya Pushcha. [Roles of the peregrine and the short-toed eagle in game management of Białowieża Forest reserve.] Unpublished Thesis, University of Belarus, Minsk

Duke GE, Evanson OA, Jegers A (1976) Meal to pellet intervals in 14 species of captive raptors. Comp Biochem Physiol 53A:1–6

Dunstone N (1993) The mink. T&AD Poyser Natural History, London
Dzięciołowski R (1969) Ciężar naszych jeleni nizinnych. [Weight of our lowland red deer.] Łowiec Polski 19:4 (in Polish)
Dzięciołowski R (1970a) Variation in red deer (*Cervus elaphus* L.) food selection in relation to environment. Ekol Pol 18:635–645
Dzięciołowski R (1970b) Foods of the red deer as determined by rumen content analyses. Acta Theriol 15:89–110
Dzięciołowski R, Pielowski Z (1993) Łoś. [The moose.] Wydawnictwo Anton-5, Warszawa (in Polish)
Dziurdzik B (1973) Key to the identification of hairs of mammals from Poland. Acta Zool Cracov 18:73–91
Engelman WE, Fritzsche J, Gunther R, Obst FJ (1985) Lurche und Kriechtiere Europas. Neuman, Leipzig
Erlinge S (1987) Predation and noncyclicity in a microtine population in southern Sweden. Oikos 50:347–352.
Erlinge S, Göransson G, Hansson L, Högstedt G, Liberg O, Nilsson IN, Nilsson T, von Schantz T, Sylvén M (1983) Predation as a regulating factor on small rodent populations in southern Sweden. Oikos 40:36–52.
Erlinge S, Göransson G, Högstedt G, Jansson G, Liberg O, Loman J, Nilsson IN, von Schantz T, Sylvén M (1984) Can vertebrate predators regulate their prey? Am Nat 123:125–133
Erlinge S, Göransson G, Högstedt G, Jansson G, Liberg O, Loman J, Nilsson IN, von Schantz T, Sylvén M (1988) More thoughts on vertebrate predator regulation of prey. Am Nat 132:148–154
Erlinge S, Agrell J, Nelson J, Sandell M (1991) Why are some microtine populations cyclic while others are not? Acta Theriol 36:63–71
Errington PL (1946) Predation and vertebrate populations. Q Rev Biol 21:144–177
Fairley JS, Ward DP, Smal CM (1987) Correction factors and mink faeces. Ir Nat J 22:334–336
Falińska K (1971) An estimate of diaspore production in the ecosystem of a mixed oak-hornbeam forest (*Querco-Carpinetum*) in the Białowieża National Park. Ekol Pol 19:525–561
Falińska K (1991) Plant demography in vegetation succession. Kluwer, Dordrecht
Faliński JB (1973) Herb layer filling by plant cormus in the *Querco-Carpinetum* community in the Białowieża National Park. Phytocoenosis 2(2):123–142
Faliński JB (1986) Vegetation dynamics in temperate lowland primeval forest. Dr W Junk, Dordrecht
Fedyushin AV, Dolbik MS (1967) Pticy Belorussii. [Birds of Belarus.] Uradzhai, Minsk
Fenchel T (1974) Intrinsic rate of natural increase: the relationship with body size. Oecologia 14:317–326
Folitarek SS (1948) Khishchnye pticy kak faktor estestvennogo otbora v prirodnykh populyaciyakh melkikh gryzunov. [Raptors as the factor of natural selection in small rodent populations.] Zh Obshchei Biologii 9(1):47–64 (in Russian)
Formozov AN (1946) Snezhnyi pokrov kak faktor sredy, ego znachenie v zhizni mlekopitayushchikh i ptic SSSR. [Snow cover as an environmental factor, its role in the lives of mammals and birds of the Soviet Union.] Izdatelstvo Moskovskogo Obshchestva Ispytatelei Prirody, Moskva (in Russian)
Formozov AN (ed) (1952) Metody ucheta chislennosti i geograficheskogo raspredeleniya nazemnykh pozvonochnykh. [Methods of censusing numbers and geographic distribution of terrestrial vertebrates.] Izdatelstvo Akademii Nauk SSSR, Moskva (in Russian)
Frąckowiak W, Gula R (1992) The autumn and spring diet of brown bear *Ursus arctos* in the Bieszczady Mountains in Poland. Acta Theriol 37:339–344
Fruziński B (1992) Dzik. [The wild boar.] Wydawnictwo Cedrus, Warszawa (in Polish)
Gaiduk VE, Bunevich AN, Blockaya ES (1985) Biotopicheskoe raspredelenie, pitanie i dinamika chislennosti zholtogorloi myshi v Belovezhskoi Pushche. [Habitat distribution, food and population dynamics of yellow-necked mouse in Białowieża Forest.] Zapovedniki Belarusii 9:121–127 (in Russian)

Gavrin VF (1953) Pticy Belovezhskoi Pushchi. [Birds of Białowieża Primeval Forest.] Unpublished report, Zapovedno-Okhotnichie Khozaistvo Belovezhskaya Pushcha, Kamenyuki (in Russian)
Gavrin VF (1969) Ekologiya ryabchika v Belovezhskoi Pushche. [Ecology of the hazel hen in Białowieża Primeval Forest.] Belovezhskaya Pushcha, Issledovaniya 3:146–172 (in Russian)
Gavrin VF, Donaurov SS (1954) Volk v Belovezhskoi Pushche. [Wolf in Białowieża Primeval Forest.] Zool Zh 33:904–924 (in Russian)
Gavrin VF, Krapivnyi AP (1968) Pitanie i chislennost lisicy v Belovezhskoi Pushche. [Food and population dynamics of the red fox in Białowieża Forest.] Okhotniche Promyslovye Zveri, Biol i Khozaistvennoe Ispolzovanie 1:245–264 (in Russian)
Genko N (1902–1903) Kharakteristika Belovezhskoi pushchi i istoricheskiya o nei dannyya. [Characteristics of Białowieża Primeval Forest and the historical data on it.] Lesnoi Zh 22(5):1014–1056, 22(6):1269–1302, 23(1):22–56 (in Russian)
Geptner VG, Sludskii AA (1972) Mlekopitayushchie Soveckogo Soyuza. Tom Vtoroi. Khishchnye (gieny i koshki). [Mammals of the Soviet Union, vol 2. Carnivores (hayenas and cats).] Izdatelstvo Vysshaya Shkola, Moskva
Gębczyńska Z (1966) Estimation of rodents numbers in a plot of *Querceto Carpinetum* forest. Acta Theriol 11:315–328
Gębczyńska Z (1976) Food habits of the bank vole and phenological phases of plants in an oak-hornbeam forest. Acta Theriol 21:223–236
Gębczyńska Z (1980) Food of the roe deer and red deer in the Białowieża Primeval Forest. Acta Theriol 25:487–500
Gębczyńska Z, Krasińska M (1972) Food preferences and requirements of the European bison. Acta Theriol 17:105–117
Gębczyńska Z, Gębczyński M, Martynowicz E (1991) Food eaten by the free-living European bison in Białowieża Forest. Acta Theriol 36:307–313.
Gibet LA (1960) Chislennost khishchnykh ptic i ee svyaz s gryzunami v stepnoi zone severnogo Kazakhstana. [Numbers of raptors and their relationships with rodents in the steppe zone of northern Kazakhstan.] Ornitologiya 3:278–291 (in Russian)
Głowaciński Z, Witkowski Z (1970) Numbers and biomass of amphibians estimated by the capture and removal method. Wiad Ekol 16(4):328–340 (in Polish with English Summary)
Golodushko BZ (1958) Materialy po pitaniyu obyknovennogo kanyuka *Buteo buteo* L. i malogo podorlika *Aquila pomarina* Brehm Belovezhskoi Pushchi. [Data on foraging by the common buzzard *Buteo buteo* L. and the lesser spotted eagle *Aquila pomarina* Brehm in the Belovezha Primeval Forest.] Tr Zapovedno-okhotnichego Khozaistva Belovezhskaya Pushcha 1:100–109 (in Russian)
Golodushko BZ (1959) Da ekalogii arla-karlika – *Aquila pennata* Gmel. u Belavezhskai Pushchy. [On ecology of the booted eagle – *Aquila pennata* Gmel. in Białowieża Primeval Forest.] Vesti Akademii Navuk Belaruskai SSR 1:114–117 (in Belarussian)
Golodushko BZ (1960) K ekologii chegloka v Belovezhskoi Pushche. [Ecology of the hobby in Białowieża Forest.] Ornitologiya 3:139–145 (in Russian)
Golodushko BZ (1961a) Chislennost amfibii i reptilii i ikh rol v pitanii kanyuka i malogo podorlika Belovezhskoi Pushchi. [Numbers of amphibians and reptiles and their role in food of the common buzzard and lesser spotted eagle of Białowieża Primeval Forest.] In: Ges ND (ed) Fauna i ekologiya nazemnykh pozvonochnykh Belorussii. [Fauna and ecology of terrestrial vertebrates of Belarus.] Izdatelstvo Ministerstva vysshego, srednego spetsialnogo i professionalnogo obrazovaniya BSSR, Minsk, pp 143–149 (in Russian)
Golodushko BZ (1961b) O roli khishchnykh ptic v biocenoze Belovezhskoi Pushchi. [Role of raptors in ecosystems of Białowieża Forest.] In: Ges ND (ed) Fauna i ekologiya nazemnykh pozvonochnykh Belorussii. [Fauna and ecology of terrestrial vertebrates of Belarus.] Izdatelstvo Ministerstva vysshego, srednego specialnogo i professionalnogo obrazovaniya BSSR, Minsk, pp 98–111 (in Russian)
Golodushko BZ (1961c) O pishchevykh otnosheniyakh khishchnykh ptic Belovezhskoi Pushchi. [Trophic relationships of raptors in Białowieża Forest.] In: Ges ND (ed) Fauna i ekologiya nazemnykh pozvonochnykh Belorussii. [Fauna and ecology of terrestrial verte-

brates of Belarus.] Izdatelstvo Ministerstva vysshego, srednego specialnogo i professionalnogo obrazovaniya BSSR, Minsk, pp 112–132

Golodushko BZ, Danilyuk II (1961) Chislennost i stacialnoe raspredelenie vorobinykh ptits Belovezhskoi pushchi. [Numbers and distribution of passerine birds of Białowieża Forest.] In: Ges ND (ed) Fauna i ekologiya nazemnykh pozvonochnykh Belorussii. [Fauna and ecology of terrestrial vertebrates of Belarus.] Izdatelstvo Ministerstva vysshego, srednego spetsialnogo i professionalnogo obrazovaniya BSSR, Minsk, pp 133–142

Golodushko BZ, Samusenko EG (1961) Pitanie vorobinogo sychika v Belovezhskoi Pushche. [Food of the pygmy owl in Białowieża Forest.] Tr IV Pribaltiiskoi ornitologicheskoi konferencii, Riga, pp 135–140

Goszczyński J (1974) Studies on the food of foxes. Acta Theriol 19:1–18

Goszczyński J (1976) Estimation of daily food ratio of *Tyto alba* Scopoli under natural conditions. Pol Ecol Stud 2(1):95–102

Goszczyński J (1977) Connections between predatory birds and mammals and their prey. Acta Theriol 22:399–430

Goszczyński J (1985) The effect of structural differentiation of ecological landscape on the predator-prey interaction. Rozprawy Naukowe i Monografie – Publications of Warsaw Agricultural University SGGW-AR, Warszawa (in Polish with English Summary)

Goszczyński J (1986a) Locomotor activity of terrestrial predators and its consequences. Acta Theriol 31:79–95

Goszczyński J (1986b) Diet of foxes and martens in central Poland. Acta Theriol 31:491–506

Goszczyński J (1989) Spatial distribution of red foxes *Vulpes vulpes* in winter. Acta Theriol 34:361–372

Haglund B (1968) Winter habits of the bear and the wolf as revealed by tracking in the snow. Viltrevy 5(6):213–361

Hanski I (1984) Food consumption, assimilation and metabolic rate in six species of shrew (*Sorex* and *Neomys*). Ann Zool Fenn 21:157–165

Hanski I (1992) Insectivorous mammals. In: Crawley MJ (ed) Natural enemies. The population biology of predators, parasites and diseases. Blackwell Scientific, Oxford, pp 163–187

Hanski I, Korpimäki E (1995) Microtine rodent dynamics in northern Europe: parameterized models for the predator-prey interaction. Ecology 76:840–850

Hanski I, Hansson L, Henttonen H (1991) Specialist predators, generalist predators, and the microtine rodent cycle. J Anim Ecol 60:353–367

Hanski I, Turchin P, Korpimäki E, Henttonen H (1993) Population oscillations of boreal rodents: regulation by mustelid predators leads to chaos. Nature 364:232–235

Hansson L (1971) Habitat, food and population dynamics of the field vole *Microtus agrestis* (L.) in south Sweden. Viltrevy 8:267–378

Hansson L (1987) An interpretation of rodent dynamics as due to trophic interactions. Oikos 50:308–318

Hansson L, Henttonen H (1985) Gradients in density variations of small rodents: the importance of latitude and snow cover. Oecologia 67:394–402

Hansson L, Henttonen H (1988) Rodent dynamics as community processes. Trends Ecol Evol 3:195–200

Harestad AS, Bunnell FL (1979) Home range and body weight – a reevaluation. Ecology 60:389–402

Hedemann O (1936) Z dawnej Białowieży. Echa wojen. [From ancient Białowieża. Repercussions of wars.] Echa leśne 13:2–3 (in Polish)

Hedemann O (1939) L'historie de la foret de Białowieża (jusqu'a 1798). Inst Badawczy Lasów Państwowych, Rozprawy i Sprawozdania, Seria A 41:1–310 (in Polish with French Summary)

Heisey DM, Fuller TK (1985) Evaluation of survival and cause-specific mortality rates using telemetry data. J Wildl Manage 49:668–674

Henschel JR, Skinner JD (1990) The diet of the spotted hyaenas *Crocuta crocuta* in Kruger National Park. Afr J Ecol 28:69–82

Henttonen H, Oksanen T, Jortikka A, Haukisalmi V (1987) How much do weasels shape microtine cycles in the northern Fennoscandian taiga? Oikos 50:353–365

Hilden O, Helo P (1981) The great grey owl *Strix nebulosa* - a bird of the northern taiga. Ornis Fenn 58:159-166
Holišová V (1965) The food of *Pitymys subterraneus* and *P. tatricus* (Rodentia, Microtidae) in the mountain zone of the *Sorbeto-Piceetum*. Zool Listy 14:15-28
Holišová V (1967) The food of *Apodemus agrarius* (Pall.). Zool Listy 16:1-14
Holling CS (1959) The components of predation as revealed by a study of small-mammal predation of the European pine sawfly. Can Entomol 91:293-320
Holling CS (1965) The functional response of predators to prey density and its role in mimicry and population regulation. Mem Entomol Soc Can 45:1-60
Holmes JC (1995) Population regulation: a dynamic complex of interactions. Wildl Res 22:11-19
Jacobs J (1974) Quantitative measurements of food selection; a modification of the forage ratio and Ivlev's electivity index. Oecologia 14:413-417
Jankowska K (1967) Seasonal changes of vegetation and net primary production on the fresh meadow *Arrhenatheretum elatoris*. Stud Naturae, Ser A 1967(1):153-173 (in Polish with English Summary)
Jarocki FP (1830) O Pusczy Białowiezkiey i o celnieyszych w niey zwierzętach. [About Białowieża Primeval Forest and its more prominent animals.] Warszawa (in Polish)
Järvinen A (1985) Predation causing extended low densities in microtine cycles: implications from predation on hole-nesting passerines. Oikos 45:157-158
Järvinen A (1990) Changes in the abundance of birds in relation to small rodent density and predation rate in Finnish Lapland. Bird Study 37:36-39
Jezierski W, Myrcha A (1975) Food requirements of a wild boar population. Pol Ekol Stud 1(2):61-83
Jędrzejewska B (1987) Reproduction in weasels *Mustela nivalis* in Poland. Acta Theriol 32:493-496
Jędrzejewska B, Jędrzejewski W (1989) Seasonal surplus killing as hunting strategy of the weasel *Mustela nivalis* - test of a hypothesis. Acta Theriol 34:347-359
Jędrzejewska B, Jędrzejewski W (1990) Antipredatory behaviour of bank voles and prey choice of weasels - enclosure experiments. Ann Zool Fenn 27:321-328
Jędrzejewska B, Jędrzejewski W (1993) Summer food of the Pygmy owl *Glaucidium passerinum* in Białowieża National Park, Poland. Ornis Fenn 70:196-201
Jędrzejewska B, Okarma H, Jędrzejewski W, Miłkowski L (1994) Effects of exploitation and protection on forest structure, ungulate density and wolf predation in Białowieża Primeval Forest, Poland. J Appl Ecol 31:664-676
Jędrzejewska B, Bunevich AN, Jędrzejewski W (1995) Observations of brown bear *Ursus arctos* in Białowieża Primeval Forest in 1948-1950. Parki Narodowe i Rezerwaty Przyrody 14(4):87-90 (in Polish with English Summary)
Jędrzejewska B, Jędrzejewski W, Bunevich AN, Miłkowski L, Okarma H (1996) Population dynamics of wolves *Canis lupus* in Białowieża Primeval Forest (Poland and Belarus) in relation to hunting by humans, 1847-1993. Mammal Rev 26:103-126
Jędrzejewska B, Jędrzejewski W, Bunevich AN, Miłkowski L, Krasiński ZA (1997) Factors shaping population densities and increase rates of ungulates in Białowieża Primeval Forest (Poland and Belarus) in the 19th and 20th centuries. Acta Theriol 42:399-451
Jędrzejewski M (1993) Występowanie ptaków w okresie zimowym w dolinach wybranych rzek Środkowo-Wschodniej Polski w latach 1984-1993. [Occurrence of birds in the winter season in some river valleys of East-Central Poland in 1984-1993.] Unpublished Master Thesis, Wyższa Szkoła Rolniczo-Pedagogiczna, Siedlce (in Polish)
Jędrzejewski W, Jędrzejewska B (1990) Effect of a predator's visit on the spatial distribution of bank voles: experiments with weasels. Can J Zool 68:660-666
Jędrzejewski W, Jędrzejewska B (1992) Foraging and diet of the red fox *Vulpes vulpes* in relation to variable food resources in Białowieża National Park, Poland. Ecography 15:212-220
Jędrzejewski W, Jędrzejewska B (1993) Predation on rodents in Białowieża primeval forest, Poland. Ecography 16:47-64
Jędrzejewski W, Jędrzejewska B (1995) Proposed National Park of Białowieża Primeval Forest. Chrońmy Przyr Ojczystą 51(3):16-36 (in Polish with English Summary)

Jędrzejewski W, Jędrzejewska B (1996) Rodent cycles in relation to biomass and productivity of ground vegetation and predation in the Palearctic. Acta Theriol 41:1–34

Jędrzejewski W, Jędrzejewska B, Szymura A (1989). Food niche overlaps in a winter community of predators in the Białowieża Primeval Forest, Poland. Acta Theriol 34:487–496

Jędrzejewski W, Jędrzejewska B, Okarma H, Ruprecht AL (1992a) Wolf predation and snow cover as mortality factors in the ungulate community of the Białowieża National Park, Poland. Oecologia 90:27–36

Jędrzejewski W, Jędrzejewska B, McNeish E (1992b) Hunting success of the weasel *Mustela nivalis* and escape tactics of forest rodents in Białowieża National Park. Acta Theriol 37:319–328

Jędrzejewski W, Jędrzejewska B, Brzeziński M (1993a) Winter habitat selection and feeding habits of polecats (*Mustela putorius*) in the Białowieża National Park, Poland. Z Saugetierkd 58:75–83

Jędrzejewski W, Schmidt K, Miłkowski L, Jędrzejewska B, Okarma H (1993b) Foraging by lynx and its role in ungulate mortality: the local (Białowieża Forest) and the Palaearctic viewpoints. Acta Theriol 38:385–403

Jędrzejewski W, Zalewski A, Jędrzejewska B (1993c) Foraging by pine marten *Martes martes* in relation to variable food resources in Białowieża National Park. Acta Theriol 38:405–426

Jędrzejewski W, Rychlik L, Jędrzejewska B (1993d) Responses of bank voles to odours of seven species of predators: experimental data and their relevance to natural predator-vole relationships. Oikos 68:251–257

Jędrzejewski W, Jędrzejewska B, Zub K, Ruprecht AL, Bystrowski C (1994a) Resource use by Tawny owls *Strix aluco* in relation to rodent fluctuations in Białowieża National Park, Poland. J Avian Biol 25:308–318

Jędrzejewski W, Szymura A, Jędrzejewska B (1994b) Reproduction and food of the Buzzard *Buteo buteo* in relation to the abundance of rodents and birds in Białowieża National Park, Poland. Ethol Ecol Evol 6:179–190

Jędrzejewski W, Jędrzejewska B, Szymura L (1995) Weasel population response, home range, and predation on rodents in a deciduous forest in Poland. Ecology 76:179–195

Jędrzejewski W, Jędrzejewska B, Szymura A, Zub K (1996a) Tawny owl (*Strix aluco*) predation in a pristine deciduous forest (Białowieża National Park, Poland). J Anim Ecol 65:105–120

Jędrzejewski W, Jędrzejewska B, Okarma H, Schmidt K, Bunevich AN, Miłkowski L (1996b) Population dynamics (1869–1994), demography, and home range of the lynx in Białowieża Primeval Forest (Poland and Belarus). Ecography 10:122–138

Jędrzejewski W, Jędrzejewska B, Okarma H, Schmidt K, Zub K, Musiani M (in press) Prey selection, functional response, and predation by wolves in Białowieża Primeval Forest (Poland). J Mammal

Juszczyk W (1987) Płazy i gady krajowe. [Amphibians and reptiles of the country, vol. 1–3.] Państwowe Wydawnictwo Naukowe, Warszawa (in Polish)

Kaczmarek M, Wasilewski A (1977) Dynamics of numbers of the leaf-eating insects and its effect on foliage production in the Grabowy Reserve in the Kampinos National Park. Ekol Pol 25:653–673.

Kaczmarek M, Wasilewski A (1978) Reduction of the numbers of *Operophthera brumata* (L.) over the outbreak and post-outbreak period in the Grabowy Reserve of the Kampinos National Park. Ekol Pol 26:141–150.

Kadochnikov NP (1953) O vzaimootnoshenii khishchnykh ptic i obshchestvennoi polevki v stepnom Azerbaidzhane. [On the relationships between birds of prey and the social vole in the steppe Azerbaijan.] Zool Zh 32:1222–1233 (in Russian)

Kaikusalo A (1982) Predatory mammals and vole populations in the fell regions of northwest Finland. Suomen Riista 29:89–92 (in Finnish with English Summary)

Karcov G (1903) Belovezhskaya Pushcha. Ee istoricheskii ocherk, sovremennoe okhotniche khozaistvo i vysochaishe okhoty v Pushche. [Białowieża Primeval Forest. Historical description, contemporary game management, and monarchial hunts in the forest.] A Marks, St Petersburg (in Russian)

Karpiński JJ (1949) O niedźwiedziach w Puszczy Białowieskiej. [On the brown bears of Białowieża Primeval Forest.] Chrońmy Przyr Ojczystą 4–6:35–41 (in Polish)

Karpiński JJ (1954) Bird fauna in the biocenosis of national park of Białowieża (investigations). Rocz Nauk Leśnych, Prace Instytutu Badawczego Leonictwa, Warszawa 5(120):3–104 (in Polish with English Summary)

Kauhala K, Kaunisto M, Helle E (1993) Diet of the raccoon dog, *Nyctereutes procyonoides*, in Finland. Z Säugetierkd 58:129–136

Kenward R, Marcstrom V, Karlbom M (1981) Goshawk winter ecology in Swedish pheasant habitats. J Wildl Manage 45:397–408

Kidd NAC, Lewis GB (1987) Can vertebrate predators regulate their prey? A reply. Am Nat 130:448–453

King CM (1989) The natural history of weasels and stoats. Christopher Helm, London

Kirby KJ, Webster SD, Antczak A (1991) Effects of forest management on stand structure and the quantity of fallen dead wood: some British and Polish examples. Forest Ecol Manage 43:167–174

Korhonen H (1988 a) Voluntary regulation of energy balance in farmed raccoon dogs. Comp Biochem Physiol 89A:219–222.

Korhonen H (1988 b) Regulation of energy economy in raccoon dogs and blue foxes: a process of dynamic interactions. Comp Biochem Physiol 91A:263–268

Korochkina LN, Kovonkov MP, Tolkach VN (1980) Belovezhskaya Pushcha. [Białowieża Primeval Forest.] Izdatelstvo Uradzhai, Minsk (in Russian)

Korpimäki E (1985) Predation causing synchronous decline phases in microtine and shrew populations in western Finland. Oikos 46:124–127

Korpimäki E (1993) Regulation of multiannual vole cycles by density-dependent avian and mammalian predation? Oikos 66:359–363

Korpimäki E (1994) Rapid or delayed tracking of multi-annual vole cycles by avian predators? J Anim Ecol 63:619–628

Korpimäki E, Norrdahl K (1989). Avian predation on mustelids in Europe. 1. Occurrence and effects on body size variation and life traits. Oikos 55:205–215

Korpimäki E, Norrdahl K (1991) Numerical and functional responses of kestrels, short-eared owls, and long-eared owls to vole densities. Ecology 72:814–826

Korpimäki E, Norrdahl K, Rinta-Jaskari T (1991) Responses of stoats and least weasels to fluctuating food abundances: is the low phase of the vole cycle due to mustelid predation? Oecologia 88:552–561

Kossak S (1988) Winter penetration and movements of a male lynx. Acta Theriol 33:354–360

Kossak S (1989) Multiple hunting by lynx and red fox and utilisation of prey by some carnivores. Acta Theriol 34:505–512

Kossak S (1995) Game animal numbers in the Białowieża Forest and proposed ways of game management. Sylwan 139(8):25–41 (in Polish with English Summary)

Kowalski K, Ruprecht AL (1984) Rząd: Nietoperze – Chiroptera. [Order: Bats – Chiroptera.] In: Pucek Z (ed) Klucz do oznaczania ssaków Polski. [Key for identification of Polish mammals, 2nd edn.] PWN – Polish Scientific Publishers, Warszawa, pp 85–138 (in Polish)

Kowalski K, Pucek Z, Ruprecht AL (1984) Rząd: Gryzonie – Rodentia. [Order: Rodents – Rodentia.] In: Pucek Z (ed) Klucz do oznaczania ssaków Polski. [Key for identification of Polish mammals, 2nd edn.] PWN – Polish Scientific Publishers, Warszawa, pp 149–240 (in Polish)

Kozlo PG (1975) Dikii kaban. [Wild boar.] Izdatelstvo Uradzhai, Minsk (in Russian)

Koźmiński Z (1923) The amphibian and reptile fauna of the Białowieża-Forest. Białowieża (Wydawnictwo Ministerstwa Rolnictwa i Dóbr Państwowych, Departament Leśnictwa, Warszawa) 2:7–118 (in Polish with English Summary)

Krasińska M (1988) Hybrids of European bison and domestic cattle. Ossolineum, Prace Habilitacyjne, Wrocław (in Polish with English Summary)

Krasiński Z (1967) Free living European bisons. Acta Theriol 12:391–405

Krasiński Z (1978) Dynamics and structure of the European bison population in the Białowieża Primeval Forest. Acta Theriol 23:3–48

Krasiński ZA, Bunevich AN, Krasińska M (1994) Characteristics of the European bison populations in the Polish and Belarussian parts of the Białowieża Forest. Parki Narodowe i Rezerwaty Przyrody 13(4):25–67 (in Polish with English Summary)

Kratochvil J (ed) (1968) History of the distribution of the lynx in Europe. Acta Sc Nat Brno 2(4):1–50

Krebs CJ (1995) Two paradigms of population regulation. Wildl Res 22:1–10

Krebs CJ, Boutin S, Boonstra R, Sinclair ARE, Smith JNM, Dale MRT, Martin K, Turkington R (1995) Impact of food and predation on the snowshoe hare cycle. Science 269:1112–1115

Kritskaya TI (1961) The role of the raccoonlike dog (*Nyctereutes procynoides*) in the biocoenose of the Manych steppe. Zool Zh 40:788–790 (in Russian with English Summary)

Kruuk H (1989) The social badger. Ecology and behaviour of a group-living carnivore (*Meles meles*). Oxford University Press, Oxford, pp 1–155

Kruuk H (1995) Wild otters. Predation and populations. Oxford University Press, Oxford

Kurskov AN (1958) Materialy k izucheniyu rukokrylykh Belovezhskoi Pushchi. [Studies on bats of Białowieża Primeval Forest.] Tr Zapov-okhotnichego Khoz Belovezhskaya Pushcha 1:120–138 (in Russian)

Kwiatkowski W (1994) Vegetation landscapes of Białowieża Forest. Phytocoenosis (Supplementum Cartographiae Geobotanicae) 6(NS):35–88

Lack D (1966) Population studies of birds. Oxford University Press, Oxford

Lebedeva LS (1956) Ekologicheskie osobennosti kabana Belovezhskoi Pushchi [Ecological characteristics of the wild boar in Białowieża Primeval Forest]. Uch Zap Mosk Gorodskogo Pedagog Inst VP Potemkina 61:105–271 (in Russian)

Levins R (1968) Evolution in changing environments. Princeton University Press, Princeton

Lewartowski Z, Wołk K (1983) Breeding avifauna of a moss-spruce forest and related habitats in the Białowieża Primeval Forest. Acta ornithol 19:97–111

Litvin KYe, Ovsyanikov NG (1990) Relationship between the reproduction and numbers of snowy owls and arctic foxes and the numbers of true lemmings on the Vrangel Island. Zool Zh 69:52–64 (in Russian with English Summary)

Lockie JD (1959) The estimation of the food of foxes. J Wildl Manage 23:224–227

Lockie JD (1961) The food of the pine marten *Martes martes* in West Ross Shire, Scotland. Proc Zool Soc Lond 136:187–195

Lockie JD (1964) Distribution and fluctuations of the pine marten, *Martes martes* (L.), in Scotland. J Anim Ecol 33:349–356

Lode T (1990) Le régime alimentaire d'un petit carnivore, le putois *Mustela putorius* dans l'ouest de la France. Gibier Faune Sauv 7:193–203

Löfgren O, Hörnfeldt B, Carlsson BG (1986) Site tenacity and nomadism in Tengmalm's owl (*Aegolius funereus* (L.)) in relation to cyclic food production. Oecologia 69:321–326

Loman I (1988) Alternative prey that decreases vole population cyclicity: a simulation study based on field data. Ecol Model 40:265–310

Lowe VPW (1980) Variation in digestion of prey by the Tawny owl (*Strix aluco*). J Zool (Lond) 192:283–293

Marcström V, Kenward RE, Engren E (1988) The impact of predation on boreal tetraonids during vole cycles: an experimental study. J Anim Ecol 57:859–872

März R (1987) Gewöll-und Rupfungskunde. Akademie-Verlag, Berlin

Martin TE (1992) Interaction of nest predation and food limitation in reproductive strategies. Curr Ornithol 9:163–197

May RM (1981) Models for single populations. In: May RM (ed) Theoretical ecology. Principles and applications, 2nd edn. Blackwell Scientific, Oxford, pp 5–77

Merkens M, Harestad AS, Sullivan TP (1991) Cover and efficacy of predator-based repellents for Townsend's vole. J Chem Ecol 17:401–412

Messier F (1991) The significance of limiting and regulating factors on the demography of moose and white-tailed deer. J Anim Ecol 60:377–393

Messier F (1995a) On the functional and numerical responses of wolves to changing prey density. In: Carbyn LN, Fritts SH, Seip DR (eds) Ecology and conservation of wolves in a changing world. Proc 2nd North Am Symp on Wolves. Canadian Circumpolar Institute, University of Alberta, Edmonton, pp 187–197

Messier F (1995b) Trophic interactions in two northern wolf-ungulate systems. Wildl Res 22:131–146
Messier F, Crete M (1985) Moose-wolf dynamics and the natural regulation of moose populations. Oecologia 65:503–512
Mikkola H (1983) Owls of Europe. T & AD Poyser, London
Miłkowski L (1969) Z doświadczeń nad inwentaryzacją grubej zwierzyny w Puszczy Białowieskiej. [Experience in censusing big game in Białowieża Primeval Forest.] Łowiec Polski 22(1361):4–5 (In Polish)
Miłkowski L (1970) Łowiectwo w Puszczy Białowieskiej. Przewodnik dla myśliwych. [Hunting in Białowieża Primeval Forest. Guide for hunters.] Wydawnictwo Wojewódzki Ośrodek Infomacji Turystycznej, Białystok (in Polish)
Miścicki S (1996) Forest regeneration and its damage by herbivorous ungulates in the Białowieża National Park. In: Paschalis P, Zajączkowski S (eds) Biodiversity protection of Białowieża Primeval Forest. Selected papers. Grant GEF 05/21685 POL. Forest Biodiversity Protection Project, Warsaw, pp 91–108
Moreno E (1985) Clave osteologica para la identificacion de los Passeriformes Ibericos. I. Aegithalidae, Remizidae, Fringillidae, Alaudidae. Ardeola 32(2):295–377
Moreno E (1986) Clave osteologica para la identificacion de los Passeriformes Ibericos. II. Hirundinidae, Prunellidae, Sittidae, Cerothidae, Troglodytidae, Cinclidae, Lanidae, Oriolidae, Corvidae, Sturnidae, Motacillidae. Ardeola 33(1–2):69–129
Moreno E (1987) Clave osteologica para la identificacion de los Passeriformes Ibericos. III. Muscicapidae. Ardeola 34(2):243–273
Morow K (1976) Food habits of moose from Augustów Forest. Acta Theriol 21:101–116
Murdoch WW, Oaten A (1975) Predation and population stability. Adv Ecol Res 9:2–132
Nasimovich AA (1948) Opyt izucheniya ekologii mlekopitayushchikh putem zimnikh troplenii. [Investigations of mammalian ecology by winter snow tracking.] Zool Zh 27(4):371–378 (in Russian with English Summary)
Naumov RL (1965) Metodika absolyutnogo ucheta ptic v gnezdovoi period na marshrutakh. [Method of absolute census of birds on transects in the breeding season.] Zool Zh 44:81–94 (in Russian)
Newton I (1979) Population ecology of raptors. T and AD Poyser, Berkhamsted
Newton I (1993) Predation and limitation of bird numbers. Curr Ornithol 11:143–198
Newton I (1994) Experiments on the limitation of bird breeding densities: a review. Ibis 136:397–411
Nicholson AJ (1933) The balance of animal populations. J Anim Ecol 2:131–178
Norrdahl K (1995) Population cycles in northern small mammals. Biol Rev 70:621–637
Nybo JO, Sonerud GA (1990) Seasonal changes in diet of hawk owl *Surnia ulula*: importance of snow cover. Ornis Fenn 67:45–51
Ognev S (1907) Poezdka v Belovezhskuyu Pushchu. [A trip to Białowieża Forest.] Okhot Vestn 19:301–306 (in Russian)
Okarma H, Jędrzejewski W (1997) Livetrapping wolves with nets. Wildl Soc Bull 25:78–82
Okarma H, Jędrzejewska B, Jędrzejewski W, Krasiński ZA, Miłkowski L (1995) The roles of predation, snow cover, acorn crop, and man-related factors on ungulate mortality in Białowieża Primeval Forest, Poland. Acta Theriol 40:197–217
Okarma H, Jędrzejewski W, Schmidt K, Kowalczyk R, Jędrzejewska B (1997) Predation of Eurasian lynx on roe deer and red deer in Białowieża Primeval Forest, Poland. Acta Theriol 42:203–224
Okarma H, Jędrzejewski W, Schmidt K, Śnieżko S, Bunevich AN, Jędrzejewska B (1998) Home ranges of wolves in Białowieża Primeval Forest, Poland, compared with other Eurasian populations. J Mammal 79
Oksanen T, Oksanen L, Norberg M (1992) Habitat use by small mustelids in north Fennoscandian tundra: a test of the hypothesis of patchy exploitation ecosystem. Ecography 15:237–244
Olszewski JL (1968) Role of uprooted trees in the movements of rodents in forests. Oikos 19:99–104

Pearson OP (1966) The prey of carnivores during one cycle of mouse abundance. J Anim Ecol 35:217–233
Pearson OP (1971) Additional measurements of the impact of carnivores on California voles (*Microtus californicus*). J Mammal 52:41–49.
Peterson RO, Page RE (1983) Wolf-moose fluctuations at Isle Royale National Park, Michigan, USA. Acta Zool Fenn 174:251–253
Petty SJ (1987) Breeding of tawny owls (*Strix aluco*) in relation to their food supply in an Upland Forest. In: Hill DJ (ed) Breeding and management in birds of prey. Proc Conf, Univeristy of Bristol, pp 167–179
Pianka ER (1973) The structure of lizard communties. Annu Rev Ecol Syst 4:53–74
Pielowski Z (1975) Ecological characteristics of hare population (*Lepus europaeus* Pallas, 1778). Rocz Akad Roln w Poznaniu, Prace habilitacyjne 65:1–33 (in Polish with English Summary)
Pielowski Z (1988) Sarna. [The roe deer, 3rd edn.] Państwowe Wydawnictwo Rolnicze i Leśne, Warszawa (in Polish)
Pikulik MM (1985) Zemnovodnye Belorussii. [Amphibians of Belarus.] Nauka i Tekhnika, Minsk, pp 1–191 (in Russian)
Pikulik MM, Kosov SV, Drobenkov SM, Letenckii AA, Khandogii AV (1987) Ocenka vliyaniya osushitelnoi melioracii nizinnykh bolot na sostoyanie amfibii i reptilii (na primere Belorusskogo Polesya, verkhove rek Yaseldy i Nareva). [Evaluation of the impact of lowland marsh drainage on the status of amphibians and reptiles (an example of upper Yaselda and Narev rivers, Belarussian Polese).] Izdatelstvo VINITI, Akademiya Nauk Belorusskoi SSR, Minsk (in Russian)
Pikulik MM, Bakharev VA, Kosov SV (1988) Presmykayushchiesya Belorussii [The reptiles of Belarus]. Nauka i Tekhnika, Minsk (in Russian)
Piotrowska M, Wesołowski T (1989) The breeding ecology and behaviour of the chiffchaff *Phylloscopus collybita* in primaeval and managed stands of Białowieża Forest (Poland). Acta Ornithol 25(1):25–76
Piotrowska M, Wołk K (1983) Breeding avifauna in coniferous forests of the Białowieża Primeval Forest. Acta Ornithol 19(3):81–95
Polish Red Data Book of Animals (1992) Państwowe Wydawnictwo Rolnicze i Leśne, Warszawa (in Polish and English)
Pollard E, Lakhani KH, Rothery P (1987) The detection of density-dependence from a series of annual censuses. Ecology 68:2046–2055
Powell RA (1982) Evolution of black-tipped tails in weasels: predator confusion. Am Nat 119:126–131
Price MV, Wasert NM, Bass TA (1984) Effects of moonlight on microhabitat use by desert rodents. J Mammal 65:353–356
Priklonsky SG (1965) Coefficient to treat the data of winter transect method of census taking of game animals by their traces. Byull Mosk O Ispyt Prir Otd Biol 30(6):5–12 (in Russian with English Summary)
Proekt organizacii i razvitiya lesnogo khozaistva Gosudarstvennogo Nacionalnogo Parka Belovezhskaya Pushcha na 1993–2012 gg, vol I. Ob'yasnitelnaya zapiska. [Project of organization and development of the forest management in the State National Park Belovezhskaya Pushcha in 1993–2012, vol. 1. Descriptive part.] (1993) Ministerstvo Lesnogo Khozaistva Respubliki Belarus, Belorusskoe gosudarstvennoe lesoustroitelnoe proizvodstvennoe obedinenie BELGOSLES, Minsk (in Russian)
Promberger C, Schröder W (eds) (1992) Wolves in Europe. Status and perspective. Proc Worksh Wolves in Europe – current status and prospects, Oberammergau, Germany, 2–5 April, Wildbiologische Gesellschaft, Munich
Pucek Z (1960) Sexual maturation and variability of the reproductive system in young shrews (*Sorex* L.) in the first calendar year of life. Acta Theriol 3:269–296
Pucek Z (1969) Trap response and estimation of numbers of shrews in removal catches. Acta Theriol 14:403–426
Pucek Z (1984) Klucz do oznaczania ssaków Polski [Key for identification of Polish mammals, 2nd edn.] PWN – Polish Scientific Publishers, Warszawa (in Polish)

Pucek Z (1991 a) History of the European bison and problems of its protection and management. In: Bobek B, Perzanowski K, Regelin W (eds) Transactions of the 18th Congress of International Union of Game Biologists, Kraków 1987. Świat, Kraków-Warszawa, pp 19–39

Pucek Z (1991 b) Conservation strategy for European Bison. Proc Int Symp Ongules/Ungulates 91, Touluse-France, 2-6 Sept, 1991, pp 589–594

Pucek Z, Olszewski J (1971) Results of extended removal catches of rodents. Ann Zool Fenn 8:37–44

Pucek Z and Sych L (1984) Rząd: Parzystokopytne - Artiodactyla. [Order: Even-toed ungulates - Artiodactyla.] In: Pucek Z (ed) Klucz do oznaczania ssaków Polski. [Key for identification of Polish mammals, 2nd edn.] PWN - Polish Scientific Publishers, Warszawa, pp 319–354 (in Polish)

Pucek Z, Bobek B, Łabudzki L, Miłkowski L, Morow K, Tomek A (1975) Estimates of density and number of ungulates. Pol Ecol Stud 1(2):121–135

Pucek Z, Jędrzejewski W, Jędrzejewska B, Pucek M (1993) Rodent population dynamics in a primeval deciduous forest (Białowieża National Park) in relation to weather, seed crop, and predation. Acta Theriol 38:199–232

Pugacewicz E (1993) The occurrence of the booted eagle (*Hieraaetus pennatus*) in the Polish part of the Białowieża Primaeval Forest. Notatki Orn 34:299–312 (in Polish with English Summary)

Pugacewicz E (1995) Population of the eagle owl (*Bubo bubo*) in the Nizina Północnopodlaska lowlands in 1984–1994. Notatki Orn 36:119–134 (in Polish with English Summary)

Pugacewicz E (1996) Birds of prey breeding in the Polish part of the Białowieża Primaeval Forest. Notatki Orn 37:173–224 (in Polish with English Summary)

Pugacewicz E (1977) Ptaki lęgowe Puszczy Białowieskiej. [Breeding birds of the Białowieża Forest.] Północnopodlaskie Towarzystwo Ochrony Ptaków, Białowieża (in Polish with English Summary)

Puławski W (1967) Osowate - Vespidae, Masaridae. [Wasps - Vespidae, Masaridae.] Klucze Oznaczania Owadów Pol 64-65:1–33 (in Polish)

Rachwald A (1992) Habitat preference and activity of the noctule bat *Nyctalus noctula* in the Białowieża Primeval Forest. Acta Theriol 37:413–422

Raczyński J, Ruprecht AL (1974) The effect of digestion on the osteological composition of owl pellets. Acta Ornithol 14:25–38

Roger M, Pascal, M, Pruniere P (1991) Facteurs correctifs de quantification du régime alimentaire du putois *Mustela putorius* L. Gibier Faune Sauv 7:343–357

Rohrig E (1991) Floral composition and its evolutionary development. In: Rohrig E, Ulrich B (eds) Ecosystems of the world 7. Temperate deciduous forests. Elsevier, Amsterdam, pp 17–23

Romanov VS, Smoktunovich EA (1986) Ocenka prostranstvennogo raspredeleniya evropeiskoi kosuli zimoi. [An estimate of the distribution of roe deer in winter.] Zapovedniki Belorussii 10:117–123 (in Russian)

Ronkainen H, Ylönen H (1994) Behaviour of cyclic bank voles under risk of mustelid predation: do females avoid copulations? Oecologia 97:377–381

Ruprecht AL (1976) New observations on bats (Chiroptera) of Białowieża. Przegl Zool 20(1):115–123 (in Polish with English Summary)

Ruprecht AL (1989) New records of the European Pond Terrapin, *Emys orbicularis* (Linnaeus, 1758) in the western part of the Białowieża Primeval Forest. Przegl Zool 33(1):125–128 (in Polish with English Summary)

Ruprecht AL, Buszko S (1985) Stone marten, *Matres foina* (Erxleben, 1777) in western part of the Białowieża Primeval Forest. Przegl Zool 29(1):105–108 (in Polish with English Summary)

Ruprecht AL, Szwagrzak A (1987) Zur Ernährung der Eulen im Westteil des Bialowieza-Urwaldes. Ökol Vögel 9:89–96

Ruprecht AL, Buchalczyk T, Wójcik JM (1983) The occurrence of minks (Mammalia: Mustelidae) in Poland. Przegl Zool 27(1):87–99 (in Polish with English Summary)

Ryabicev VK, Ryzhanovskii VN, Shutov SV (1976) Vliyanie khishchnikov na effektivnost razmnozheniya ptic na Yamale pri depressii gryzunov. [The impact of predators on breeding success of birds during rodent population crash in Yamal.] Ekologiya 1976(4):103–104 (in Russian)

Ryszkowski L, Goszczyński J, Truszkowski J (1973) Trophic relationships of the common vole in cultivated fields. Acta Theriol 18:125–165

Saarikko J (1992) Risk of predation and foraging activity in shrews. Ann Zool Fenn 29:291–299

Sablina TB (1955) Kopytnye Belovezhskoi Pushchi. [Ungulates of Białowieża Primeval Forest.] Tr Inst Morfol Zhivotn AN Severcova, Akad Nauk SSSR 15:1–191 (in Russian)

Samusenko EG, Golodushko BZ (1961) Pitanie enotovidnoi sobaki v Belorussii. [The food of the raccoon dog in Belarus.] In: Ges ND (ed) Fauna i ekologiya nazemnykh pozvonochnykh Belorussii. [Fauna and ecology of terrestrial vertebrates of Belarus.] Izdatelstvo Ministerstva vysshego, srednego specialnogo i professionalnogo obrazovaniya BSSR, Minsk, pp 71–82 (in Russian)

Schmidt K, Jędrzejewski W, Okarma H (1997) Spatial organization and social relations in the Eurasian lynx population in Białowieża Primeval Forest, Poland. Acta Theriol 42:289–312

Schmidt K (in press) Mobility and space use by females with young and dispersal of juveniles in the Eurasian lynx. Acta Theriol

Serzhanin IN (1961) Mlekopitayushchie Belorussii. [Mammals of Belarus.] Izdatelstvo Akademii Nauk Belorusskoi SSR, Minsk (in Russian)

Severcov AS, Surova GS (1979) Effect of predators on the population of *Rana temporaria* tadpoles. Zool Zh 58:1374–1379 (in Russian with English Summary)

Shostak SV, Bunevich AN (1986) Vliyanie rysi na populyaciyu evropeiskogo blagorodnogo olenya. [Influence of lynx on European red deer population.] IV Sezd Vses Teriologicheskogo Obshchestva, Akad Nauk SSSR, Moskva, 3:76–77

Sidorovich VE (1993) Reproductive plasticity of the American mink *Mustela vison* in Belarus. Acta Theriol 38:175–183

Sidorovich VE, Jędrzejewska B, Jędrzejewski W (1996) Winter distribution and abundance of predatory mustelids and beavers in the river valleys of Białowieża Primeval Forest. Acta Theriol 41:155–170

Sikorski MD (1989) System kojarzenia się i charakterystyka sukcesu reprodukcyjnego w naturalnej populacji nornicy rudej. [Mating system and reproduction success in a free-living population of bank vole.] Unpublished PhD Thesis, Mammal Research Institute, Białowieża (in Polish)

Sinclair ARE (1989) Population regulation in animals. In: Cherrett JM (ed) Ecological concepts. Blackwell Scientific, Oxford, pp 197–241

Sinclair ARE, Pech RP (1996) Density dependence, stochasticity, compensation and predator regulation. Oikos 75:164–173

Sinclair ARE, Olsen PD, Redhead TD (1990) Can predators regulate small mammal populations? Evidence from house mouse outbreaks in Australia. Oikos 59:382–392

Skogland T (1991) What are the effects of predators on large ungulate populations? Oikos 61:401–411

Sokołowski AW (1993) Phytosociological characteristics of forest communities in the Białowieża National Park. Parki Narodowe i Rezerwaty Przyrody 12(3):5–190 (in Polish with English Summary)

Sokołowski J (1979) Ptaki Polski. [Birds of Poland.] Wydawnictwa Szkolne i Pedagogiczne, Warszawa (in Polish)

Solomon ME (1949) The natural control of animal populations. J Anim Ecol 18:1–35

Sonerud GA (1986) Effect of snow cover on seasonal changes in diet, habitat, and regional distribution of raptors that prey on small mammals in boreal zones of Fennoscandia. Holarct Ecol 9: 33–47

Southern HN (1970) The natural control of a population of Tawny owls (*Strix aluco*). J Zool (Lond) 162:197–285

Southern HN, Lowe VP (1968) The pattern of distribution of prey and predation in Tawny owl territories. J Anim Ecol 37:75–97

Sumiński P (1989) Borsuk. [The badger, 3rd edn.] Państwowe Wydawnictwo Rolnicze i Leśne, Warszawa (in Polish)

Surova GS (1990) The press of predators on larvae of *Rana arvalis* in nature. Zool Zh 69(10):86–97 (in Russian with English Summary)

Śliwiński P (1987) Elementy ekologii populacji borsuka na terenie LZD Rogów. [Aspects of ecology of badger population in Rogów forest.] Unpublished Master Thesis, Szkoła Główna Gospodarstwa Wiejskiego - Akademia Rolnicza, Warszawa

Tabachnick BG, Fidell LS (1983) Using multivariate statistics. Harper & Row, New York

Temeles EJ (1985) Sexual size dimorphism of bird-eating hawks: the effect of prey vulnerability. Am Nat 125:485–499

Tolkach VN, Ramlav EA (1973) Semenonoshenie sosny i eli v Belovezhskoi pushche. [Seed production by pine and spruce in Białowieża Primeval Forest.] Belovezhskaya Pushcha, Issledovaniya 7:51–61 (in Russian)

Tomiałojć L (1980) The combined version of the mapping method. In: Oelke H (ed) Bird census work and nature conservation. Proc 6th Int Conf Bird Census Work, University of Göttingen, Göttingen, pp 92–106

Tomiałojć L (1993) Breeding ecology of the blackbird *Turdus merula* studied in the primaeval forest of Białowieża (Poland). Part 1. Breeding numbers, distribution and nest sites. Acta Ornithol 27:131–157

Tomiałojć L (1994) Breeding ecology of the blackbird *Turdus merula* studied in the primaeval forest of Białowieża (Poland). Part 2. Reproduction and mortality. Acta Ornithol 29:101–121

Tomiałojć L (1995) The birds of the Białowieża Forest - additional data and summary. Acta Zool Cracov 38(3):363–397

Tomiałojć L, Wesołowski T (1990) Bird communities of the primaeval temperate forest of Białowieża, Poland. In: Keast A (ed) Biogeography and ecology of forest bird communities. SPB Academic, The Hague, pp 141–165

Tomiałojć L, Wesołowski T (1994) Die Stabilität der Vogelgemeinschaft in einem Urwald der gemässigten Zone: Ergebnisse einer 15jährigen Studie aus dem Nationalpark von Białowieża (Polen). Ornithol Beob 91:75–110

Tomiałojć L, Wesołowski T (1996) Structure of a primaeval forest bird community during 1970s and 1990s (Białowieża National Park, Poland). Acta Ornithol 31:133–154

Tomiałojć L, Wesołowski T, Walankiewicz W (1984) Breeding bird community of a primaeval temperate forest (Białowieża National Park, Poland). Acta Ornithol 20:241–310

Traczyk T, Traczyk H, Moszyńska B (1973) Herb layer production of two pinewood communities in the Kampinos National Park. Ekol Pol 21:37–55

Tubbs CR (1986) The New Forest. W Collins, London

Tukey JW (1977) Exploratory data analysis. Addison-Wesley, Reading, Massachusetts

Utenkova AP, Nazarova NS (1968) Raspredelenie dozhdevykh chervei v pochvakh Belovezhskoi Pushchi. [Distribution of earthworms in soils of Białowieża Forest.] Belovezhskaya Pushcha, Issledovaniya 2:155–160 (in Russian)

Utenkova AP, Strelkov AZ (1971) Uchet opada drevesnykh porod v lesakh Belovezhskoi Pushchi. [Survey of organic matter falling from trees in the forests of Białowieża Primeval Forest.] Belovezhskaya Pushcha, Issledovaniya 4:108–120 (in Russian)

Van Ballenberghe V, Ballard WB (1994) Limitation and regulation of moose populations: the role of predation. Can J Zool 72:2071–2077

Velleman PF, Hoaglin DC (1981) Applications, basics, and computing of exploratory data analysis. Duxbury, Boston

Virkkala R (1992) Fluctuations of vole-eating birds of prey in northern Finland. Ornis Fenn 69:97–100

Walankiewicz W (1991) Do secondary cavity-nesting birds suffer more from competition for cavities or from predation in a primeval deciduous forest? Nat Areas J 11(4):203–212

Walton KC (1968) The distribution of the polecat, *Putorius putorius* in Great Britain, 1963–67. J Zool (Lond) 155:237–240

Weber D (1988a) Die aktuelle Verbreitung des Iltisses (*Mustela putorius* L.) in der Schweiz. Rev Suisse Zool 95:1041–1056

Weber D (1988b) Experiments on microhabitat preference of polecats. Acta Theriol 33:403–413

Weber D (1989) Foraging in the polecats *Mustela putorius* L. in Switzerland: the case of a specialist anuran predator. Z Saugetierkd 54:377–392

Wesołowski T (1981) Population restoration after removal of wrens (*Troglodytes troglodytes*) breeding in Primaeval Forest. J Anim Ecol 50:809–814

Wesołowski T (1983) The breeding ecology and behaviour of wrens *Troglodytes troglodytes* under primaeval and secondary conditions. Ibis 125:499–515

Wesołowski T (1985) The breeding ecology of the wood warbler *Phylloscopus sibilatrix* in primaeval forest. Ornis Scand 16:49–60

Wesołowski T (1995) Ecology and behaviour of white-backed woodpecker (*Dendrocopos leucotos*) in a primaeval temperate forest (Białowieża National Park, Poland). Vogelwarte 38(2):61–75

Wesołowski T, Stawarczyk T (1991) Survival and population dynamics of Nuthatches *Sitta europaea* breeding in natural cavities in a primeval temperate forest. Ornis Scand 22:143–154

Wesołowski T, Tomiałojć L (1997) Breeding bird dynamics in a primaeval temperate forest: long-term trends in Białowieża National Park (Poland). Ecography 20:432–453

Wesołowski T, Tomiałojć L, Stawarczyk T (1987) Why low numbers of *Parus major* in Białowieża Forest - removal experiments. Acta Ornithol 23:303–316

Więcko E (1984) Puszcza Białowieska. [Białowieża Primeval Forest.] PWN - Polish Scientific Publishers, Warszawa (in Polish)

Włodek K, Krzywiński A (1986) Zu Biologie und Verhalten des Marderhundes (*Nyctereutes procyonoides*) in Polen. Z Jagdwiss 32:203–215

Wołk K (1985) Occurrence of birds in the annual cycle versus the structure of forest in the Białowieża National Park. Parki Narodowe i Rezerwaty Przyrody 6(1):57–75 (in Polish with English Summary)

Wołk E, Kozłowski J (1989) Changes of body weight and hematological parameters in a fluctuating population of *Apodemus flavicollis*. Acta Theriol 34:439–464

Wójcik AM (1993) Genetic variation in a fluctuating population of the yellow-necked mouse *Apodemus flavicollis*. Acta Theriol 38:273–290

Wójcik JM, Wołk K (1985) The daily activity rhythm of two competitive rodents: *Clethrionomys glareolus* and *Apodemus flavicollis*. Acta Theriol 30:241–258

Yalden DW (1985) Dietary separation of owls in the Peak District. Bird Study 32:122–131

Ylönen H (1989) Weasels *Mustela nivalis* suppress reproduction in cyclic bank voles *Clethrionomys glareolus*. Oikos 55:138–140

Ylönen H, Jędrzejewska B, Jędrzejewski W, Heikkila J (1992) Antipredatory behaviour of *Clethrionomys* voles - 'David and Goliath' arms race. Ann Zool Fennici 29:207–216

Zalewski A (1996) Choice of age classes of bank voles *Clethrionomys glareolus* by pine marten *Martes martes* and tawny owl *Strix aluco* in Białowieża National Park. Acta Oecol 17:233–244

Zalewski A (1997) Patterns of resting site use by pine marten, *Martes martes*, in Białowieża National Park (Poland). Acta Theriol 42:153–168

Zalewski A, Jędrzejewski W, Jędrzejewska B (1995) Pine marten home ranges, numbers and predation on vertebrates in a deciduous forest (Białowieża National Park, Poland). Ann Zool Fenn 32:131–144

Zheltukhin AS (1984) Daily activity and sizes of home ranges of the European lynx in the southern taiga of the Upper Volga river region. Byull Mosk O Ispyt Prir Otd Biol 89(1):54–61 (in Russian with English Summary)

Zemanek M (1972) Food and feeding habits of rodents in a deciduous forest. Acta Theriol 17:315–325

Subject Index

Ecological Studies

Volumes published since 1992

Volume 94
The Ecology of Aquatic Hyphomycetes (1992)
F. Bärlocher (Ed.)

Volume 95
Palms in Forest Ecosystems of Amazonia (1992)
F. Kahn and J.-J. DeGranville

Volume 96
Ecology and Decline of Red Spruce in the Eastern United States (1992)
C. Eagar and M.B. Adams (Eds.)

Volume 97
The Response of Western Forests to Air Pollution (1992)
R.K. Olson, D. Binkley, and M. Böhm (Eds.)

Volume 98
Plankton Regulation Dynamics (1993)
N. Walz (Ed.)

Volume 99
Biodiversity and Ecosystem Function (1993)
E.-D. Schulze and H.A. Mooney (Eds.)

Volume 100
Ecophysiology of Photosynthesis (1994)
E.-D. Schulze and M.M. Caldwell (Eds.)

Volume 101
Effects of Land Use Change on Atmospheric CO_2 Concentrations: South and South East Asia as a Case Study (1993)
V.H. Dale (Ed.)

Volume 102
Coral Reef Ecology (1993)
Y.I. Sorokin

Volume 103
Rocky Shores: Exploitation in Chile and South Africa (1993)
W.R. Siegfried (Ed.)

Volume 104
Long-Term Experiments With Acid Rain in Norwegian Forest Ecosystems (1993)
G. Abrahamsen et al. (Eds.)

Volume 105
Microbial Ecology of Lake Plußsee (1993)
J. Overbeck and R.J. Chrost (Eds.)

Volume 106
Minimum Animal Populations (1994)
H. Remmert (Ed.)

Volume 107
The Role of Fire in Mediterranean-Type Ecosystems (1994)
J.M. Moreno and W.C. Oechel (Eds.)

Volume 108
Ecology and Biogeography of Mediterranean Ecosystems in Chile, California and Australia (1994)
M.T.K. Arroyo, P.H. Zedler, and M.D. Fox (Eds.)

Volume 109
Mediterranean-Type Ecosystems. The Function of Biodiversity (1995)
G.W. Davis and D.M. Richardson (Eds.)

Volume 110
Tropical Montane Cloud Forests (1995)
L.S. Hamilton, J.O. Juvik, and F.N. Scatena (Eds.)

Volume 111
Peatland Forestry. Ecology and Principles (1995)
E. Paavilainen and J. Päivänen

Volume 112
Tropical Forests: Management and Ecology (1995)
A.E. Lugo and C. Lowe (Eds.)

Volume 113
Arctic and Alpine Biodiversity. Patterns, Causes and Ecosystem Consequences (1995)
F.S. Chapin III and C. Körner (Eds.)

Volume 114
Crassulacean Acid Metabolism. Biochemistry, Ecophysiology and Evolution (1996)
K. Winter and J.A.C. Smith (Eds.)

Volume 115
Islands. Biological Diversity and Ecosystem Function (1995)
P.M. Vitousek, L.L. Loope, and H. Adsersen (Eds.)

Ecological Studies

Volumes published since 1992

Volume 116
High Latitude Rainforests and Associated Ecosystems of the West Coast of the Americas: Climate, Hydrology, Ecology and Conservation (1996)
R.G. Lawford, P. Alaback, and E.R. Fuentes (Eds.)

Volume 117
Anticipated Effects of a Changing Global Environment on Mediterranean-Type Ecosystems (1995)
J. Moreno and W.C. Oechel (Eds.)

Volume 118
Impact of Air Pollutants on Southern Pine Forests (1996)
S. Fox and R.A. Mickler (Eds.)

Volume 119
Freshwater Ecosystems of Alaska (1997)
A.M. Milner and M.W. Oswood (Eds.)

Volume 120
Landscape Function and Disturbance in Arctic Tundra (1996)
J.F. Reynolds and J.D. Tenhunen (Eds.)

Volume 121
Biodiversity and Savanna Ecosystem Processes. A Global Perspective (1996)
O.T. Solbrig, E. Medina, and J.F. Silva (Eds.)

Volume 122
Biodiversity and Ecosystem Processes in Tropical Forests (1996)
G.H. Orians, R. Dirzo, and J.H. Cushman (Eds.)

Volume 123
Marine Benthic Vegetation. Recent Changes and the Effects of Eutrophication (1996)
W. Schramm and P.H. Nienhuis (Eds.)

Volume 124
Global Change and Arctic Terrestrial Ecosystems (1997)
W.C. Oechel (Ed.)

Volume 125
Ecology and Conservation of Great Plains Vertebrates (1997)
F.L. Knopf and F.B. Samson (Eds.)

Volume 126
The Central Amazon Floodplain: Ecology of a Pulsing System (1997)
W.J. Junk (Ed.)

Volume 127
Forest Decline and Ozone: A Comparison of Controlled Chamber and Field Experiments (1997)
H. Sandermann, A.R. Wellburn, and R.L. Heath (Eds.)

Volume 128
Productivity and Sustainability of Southern Forest Ecosystems (1997)
R.A. Mickler and S. Fox (Eds.)

Volume 129
Pelagic Nutrient Cycles: Herbivores as Sources and Sinks (1997)
T. Andersen

Volume 130
Vertical Food Web Interactions: Evolutionary Patterns and Driving Forces (1997)
K. Dettner, G. Bauer, and W. Völkl (Eds.)

Volume 131
The Structuring Role of Submerged Macrophytes in Lakes (1998)
E. Jeppesen et al. (Eds.)

Volume 132
Vegetation of the Tropical Pacific Islands (1998)
D. Mueller-Dombois and F.R. Fosberg

Volume 133
Aquatic Humic Substances: Ecology and Biogeochemistry (1998)
D.O. Hessen and L.J. Tranvik (Eds.)

Volume 134
Oxidant Air Pollution Impacts in the Montane Forests of Southern California (1998)
P.R. Miller and J.R. McBride (Eds.)

Volume 135
Predation in Vertebrate Communities: The Białowieża Primeval Forest as a Case Study (1998)
B. Jędrzejewska and W. Jędrzejewski

MIX
Papier aus verantwortungsvollen Quellen
Paper from responsible sources
FSC® C105338

If you have any concerns about our products,
you can contact us on
ProductSafety@springernature.com

In case Publisher is established outside the EU,
the EU authorized representative is:
Springer Nature Customer Service Center GmbH
Europaplatz 3, 69115 Heidelberg, Germany

Printed by Libri Plureos GmbH
in Hamburg, Germany